Prokaryotic Antimicrobial Peptides

Djamel Drider · Sylvie Rebuffat
Editors

Prokaryotic Antimicrobial Peptides

From Genes to Applications

Springer

Editors
Djamel Drider
Nantes-Atlantic National College
of Veterinary Medicine
Food Science and Engineering (ONIRIS)
Nantes, France
djamel.drider@onris-nantes.fr

Sylvie Rebuffat
Muséum National d'Histoire Naturelle
Laboratory of Communication Molecules
and Adaptation of Microorganisms
Paris, France
rebuffat@mnhn.fr

ISBN 978-1-4419-7691-8 e-ISBN 978-1-4419-7692-5
DOI 10.1007/978-1-4419-7692-5
Springer New York Dordrecht Heidelberg London

Library of Congress Control Number: 2011921930

Printed on acid-free paper

Springer is part of Springer Science+Business Media (www.springer.com)

Contents

Contributors

Hikmate Abriouel
Health Sciences Department, University of Jaen, Jaen, Spain

Yanath Belguesmia
Nantes-Atlantic National College of Veterinary Medicine, Food Science and Engineering, Nantes, France

Nabil Ben Omar
Health Sciences Department, University of Jaen, Jaen, Spain

A. Brand
Department of Microbiology, Stellenbosch University, Stellenbosch, South Africa

Kim A. Brogden
Department of Periodontics and Dows Institute for Dental Research, College of Dentistry, The University of Iowa, Iowa City, IA, USA

Mathieu Chauleau
Institut de Biologie Physico-Chimique, Paris, France

Nour-Eddine Chihib
Laboratoire ProBioGEM, UPRES-EA 1026, Polytech-Lille/IUT A, Université des Sciences et Technologies de Lille, Villeneuve d'Ascq, France

M.L. Chikindas
Department of Food Science, Rutgers, The State University of New Jersey, New Brunswick, NJ, USA

Luis M. Cintas
Departamento de Nutrición, Bromatología y Tecnología de los Alimentos, Facultad de Veterinaria, Universidad Complutense de Madrid (UCM), Madrid, Spain

Paul D. Cotter
Teagasc, Moorepark Food Research Centre, Fermoy, Cork, Ireland; Alimentary Pharmabiotic Centre, University College Cork, Cork, Ireland

Anne de Jong
Department of Molecular Genetics, University of Groningen, Groningen Biomolecular Sciences and Biotechnology Institute, Haren, The Netherlands

Miklos de Zamaroczy
Institut de Biologie Physico-Chimique, Paris, France

Delphine Destoumieux-Garzón
CNRS, Ifremer, IRD, Université Montpellier 2, UMR5119, Laboratoire Ecosystèmes Lagunaires, Montpellier, France

L.M.T. Dicks
Department of Microbiology, Stellenbosch University, Stellenbosch, South Africa

Djamel Drider
Department of Research into Sanitary Risks and Biotechnology of Reproduction, UPSP 5301 DGER, Nantes-Atlantic National College of Veterinary Medicine, Food Science and Engineering (ONIRIS), La Chantrerie Cedex 03, BP 40706, 44307 Nantes, France

Antonio Gálvez
Health Sciences Department, Microbiology Division, Faculty of Experimental Sciences, University of Jaen, Jaen, Spain

Helen Sophie Haugen
Department of Molecular Biosciences, University of Oslo, Oslo, Norway

Pablo E. Hernández
Departamento de Nutrición, Bromatología y Tecnología de los Alimentos, Facultad de Veterinaria, Universidad Complutense de Madrid (UCM), Madrid, Spain

Carmen Herranz
Departamento de Nutrición, Bromatología y Tecnología de los Alimentos, Facultad de Veterinaria, Universidad Complutense de Madrid (UCM), Madrid, Spain

T.D.J. Heunis
Department of Microbiology, Stellenbosch University, Stellenbosch, South Africa

Colin Hill
Department of Microbiology, University College Cork, College Road, Cork, Ireland

Shun Iwatani
Laboratory of Microbial Technology, Division of Microbial Science and Technology, Department of Bioscience and Biotechnology, Faculty of Agriculture, Kyushu University, Fukuoka, Japan

Teymur Kazakov
The Waksman Institute for Microbiology, Rutgers, The State University of New Jersey, Piscataway, NJ, USA

Per Eugen Kristiansen
Department of Molecular Biosciences, University of Oslo, Oslo, Norway

Anneke Kuipers
BiOMaDe Technology Foundation, Groningen, The Netherlands

Oscar P. Kuipers
Department of Molecular Genetics, University of Groningen, Groningen Biomolecular Sciences and Biotechnology Institute, Groningen, The Netherlands

Rosario Lucas
Health Sciences Department, University of Jaen, Jaen, Spain

Leah A. Martin-Visscher
Department of Chemistry, University of Alberta, Edmonton, Alberta, Canada

Gert N. Moll
BiOMaDe Technology Foundation, Groningen, The Netherlands

Karim Naghmouchi
Université de Tunis-El Manaar-Faculté des Sciences, Tunis, Tunisia

Ingolf F. Nes
Laboratory of Gene Technology, Department of Chemistry, Biotechnology and Food Science, Norwegian University of Life Sciences, Aas, Norway

Jon Nissen-Meyer
Department of Molecular Biosciences, University of Oslo, Oslo, Norway

K. Sutyak Noll
Department of Food Science, Rutgers, The State University of New Jersey, New Brunswick, NJ, USA

Camilla Oppegård
Department of Molecular Biosciences, University of Oslo, Oslo, Norway

Jean Peduzzi
CNRS, Muséum National d'Histoire Naturelle, Paris, France

Mary C. Rea
Teagasc, Moorepark Food Research Centre, Fermoy, Cork, Ireland

Sylvie Rebuffat
Laboratory of Communication Molecules and Adaptation of Microorganisms, Muséum National d'Histoire Naturelle - CNRS, Paris, France

Margaret A. Riley
Department of Biology, University of Massachusetts Amherst, Amherst, MA, USA

Rick Rink
BiOMaDe Technology Foundation, Groningen, The Netherlands

Per Rogne
Department of Molecular Biosciences, University of Oslo, Oslo, Norway

R. Paul Ross
Teagasc, Moorepark Food Research Centre, Fermoy, Cork, Ireland

Lucila Saavedra
Centro de Referencia para Lactobacilos (CERELA-CONICET), San Miguel de Tucumán, Tucumán, Argentina

Ekaterina Semenova
The Waksman Institute for Microbiology, Rutgers, The State University of New Jersey, Piscataway, NJ, USA

Fernando Sesma
Centro de Referencia para Lactobacilos (CERELA-CONICET), San Miguel de Tucumán, Tucumán, Argentina

Konstantin Severinov
The Waksman Institute for Microbiology, Rutgers, The State University of New Jersey, Piscataway, NJ, USA;
Department of Molecular Biology and Biochemistry, Rutgers, The State University of New Jersey, Piscataway, NJ, USA;
Institutes of Molecular Genetics and Gene Biology, Russian Academy of Sciences, Moscow, Russia

Kenji Sonomoto
Laboratory of Microbial Technology, Division of Applied Molecular Microbiology and Biomass Chemistry, Department of Bioscience and Biotechnology, Faculty of Agriculture, Graduate School, Kyushu University, Fukuoka, Japan

Alessandro Tossi
Department of Life Sciences, University of Trieste, Trieste, Italy

Marco J. van Belkum
Department of Chemistry, University of Alberta, Edmonton, Alberta, Canada

Auke J. van Heel
Department of Molecular Genetics, University of Groningen, Groningen Biomolecular Sciences and Biotechnology Institute, Haren, The Netherlands

D.A. van Staden
Department of Microbiology, Stellenbosch University, Stellenbosch, South Africa

Gaëlle Vassiliadis
CNRS, Muséum National d'Histoire Naturelle, Paris, France

John C. Vederas
Department of Chemistry, University of Alberta, Edmonton, Alberta, Canada

Takeshi Zendo
Laboratory of Microbial Technology, Division of Microbial Science and Technology, Department of Bioscience and Biotechnology, Faculty of Agriculture, Graduate School, Kyushu University, Fukuoka, Japan

Part I
Introduction: History, Current Knowledge, and Future Research on Antimicrobial Peptides

Chapter 1
History, Current Knowledge, and Future Directions on Bacteriocin Research in Lactic Acid Bacteria

Ingolf F. Nes

All organisms, both eukaryotic organisms and bacteria, are able to produce ribosomally antimicrobial peptides. In bacteria, such compounds are referred to as bacteriocins. The history of bacteriocins goes back to the early 1920s. One has experienced many disappointments in the efforts how to put these compounds into practical use despite being one of the most promising groups of antimicrobial agents to fight bacterial pathogens. However, today, we see new possibilities how to take advantage of such peptides for the benefit of man and animals. Bacteriocin production has become an important property of probiotic bacteria, and targeted use of bacteriocins to fight certain pathogens may have a future.

We should separate bacteriocins from our traditional peptide antibiotics. First, the peptide antibiotics differ from ribosomally synthesized peptides because peptide antibiotics are synthesized by enzymes. Second, bacteriocins are targeted at a narrow spectrum of bacteria often within the species of the producer or closely related ones, while the classical antibiotics are active against broad spectra of bacteria. Another feature that separates bacteriocins from antibiotics is their potency against susceptible bacteria; bacteriocins are unique because they can kill bacteria at nanomolar concentrations, while antibiotics are needed in much higher concentrations.

For many reasons, it is meaningful to separate the bacteriocins of gram-positive and gram-negative bacteria, and this short overview focuses on bacteriocins from gram-positive bacteria. It is most fruitful to divide G+ bacteriocins into two major groups: the heat-stable lantibiotics (Class I) and the nonmodified (some minor modifications may exist) and heat-stable bacteriocins (Class II). These two major classes are further divided into subclasses (Chatterjee et al. 2005; Cotter et al. 2005; Nes et al. 2006). Numerous excellent reviews on bacteriocins have been published

I.F. Nes (✉)
Laboratory of Gene Technology, Department of Chemistry, Biotechnology and Food Science, Norwegian University of Life Sciences, Aas 1432, Norway
e-mail: Ingolf.nes@umb.no

D. Drider and S. Rebuffat (eds.), *Prokaryotic Antimicrobial Peptides: From Genes to Applications*, DOI 10.1007/978-1-4419-7692-5_1,

in recent years (Breukink and de Kruijff 2006; Chatterjee et al. 2005; Cotter et al. 2005; Diep et al. 2009; Drider et al. 2006; Nes et al. 2007; Nissen-Meyer et al. 2009; Oppegard et al. 2007; Willey and van der Donk 2007).

The focus of bacteriocins in gram-positive bacteria has for the most part been on lactic acid bacteria (LAB) due to their apparent importance in food and feed fermentation, and also by being considered as GRAS organisms by FDA, and not least because of good funding in the 1990s and into the twenty-first century by the European Union. An important reason for research on bacteriocin has been and still is their extreme potency as antimicrobials as observed with some bacteriocins that are active at nanomolar concentrations against a number of bacteria including pathogens such as *Listeria monocytogenes*. Some bacteriocins exhibit activity against multidrug-resistant nosocomial pathogens such as methicillin-resistant *Staphylococcus aureus* (MRSA) and vancomycin-resistant enterococcus (VRE). Thus, it is believed that they could have a potential in medical and veterinary applications.

Fermented food and feed and plant material have been a source for isolation of bacteriocin-producing LAB, but intestinal and fecal sampling of LAB from animals and humans has become an increasingly important source for such bacteria due to an increased awareness of their importance as probiotic bacteria.

Modification and Structure of Bacteriocins

Lantibiotics are gene-encoded peptides that contain intramolecular ring structures by the formation the thioether bridges between dehydrated serine or threonine and cysteines that confer lanthionine and methyllanthionine residues, respectively. Additional but less frequent modifications have been identified in some lantibiotics. Such modifications include lysinoalanine, 2-oxybutyrate, 2-oxopropionate, 2-hydroxypropionate, β-hydroxy-aspartate, S-aminovinyl-D-cysteine, S-aminovinyl-D-methylcysteine, and D-alanine.

After the discovery of lantibiotic bacteriocins that goes back to the early 1920s, it took more than 50 years for the structure of nisin, the first identified lantibiotic, to be determined, i.e., in 1971 (de Klerk and Smit 1967; Gross and Morell 1971). In the 1960s, it was reported that lactobacilli produced antimicrobial substances different from the organic acids (de Klerk and Smit 1967; Sabine 1963; Tramer 1966). It was a slow start, and the area of lantibiotic research did not take off before the 1970s. Since then, numerous lantibiotics have been identified and characterized with respect to structure, mode of action, genetics, regulation, synthesis, and modification. Many excellent and comprehensive review articles have been published in recent years on these topics of lantibiotics (Bonelli et al. 2006; Chatterjee et al. 2005; Dufour et al. 2007; Guder et al. 2000; Pag and Sahl 2002; Twomey et al. 2002; Willey and van der Donk 2007; Xie and van der Donk 2004).

There is presently focus on development of bioengineered lantibiotics, to reveal the location of essential and variable domains therein and to create derivatives with broader specificity, increased stability, and even higher activities against specific target organisms for both in vivo and in vitro use.

In vitro modification systems have successfully been used to introduce thioether rings into other biologically active peptides. These enzymes have been the focus of recent bioengineering studies.

Of particular significance with respect to posttranlational modification is the new bacteriocin thuricin CD, a strong anticlostridial bacteriocin that is particularly effective against *Clostridium difficile* (Rea et al. 2010), produced by *Bacillus thuringiensis*. This two-peptide bacteriocin is quite unique not just because of its anti-clostridial activity but also because both peptides feature three posttranslationally modified sulfur to alpha carbon in thioether linkages. It should also be added that such modification has previously been identified in the circular bacteriocin.

Also, class II bacteriocins can be structured by inducing certain posttranslational changes, and the most common modification is the conserved N-terminal cysteine-bridge formation that has shown to be of crucial importance for the antimicrobial activity of class IIa bacteriocin (Eijsink et al. 1998).

Also, circular bacteriocins are posttranslationally modified through a head-to-tail backbone covalent linkage (Maqueda et al. 2008).

Structures of many bacteriocins have been resolved by NMR analysis. Such studies include both classes of lantibiotics, Class II bacteriocins (Kristiansen et al. 2005; Oppegard et al. 2007; Rogne et al. 2009; Sprules et al. 2004) and cyclic bacteriocins (Gonzalez et al. 2000; Martin-Visscher et al. 2009). Combined with functional analysis, important structural features important for the antimicrobial activity have been determined. In addition, immunity proteins have also been structurally determined (Johnsen et al. 2005; Martin-Visscher et al. 2008). Hopefully, these structural studies combined with functional studies will bring together how the bacteriocins work and how the immunity interacts with its bacteriocin to prevent self-destruction of the host.

Genetics

The genes required for biosynthetic machinery of lantibiotics are complex and are often organized in operons. Together with the structural gene(s) (*lanA*), genes encoding modification enzymes, externalization system of the bacteriocins as well as immunity genes to protect the producer for self-destruction are needed. In addition, it has been shown that the production of some lantibiotics is also regulated by a two-component regulatory system (Kleerebezem et al. 2001), although alternative regulatory systems are identified in a few lantibiotics as seen for lactocin S (Rawlinson et al. 2005).

The modifications are introduced either by one biosynthetic enzyme (LanM) or by a dehydratase (LanB) in combination with a cyclase (LanC). The structure of NisC has been resolved; the reaction mechanism of LctM has been studied, and the amino-acid residues in the active site were identified by mutagenesis studies

(Chatterjee et al. 2006; Li and van der Donk 2007; Li et al. 2006; Rink et al. 2007; You and van der Donk 2007).

Class II bacteriocins need usually only four genes for synthesis that constitute a structural bacteriocin gene, a dedicated immunity gene, and a transporter and its accessory gene.

As seen with some lantibiotics, class II bacteriocins can also be regulated by a two- or three-component regulatory system, and under such circumstances, three more genes are needed: a gene that encodes a peptide pheromone, and genes for a sensor (histidine protein kinase) and a DNA-binding protein that activates gene expression (response regulator) (Nes et al. 1996; Nes and Eijsink 1999). Often, several bacteriocins are clustered together and share both the transport and regulatory system as seen in the plantaricin system (Diep et al. 2009).

Mode of Action and Targets

Numerous mode-of-action studies have been published through all these years, and the membrane has been the target for most of these studies. With few exceptions, bacteriocins cause membrane permeabilization, triggering leakage of intracellular compounds and dissipation of the proton-motive force. Such studies have been carried out with both whole cells and various liposome and vesicle systems. Most of the permeabilization studies suffered by using higher concentrations of bacteriocins than what is needed to kill living target cells. The most obvious conclusion from such studies is that the observed leakage of intracellular components was most likely due to detergent effects and other secondary effects produced by high bacteriocin concentrations and was not associated with the in vivo mode of action.

The identification of lipid II as target of nisin was of crucial importance to understand the antimicrobial activity of this bacteriocin (Brotz et al. 1998). It was a breakthrough when it was shown that nisin actually binds to lipid (Breukink and de Kruijff 2006; Brotz et al. 1998; Pag and Sahl 2002; Schneider and Sahl 2010). Recently, it has been shown that lipid II is the target not only for certain lantibiotics but also for plectasin, a fungal defensin that acts by binding to the bacterial cell-wall precursor (Schneider et al. 2010).

Some class II bacteriocins have been shown to need a specific target for activity. Already in the early 1990s, it was suggested that lactococcin A, a class II bacteriocin, kills susceptible bacteria through a target. This suggestion was based on a study that proposed a membrane-associated protein specific for lactococci acts as receptor for lactococcin A (van Belkum et al. 1991). But this notion was not proved before 2007, when it was experimentally demonstrated that a membrane protein component of the man-PTS system was the actual target site of lactococcin A (Diep et al. 2007). Also, class IIa bacteriocins target the same membrane component of the mannose PTS system, but the recognition specificity is different from that of lactococcin A (Arous et al. 2004; Diep et al. 2007; Gravesen et al. 2002; Héchard et al. 2001; Kjos et al. 2009). It has now been shown that amino-acid sequence of

ManC membrane protein is required for the target specificity of the class IIa bacteriocins (Kjos and Diep unpublished results).

Not only bacteriocins from gram-positive bacteria are using the mannose-PTS as a target for killing but it has been demonstrated that certain microcins from gram-negative bacteria also need this membrane protein(s). A study of microcin E492 produced by *Klebsiella pneumoniae* has shown that a mannose permease is involved in the antimicrobial activity, too. The work concluded that the toxicity is strictly dependent on the presence of ManYZ, the inner membrane protein complex involved in mannose uptake (Bieler et al. 2006).

In the future, we will certainly disclose new targets for other bacteriocins, and there are good reasons to believe that all bacteriocins do have specific targets/receptors/docking molecule.

Application

Many foodborne lactic acid bacteria produce bacteriocins, and several of the most recognized bacteriocins have been isolated from such LAB obtained from fermented food, just to mention, nisin, lactococcin A, and lacticin 3147. Therefore, we can conclude that bacteriocins have a long history of use in food production, on one hand. On the other hand, implementation of either bacteriocin-producing starter cultures or use of cell-free bacteriocin supplements in foods has been much less. But both kinds of products are available for commercial applications. It is surprising that the industry has not used this opportunity more, but this may be due to too high expectation of such compounds that set back such efforts. Several attractive applications have been investigated, but only a few are presently in use.

In recent years, probiotics have become very popular, and LAB are important probiotic players among them. Bacteriocin production has always been considered a beneficial and probiotic feature of LAB, but it has been questioned if bacteriocins are produced and work within the intestinal tract of the host. In a recent work, it has been shown that bacteriocin-producing *Lactobacillus salivarius* protected mice against infection with the invasive foodborne pathogen *L. monocytogenes* (Corr et al. 2007). It was conclusively confirmed that bacteriocin production was the primary mediator of protection against *Listeria* infection. This study supports the idea that bacteriocin production should be included as an important trait and a criterion for selection of probiotic bacteria.

Medical application of antimicrobial peptides, particularly peptides from eukaryotic organisms, has been under investigation for a long time. Bacteriocins have not attracted the same attention mainly because their activities are more bacterial species-specific than that of their eukaryotic counterparts, which often act on both gram-positive and gram-negative bacteria (Zasloff 2002). However, bacteriocins are much more potent against the ones they strike, and they act at nanomolar concentrations, while eukaryotic peptides have to be used at micromolar range.

In order to benefit the broad specificity of the eukaryotic peptides with the high potency of bacteriocins, it may be useful to combine these two activities. It has been demonstrated that by combining pediocin-like bacteriocins (Class IIa) with eukaryotic peptides, a synergistic effect can be obtained (Luders et al. 2003).

Future Trends

There has been a continuous effort to isolate new bacteriocins. The most common approach is to isolate new bacteria, test for antimicrobial activity and then isolate and characterize the activity. This approach is limited by the susceptibility of the indicators used. Also, the growth conditions used for screening new isolates are limiting factors to uncover potential bacteriocin producers, since bacteriocin production can be under the control of a quorum-sensing mechanism (Nes et al. 1996) and the expression can also be off or on only under very narrowly defined conditions (Diep et al. 2000). By the advent of genome sequencing, an increasing number of bacterial genomes are published, and now more than 1,000 genomes are available. Owing to reduced cost of genome sequencing, genome sequencing will be the preferred approach to obtain new bacteriocins and their genes. Development of in silico approaches to identify bacteriocin operons is becoming available, including bacteriocin databases (de Jong et al. 2006, 2010; Hammami et al. 2010). By using such tools, new bacteriocins have already been identified (Begley et al. 2009; Diep et al. 2006; Lawton et al. 2007).

New features will be found and included to improve algorithms for identification of new bacteriocin genes in genome DNA sequence data banks. Recently, a new group of membrane-bound bacteriocin immunity proteins has been identified, and this information has been used to search for novel bacteriocins in sequenced genomes, and seven new bacteriocin-like loci have been identified in gram-positive bacteria (Kjos et al. 2010).

Target identification is a promising area of bacteriocin research. Identification of bacteriocin targets will permit a more detailed molecular analysis and insight into the mechanism of action and rational designs to improve and produce antimicrobial peptides with a broader target specificity and high potency. It is of particular interest that mannose-PTS is target for bacteriocins from both gram-negative and gram-positive bacteria because such PTS systems are unique to bacteria and accordingly bacteriocins should not affect eukaryotic cells. It is tempting to refer to Dr. Erni's commentarial note "Development of The Mannose Transporter Complex: an Open Door for the Macromolecular Invasion of Bacteria" (Erni 2006). His view has been further supported and extended by the finding that the same transport system is the target for Class II bacteriocins in gram-positive bacteria, too.

Peptide and gene sequence information of bacteriocins have laid ground for protein engineering aimed at production of peptides with new properties. Lantibiotics are structurally more complex than Class II bacteriocins and have attracted most focus for protein engineering (Kuipers et al. 1996). Such studies have also included changes of the dehydrated amino acid as well as lanthionines (thioether bridge) in new peptides. New molecules with increased solubility in water (Rollema et al. 1995) and increased stability and sporicidal activity have been achieved (Liu and Hansen 1992). Hopefully, the future will bring forward new and more efficient antimicrobial peptides based on our knowledge of biologically produced peptides. The future looks bright for both medical and food applications of bacteriocins.

References

Arous S, Dalet K, Hechard Y (2004) Involvement of the mpo operon in resistance to class IIa bacteriocins in *Listeria monocytogenes*. FEMS Microbiol Lett 238:37–41

Begley M, Cotter PD, Hill C, Ross RP (2009) Identification of a novel two-peptide lantibiotic, lichenicidin, following rational genome mining for LanM proteins. Appl Environ Microbiol 75:5451–5460

Bieler S, Silva F, Soto C, Belin D (2006) Bactericidal activity of both secreted and nonsecreted microcin E492 requires the mannose permease. J Bacteriol 188:7049–7061

Bonelli RG, Wiedemann I, Sahl H-G (2006) "Lantibiotics" in Handbook of Biological Active Peptides, page-97-105. Academic Press, San Diego, CA, USA

Breukink E, de Kruijff B (2006) Lipid II as a target for antibiotics. Nat Rev Drug Discov 5:321–332

Brotz H, Josten M, Wiedemann I, Schneider U, Gotz F, Bierbaum G, Sahl HG (1998) Role of lipid-bound peptidoglycan precursors in the formation of pores by nisin, epidermin and other lantibiotics. Mol Microbiol 30:317–327

Chatterjee C, Paul M, Xie L, van der Donk WA (2005) Biosynthesis and mode of action of lantibiotics. Chem Rev 105:633–684

Chatterjee C, Patton GC, Cooper L, Paul M, van der Donk WA (2006) Engineering dehydro amino acids and thioethers into peptides using lacticin 481 synthetase. Chem Biol 13:1109–1117

Corr SC, Li Y, Riedel CU, O'Toole PW, Hill C, Gahan CG (2007) Bacteriocin production as a mechanism for the antiinfective activity of *Lactobacillus salivarius* UCC118. Proc Natl Acad Sci U S A 104:7617–7621

Cotter PD, Hill C, Ross RP (2005) Bacteriocins: developing innate immunity for food. Nat Rev Microbiol 3:777–788

de Jong A, van Hijum SA, Bijlsma JJ, Kok J, Kuipers OP (2006) BAGEL: a web-based bacteriocin genome mining tool. Nucleic Acids Res 34:W273–W279

de Jong A, van Heel AJ, Kok J, Kuipers OP (2010) BAGEL2: mining for bacteriocins in genomic data. Nucleic Acids Res 1(38 suppl):W647–W651

de Klerk HC, Smit JA (1967) Properties of a *Lactobacillus fermenti* bacteriocin. J Gen Microbiol 48:309–316

Diep DB, Axelsson L, Grefsli C, Nes IF (2000) The synthesis of the bacteriocin sakacin A is a temperature-sensitive process regulated by a pheromone peptide through a three-component regulatory system. Microbiology 146(Pt 9):2155–2160

Diep DB, Godager L, Brede D, Nes IF (2006) Data mining and characterization of a novel pediocin-like bacteriocin system from the genome of *Pediococcus pentosaceus* ATCC 25745. Microbiology 152:1649–1659

Diep DB, Skaugen M, Salehian Z, Holo H, Nes IF (2007) Common mechanisms of target cell recognition and immunity for class II bacteriocins. Proc Natl Acad Sci U S A 104:2384–2389

Diep DB, Straume D, Kjos M, Torres C, Nes IF (2009) An overview of the mosaic bacteriocin pln loci from *Lactobacillus plantarum*. Peptides 30:1562–1574

Drider D, Fimland G, Héchard Y, McMullen LM, Prévost H (2006) The continuing story of class IIa bacteriocins. Microbiol Mol Biol Rev 70:564–582

Dufour A, Hindre T, Haras D, Le Pennec JP (2007) The biology of lantibiotics from the lacticin 481 group is coming of age. FEMS Microbiol Rev 31:134–167

Eijsink VG, Skeie M, Middelhoven PH, Brurberg MB, Nes IF (1998) Comparative studies of class IIa bacteriocins of lactic acid bacteria. Appl Environ Microbiol 64:3275–3281

Erni B (2006) The mannose transporter complex: an open door for the macromolecular invasion of bacteria. J Bacteriol 188:7036–7038

Gonzalez C, Langdon GM, Bruix M, Galvez A, Valdivia E, Maqueda M, Rico M (2000) Bacteriocin AS-48, a microbial cyclic polypeptide structurally and functionally related to mammalian NK-lysin. Proc Natl Acad Sci U S A 97:11221–11226

Gravesen A, Ramnath M, Rechinger KB, Andersen N, Jansch L, Hechard Y, Hastings JW, Knochel S (2002) High-level resistance to class IIa bacteriocins is associated with one general mechanism in *Listeria monocytogenes*. Microbiology 148:2361–2369

Gross E, Morell JL (1971) The structure of nisin. J Am Chem Soc 93:4634–4635

Guder A, Wiedemann I, Sahl HG (2000) Posttranslationally modified bacteriocins – the lantibiotics. Biopolymers 55:62–73

Hammami R, Zouhir A, Le Lay C, Ben Hamida J, Fliss I (2010) BACTIBASE second release: a database and tool platform for bacteriocin characterization. BMC Microbiol 10:22

Héchard Y, Pelletier C, Cenatiempo Y, Frere J (2001) Analysis of sigma(54)-dependent genes in *Enterococcus faecalis*: a mannose PTS permease (EII(Man)) is involved in sensitivity to a bacteriocin, mesentericin Y105. Microbiology 147:1575–1580

Johnsen L, Dalhus B, Leiros I, Nissen-Meyer J (2005) 1.6-Angstroms crystal structure of EntA-im. A bacterial immunity protein conferring immunity to the antimicrobial activity of the pediocin-like bacteriocin enterocin A. J Biol Chem 280:19045–19050

Kjos M, Nes IF, Diep DB (2009) Class II one-peptide bacteriocins target a phylogenetically defined subgroup of mannose phosphotransferase systems on sensitive cells. Microbiology 155:2949–2961

Kjos M, Snipen L, Salehian Z, Nes IF, Diep DB (2010) The abi proteins and their involvement in bacteriocin self-immunity. J Bacteriol 192:2068–2076

Kleerebezem M, Kuipers OP, de Vos WM, Stiles ME, Quadri LE (2001) A two-component signal-transduction cascade in *Carnobacterium piscicola* LV17B: two signaling peptides and one sensor-transmitter. Peptides 22:1597–1601

Kristiansen PE, Fimland G, Mantzilas D, Nissen-Meyer J (2005) Structure and mode of action of the membrane-permeabilizing antimicrobial peptide pheromone plantaricin A. J Biol Chem 280:22945–22950

Kuipers OP, Bierbaum G, Ottenwalder B, Dodd HM, Horn N, Metzger J, Kupke T, Gnau V, Bongers R, van den Bogaard P, Kosters H, Rollema HS, de Vos WM, Siezen RJ, Jung G, Gotz F, Sahl HG, Gasson MJ (1996) Protein engineering of lantibiotics. Antonie Van Leeuwenhoek 69:161–169

Lawton EM, Cotter PD, Hill C, Ross RP (2007) Identification of a novel two-peptide lantibiotic, haloduracin, produced by the alkaliphile *Bacillus halodurans* C-125. FEMS Microbiol Lett 267:64–71

Li B, van der Donk WA (2007) Identification of essential catalytic residues of the cyclase NisC involved in the biosynthesis of nisin. J Biol Chem 282:21169–21175

Li B, Yu JP, Brunzelle JS, Moll GN, van der Donk WA, Nair SK (2006) Structure and mechanism of the lantibiotic cyclase involved in nisin biosynthesis. Science 311:1464–1467

Liu W, Hansen JN (1992) Enhancement of the chemical and antimicrobial properties of subtilin by site-directed mutagenesis. J Biol Chem 267:25078–25085

Luders T, Birkemo GA, Fimland G, Nissen-Meyer J, Nes IF (2003) Strong synergy between a eukaryotic antimicrobial peptide and bacteriocins from lactic acid bacteria. Appl Environ Microbiol 69:1797–1799

Maqueda M, Sanchez-Hidalgo M, Fernandez M, Montalban-Lopez M, Valdivia E, Martinez-Bueno M (2008) Genetic features of circular bacteriocins produced by gram-positive bacteria. FEMS Microbiol Rev 32:2–22

Martin-Visscher LA, Sprules T, Gursky LJ, Vederas JC (2008) Nuclear magnetic resonance solution structure of PisI, a group B immunity protein that provides protection against the type IIa bacteriocin piscicolin 126, PisA. Biochemistry 47:6427–6436

Martin-Visscher LA, Gong XD, Duszyk M, Vederas JC (2009) The three-dimensional structure of Carnocyclin A reveals that many circular bacteriocins share a common structural motif. J Biol Chem 284:28674–28681

Nes IF, Eijsink VGH (1999) Regulation of Group II Peptide Bacteriocin Synthesis by Quorum - Sensing Mechansims p. In: Dunny GM, Winans SC (eds) Cell-Cell Signaling in Bacteria. ASM Press, Washington, D.C, pp 175–192

Nes IF, Diep DB, Havarstein LS, Brurberg MB, Eijsink V, Holo H (1996) Biosynthesis of bacteriocins in lactic acid bacteria. Antonie Van Leeuwenhoek 70:113–128

Nes IF, Brede DA, Holo H (2006) The non-lantibiotic heat-stable bacteriocins in gram positive bacteria. In: Kastin AJ (ed) Handbook of Biological Active Peptides. Academic Press, San Diego, CA, USA

Nes IF, Diep DB, Holo H (2007) Bacteriocin diversity in *Streptococcus* and *Enterococcus*. J Bacteriol 189:1189–1198

Nissen-Meyer J, Rogne P, Oppegard C, Haugen HS, Kristiansen PE (2009) Structure-function relationships of the non-lanthionine-containing peptide (class II) bacteriocins produced by gram-positive bacteria. Curr Pharm Biotechnol 10:19–37

Oppegard C, Rogne P, Emanuelsen L, Kristiansen PE, Fimland G, Nissen-Meyer J (2007) The two-peptide class II bacteriocins: structure, production, and mode of action. J Mol Microbiol Biotechnol 13:210–219

Pag U, Sahl HG (2002) Multiple activities in lantibiotics–models for the design of novel antibiotics? Curr Pharm Des 8:815–833

Rawlinson EL, Nes IF, Skaugen M (2005) Identification of the DNA-binding site of the Rgg-like regulator LasX within the lactocin S promoter region. Microbiology 151:813–823

Rea MC, Sit CS, Clayton E, O'Connor PM, Whittal RM, Zheng J, Vederas JC, Ross RP, Hill C (2010) Thuricin CD, a posttranslationally modified bacteriocin with a narrow spectrum of activity against *Clostridium difficile*. Proc Natl Acad Sci U S A 107:9352–9357

Rink R, Kluskens LD, Kuipers A, Driessen AJ, Kuipers OP, Moll GN (2007) NisC, the cyclase of the lantibiotic nisin, can catalyze cyclization of designed nonlantibiotic peptides. Biochemistry 46:13179–13189

Rogne P, Haugen C, Fimland G, Nissen-Meyer J, Kristiansen PE (2009) Three-dimensional structure of the two-peptide bacteriocin plantaricin JK. Peptides 30:1613–1621

Rollema HS, Kuipers OP, Both P, de Vos WM, Siezen RJ (1995) Improvement of solubility and stability of the antimicrobial peptide nisin by protein engineering. Appl Environ Microbiol 61:2873–2878

Sabine DB (1963) An antibiotic-like effect of *Lactobacillus acidophilus*. Nature 199:811

Schneider T, Sahl H-G (2010) An oldie but a goodie – cell wall biosynthesis as antibiotic target pathway. Int J Med Microbiol 300:161–169

Schneider T, Kruse T, Wimmer R, Wiedemann I, Sass V, Pag U, Jansen A, Nielsen AK, Mygind PH, Raventos DS, Neve S, Ravn B, Bonvin AM, De Maria L, Andersen AS, Gammelgaard LK, Sahl HG, Kristensen HH (2010) Plectasin, a fungal defensin, targets the bacterial cell wall precursor Lipid II. Science 328:1168–1172

Sprules T, Kawulka KE, Gibbs AC, Wishart DS, Vederas JC (2004) NMR solution structure of the precursor for carnobacteriocin B2, an antimicrobial peptide from *Carnobacterium piscicola*. Eur J Biochem 271:1748–1756

Tramer J (1966) Inhibitory effect of *Lactobacillus acidophilus*. Nature 211:204–205

Twomey D, Ross RP, Ryan M, Meaney B, Hill C (2002) Lantibiotics produced by lactic acid bacteria: structure, function and applications. Antonie Van Leeuwenhoek 82:165–185

van Belkum MJ, Kok J, Venema G, Holo H, Nes IF, Konings WN, Abee T (1991) The bacteriocin lactococcin A specifically increases permeability of lactococcal cytoplasmic membranes in a voltage-independent, protein-mediated manner. J Bacteriol 173:7934–7941

Willey JM, van der Donk WA (2007) Lantibiotics: peptides of diverse structure and function. Annu Rev Microbiol 61:477–501

Xie L, van der Donk WA (2004) Post-translational modifications during lantibiotic biosynthesis. Curr Opin Chem Biol 8:498–507

You YO, van der Donk WA (2007) Mechanistic investigations of the dehydration reaction of lacticin 481 synthetase using site-directed mutagenesis. Biochemistry 46:5991–6000

Zasloff M (2002) Antimicrobial peptides of multicellular organisms. Nature 415:389–395

Chapter 2
Bacteriocin-Mediated Competitive Interactions of Bacterial Populations and Communities

Margaret A. Riley

Abstract Explaining the coexistence of competing species is a major challenge in community ecology. In bacterial systems, competition is often driven by the production of bacteriocins; narrow spectrum proteinaceous toxins that serve to kill closely related species providing the producer better access to limited resources. Bacteriocin producers have been shown to competitively exclude sensitive, non-producing strains. However, the interaction dynamics between bacteriocin producers, each lethal to its competitor, are largely unknown. Several recent studies have revealed some of the complexity of these interactions, employing a suite of in vitro, in vivo, and in silico bacterial model systems. This chapter describes the current state of knowledge regarding the population and community ecology of this potent family of toxins.

Introduction

Bacteria engage in a never-ending arms race in which they compete for limited resources and niche space. The outcome of this intense interaction is the evolution of a diverse and powerful arsenal of biological weapons. Most species of bacteria produce one, and usually many more, of these potent biocontrol agents, including classical antibiotics, lytic agents, lysozymes, and bacteriocins (Cascales et al. 2007). The microbial weapons of choice, as assessed by the frequency with which they are encountered in natural populations of bacteria and in their diversity of forms, are the bacteriocins.

Bacteriocins are loosely defined as biologically active protein moieties with a bacteriocidal mode of action (Tagg et al. 1976; James et al. 1991). Two main features distinguish the majority of bacteriocins from classical antibiotics: bacteriocins are

M.A. Riley (✉)
Department of Biology, University of Massachusetts Amherst,
611 North Pleasant Street, Amherst, MA 01003, USA
e-mail: riley@bio.umass.edu

D. Drider and S. Rebuffat (eds.), *Prokaryotic Antimicrobial Peptides: From Genes to Applications*, DOI 10.1007/978-1-4419-7692-5_2,

ribosomally synthesized and have a relatively narrow killing spectrum (Riley and Wertz 2002). Indeed, bacteriocins are often only toxic to bacteria closely related to the producing strain. The bacteriocin family includes a diversity of proteins in terms of size, microbial target, mode of action, release, and immunity mechanisms and can be divided into two main groups: those produced by Gram-negative and Gram-positive bacteria (Gordon et al. 2006; Heng et al. 2007).

Their production occurs across all major groups of Eubacteria and the Archaebacteria (Webster 1991). Within a species, tens or even hundreds of different kinds of bacteriocins are produced (James et al. 2002; Riley and Gordon 1992). Colicins, bacteriocins produced by *Escherichia coli*, are found in 30–50% of the strains isolated from human hosts and are often referred to as virulence factors (Riley and Gordon 1992). Much higher levels of bacteriocin production have been found in some Gram-negative bacteria, such as *Pseudomonas aeruginosa*, in which >90% produce bacteriocins (Michel-Briand and Baysse 2002).

Despite high levels of bacteriocin diversity, these proteins share many general characteristics (James et al. 2002; De Vuyst et al. 1994). They are generally high-molecular weight protein antibiotics that kill closely related strains or species. The bacteriocin gains entry into the target cell by recognizing specific cell surface receptors and then kills the cell by forming ion-permeable channels in the cytoplasmic membrane, by nonspecific degradation of cellular DNA, by inhibition of protein synthesis through the specific cleavage of 16s rRNA, or by cell lysis.

Without question, bacteriocins serve some function in microbial communities. This statement follows from the detection of bacteriocin production in all surveyed lineages of prokaryotes. Equally compelling is the inference of strong positive selection acting on some bacteriocins (Tan and Riley 1996; Riley 1998). Finally, there is a well-developed body of theory and empirical data that details the potential role of bacteriocins which play in maintaining microbial diversity at the population and community levels (Chao and Levin 1981; Frank 1994; Gordon and Riley 1999; Czárán et al. 2002; Kerr et al. 2002). Such empirical observations and theoretical results argue that these toxins play a critical role in mediating microbial interactions. What remains in question is what, precisely, that ecological role is. Bacteriocins may serve as anticompetitors enabling the invasion of a strain or a species into an established microbial population or community. They may also play a defensive role and act to prohibit the invasion of other strains or species into an occupied niche or limit the advance of neighboring cells. An additional role has recently been proposed for bacteriocins produced by Gram-positive bacteria, that of regulating quorum sensing (Miller and Bassler 2001). This chapter describes the current state of knowledge regarding the population and community ecology of this potent family of toxins.

Colicins: The Model Bacteriocin

The most extensively studied bacteriocins are the colicins, which are produced by *E. coli* (Pugsley 1985; Pugsley and Oudega 1987; James et al. 2002; Cascales et al. 2007). Colicins were first identified almost 100 years ago as a heat-labile product present in cultures of *E. coli* V and toxic to *E. coli* S. They were given the name of colicin to identify the producing species (Gratia 1925). Fredericq demonstrated that colicins were proteins and that they had a limited range of activity due to the presence or absence of specific receptors on the surface of sensitive cells (Fredericq and Levine 1947).

Colicins are archetypical of a large subfamily of bacteriocins found primarily in the family *Enterobacteriaceae*. One of the defining features of colicin-like toxins is that they are composed of three functional domains: a central binding domain that recognizes and adheres to specific receptor sites on the surfaces of target cells, an amino-terminal translocation domain responsible for entry into the cell, and a carboxy-terminal killing domain that actually kills the cell. Colicins kill target cells through a variety of mechanisms. Nomura showed that colicins E1 and K inhibit macromolecular synthesis without the arrest of respiration, colicin E2 causes DNA breakdown, and colicin E3 stops protein synthesis (Nomura and Witten 1967). In each case, he showed that the lethal action is reversed by treatment with trypsin. Since his pioneering work, colicins were shown to kill their targets by either membrane permeabilization or nucleic acid degradation (Braun et al. 1994; Riley and Wertz 2002; Smarda and Smajs 1998). Colicins are classified according to the nature of the killing domain. The nuclease group includes colicins that degrade DNA, rRNA, or tRNA. The pore former colicins kill by the formation of voltage-gated channels in the cytoplasmic membrane. The third group contains colicins that affect the peptidoglycan cell wall. Colicin operons typically contain three genes: the toxin-encoding gene; an immunity gene, whose product specifically binds to and confers protection against the encoded toxin; and a lysis gene, whose product contributes to the release of toxin into the environment.

Recent surveys of *E. coli*, *Salmonella enterica*, *Hafnia alvei*, *Citrobacter freundii*, *Klebsiella oxytoca*, *Klebsiella pneumoniae*, and *Enterobacter cloacae* reveal levels of bacteriocin production ranging from 3 to 26% of environmental isolates (Gordon 2006; Riley 2002). Colicins are found in 30–50% of the strains isolated from human hosts (Riley and Gordon 1992). Much higher levels of bacteriocin production have been found in some Gram-negative bacteria, such as *P. aeruginosa*, in which >90% of both environmental and clinical isolates produce bacteriocins (Michel-Briand and Baysse 2002).

Until recently, little was known about the phylogenetic breadth over which bacteriocins kill. To produce such an estimate, we took advantage of a recently determined molecular phylogeny of enteric bacteria (Wertz et al., 2003). The frequency of killing within each taxon for each bacteriocin was mapped onto the enteric phylogeny (Fig. 2.1) (Riley et al. 2003). Not surprising, these data reveal that bacteriocins usually kill members of their own species. However, a surprisingly high level of interspecific killing was observed, with almost half of the bacteriocins killing in more than one taxon. Further, the relationship between killing ability and phylogenetic distance is nonlinear. In other words, some bacteriocin producers kill distantly related bacteria but not their closest relatives. This nonlinear relationship is seen in Fig. 2.1, in which 18 of 36 columns indicate killing outside of the producer strains own species (off the diagonal). The observation of a broad killing range for numerous enteric bacteriocins requires that the ecological role proposed for bacteriocins be reconsidered. It may well be that they serve a broader, community

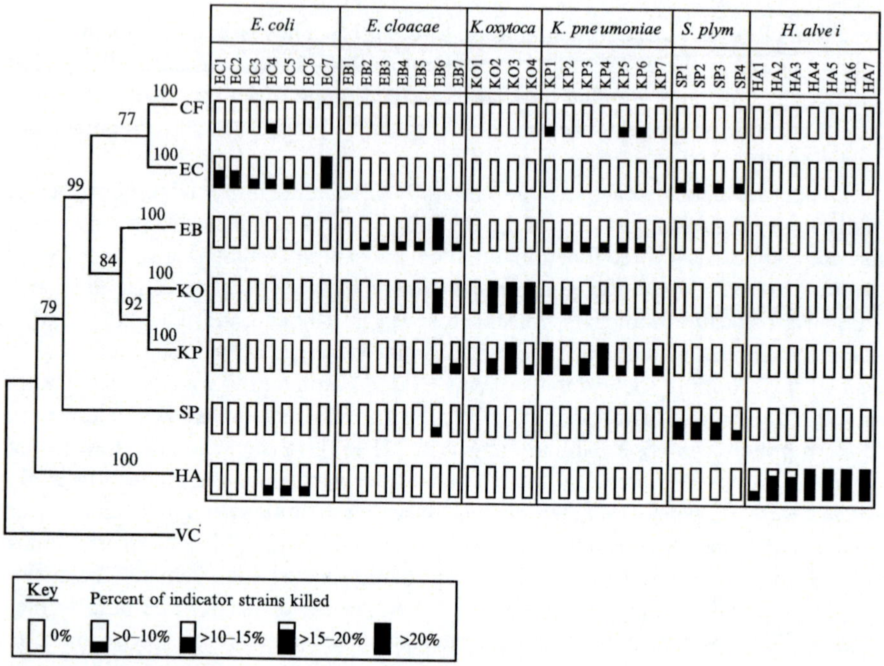

Fig. 2.1 Enteric bacteriocin phylogenetic killing range. The frequency of bacteriocin killing within each of the seven enteric taxa is mapped onto a composite molecular phylogeny of enteric bacteria (adopted from Wertz and Riley 2003). The bacteriocins assayed for killing breadth are indicated across the *top* (EC = *Escherichia coli*, CF = *Citrobacter freundii*, KO = *Klebsiella oxytoca*, KP = *Klebsiella pneumoniae*, EB = *Enteroterobacter cloacae*, HA = *Hafnia alvei*, SP = *Serratia plymuthica*, VC = *Vibrio cholera*). Each column provides the frequency of killing for each bacteriocin assayed against 40 indicator strains for each taxa in the molecular phylogeny

level role than has been envisioned to date. This finding complements recent theoretical work, which suggests that bacteriocins (and other microbial defense systems) may be responsible for maintaining much of the extraordinary diversity of microbes observed in nature (Czárán et al. 2002; Kerr et al. 2002).

The Ecological Role of Colicins

Despite the pervasive role of toxin production in the microbial world, little is known about the ecology of this form of competition. Previous theoretical and empirical studies have suggested that toxin production serves as a strategy to obtain access to nutrients (Chao and Levin 1981; Ivanovska and Hardwick 2005; Riley and Gordon 1992). However, a more recent study testing competitive interactions between toxin producers and sensitive yeast strains under low and high nutrient conditions concluded that toxin producers only out-compete sensitive cells in high nutrient environments (Wloch-Salamon et al. 2008). This observation supports the theoretical prediction that toxin production has evolved to occur as a competitive strategy under conditions of abundant resources (Frank 1994). Both studies suggest that toxin production may be more important in the invasion of niches, than in obtaining nutrients (Brown and Taddei 2009).

One of the more compelling models of toxin-mediated bacterial competition employs *E. coli* and their colicins. Some of the earliest ecological studies were inconclusive and sometimes even contradictory (Ikari et al. 1969). More recently, however, a more robust theoretical and empirical base has been established defining the conditions that favor maintenance of toxin-producing bacteria in both population and community settings. Chao and Levin (1981) showed that the conditions for invasion of a colicin-producing strain are much broader in a spatially structured environment (i.e., one that restricts the movement of strains and nutrients), than in an unstructured one (i.e., where mass action prevails). In an unstructured environment with mass action, a small population of producers cannot invade an established population of sensitive cells (Durrett and Levin 1997). This failure occurs because producers pay a price for toxin production (energetic costs of plasmid carriage and lethality of production), while the benefits (the resources made available by killing sensitive organisms) are distributed at random. In a physically structured environment, such as on the surface of an agar plate, the strains grow as separate colonies and toxin diffuses out from a colony of producers, killing sensitive neighbors (Kerr et al. 2002). The access to resources that would have otherwise been consumed by the sensitive cells, as well as the nutrients from the killed individuals, is made available to the producing colony owing to its proximity. Therefore, killers can increase in frequency even when initially rare.

Several modeling efforts have incorporated additional biological reality. Two such efforts introduced a third species, one that is resistant to the toxin but cannot itself produce the toxin (Nakamaru and Iwasa 2000). Resistance can be conferred through mutations in either the binding site or the translocation machinery required for a bacteriocin to enter the target cell. Acquisition of an immunity gene will also confer resistance to its cognate bacteriocin. It is assumed that there is a cost to resistance and that this cost is less than the cost of toxin production borne by the killer strain (Riley and Wertz 2002). Owing to this third member, pair-wise interactions among the strains have the nontransitive structure of the childhood game of rock–scissors–paper (Karolyi et al. 2005; Kerr et al. 2002). The producer strain beats the sensitive strain, owing to the toxin's effects on the latter. The sensitive strain beats the resistant strain because only the latter suffers the cost of resistance. And the resistant strain wins against the producer because the latter bears the higher cost of toxin production and release while the former pays only the cost of resistance.

Kerr et al. (2002) investigated this colicin-mediated nontransitive interaction. Their study revealed that in environments where interactions and dispersal are not solely local, the resistant strain overtook the community during the course of the experiment. In contrast, in a structured environment, this game permits a quasi-stable global equilibrium, one in which all three strains can persist with nearly constant global abundance (Laird and Schamp 2008; Neumann and Schuster 2007).

A third environment, mixed plate, which is intermediate between structured and unstructured, revealed that growth on a surface is not the key factor, as resistance overtook the other strains on this plate also. Additional complexity was incorporated by the addition of a second producing strain. In one such study, two producing strains were grown in competition in a static environment, on plates. In this study, each colicin was able to induce its counterpart's production. The static plate experiment showed that in a structured environment that allow only local interactions coexistence results in absolute spatial isolation of the two strains. These results support the prediction that balanced chasing in a spatially structured, nonhierarchical community may result in the maintenance of diversity.

Surveys of colicin production in natural populations suggest that natural populations of *E. coli* partially match the predictions of these ecological models: *E. coli* producer strains are found at frequencies ranging from 10 to 50% (Riley and Gordon 1992; Gordon and Riley 1999; Gordon and O'Brien 2006; Barnes et al. 2007). Resistant strains are even more abundant, being found at frequencies of 50–90%. In fact, most strains are resistant to all co-segregating colicins. Finally, there is a small population of sensitive cells. The simultaneous presence of both colicin-producing and -resistant strains raises an interesting paradox. How can we explain the persistence of colicin-producing strains when their putative targets are relatively rare? The models predict this distribution of phenotypes results from frequent horizontal transfer of resistance and the significant cost associated with colicin production (Barnes et al. 2007). In other words, if a strain can gain resistance and lose production, they will over time – just as was observed in *E. coli* isolated from field mouse population over a period of 3 months (Gordon et al. 1998).

Some of the most recent investigations into colicin ecology and evolution have examined how the presence of a colicin impacts levels and types of gene expression. The first such study (Vreizen et al. 2009) followed the population dynamics of a colicin-producing strain of *E. coli* exposed to serial transfer conditions, with no competitors, over 253 generations. To this end, an in vitro serial transfer experiment was conducted in which a colicinogenic strain and its nonproducing ancestor were evolved for 253 generations. As was previously observed in a similar experiment carried out in vitro, the strain evolved a reduction in killing activity and a corresponding increase in fitness, relative to the ancestral strain. This result is not surprising, given that colicin production is quite costly due to (1) the cost of replication and maintenance of the colicin plasmid (Feldgarden et al., 1995), (2) the cost of colicin production, and (3) the occurrence of autolysis of the producing cells (Braun et al. 1994). When colicins are produced in an environment with no competitors, as is the case in this serial transfer experiment, it is reasonable to expect that the costs of colicin production will outweigh its benefits. We speculate that the lack of competitive interactions results in strong selection for a decrease or elimination of colicin production. In the in vitro study described above, molecular characterization of the evolved strains revealed no changes in the DNA sequence of the evolved colicin plasmids. However, a set of chromosomally encoded loci experienced changes in gene expression that were positively correlated with the reduction in killing. Further studies are required to more accurately identify the precise molecular changes that result in the observed reduction in killing.

Perhaps these future studies will also reveal why selection results in changes in gene expression, rather than simple elimination of the plasmid itself.

Experimental and theoretical work on the ecology of bacteriocin-mediated allelopathy highlights the importance of cell–cell interactions and spatial structure in mediating the outcome of competition. One recent study explored this hypothesis using two *E. coli* colicin producers (colicins E2 and E9), which are sometimes found together in the same host (Gordon et al. 1998). This study demonstrated that each colicin has the ability to induce its counterpart's production (Fig. 2.2). Over 50 years ago, it was found that colicin induction could be enhanced by the addition of mutagenic agents, such as mitomycin C and UV light (Herschman and Helinski 1967). Induction by DNA damaging agents was later linked to the SOS motifs, conserved in all the promoter regions of colicins (Gillor et al. 2008). Interestingly, transcriptional response of an *E. coli* strain to damage induced by a DNA degrading colicin showed strong induction of the LexA-regulated SOS system (Walker et al. 2004). These data suggest that the colicin mutual induction presented in this study results from the DNase toxins induction of the SOS response that in turns control colicin production.

The next step in these studies involved the introduction of two colicin-producing strains in a structured environment that allow only local interactions, such coexistence results in absolute spatial isolation of the two strains. These results support the prediction that balanced chasing in a spatially structured, nonhierarchical community may result in the maintenance of diversity. This was tested in vivo using a murine model by allowed pairs of co-caged mice, each carrying a single enteric

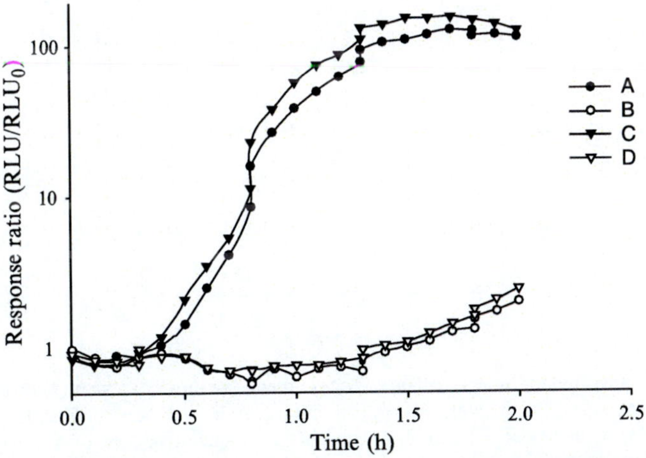

Fig. 2.2 Mutual colicin induction. The proteins of isogenic strains carrying colicin E2 or E7 plasmids and a colicin-free control strain were crudely extracted and used to induce reporter strains carrying *ce2a* and *ce7a* promoters (colicin E2 and E7 promoters) fused to *Photorhabdus luminescence luxCDABE* reporter operon. Colicin E2 crude protein extract was used to induce pDEW-E7 reporter vector (A; *filled circle*), while colicin E7 extract was used to induce pDEW-E2 reporter vector (B; *filled triangle*); the colicin-free strain was tested with the pDEW-E7 (C; *open circle*), and pDEW-E2 (D; *open triangle*) reporter vectors

strain producing one of the tested colicins, to interact freely and monitored their colon enteric residents for almost 4 months. Unlike previous reports in which colicin producers, nonproducers, and resistant strains competitively replaced one another (Kirkup and Riley 2004); in the current setting, no such invasion was observed. Throughout the experiment, the colicin-producing strains persisted, each in its host's colon, excluding the invasion of the competing producer population. Finally, in an attempt to generalize these observations, simulations were undertaken with competing bacteriocin-producing populations assuming mutual induction. The mathematical model further supported the empirical results, demonstrating that spatially distributed, cross-inducing producer populations can mutually exclude one another and thus maintain a steady-state coexistence (Fig. 2.3).

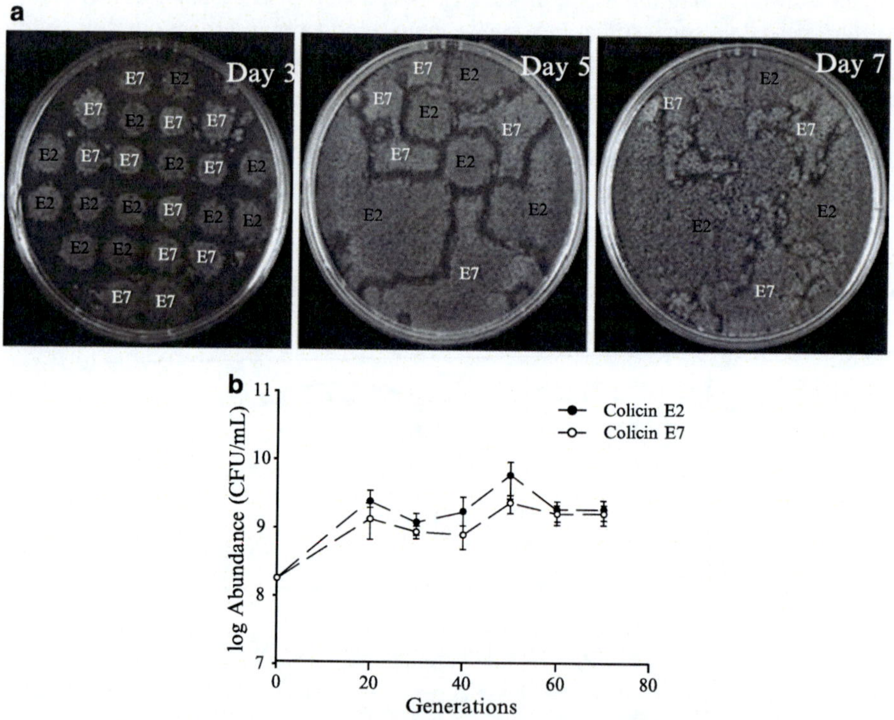

Fig. 2.3 Community dynamics in a structured environment. We initiated a static plate environment by randomly depositing 24 droplets from pure culture of *E. coli* strain BZB1011 carrying a plasmid encoding either colicin E2 or E7. The changing spatial pattern of the community is photographed over time (**a**) showing the spread of the strains droplets (day 3) to lawns bordered by clearing zone (day 5) that was later colonized by strains resistant to both colicins (day 7). Upon analysis of the cells concentration (**b**), the abundance of the *E. coli* harboring colicin E2 (*filled circles*) and E7 (*open circles*) encoding plasmids was shown to remain invariable throughout the experiment. Data points are the mean of two impendent experiments each performed in duplicate, the *bars* represent the standard deviation of the average cell concentration. The data points depicted on the *X*-axis are separated by 24 h and approximately ten bacteria generations

Such studies suggest that cross-induction in structured environments controls the invasion of susceptible bacteriocin producers as the established community acts to increase the local concentration of the toxin to a lethal dose in order to prevent invasion. The outcome of such an interaction pattern on the relative cost and benefit of investment in allelopathy has strong implications on an evolutionary scale. It has been shown that when toxin producers are locally scarce, they are unable to generate sufficient toxins to compensate for the cost of production (Chao and Levin 1981; Gardner et al. 2004). In contrast, if toxin producers are induced by their competitors/invaders, then bacteriocin production confers a fitness advantage, as the gain from a given investment in killing is maximized, being directed against an evident adversary. This will favor the regulation of bacteriocin that is readily susceptible to induction by a competing toxin producer, which should, in turn, become widespread, and thereby maintain and even enhance bacteriocin diversity.

The Colicin Mouse Model

A variety of mouse models have been employed to provide a more realistic evaluation of the role of bacteriocins in mediating bacterial strain dynamics. Although none of these models are able to control precisely the bacterial communities in mice or measure the impact of host responses, they do provide a clear indication of population dynamics of strains that have differing abilities to produce or resist colicins. One of the very first such studies demonstrated that bacteriocin production improves the establishment success of the producing strains in the mouse colon (McCormick et al. 1989). In this case, a pair of *E. coli* strains, isogenic save for the production of microcin V, were introduced into the large intestine of streptomycin-treated mice. When the two strains were fed together, the microcin-deficient strain was eliminated from the large intestine.

Additional realism was incorporated into such studies with the introduction of the same three strains employed by Kerr et al. (2002) (producer, sensitive, and resistant), which were monitored after introduction into a mouse colon (Kirkup and Riley 2004). These experiments revealed exactly the same nontransitive interactions described in the in vitro studies of Kerr et al. (2002). When a mouse harbored a sensitive strain, an introduced colicin-producing strain was able to invade. When a colicin-producing strain was resident, an introduced R strain was able to invade. In both experimental systems, the nontransitive nature of colicin-mediated dynamics was further revealed (Kirkup and Riley 2004).

Further mouse-based studies of colicin ecology focused solely on the colicin-producing strains (Majeed et al., 2011). One such study monitored the evolution of killing phenotype of a colicin-producing strain maintained in a mouse colon. Prior to introduction of the colicin-producing strains in the mice, the native Gram-negative bacterial community was eliminated with antibiotic application. Thus, there was a little or no competition with close relatives in this experimental design. The producing strains evolved significantly reduced killing activity over the 4 months of sampling.

More recent studies employed the same mouse model and incorporated additional colicin diversity into the experimental design (Gillor et al. 2009). Mice were inoculated with a single enteric strain producing one colicin. The mice were then paired in cages and allowed to interact freely, while their colon-based enteric residents were monitored for almost 4 months. Unlike previous reports in which colicin producers, nonproducers, and resistant strains competitively replaced one another (Kirkup and Riley 2004), in the current setting, no such invasion was observed (Fig. 2.4). Throughout the experiment, the colicin-producing strains persisted, each in its host's colon, excluding the invasion of the competing producer population.

These results may be subject to spatial dispersal; it has been previously suggested that limited dispersal favors toxin producers, while high dispersal, although still beneficial to toxin producers, is nevertheless insufficient to compensate for the cost of carrying toxin-encoding agents (Cascales et al. 2007; Wloch-Salamon et al. 2008; Inglis et al. 2009). Enterobactereaceae adhere to the colon epithelial cells, forming a stable biofilm (Everett et al. 2004), in other words, a structured environment in which cell–cell interactions are localized. Thus, in an unstructured environment at high dispersal, the interactions between bacteriocin producers may increase, and due to mutual induction, lead to elevated toxin production and release. As colicin production is costly and release is lethal to the producing cell, the competitive advantage of the producer's population will consequently be reduced. However, in a structured environment at low dispersal, competitive interaction will be localized and only a small percentage of the producer's population will be induced to fend off invading cells.

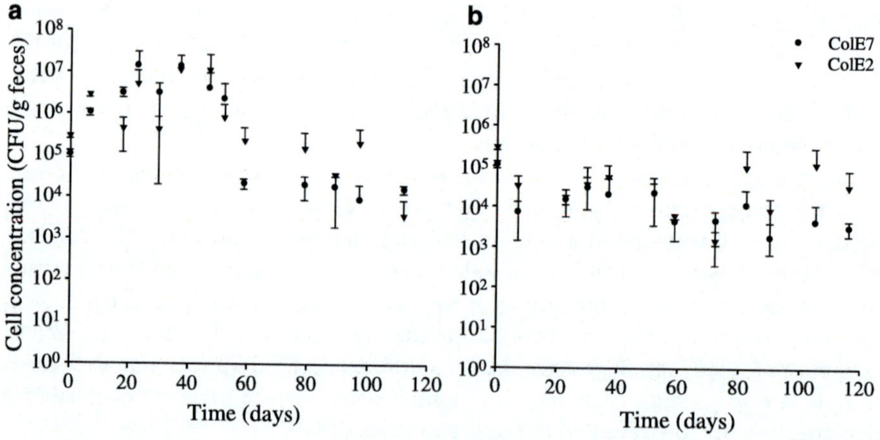

Fig. 2.4 Effect of competition on bacterial population size in mice. Bacterial density (CFU per gram fecal matter) monitored over time in mice in control (**a**) or experimental (**b**) cages. The control cages hosted mice harboring either *E. coli* strain BZB1011 bearing pDEW-E2 or mice harboring *E. coli* strain BZB1011 bearing pDEW-E7. The experimental cages contained one mouse established with the *E. coli* strain BZB1011 bearing pDEW-E2 and one mouse with *E. coli* strain BZB1011 bearing pDEW-E7. Each point represents the mean CFU/g feces averaged for strains bearing pDEW-E2 (*filled triangle*) or pDEW-E7 (*filled circle*) recovered from the mice. *Bars* represent the standard error for each point

In this final in vivo study, the competitive interactions between two populations of colicin producers in the mouse colon were resolved in mutual exclusion. This result suggests that the established biofilm of each of the producing cells could successfully prevent the invasion of cells producing a different colicin, both competing in a structured environment at low dispersal. It is assumed that the interaction between the populations was localized such that only a small part of the established population was induced by the invaders, just enough to prevent their advance. The mathematical model demonstrating mutual exclusion to be a robust result when bacteriocin producers interact locally, cross-inducing one another, further supported this observation.

These experimental and theoretical studies of the ecology of bacteriocin-mediated allelopathy highlight the importance of cell–cell interactions and spatial structure in mediating the outcome of competition. The data suggest that cross-induction in structured environments controls the invasion of susceptible bacteriocin producers as the established community acts to increase the local concentration of the toxin to a lethal dose in order to prevent invasion. The outcome of such an interaction pattern on the relative cost and benefit of investment in allelopathy has strong implications on an evolutionary scale. It has been shown that when toxin producers are locally scarce, they are unable to generate sufficient toxins to compensate for the cost of production (Chao and Levin 1981; Gardner et al. 2004). In contrast, if toxin producers are induced by their competitors/invaders, then bacteriocin production confers a fitness advantage, as the gain from a given investment in killing is maximized, being directed against an evident adversary. This will favor the regulation of bacteriocin that is readily susceptible to induction by a competing toxin producer, which should, in turn, become widespread, and thereby maintain and even enhance bacteriocin diversity.

Conclusions

There has been a virtual explosion in studies of the probiotic and antibiotic use of bacteriocins (Breukink and de Kruijff 1999; Audisio et al. 2005; Hillman et al. 2000; Aroutcheva et al. 2001; Avonts and De Vuyst 2001; Brashears et al. 2003; Gillor et al. 2004; Joerger 2003; Aslim and Kilic 2006; Corr et al. 2007; Diez-Gonzalez 2007; Falagas et al. 2007). As the potential health benefits of bacteriocins are being realized, it is critical to maintain a strong link between the ecological and applied studies. We learned far too late the cost (in terms of human morbidity and mortality) of our overuse and misuse of classical antibiotics. This failure to understand how we were creating ideal conditions for the rapid evolution and spread of antibiotic resistance has, unfortunately, resulted in one of the most significant human health challenges, the rise of human pathogens able to resist many of our most powerful antibiotic drugs. We are in a far better position to employ our growing understanding of bacteriocin ecology to develop sound and resilient methods with which to apply the power of bacteriocins for humans needs. As this review clearly illustrates, the studies of bacteriocin ecology and evolution have kept pace with, and in some cases advanced far beyond, our more limited forays into their medical applications. Clearly, we have far to go in developing a generalized model of bacteriocin ecology. With each new finding, we learn how complex are the roles these toxins serve in bacterial populations and communities. Even more challenging, it is almost certainly the case that such roles will differ for different species and even for the same species experiencing different environmental pressures.

Acknowledgments This work was supported by National Institutes of Health grants R01GM068657-01A2 and R01A1064588-01A2 to M.A.R.

References

Aroutcheva AA, Simoes JA, Faro S (2001) Antimicrobial protein produced by vaginal *Lactobacillus acidophilus* that inhibits *Gardnerella vaginalis*. Infect Dis Obstet Gynecol 9:33–39

Aslim B, Kilic E (2006) Some probiotic properties of vaginal lactobacilli isolated from healthy women. Jpn J Infect Dis 59:249–253

Audisio MC, Terzolo HR, and Apella MC (2005) Bacteriocin from honeybee beebread Enterococcus avium, active against Listeria monocytogenes. Appl Environ Microbiol 71, 3373–3375

Avonts L, De Vuyst L (2001) Antimicrobial potential of probiotic lactic acid bacteria. Meded Rijksuniv Gent Fak Landbouwkd Toegep Biol Wet 66:543–550

Barnes B, Sidhu H, Gordon DM (2007) Host gastro-intestinal dynamics and the frequency of colicin production by *Escherichia coli*. Microbiology 153:2823–2827

Brashears MM, Jaroni D, Trimble J (2003) Isolation, selection, and characterization of lactic acid bacteria for a competitive exclusion product to reduce shedding of *Escherichia coli* O157: H7 in cattle. J Food Prot 66:355–363

Braun V, Pilsl H, and Gross P (1994) Colicins: structures, modes of action, transfer through membranes, and evolution. Arch Microbiol 161, 199–206

Breukink E, de Kruijff B (1999) The lantibiotic nisin, a special case or not? Biochim Biophys Acta 1462:223–234

Brown SP, Inglis RF, Taddei F (2009) Evolutionary ecology of microbial wars: within-host competition and (incidental) virulence. Evol Appl 2:32–39

Cascales E, Buchanan SK, Duche D, Kleanthous C, Lloubes R, Postle K et al (2007) Colicin biology. Microbiol Mol Biol Rev 71:158–229

Chao L, Levin BR (1981) Structured habitats and the evolution of anticompetitor toxins in bacteria. Proc Natl Acad Sci USA 78:6324–6328

Corr SC, Li Y, Riedel CU, O'Toole PW, Hill C, Gahan CGM (2007) Bacteriocin production as a mechanism for the antfinfective activity of *Lactobacillus salivarius* UCC118. Proc Natl Acad Sci USA 104:7617–7621

Czárán TL, Hoekstra RF, and Pagie L (2002) Chemical warfare between microbes promotes biodiversity. PNAS 99, 786–790.

De Vuyst LVEJ (1994) Bacteriocins of Lactic Acid Bacteria: Microbiology, Genetics and Applicationslactic acid bacteria. (London, Blackie Academic & Professional)

Diez-Gonzalez F (2007) Use of bacteriocin in livestock. In: Riley MA, Gillor O (eds) Research and applications in bacteriocins. Horizon Bioscience, Norfolk, pp 117–129

Durrett R, Levin S (1997) Allelopathy in spatially distributed populations. J Theor Biol 185:165–171

Everett ML, Palestrant D, Miller SE, Bollinger RR, Parker W (2004) Immune exclusion and immune inclusion: a new model of host-bacterial interactions in the gut. Clin Appl Immunol Rev 4:321–332

Falagas ME, Betsi GI, Athanasiou S (2007) Probiotics for the treatment of women with bacterial vaginosis. Clin Microbiol Infect 13:657–664

Feldgarden M, Golden S, Wilson H, and Riley MA (1995) Can phage defence maintain colicin plasmids in Escherichia coli? Microbiology 141 (Pt 11), 2977–2984

Frank SA (1994) Spatial polymorphism of bacteriocins and other allelopathic traits. Evol Ecol 8:369–386

Fredericq P, and Levine M (1947) Antibiotic Interrelationships among the Enteric Group of Bacteria. J Bacteriol 54:785–792

Gardner A, West SA, Buckling A (2004) Bacteriocins, spite and virulence. Proc Biol Sci 271:1529–1535

Gillor O, Giladi I, and Riley MA (2009) Persistence of colicinogenic Escherichia coli in the mouse gastrointestinal tract. BMC Microbiol 9, 165

Gillor O, Kirkup BC, Riley MA (2004) Colicins and microcins: the next generation antimicrobials. Adv Appl Microbiol 54:129–146

Gillor O, Vriezen JAC, Riley MA (2008) The role of SOS boxes in enteric bacteriocin regulation. Microbiology 154:1783–1792

Gordon DM, O'Brien CL (2006) Bacteriocin diversity and the frequency of multiple bacteriocin production in *Escherichia coli*. Microbiology 152:3239–3244

Gordon DM, Riley MA (1999) A theoretical and empirical investigation of the invasion dynamics of colicinogeny. Microbiology 145:655–661

Gordon DM, Riley MA, Pinou T (1998) Temporal changes in the frequency of colicinogeny in *Escherichia coli* from house mice. Microbiology 144:2233–2240

Gordon EA (2006) The diversity of Bacteriocins in Gram-negative bacteria. In Bacteriocins: Ecology and Evolution (Berlin, Springer), pp. 5–18

Gratia, A. (1925). C R Soc Biol 93, 1040

Heng NC, Burtenshaw GA, Jack RW, and Tagg JR (2007) Ubericin A, a class IIa bacteriocin produced by Streptococcus uberis. Appl Environ Microbiol 73:7763–7766

Herschman HR, Helinski DR (1967) Comparative study of the events associated with colicin induction. J Bacteriol 94:691–699

Hillman JD, Brooks TA, Michalek SM, Harmon CC, Snoep JL, van Der Weijden CC (2000) Construction and characterization of an effector strain of *Streptococcus mutans* for replacement therapy of dental caries. Infect Immun 68:543–549

Ikari NS, Kenton DM, Young VM (1969) Interaction in germfree mouse intestine of colicinogenic and colicin-sensitive microorganisms. Proc Soc Exp Biol Med 130:1280–1284

Inglis RF, Gardner A, Cornelis P, Buckling A (2009) Spite and virulence in the bacterium *Pseudomonas aeruginosa*. Proc Natl Acad Sci USA 106:5703–5707

Ivanovska I, Hardwick JM (2005) Viruses activate a genetically conserved cell death pathway in a unicellular organism. J Cell Biol 170:391–399

James R, Lazdunski C, and Pattus F (1991) Bacteriocins, Microcins and Lantibiotics. In NATO ASI series (New York, Springer-Verlag), p. 519

James R, Penfold CN, Moore GR, Kleanthous C (2002) Killing of *E. coli* cells by E group nuclease colicins. Biochimie 84:381–389

Joerger RD (2003) Alternatives to antibiotics: bacteriocins, antimicrobial peptides and bacteriophages. Poult Sci 82:640–647

Karolyi G, Neufeld Z, Scheuring I (2005) Rock-scissors-paper game in a chaotic flow: the effect of dispersion on the cyclic competition of microorganisms. J Theor Biol 236:12–20

Kerr B, Riley MA, Feldman MW, Bohannan BJ (2002) Local dispersal promotes biodiversity in a real-life game of rock-paper-scissors. Nature 418:171–174

Kirkup BC, Riley MA (2004) Antibiotic-mediated antagonism leads to a bacterial game of rock-paper-scissors in vivo. Nature 428:412–414

Laird RA, and Schamp BS (2008) Does local competition increase the coexistence of species in intransitive networks? Ecology 89:237–247

Majeed H, Gillor O, Kerr B, and Riley MA (2011) Competitive interactions in Escherichia coli populations: the role of bacteriocins. ISME J 5:71–81

McCormick BA, Franklin DP, Laux DC, and Cohen PS (1989) Type 1 pili are not necessary for colonization of the streptomycin-treated mouse large intestine by type 1-piliated Escherichia coli F-18 and E. coli K-12. Infect Immun 57:3022–3029

Michel-Briand Y, and Baysse C (2002) The pyocins of Pseudomonas aeruginosa. Biochimie 84:499–510.

Miller MB, and Bassler BL (2001) Quorum sensing in bacteria. Annu Rev Microbiol 55:165–199.

Nakamaru M, Iwasa Y (2000) Competition by allelopathy proceeds in traveling waves: colicin-immune strain aids colicin-sensitive strain. Theor Popul Biol 57:131–144

Neumann G, Schuster S (2007) Continuous model for the rock-scissors-paper game between bacteriocin producing bacteria. J Math Biol 54:815–846

Nomura M, and Witten C (1967) Interaction of colicins with bacterial cells. 3. Colicin-tolerant mutations in Escherichia coli. J Bacteriol 94:1093–1111

Pugsley AP (1985) *Escherichia coli* K12 strains for use in the identification and characterization of colicins. J Gen Microbiol 131:369–376

Pugsley AP, and Oudega, B (1987) Methods for studying colicins and their plasmids. In Plasmids, a Practical Approach, K. Hardy, ed. (Oxford, IRL Press), pp. 105–161

Riley MA (1998) Molecular mechanisms of bacteriocin evolution. Annu Rev Genet 32:255–278

Riley MA, et al (2002). Bacteriocin mediated competitive interactions of bacterial populations and communities. Journal of Evolutionary Biology 16:690-697

Riley MA, Goldstone CM, Wertz JE, and Gordon D. (2003). A phylogenetic approach to assessing the targets of microbial warfare. J Evol Biol 16, 690–697

Riley MA, Gordon DM (1992) A survey of Col plasmids in natural isolates of *Escherichia coli* and an investigation into the stability of Col-plasmid lineages. J Gen Microbiol 138:1345–1352

Riley MA, and Wertz JE (2002) Bacteriocins: evolution, ecology, and application. Annu Rev Microbiol 56:117–137

Smarda J, and Smajs D (1998) Colicins–exocellular lethal proteins of Escherichia coli. Folia Microbiol (Praha) 43:563–582

Tagg JR, Dajani AS, and Wannamaker LW (1976) Bacteriocins of gram positive bacteria. Bacteriol Rev 40:722–756

Tan Y, and Riley MA (1996) Rapid invasion by colicinogenic Escherichia coli with novel immunity functions. Microbiology 142 (Pt 8), 2175–2180

Vriezen VAR (2009) The Evolution of reduced microbial killing. Genome Biology and Evolution 400–408

Walker D, Rolfe M, Thompson A, Moore GR, James R, Hinton JCD, Kleanthous C (2004) Transcriptional profiling of colicin-induced cell death of *Escherichia coli* MG1655 identifies potential mechanisms by which bacteriocins promote bacterial diversity. J Bacteriol 186: 866–869

Webster RE (1991) The tol gene products and the import of macromolecules into Escherichia coli. Mol Microbiol 5:1005–1011

Wertz JE, Goldstone C, Gordon DM, and Riley MA (2003) A molecular phylogeny of enteric bacteria and implications for a bacterial species concept. J Evol Biol 16:1236–1248

Wloch-Salamon DM, Gerla D, Hoekstra RF, de Visser J (2008) Effect of dispersal and nutrient availability on the competitive ability of toxin-producing yeast. Proc R Soc Lond B Biol Sci 275:535–541

Part II
Classification of Prokaryotic Antimicrobial Peptides

Chapter 3
Classification of Bacteriocins from Gram-Positive Bacteria

Mary C. Rea, R. Paul Ross, Paul D. Cotter, and Colin Hill

Abstract Bacteriocins are ribosomally synthesised antimicrobial peptides produced by bacteria, including many Gram-positive species. The classification of bacteriocins from Gram-positive bacteria is complicated by their heterogeneity and thus, as the number of Gram-positive bacteriocins identified has continued to increase, classification schemes have had to continuously evolve. Here, we review the various classification approaches, both historical and current, their underlying scientific basis and their relative merit, and suggest a rational scheme given the state of the art.

Introduction

Bacteriocins are ribosomally synthesised antimicrobial peptides produced by one bacterium that are active against other bacteria. It has been hypothesised that all bacteria produce at least one bacteriocin, and it is merely the lack of suitable expression and detection systems to detect the antimicrobial activity which prevents its detection (Tagg et al. 1976). While this may not be strictly true, it is apparent that bacteriocin producers are widespread in nature. Their frequent presence is probably a reflection of their importance in providing competitive advantages in complex microbial niches (O'Shea et al. 2009). Indeed, it has been suggested that many strains only produce bacteriocins in nutritionally deprived, overcrowded environments (Riley and Wertz 2002). The term "bacteriocine" was first coined more than 50 years ago to describe the prototype Gram-negative bacteriocin, colicin (Jacob et al. 1953). However, the first reference to bacteriocin-mediated inhibition dates from over 130 years ago when Pasteur and Joubert (1877) reported the inhibition of *Bacillus anthracis* by bacteria isolated from urine samples.

P.D. Cotter (✉)
Teagasc, Moorepark Food Research Centre, Fermoy, Cork, Ireland
and
Alimentary Pharmabiotic Centre, University College Cork, Cork, Ireland
e-mail: paul.cotter@teagasc.ie

D. Drider and S. Rebuffat (eds.), *Prokaryotic Antimicrobial Peptides: From Genes to Applications*, DOI 10.1007/978-1-4419-7692-5_3,

In 1976, Tagg et al. defined bacteriocins as proteinaceous compounds that kill closely related bacteria, but it is now recognised that bacteriocins can also have activity across genera (broad spectrum). Indeed, the antimicrobial sensitivity of a target strain to any given bacteriocin may depend on the ecological conditions under which it is grown as variations in salt concentration, pH, the presence of membrane disrupting molecules or inducing cultures, and a large number of other environmental parameters can have a significant impact (Ganzle et al. 1999; Rojo-Bezares et al. 2007).

With the continuing discovery and characterisation of new bacteriocins, including many produced by Gram-positive bacteria, it has become apparent that these are a very heterogeneous group of compounds. Early criteria employed to define Gram-positive bacteriocins evolved from approaches employed to define colicins and were originally outlined by Tagg et al. (1976) which are given as follows: (1) a narrow spectrum of inhibitory activity against closely related species, (2) the presence of an essential, biologically active protein moiety, (3) a mode of action that is bactericidal in nature, (4) attachment to specific cell receptors, (5) genetic determinants for production are plasmid borne and (6) production leads to the death of the producing cell. However, due to the increased understanding of the biosynthesis, structure and mode of action of bacteriocins, it is now generally accepted that these criteria no longer accurately describe the myriad of bacteriocins produced by Gram-positive bacteria as many have a broad spectrum of inhibition, are not always plasmid encoded and contain self-protective immunity proteins. We would suggest the following criterion is a more accurate description of the modern usage of the term "bacteriocin": *modified or unmodified peptide antimicrobials produced by bacteria which are protected by a dedicated immunity system.*

Classification Schemes

With respect to bacteriocins produced by Gram-positive bacteria, those produced by lactic acid bacteria (LAB), including the genera *Lactococcus*, *Lactobacillus*, *Leuconostoc*, *Enterococcus*, *Pediococcus*, *Carnobacterium* and some members of the genus *Streptococcus*, have been the focus of particular attention. This is primarily due to the fact that the food grade nature of many of these producers means that their associated peptides can potentially be employed in a variety of different ways, including food preservation, food safety and in human and veterinary medicine (Cotter et al. 2005; Piper et al. 2009). The *Lactococcus lactis* produced Nisin A is the most extensively characterised of all bacteriocins, and the significant interest in bacteriocin-related research can largely be traced to the successful development of nisin from the initial observation of inhibitory activity by a strain of *L. lactis* against *Lactobacillus bulgaricus* (Rogers and Whittier 1928), through to its regulatory approval as a bio-protective agent for food (Delves-Broughton 2005). Due to the extensive focus on LAB bacteriocins, a number of classification schemes have been suggested which are largely applicable to other Gram-positive bacteriocins.

Table 3.1 Original classification scheme of Klaenhammer (1993)

Group	Description	Distinctive features
Class I	Post-translationally modified bacteriocins	Contain the unusual amino acids lanthionine, β-methyl lanthionine and dehydrated residues
Class II	Unmodified peptides	Small (<10 kDa) heat-stable membrane-active peptides
Class III	Unmodified proteins	Large (>30 kDa) heat-labile proteins
Class IV	Complex proteins	Contain lipid or carbohydrate moieties

Early attempts to classify LAB bacteriocins involved placing individual bacteriocins into one of eight groups based on their heat resistance, host range, trypsin sensitivity and the degree of cross-reactivity between various bacteriocin and host combinations (Geis et al. 1983; Kozak et al. 1978). This approach was superseded by that developed by Klaenhammer (1993), who grouped bacteriocins into four distinct classes with further subclasses. These groupings have formed the basis of all subsequent classification schemes for bacteriocins of Gram-positive bacteria (Table 3.1). Klaenhammer suggested four classes of bacteriocins as follows:

- Class I or lantibiotics: defined as small membrane-active peptides (<5 kDa) containing the unusual amino acids lanthionine or β-methyl lanthionine (hence the name lantibiotics) and dehydrated residues. Nisin is the most characterised of the Class I bacteriocins.
- Class II: defined as small heat-stable non-lanthionine containing membrane-active peptides characterised by the presence of a Gly-Gly processing site in the bacteriocin's precursor, the presence of amphiphilic helices with varying amounts of hydrophobicity and moderate-to-high heat stability. These were further subdivided into three subgroups:
  - Subclass IIa: *Listeria*-active peptides with a consensus sequence in the N-terminal of –Tyr-Gly-Asn-Gly-Val-Xaa-Cys-
  - Subclass IIb: Poration complexes consisting of two proteinaceous peptides for activity
  - Subclass IIc: Thiol-activated peptides requiring reduced cysteine residues for activity
- Class III: Large heat-labile proteins, often with enzymatic activity.
- Class IV: Complex proteins composed of one or more chemical moieties, either lipid or carbohydrate.

All subsequent classification schemes accept the first two major classes of LAB/Gram-positive bacteriocins, the Class I lanthionine and Class II non-lanthionine containing peptides (Cotter et al. 2005; Diep and Nes 2002; Heng et al. 2007; Klaenhammer 1993; Pag and Sahl 2002; Willey and van der Donk 2007), and most also agree that since Class IV remains unpopulated to date, it can be omitted from more recent schema.

With respect to the Class I peptides, it is important to emphasise that the presence of lanthionine residues in a peptide does not automatically make it a lantibiotic, since not all such peptides necessarily possess antimicrobial activity (Jack et al. 1995).

For example, SapB, AMfs and SapT secreted by *Streptomyces coelicolor*, *Streptomyces griseus* and *Streptomyces tendae*, respectively, are lanthionine containing peptides which play an important role in aerial mycelium formation in *Streptomyces* without any known antimicrobial activity (Kodani et al. 2004; Kodani et al. 2005; Ueda et al. 2002). It was also suggested that the Class II bacteriocins could also be divided on the basis of particular chemical groups, such as thiolbiotics (having an active –SH group) or cystibiotics (with a cysteine residue). However, this suggestion has not been widely adopted.

Diep and Nes (2002) broadly retained the classification scheme of Klaenhammer (1993), with the exception of the aforementioned exclusion of class IV (Diep and Nes 2002). The scheme proposed by Pag and Sahl (2002) dealt with lantibiotics only and this scheme was further modified by Willey and van der Donk (2007). Cotter et al. (2005) suggested that there are only two classes of Gram-positive bacteriocins: the Class I lantibiotics (further subdivided on the basis of the sequences of the unmodified pre-peptide) and Class II peptides which can be subdivided into four subclasses (IIa, pediocin-like; IIb, two-peptide; IIc, cyclic; and IId, non-pediocin unmodified peptides). This scheme reclassified the Class III bacteriocins as bacteriolysins, since they are lytic enzymes rather than peptides and as such should not be classified as bacteriocins. Heng et al. (2007) agreed broadly with this scheme but suggested a further modification in that cyclic bacteriocins which should be moved to a new Class IV and they also favoured the retention of Class III containing the large proteins such as the lysins. Class II bacteriocins have more recently been further subgrouped on the basis of their amino acid sequences (Martin-Visscher et al. 2009; Nissen-Meyer et al. 2009). Similarly, enterococcal bacteriocins have also been classified on the basis of homology of their amino acid sequences (Franz et al. 2007).

These various schemes are expanded on, and updated, below.

Class I: Post-translationally Modified Bacteriocins

Traditionally, the post-translationally modified bacteriocin group has only contained the lantibiotics. However, as outlined below, the identification of a number of additional post-translationally modified bacteriocins has confused this issue. As a consequence, we propose that Class I be further divided into Class Ia (the lantibiotics), Ib (the labyrinthopeptins) and Ic (the sactibiotics). These are each expanded upon below.

Class Ia (Lantibiotics)

The term lantibiotic, derived from *lan*thionine containing *antibiotics*, is generally applied to small peptides (<5 kDa; 19–28 amino acids in length) which undergo

Table 3.2 Classification schemes previously proposed for the lantibiotics

Scheme	Subclass		
Jung (1991)	Type A Elongated, screw-shaped amphiphilic membrane acting cationic peptides	Type B Globular, short peptides active through cell wall enzyme inhibition	
Klaenhammer (1993)	No subclassification		
Piper et al. (2009) and Cotter et al. (2005)	Twelve subclasses separated on the basis of structural propeptide sequence similarities (see also Table 3.3)		
Willey and van der Donk (2007) and Pag and Sahl (2002)	Class I Modification – LanB/C Export – LanT Leader cleavage – LanP	Class II Modification – LanM Export/cleavage – LanT	Class III No significant antimicrobial activity includes SapB, AmfS and SapT
Heng et al. (2007)	Type A Linear structure	Type B Globular structure	Type C Two component

extensive post-translational modification and contain the eponymous unusual amino acids lanthionine (Lan) and/or β-methyllanthionine (meLan) as well as other unusual amino acids, most frequently but not only dehydroalanine (Dha) and dehydrobutyrine (Dhb). A number of classification schemes for subdividing the lantibiotics have been suggested over the last ~20 years which reflect the complexity of the structure and function of these peptides (Table 3.2). Lantibiotics have been classified by structure into Type A or Type B (Jung 1991), by the amino acid sequence of the unmodified pre-peptide (Cotter et al. 2005), by the enzymes utilised for modification and secretion of the active peptide (Pag and Sahl 2002; Willey and van der Donk 2007) and by whether they require a single or two peptides for optimal activity (Heng et al. 2007).

In 1991, two categories of lantibiotics were suggested, linear or globular using the limited number of structures available (Fig. 3.1). The linear or Type A lantibiotics included nisin, subtilin, Pep5 and epidermin and were grouped on the basis of being elongated, amphiphilic, screw shaped, cationic lantibiotics (up to 34 amino acid residues long) which are similar in the arrangement of their (me) Lan bridges. These were thought to share a similar mode of action with cell death resulting from the formation of pores resulting in the dissipation of membrane potential and the efflux of small molecules from the target cell. In contrast, globular lantibiotics included mercsacidin, cinnamycin, mutacin II and lacticin 481 were grouped due to having more compact structures (up to 19 residues in length) and were at first defined as lantibiotics whose mode of action was mediated through enzyme inhibition of cell wall biosynthesis (Jung 1991). However, owing to a much improved understanding of the mode of action of lantibiotics, this division based on functional

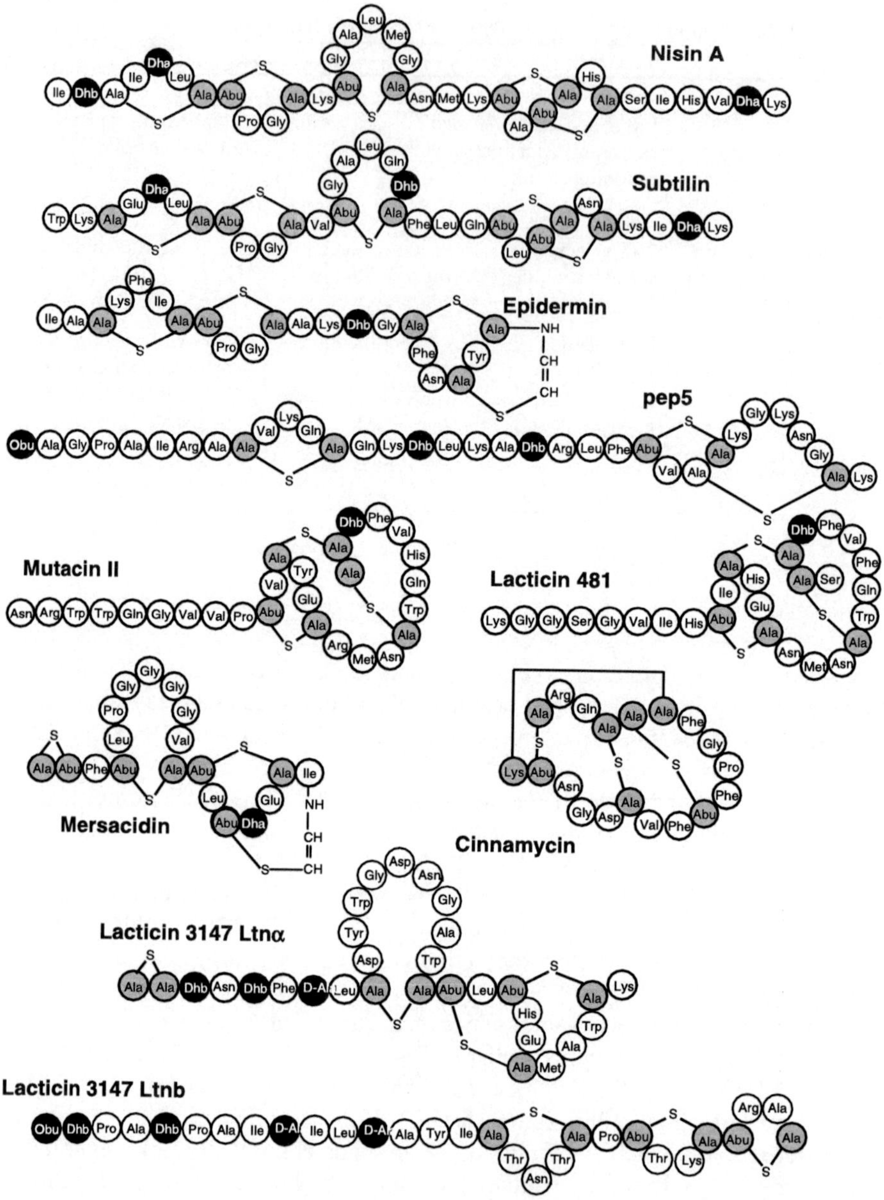

Fig. 3.1 Structures of some representative lantibiotics. Residues involved in (β-methyl) lanthionine structures are *shaded in grey* (Ala-S-Ala lanthionine; Abu-S-Ala, β-methyllanthionine), other modified residues are in *grey* (*Dha* dehydroalanine, *Dhb* dehydrobutyrine, *D-Ala* D-alanine, *Obu* oxobutyrate)

attributes has become blurred. This is illustrated by the fact that nisin, in addition to being a pore former and therefore a Type A on that basis, also inhibits peptidoglycan synthesis in Gram-positive target cells by binding to the peptidoglycan precursor lipid II and thus can also be classified as Type B (Breukink et al. 1999). It has also been revealed that Type A epidermin is not capable of pore formation in some target cells and in those circumstances its activity is a reflection of its ability to inhibit cell wall formation only, a Type B activity (Bonelli et al. 2006). The "Jung" classification scheme also pre-dated the identification of two peptide lantibiotics such as lacticin 3147, which are active as a consequence of the synergistic activity of both component peptides, Ltn-α and Ltn-β. Investigations to date have revealed that Ltn-α first targets lipid II in the cytoplasmic membrane forming a lipid II–Ltn-α complex which then recruits the Ltn-β to form a three-component complex which has the dual function of inhibiting both cell wall biosynthesis and causing subsequent pore formation (Morgan et al. 2005; Wiedemann et al. 2006). Thus, it could be argued that lacticin 3147 is composed of both Type A and Type B peptides. A similar mode of action is proposed for a number of other related two-peptide lantibiotics (Lawton et al. 2007; Oman and van der Donk 2009).

Given that the distinction between Type A and Type B bacteriocins has become blurred, this approach to classification has been superseded by schemes based on the nature of the modification enzymes employed or on the sequence of the ribosomally synthesised inactive peptides prior to modification. Pag and Sahl (2002) suggested two schemes for subdividing lantibiotics into groups based on (1) their primary structures and bridging patterns (creating groups represented by the prototypical lantibiotics nisin, epidermin, pep5, lacticin 481, mersacidin and cinnamycin) or (2) their biosynthetic and export machinery (leader peptide of either FNLD or GG type; modification enzymes of the LanB, LanC or LanM type, and whether export and processing required LanP, LanT or LanT(P) proteins). This latter approach was further developed by Willey and van der Donk (2007) who suggested three subdivisions based on the pathway by which maturation of the peptide occurs and also on the presence or absence of antibiotic activity (Table 3.2), as follows:

- Subclass I: Linear peptides, the best characterised member being nisin, whose pre-peptides are modified by two distinct enzymes LanB and LanC which dehydrate the Thr and Ser residues and mediates cyclisation, respectively. They have a dedicated ABC transporter (LanT). LanP, subtilisin-like sereine proteinases, cleave the leader peptides.
- Subclass II: Often with a more globular structure than the Subclass I lantibiotics, these peptides are modified by large LanM proteins (900–1,000 residues) – modification enzymes which exhibit both dehydratase and cyclase activity. In addition, a single multifunctional protein with a conserved N-terminal cysteine protease domain, also designated LanT, has a dual function of secretion and processing. Included in this class are lactocin S and the two-component lantibiotics such as lacticin 3147. Such is the homology between the genes encoding these modification enzymes, it has been possible to perform rational genome mining for LanM proteins to screen for novel lantibiotic producers (Begley et al. 2009).
- Subclass III: This class of lantibiotic-like peptides, grouped on the basis of related modification enzymes, is unusual in that the peptides have no associated antimicrobial ability (and as such would not fulfil the definition of lantibiotic suggested earlier in this review). Three have been described to date, namely

SapB, AmfS and SapT, which function in aerial mycelium production in *Streptomyces* (Kodani et al. 2004; Kodani et al. 2005; Ueda et al. 2002). The proposed modification enzymes for SapB and AmfS bear homology to the C-terminal domain of the LanM enzymes while lacking the zinc ligands which are important for LanM-catalysed cyclisation.

- Subclass IV: This class of lantibiotics was recently described by Goto et al. in 2010 and is grouped on the basis of an association with yet another novel class of lanthionine synthetases, designated LanL. Investigations with a LanL, encoded within the genome of a *Streptomyces venezuelae* strain, revealed that the enzyme generates dehydroamino acids via phosphorylation of Ser/Thr residues by a protein kinase domain and the subsequent elimination of a phosphate by a Ser/Thr lyase domain. LanLs also contain a C-terminal LanC-like cyclisation domain. The fact that the associated modified peptide also lacks antimicrobial activity leads to the coining of the term "lantipeptides" to describe "compounds that by structure and biosynthetic strategy are clearly related to lantibiotics but that are not known to possess antimicrobial activity" (Goto et al. 2010).

It is also possible to further subdivide lantibiotics (i.e. those with antimicrobial activity) into 12 subclasses on the basis of the sequences of the unmodified forms of the associated propeptides. This approach was first employed for LAB lantibiotics by Twomey et al. (2002) and has since been expanded to include all Gram-positive lantibiotics (Cotter et al. 2005; Piper et al. 2009). The 12 subclasses of lantibiotics based on their amino acid sequence, updated from Piper et al. (2009) to include lichenicidin (Begley et al. 2009) and clausin (Bressolier et al. 2007), are outlined in Table 3.3. In addition, microbisporicin, a potent lantibiotic active against multiresistant pathogens, has been described by Castiglione et al. (2008). The two associated peptides are, on the basis of structural analysis, obviously representative of a new lantibiotic subfamily. However, they have been excluded from the current classification system until such time as the corresponding structural and modification gene(s) have been published (Castiglione et al. 2008).

Class Ib (Labyrinthopeptins)

More recently, a new class of post-translationally modified peptides, designated the labyrinthopeptins (Meindl et al. 2010), has been identified. The newly identified labyrinthopeptins, so named as a consequence of their "labyrinthine" structure, are distinguished by the presence of labionin, a previously unidentified carbacyclic, post-translationally modified amino acid. Two labionins are located in *Actinomadura namibiensis* DSM 6313-produced labyrinthopeptins A2 and A1 (as well as the A1 derivative, A3) and are derived from Ser-Xxx-Xxx-Ser-Xxx-Xxx-Xxx-Cys motifs in the corresponding propeptides. These modifications are

Table 3.3 Lantibiotic classification (updated from Piper et al. 2009) on the basis of structural propeptide amino acid sequences

Table 1. Lantibiotic classification

Subclass	Group	Lantibiotic	Unmodified sequence of structural peptide
I (BC)	Planosporicin	Planosporicin	ITSVSWCTCTCEGGGSGCSHCCPG
	Nisin	Nisin A	ITSISLCTPGCKTGALMGCNMKTATCHCSIHVSK
		Nisin Z	ITSISLCTPGCKTGALMGCNMKTATCNCSIHVSK
		Nisin F	ITSSSLCTPGCKTGALMGCNMKTATCNCSVHVSK
		Nisin U	ITSKSLCTPGCKTGILMTCPLKTATCGC
		Nisin U2	VTSKSLCTPGCKTGILMTCPLTTATCGC
		Nisin Q	ITSISLCTPGCKTGALQGCNLKTATCNCSVHVSK
		Subtilin	WKSESLCTPGCVTGALQTCFLQTLTCNCKISK
		Ericin A	VLSKSLCTPGCITGPLQTCYLCFPTFAKC
	Epidermin	Epidermin	IASKFICTPGCAKTGSFNSYCC
		Epidermin'	IASKFLCTPGCAKTGSFNSYCC
		Gallidermin	VASKFLCTPGCAKTGSFNSYCC
		Staphylococcin T	VASKFLCTPGCAKTGSFNSYCC
		NY266	FKSWSFCTPGCAKTGSFNSYCC
		Mutacin 1140/Mutacin III	FKSWSLCTPGCARTGSFNSYCC
		Clausin	FTSVSFCTPGCGETGSFNSFCC
		Mutacin I	FSSLSLCSLGCTGVKNPFNSYCC
	Streptin	Streptin	VGSRYLCTPGSCVVKLVCFTTTVK
	Pep5	Pep5	TAGPAIRASVKQCQKTLKATRLFTVSCKGKNGCK
		Epicidin 280	SLGPAIKATRQVCPKATRFVTVSCKKSDCQ
		Epilancin K7	SASVLKTSIKVSKKYCKGVTLTCGCNITGGK
II(M)	Lacticin 481	Macedocin	GKNGVKTISHECHLNTWAFLATCCS
		Lacticin 481	KGGSGVIHTISHECNMNSWQFVFTCCS
		MukA1	MGKGAVGTISHECRYNSWAFLATCCS
		MukA2	AGNGVIRTITQGCRMPNNMQVLFTC
		MukA3	MGKGAVGTISHECRYNSWAFLATCCS
		MukA'	MGKGAVGTISHECRYNSWAFLATCCS
		Salivaracin B	GGGVIGTISHQCRMNSWQFLFTCCS
		Lacticin J46	KGGSGVIHTISHEVIYNSWNFVFTCCS
		Mutacin II	NRWWQGVVPTVSYECRMNSWQHVFTCC
		Butyrivibriocin	GNGVIKTISHECHMNTWQFIFTCCS
		RumA	GNGVLKTISHECNMNTWQFLFTCC
		Streptococcin SA-F22	GKNGVFKTISHECHLNTWAFLATCCS
		Nukacin ISK-1	KKKSGVIPTVSHDCHMNSFQFVFTCCS
		Variacin	GSGVIPTISHECHMNSFQFVFTCCS
		Salivaricin A	KRGSGWIATITDDCPNSVFVCC
		Salivaricin A1	KKGSGWFATITDDCPNSVFVCC
		Bovicin HJ50	ADRGWIKTLTKDCPNVISSICAGTIITACKNCA
	Mersacidin	RumB	GEPRVSDGSQFCTSTKECNWGTIMFVCC
		Plantaricin C	KKTKKNXSGDICTLTSECDHLATWVCX
		Mersacidin	CTFTLPGGGGVCTLTSECIC
		Michiganin	SGWLCTLTIQCGIIAR
		Actagardine	CSGWVCTLTIECGTVICAC
		C55a	CSTNTFSLSDYWGNKGNWCTATHECMSWCK
		Ltn α	CSTNTFSLSDYWGNNGAWCTLTHECMAWCK
		SmbB	IGTTVVNSTFSIVLGNKGVICTVTVECMRNCSK
		bhtA-alpha	IGTTVVNSTFSIVLGNKGVICTVTVECMRNCQ
		Plwa	KCKWWNISCDLGNNGHVCTLSHECQVSCN
		BhaA1	CAWYNISCRLGNKGAYCTLTVECMPSCN
		BliA1	TITLSTCAILSKPLGNNGYLCTVTKECMPSCN
	LtnA2	Ltnβ	TTPATPAISILSAYISTNTCPTTKCTRAC
		C55b	GTPLALLGGAATGVIGYISNQTCPTTACTRAC
		Plwb	GDPEARSGIPCTIGAAVAASIAVCPTTKCSKRCGKRKK
		SmbA/bhtA-beta	STPACAIGVVGITVAVTGISTACTRCINK
		BhaA2	GDVHAQTTWPCATVGVSVALCPTTKCTSQC
		BliA2	NDVNPETTPATTSSWTCITAGVTVSASLCPTTKCTSTC
	Cytolysin	CylLl	GDVAETTPVCAACGWVGGGIFTGVTVVVSLKHC
		CylLs	GDVAETTPACFTIGIGVGGLFSAKC
	Lactosin S	LasA	TPVIASVAVSMELLPTASVLYSDVAGCFKYSAKHHC
	Cinnamycin	Cinnamycin	CRQSCSFGPFTFVCDGNTK
		Duramycin	CKQSCSFGPFTFVCDGNTK
		Duramycin B	CRQSCSFGPFTFVCDGNTK
		Duramycin C	CANSCSYGPFTFVCDGNTK
		Ancovenin	CVQSCSFGPLTWSCDGNTK
	Sublancin	Sublancin 168	LGKAQCAALWLQCASGGTIGCGGGAVACQNYRQFCR
III		RamS	TGSRASLLLCGDSSLSITTCN
		AmfS	TGSQVSLLVCEYSSLSVVLCP
		SapT	YTQGCSGLCTIVICATVVICG
IV(L)		Venezuelin	ATCECVGLLTLLNTVCIGISCA

Residues which are 100% conserved across subclasses are emphasised by *black boxes* and residues conserved 100% within subclasses are emphasised by *grey boxes* thought to be catalysed by LabKC, a bifunctional protein with an N-terminal serine/threonine kinase function and a C-terminal lanthionine cyclase function. In addition to being significant as a consequence of containing these unusual structures, the labyrinthopeptins are also noteworthy by virtue of their being active against the herpes simplex virus and exhibiting potential as agents for the treatment of neuropathic pain.

Class Ic (Sactibiotics)

During the early years of the last decade, the elucidation of the structure of subtilosin A created an interesting conundrum. Subtilosin A, produced by *Bacillus subtilus*, is a cyclic peptide but differs significantly from all other circular bacteriocins (subclass IId) in that it is considerably smaller in size and is extensively post-translationally modified, with cross-linkages formed between the sulphurs of three cysteine residues and the α-carbon of two phenylalanines and one threonine (Kawulka et al. 2003; Marx et al. 2001). Initially, the existence of this unusual peptide was not normally taken into consideration when developing new classification schemes, although Martin-Visscher et al. (2009) proposed that subtilosin A should be put in a class of its own. The recent identification of another Gram-positive bacteriocin, the two-peptide thuricin CD, which also contains cysteine to α-carbon bridges (Rea et al. 2010) means that this emerging group of bacteriocins can no longer be omitted from classification schemes. Thuricin CD is a two-peptide bacteriocin (Trnα and Trnβ) produced by *Bacillus thuringiensis* 6431 which is active against *Clostridium difficile*. It shows some similarity to subtilosin A by virtue of the size of the respective peptides and the existence of intramolecular cross-linkages between three cysteines in each peptide and the α-carbons of Ser_{21}, Thr_{25} and Thr_{28} in Trn-α and Thr_{21}, Ala_{25} and Tyr_{28} in trn-β, respectively. Although the Trn-α and Trn-β peptides share a degree of homology (57% sequence similarity and 47.4% identity) with one another, they do not share any significant degree of homology with subtilosin A (Table 3.4). Genetic sequencing of the thuricin operon identified two complete and overlapping ORFs designated *trnC* (1,521 bp) and *trnD* (1,245 bp), ten nucleotides downstream of the structural genes (Rea et al. 2010). Homology searches revealed that TrnC and TrnD belong to the radical *S*-adenosylmethionine (SAM or AdoMet) superfamily of proteins. These enzymes generally generate catalytic radicals via the reductive cleavage of AdoMet by

Table 3.4 Alignment of sactibiotics (Class Ic)

PEPTIDE	AMINO ACID SEQUENCE
Trn-α	G W N A C V I G C I G S C V I S E G I G S L V G T A F T L G
Trn-β	G W V A C V G A C G T V C L A S G G V G T E F A A A S Y F L
Subtilosin A	N K G C A T C S I G A A C L V D G P I P D F E I A G A T G L F G L W G

Amino acid residues conserved through Trn-α and Trn-β are *shaded in black*. Amino acid residues conserved between thuricin peptides and subtilosin A are *shaded in grey* combining SAM and an unusual iron sulphur cluster [4Fe–4S] (Fontecave et al. 2004). A conserved single hallmark signature motif (C–X_3–C–X_2–C) that coordinates the [4Fe–4S] cluster was identified within the N-termini of both putative enzymes. TrnC exhibited sequence similarity to conserved domains of the Fe–S oxidoreductase arylsulfatase regulatory proteins (AslB) and TrnD had sequences similar to the MoaA family of molybdenum co-factor enzymes. Radical SAM enzymes rarely occur in bacteriocin gene clusters, but are included in those of subtilosin A and propioicin F (Brede et al. 2004; Zheng et al. 2000). Sequence alignments indicated that AlbA of subtilosin A exhibits only 19 and 17% identity to TrnC and TrnD, respectively, although AlbA exhibits some features of each individual thuricin CD proteins. As a consequence of the nature of the linkages that are found in both subtilosin A and thuricin CD, we propose that bacteriocins containing these structures be known as sactibiotics (*s*ulphur to *α* *c*arbon linkage-containing an*tibiotics*).

Class II: Unmodified Bacteriocins

The Class II group of bacteriocins are a heterogenous group of peptides (<10 kDa) consisting of standard amino acid residues (which may be linked by disulphide bridges or cyclised at the N and C terminus). These were subdivided into three classes by Klaenhammer (1993); IIa pediocin-like, IIb two-component and IIc thiol-activated. Others have suggested alternative groupings but all classifications have identified two major classes, namely, the Class IIa "pediocin-like" or *Listeria*-active, and the Class IIb two-peptide bacteriocins. A more recent approach has been to change the third subclass, IIc, such that it now contains all other class II bacteriocins (Drider et al. 2006; Eijsink et al. 2002; Nes et al. 1996) while Cotter et al. (2005), in an effort to introduce more order into the classification scheme, have suggested four subdivisions; retaining the IIa and IIb subclasses while introducing two new subdivisions. These would be the Class IIc (cyclic bacteriocins) and Class IId (non-pediocin single linear peptides). Nissen-Meyer et al. (2009) retain this classification scheme in their review of the structure–function relationships of Class II bacteriocins.

Class IIa (Pediocin-Like Bacteriocins)

Pediocin-like bacteriocins are produced by a wide range of Gram-positive bacteria. In recent years, they have been recognised as having potential as natural food preservatives and in possible biomedical applications through their inhibition of spoilage/pathogenic bacteria such as *Listeria monocytogenes* and *Staphylococcus aureus* and spore forming bacteria such as *Bacillus cereus* and *Clostridium perfringens* (Drider et al. 2006). These potential applications and the fact that they are often produced by food grade organisms have resulted in a resurgence of interest in these peptides. There are now 28 different bacteriocins in Class IIa (Nissen-Meyer et al. 2009). One characteristic that unites all members is their narrow spectrum of activity and their high potency against *L. monocytogenes.* They range in size from 55 amino acids for acidocin A (Tahara et al. 1996) to 37 amino acids in the case of sakacin G (Simon et al. 2002), and share a conserved N-terminal region containing the amino acids Y-G-N-G-V-X_1-C-X_2-K/N-X_3-X_4-C (where X is any amino acid). This hydrophilic, cationic sequence is sometimes referred to as the "pediocin box." The N-terminal region of all pediocin-like bacteriocins identified to date contains two cysteines joined by a disulphide bond with X_{1-4} representing polar uncharged or charged residues (Papagianni and Anastasiadou 2009). They are otherwise unmodified except for the cleavage of the leader peptide from its precursor which frequently occurs at a specific Gly-Gly processing site and is preformed in conjunction with export from the cell by a dedicated ABC transporter and its accessory protein (Ennahar et al. 2000). The hydrophobic and/or amphiphilic C-terminal region is less conserved, and the use of sequence alignments of the mature peptides as the basis for further division of class IIa bacteriocins into four subgroups has been proposed by Nissen-Meyer et al. (2009) as outlined in Table 3.5. There is a flexible hinge at the conserved Asp17 in peptides of subgroup 1 and also at Asn17/Asp17 in peptides of subgroups 2, 3 and 4. It is proposed that the conserved region within the pediocin box is responsible for the non-specific binding to the cell surface while the less conserved region of C-terminal domains have been shown to determine target specificity (Johnsen et al. 2005). To date, the three-dimensional (3D) structure of four Class IIa bacteriocins has been determined using nuclear magnetic resonance (NMR); including carnobacteriocin B2 (Sprules et al. 2004), curvacin A (Haugen et al. 2005), leucocin A (Fregeau Gallagher et al. 1997) and sakacin P (and a sakacin P variant) (Uteng et al. 2003). It has been shown that the pediocin-like bacteriocins do indeed consist of two domains which are separated by a flexible hinge, and variation in the 3D structure is suggested as a basis for the differences in specificity between these bacteriocins. The three class IIa bacteriocins with a second disulphide bridge (pediocin PA-1, enterocin A and sakacin G) have been shown to display increased antimicrobial potencies at high temperatures and broader inhibition spectra of activity than their one disulphide bridge containing counterparts (sakacin P and curvacin A) (Eijsink et al. 1998; Fimland et al. 2000). The maintenance of antimicrobial potency at elevated temperatures has been investigated in greater depth. It was noted that at higher temperatures peptides with a single disulphide bond, such as sakacin P or enterocin P, experience partial disruption of the helical structure, whereas those with a second disulphide bond at the C-terminal like PA-1 retain their tertiary structure (Kaur et al. 2004). This study also established that the retention of the α-helix in the C-terminal region is critical for activity. This activity is mediated by the induction of membrane permeability of the target cell membrane, probably by the formation of ion selective pores causing dissipation of the proton motive force and depletion of intracellular ATP (Chikindas et al. 1993; Drider et al. 2006; Herranz et al. 2001a, b).

Table 3.5 Multiple sequence alignment of pediocin-like (Class IIa) bacteriocins adapted from Nissen-Meyer et al. 2009

SUBGROUP	AMINO ACID SEQUENCE OF STRUCTURAL PEPTIDE
Subgroup I	* **
Coagulin	K Y Y G N G V T C G K H S C S V D W G K A T T C I I N N G A M A W A T G G H Q G T H K C
Divergicin M35	K Y Y G N G V Y C N S K K O W V D W G T A Q G C I D - - V V I G Q L G G G I P G K G K C
Divergicin V41	K Y Y G N G V Y C N S K K O W V D W G Q A H G C I G Q T V V G G W L G G A I P G - - K C
Enterocin A	T T H S G K Y Y G N G V Y C T K N K C T V D W A K A T T C I A G M S I G G F L G G A I P G - - K C
Leucocin C	K N Y G N G V H C T K K G C S V D W G Y A W T N I A N N S V M N G L T G G N A G W H N
Listeriocin 743 A	K S Y G N G V H C N K K K C W V D W G S A I S T I G N N S A A N W A T G G A A G W K S
Munticin	K Y Y G N G V S C N K K G C S V D W G K A I G I I G N N S A A N L A T G G A A G W S K
Munticin KS	K Y Y G N G V S C N K K G C S V D W G K A I G I I G N N S A A N L A T G G A A G W K S
Pediocin PA-1	K Y Y G N G V T C G K H S C S V D W G K A T T C I I N N G A M A W A T G G H Q G N H K C
Pisciocin CS 526	K Y Y G N G L S x N K K G x T V D W G T A I G I I G N N A A A N x A T G G A A G x N K
Piscicolin 126	K Y Y G N G V S C N K N G C T V D W S K A I G I I G N N A A A N L T T G G A A G W N K G
Sakacin P	K Y Y G N G V H C G K H S C T V D W G T A I G N I G N N A A A N W A T G G N A G W N K
Sakacin 5X	K Y Y G N G L S C N K S G C S V D W S K A I S I I G N N A V A N L T T G G A A G W K S
Subgroup II	**
Lactococcin MMFII	T S Y G N G V H C N K S K O W I D V S E L E T Y K A G T V S N P K D I L W
Leucoccin A	K Y Y G N G V H C T K S G C S V N W G E A E S A G V H R L A N G G N G F W
Mesentericin Y105	K Y Y G N G V H C T K S G C S V N W G E A A S A G I H R L A N G G N G F W
Plantaricin C19	K Y Y G N G L S C S K K G C T V N W G Q A F S C G V N R V A T A G H G K x
Plantaricin 423	K Y Y G N G V T C G K H S C S V N W G Q A F S C S V S H L A N F G H G K C
Sakacin G	K Y Y G N G V S C N S H G C S V N W G Q A W T C G V N H L A N G G H G V C
Subgroup III	**
Curvicin A	A R S Y G N G V Y C N N K K O W V N R G E A T O S I I G G M I S G W A S G L A G M
Carnobacteriocin BM1	A I S Y G N G V Y C N K E K O W V N K A E N K Q A I T G I V I G G W A S S L A Q M G H
Enterocin P	A T R S Y G N G V Y C N N S K C W V N W G E A K E N I A G I V I S G W A S G L A Q M G H
Subgroup IV	* **
Bacteriocin 31	A T Y Y G N G L Y C N K O K O W V D W N K A S R E I G K I I V N G W V Q H G P W A P R
Bacteriocin RC714	A T Y Y G N G L Y C N K E K O W V D W N Q A K G E I G K I I V N G W V N H G P W A P
Bacteriocin T8	A T Y Y G N G L Y C N K E K O W V D W N Q A K G E I G K I I V N G W V N H G P W A P R R
Penocin A	K K Y Y G N G V H C G K K T O Y V D W G Q A T A S I G K I I V N G W T Q H G P W A H R
Enterocin SE-K4	A T Y Y G N G V Y C N K Q K O W V D W S R A R S E I I D R G V K A Y V N G F T K V L G
Carnobacteriocin B2	V N Y G N G V S C S K T K C S V N W G Q A F Q E R Y T A G I N S F V S G V A S G A G S I G R R P

Sequences 100% conserved between all subgroups are *highlighted in black*. Sequences 100% conserved between members of the same subgroup are *highlighted in grey*. **Proposed site of flexible hinge in each peptide at residue 17 Asp or Asn where residue one is before the well-conserved YGNGY motif*

Class IIb: Two-Peptide Unmodified Bacteriocins

There are currently 16 two-peptide unmodified bacteriocins (Nissen-Meyer et al. 2009). These are primarily of LAB origin, but this subclass also contains the anti-botulinum bacteriocin brochochin-C produced by *Brochothrix campestris*. The number of amino acid residues within these peptides varies from <40 amino acid residues, for example, plantaricin S and lactococcin G with 26/24 and 39/35 residues in its two peptides, respectively (Jimenez-Diaz et al. 1995; Nissen-Meyer et al. 1992), to larger peptides such as brochocin-C, sakacin T, gassericin T and lactacin F, in which case at least one of the peptides is >50 amino acid residues in length (Kawai et al. 2000; McCormick et al. 1998; Muriana and Klaenhammer 1991; Vaughan et al. 2001). In all cases, the genes encoding the two bacteriocin components are found side-by-side on the same operon, while the gene encoding the corresponding immunity protein, which provides protection to both peptides, is immediately adjacent. The gene encoding the dedicated ABC transporter is also located within the same, or an adjacent, operon (Nes et al. 2002; Oppegard et al. 2007). The bacteriocin peptides have little or no activity in isolation and need to physically interact to form an active complex (Oppegard et al. 2007). Mode of action studies with a number of class IIb bacteriocins has revealed that permeabilisation of the target cell membrane occurs on contact with the peptides, leading to an efflux of small molecules and ultimately cell death (Nissen-Meyer et al. 2009). While amino acid sequences and structures of the two-peptide bacteriocins vary, a common GxxxG-motive present on both peptides is conserved (Table 3.6). On rare occasions, some additional homology is observed, e.g. enterocin 1071A and lactococcin Gα share 64% similarity, enterocin 1071B shares 61% similarity with lactococcin Gβ and lactococcin Qβ (Zendo et al. 2006) and enterocin L50 A and B share 72% similarity to each other (Criado et al. 2006; Garneau et al. 2002; Zendo et al. 2006). In general, peptides are only active with their complementary peptides from the same two-component complex, but enterocin 1071, lactococcin G and lactococcin Q are exceptional in that, as a consequence of the similarity between the various peptides, components from two different bacteriocins can function together without compromising the antimicrobial activity (Oppegard et al. 2007). To date, the three-dimensional structure of three unmodified two-component bacteriocins have been elucidated; plantaricin EF, plantaricin JF and lactococcin G (Fimland et al. 2008; Rogne et al. 2008; Rogne et al. 2009). In all three cases, the peptides were shown to form amphiphilic α-helixes when exposed to membrane-mimicking environments. It had been demonstrated that in lactococcin G the helix–helix interaction motif GxxxG is vital to its antimicrobial activity (Senes et al. 2004). Using plantaricin JK as a model, variants were produced whereby replacement of the Gly residues of the GxxxG motif with large residues caused loss of antimicrobial activity. As a result, it was proposed that the two peptides interact through their GxxxG motives and that replacement of the Gly residues interrupted this interaction (Rogne et al. 2009). It has been suggested that all three Class IIb bacteriocins for which the three-dimensional structures have been elucidated form

Table 3.6 Sequence alignments of two-peptide Class II bacteriocins (adapted from Nissen-Meyer et al. 2009)

BACTERIOCIN	PEPTIDE	AMINO ACID SEQUENCE
A		
ABP-118	ABP-α	K R G P N C V G N F L G G L F A G A A A G V P L G P A G I V G G A N L G M V G G A L T C L
	ABP-β	K N G Y G G S G N R W V H C G A G I V G G A L I G A I G G P W S A V A G G I S G G F T S C R
Brochocin C	BrcA	Y S S K D C L K D I G K G I G A G T V A G A A G G G L A A G L G A I P G A F V G A H F G V I G G S A A C I G G L L G N
	BrcB	K I N W G N V G G S C V G G A V I G G A L G G L G G A G G G C I T G A I G S I W D Q W
Lacticin F	LafA	R N N W Q T N V G G A V G S A M I G A T V G G T I C G P A C A V A G A H Y L P I L W T G V T A A T G G F G K I R K
	LafX	N R W G D T V L S A A S G A G T G I K A C K S F G P W G M A I C G V G G A A I G G Y F G Y T H N
Lactocin 705	705α	G M S G Y I Q G I P D F L K G Y L H G I S A A N K H K K G R L G Y
	705β	G F W G G L G Y I A G R V G A A Y G H A Q A S A N N H H S P I N G
Mutacin IV	NlmA	K V S G G E A V A A I G I C A T A S A A I G G L A G A T L V T P Y C V G T W G L I R S H
	NlmB	D K Q A A D T F L S A V G G A A S G F T Y C A S N G V W H P Y I L A G C A G V G A V G S V V F P H
Plantaricin E/F	PlnE	F N R G G Y N F G K S V R H V V D A I G S V A G I R G I L K S I R
	PlnF	V F H A Y S A R G V R N N Y K S A V G P A D W V I S A V R G F I H G
Plantaricin J/K	PlnJ	G A W K N F W S S L R K G F Y D G E A G R A I R R
	Pln K	R R S R K N G I G Y A I G Y A F G A V E R A V L G G S R D Y N K
Plantaricin NC8	Plc-α	D L T T K L W S S W G Y Y L G K K A R W N L K H P Y V Q F
	Plc-β	S V P T S V Y T L G I K I L W S A Y K H R K T I E K S F N K G F Y H
		* * * * *
Plantaricin S	Pls-α	R N K L A Y N M G H Y A G K A T I F G L A A W A L L A
	Pls-β	K K K K Q S W Y A A A G D A I V S F G E G F L N A W
		* * * * *
Salivaricin P	Sln 1	K R G P N C V G N F L G G L F A G A A A G V P L G P A G I V G G A N L G M V G G A L T C L
	Sln 2	K N G Y G G S G N R W V H C G A G I V G G A L I G A I G G P W S A V A G G I S G G F A S C H
Thermophilin 13	ThmA	Y S G K D C L K D M G G Y A L A G A G S G A L W G A P A G G V G A L P G A F V G A H V G A I A G G F A C M G G M I G N K F N
	ThmB	Q I N W G S V V G H C I G G A I I G G A F S G G A A A G V G C L V G S G K A I I N G L
B		
Lactococcin G	LcnG-α	G T W D D I G Q G I G R V A Y W V G K A M G N M S D V N Q A S R I N R K K K H
Lactococcin Q	LcnQ-a	S I W G D I G Q G V G K A A Y W V G K A M G N M S D V N Q A S R I N R K K K H
Enterocin 1071	Ent-α	E S V F S K I G N A V G P A A Y W I L K G L G N M S D V N Q A D R I N R K K - H
Lactococcin G	LcnG-β	K K W G W L A W V D P A Y E F I K G F G K G A I K E G N K D K W K N I
Lactococcin Q	LcnQ-β	K K W G W L A W V E P A G E F L K G F G K G A I K E G N K D K W K N I
Enterocin 1071	Ent-β	G P G K W L P W L Q P A Y D F V T G L A K G I G K E G N K N K W K N V

(A) Identical residues are *shaded in black*. GxxxG motives are *shaded in grey*. SxxxS and AxxxA motives present in Pls-b and Plc-b are marked by ***** in the absence of GxxxG. (B) α and β sequences of LcnG, LcnQ and Ent; residues that are identical in all three bacteriocins are *shaded in black*. Those that are conserved in two/three are *shaded in grey* a membrane spanning helix–helix structure that may interact with a membrane bound receptor (Rogne et al. 2009) and thus these bacteriocins function in a manner similar to the single peptide Class IIa and Class IId bacteriocins. The availability of structure-related data may be of critical importance with respect to further classification of these peptides. Following structural analysis of the brochocin C α- and β-peptides, it was proposed that they possess a β-sheet structure rather than α-helices (Garneau et al. 2003). In turn, Oppegard et al. (2007) suggested that since thermophilin 13 shares a high degree of sequence similarity with brochocin C, its components may also have a β-sheet structure while the ABP-118 peptides have large hydrophobic segments with cysteine at both ends (Flynn et al. 2002) and are not likely to easily form amphililic α-helices. Therefore, the presence of two quite different structures among subclass IIb bacteriocins may in the future permit further structure-based subclassification.

Class IIc: Circular Bacteriocins

As with other bacteriocins, class IIc peptides are ribosomally synthesised and thus are distinct from enzymatically synthesised cyclic antimicrobial peptides such as gramicidin-S and mycosubtilin (Duitman et al. 1999; Maqueda et al. 2008; Mogi and Kita 2009). The class IIc precursor proteins undergo post-translational modifications resulting in the covalent linkage of their N- and C-termini to create a circular backbone. This subclass thus could arguably also include subtilosin A, but as a consequence of its associated structures, it has been positioned in a new class Ib category as noted above. NMR and X-ray diffraction approaches have been employed to elucidate the structure of two such peptides, carnocyclin A and AS-48, and has revealed the presence of regularly repeated α-helical motifs (Jimenez et al. 2005; Kawulka et al. 2003; Martin-Visscher et al. 2009). These peptides are generally heat stable, show significant resistance to proteolytic digestion and have been shown to display antilisterial activity. They have been placed in class IIc by Cotter et al. (2005) while Kemperman et al. (2003) recommended that an additional class should be added to accommodate circular bacteriocins. Furthermore, Franz et al. (2007) placed the cyclic bacteriocins, enterocin AS-48 and AS-48RJ, into a distinct, separate class when classifying enterococcal bacteriocins (Cotter et al. 2005; Franz et al. 2007; Kemperman et al. 2003) while Maqueda et al. (2008) and Heng et al. (2007) propose that the circular bacteriocins should be moved to a separate class IV (Heng et al. 2007; Maqueda et al. 2008). To date, eight cyclic bacteriocins have been characterised excluding subtilosin A and thuricin CD (Table 3.7). Six are LAB associated (Gassericin A, Reutericin 6, Enterocin AS-48, Enterocin 4, Carnocyclin A and Lactocyclicin Q) while two are non-LAB bacteriocins (cirularin A and butyrivibriocin AR10 produced by *Clostridium beijerinckii* and *Butryrivibrio fibrisolvens*, respectively). They have been subdivided into two subgroups on the basis of amino acid sequence identity (Cotter et al. 2005; Martin-Visscher et al. 2009).

Table 3.7 Circular bacteriocins

Bacteriocin	Producing strain	Structure	References
Acidocin B	*Lactobacillus acidophilus*	Not determined	Leer et al. (1995)
Butyrivibriocin AR10	*Butryrivibrio fibrisolvens AR10*	Not determined	Kalmokoff et al. (2003)
Carnocyclin A	*Carnobacterium maltaromaticum*	Compact globular bundle composed of four helices	Martin-Visscher et al. (2008, 2009), Gong et al. (2009)
Circularin A	*Clostridium beijerinckii*	Not determined	Kemperman et al. (2003)
Gassericin A	*Lactobacillus gasseri LA39*	Not completely resolved. Primarily α-helices	Kawai et al. (2004)
Lactocyclicin Q	*Lactococcus sp QU 12*	Predicted to consist of four α-helices	Sawa et al. (2009)
Reutericin 6	*Lactobacillus reuteri LA6*	Not completely resolved. Primarily α-helices	Kawai et al. (2004)
Uberolysin	*Streptococcus uberis*	Not determined	Wirawan et al. (2007)

Cotter et al. (2005) proposed two divisions of the class IIc bacteriocins, namely, class IIc(i) containing enterocin AS-48 and the non-LAB circularin A and IIc(ii) containing gassericin A, reutericin 6 and AR10 produced by *B. fibrisolvens*. In contrast, Martin-Visscher et al. (2009) while still basing the subdivisions on amino acid sequences placed the disparate carnocyclin A, lactocyclicin Q, AS-48, circularin A and uberolysin in subclass IIc(i) and the highly homologous bacteriocins gassericin A, reutericin 6 and AR10 in subclass IIc(ii). The sequence alignments as shown in Table 3.8 are adapted from Martin-Visscher et al. (2009), where they show that the greatest similarity existed among N-terminal domains and predict that this region contains two distinct helices. Sequence analysis of group 2 containing the homologous bacteriocins gassericin A, reutericicn 6 and butryivibriocin AR10 predicted four helices (Table 3.8).

Class IId: Unmodified, Linear, Non-pediocin-Like Bacteriocins

According to the various classification schemes, the unmodified, linear, non-pediocin-like, one peptide bacteriocins are placed in class IId (Cotter et al. 2005; Nissen-Meyer et al. 2009) or IIc (Heng et al. 2007), simply because they do not conform to any of the criteria now set down in the currently available classification schemes (Cotter et al. 2005; Heng et al. 2007; Nissen-Meyer et al. 2009). As a result, this group of bacteriocins encompasses a heterogeneous compilation of antimicrobial peptides from a wide variety of strains from various ecological niches of which lactococcin A, the first to be isolated, is the best characterised. Nissen-Meyer et al. (2009) have classified 31 class IId bacteriocins which are primarily LAB in

Table 3.8 Alignment of Class IIc circular bacteriocins (adapted from Martin-Visscher et al. 2009)

GROUP	AMINO ACID SEQUENCE
Subgroup 1	
Cc1A	- L V A Y G I A Q G T A E K V V S L I N A G - - - L T V G S I I S - I L G - G - - - - - V T V G L S G V F T A V K A A I A K Q G I K K A I Q L
LactQ1	L I D H L G A P R W A V D T I L G A I A V G - - - - N L A S W V L - A L V P G - P G W A V K A G L A T A A A I V K H - - - - Q G K A A A A A W
AS-48	M A K E F G I P A A V A G T V L N V V E A G G W V T T I V S I L T - A V G S G G L S L L A A A G R E S I K A Y L K K E I K K K G K R A V I A W
CirA	V A G A L G V Q T A A A T T I V N V I L N A G T L V T V L G I I A - S I A S G G A G T L M T I G W A T F K A T V Q K - L A K Q S M A R A I A Y
Uber	L A G Y T G I A S G T A K K V V D A I D K G A A A F V I I S I I S T V I S A G A L G - A V S A S A D F I I L T V K N Y I S R N L K A Q A V I W
Subgroup 2	
GassA	I Y W I A D Q F G I H L A T G T A R K L L D A M A S G A S L G T A F A A I L G V T L P A W A L A A A G A L G A T A A
Reut 6	I Y W I A D Q F G I H L A T G T A R K L L D A M A S G A S L G T A F A A I L G V T L P A W A L A A A G A L G A T A A
B-AR10	I Y F I A D K M G I Q L A P A W Y Q D I V N W V S A G G T L T T G F A I I V G V T V P A W I A E A A A A F G I A S A

Subgroup 1: Amino acids conserved in at least 75% of members are *shaded in black*; amino acids conserved in at least 40% of members are *shaded in grey*.
Subgroup 2: Amino acid residues conserved throughout all the members of Subgroup 2

origin but have also been isolated from *Staphylococcus* and *Weissella* sp. (Nissen-Meyer et al. 2009). Heng et al. (2007) have also included bacteriocins from *Propionibacterium* sp. among this class.

Bacteriolysins (Formerly Class III Bacteriocins)

Bacteriolysins (formerly class III bacteriocins) are large, heat-labile antimicrobial proteins. They have a domain-type structure, in which different domains have functions for translocation, receptor binding and lethal activity. They include five domains produced by LAB which have been genetically characterised, including helveticin J from *Lactobacillus helveticus*, zoocin A from *Streptococcus zooepidermicus*, enterolysin A produced by *Enterococcus faecalis*, millericin B produced by *Streptococcus milleri* and linocin M18 produced by a strain of *Brevibacterium linens* (Beukes et al. 2000; Hickey et al. 2003; Joerger and Klaenhammer 1986; Simmonds et al. 1997; Valdes-Stauber and Scherer 1994). Of the non-LAB proteins that fall into this category, lysostaphin is the best known (Browder et al. 1965; Schindler and Schuhardt 1964).

Table 3.9 Updated classification scheme for Gram-positive bacteriocins and bacteriocin-like peptides and proteins

Class	Description	Further subdivisions
Class I	*Modified peptides*	
	(a) Lantibiotics and lantipeptides	Subclasses I and II (modified by LanBC and LanM proteins) can also be further divided into 12 further subclasses based on amino acid sequences Subclasses III and IV (modified by RamC-like and LanL proteins, respectively)
	(b) Labyrinthopeptins	
	(c) Sactibiotics	Two subclasses: single- and two-peptide bacteriocins
Class II	*Non-modified peptides*	
	(a) Pediocin-like	Four subclasses I–IV
	(b) Two-peptide bacteriocins	Two subclasses, i.e. A and B
	(c) Circular bacteriocins	Two subclasses, i.e. 1 and 2
	(d) Linear non-pediocin-like one peptide bacteriocins	
Bacteriolysins (formerly Class III)	*Non-bacteriocin lytic proteins*	

Conclusions

Classification schemes for bacteriocins have evolved over the last 25 years, reflecting the identification of new bacteriocins and the sophistication of both molecular and chemical tools available to modern-day researchers. For instance, novel sequence data now available for many bacteriocins has allowed classification schemes to become more refined. As more information regarding the composition, three-dimensional structure and mode of action of bacteriocins becomes available, the classification schemes will without doubt have to be modified to reflect this. Here, we have endeavoured to combine existing classification approaches with new information. The most notable divergence from previous classification schemes reflects the identification of lantibiotic-like peptides that lack antimicrobial activity, i.e. the lantipeptides, and of modified bacteriocins which are obviously neither lantibiotics nor lantipeptides. Taking this approach, Class I bacteriocins have now been divided into three subclasses, namely, lantibiotics/lantipeptides, which can be further subdivided based on the amino acid sequences of the propeptides and modification enzymes, labyrinthopeptides and sactibiotics. While Class IIa bacteriocins continue to be broadly defined on the basis of the conserved N-terminal region, the "pediocin box," and subdivided through sequence alignments of the less conserved C-terminal region, it is likely that as the three-dimensional structure of more of these peptides is elucidated, further subdivisions will be proposed. This is likely to be also the case with the two-peptide, unmodified bacteriocins (Class IIb) and circular bacteriocins (Class IIc). A summary of the classification scheme that we propose is depicted in Table 3.9.

References

Begley M, Cotter PD, Hill C, Ross RP (2009) Identification of a novel two-peptide lantibiotic, lichenicidin, following rational genome mining for LanM proteins. Appl Environ Microbiol 75:5451–5460

Beukes M, Bierbaum G, Sahl HG, Hastings JW (2000) Purification and partial characterization of a murein hydrolase, millericin B, produced by *Streptococcus milleri* NMSCC 061. Appl Environ Microbiol 66:23–28

Bonelli RR, Schneider T, Sahl HG, Wiedemann I (2006) Insights into in vivo activities of lantibiotics from gallidermin and epidermin mode-of-action studies. Antimicrob Agents Chemother 50:1449–1457

Brede DA, Faye T, Johnsborg O, Odegard I, Nes IF, Holo H (2004) Molecular and genetic characterization of propionicin F, a bacteriocin from *Propionibacterium freudenreichii*. Appl Environ Microbiol 70:7303–7310

Bressolier P, Brugo MA, Robineau P, Schmitter M, Sofeir M, et al. 2007. Peptide compound with biological activity, its preparation and application. WIPO PCT/IB2007/0022003. France

Breukink E, Wiedemann I, van Kraaij C, Kuipers OP, Sahl H, de Kruijff B (1999) Use of the cell wall precursor lipid II by a pore-forming peptide antibiotic. Science 286:2361–2364

Browder HP, Zygmunt WA, Young JR, Tavormina PA (1965) Lysostaphin: enzymatic mode of action. Biochem Biophys Res Commun 19:383–389

Castiglione F, Lazzarini A, Carrano L, Corti E, Ciciliato I, Gastaldo L, Candiani P, Losi D, Marinelli F, Selva E, Parenti F (2008) Determining the structure and mode of action of microbisporicin, a potent lantibiotic active against multiresistant pathogens. Chem Biol 15:22–31

Chikindas ML, Garcia-Garcera MJ, Driessen AJ, Ledeboer AM, Nissen-Meyer J, Nes IF, Abee T, Konings WN, Venema G (1993) Pediocin PA-1, a bacteriocin from *Pediococcus acidilactici* PAC1.0, forms hydrophilic pores in the cytoplasmic membrane of target cells. Appl Environ Microbiol 59:3577–3584

Cotter PD, Hill C, Ross RP (2005) Bacteriocins: developing innate immunity for food. Nat Rev Microbiol 3:777–788

Criado R, Diep DB, Aakra A, Gutierrez J, Nes IF, Hernandez PE, Cintas LM (2006) Complete sequence of the enterocin Q-encoding plasmid pCIZ2 from the multiple bacteriocin producer *Enterococcus faecium* L50 and genetic characterization of enterocin Q production and immunity. Appl Environ Microbiol 72:6653–6666

Delves-Broughton J (2005) Nisin as a food preservative. Food Australia 57:525–527

Diep DB, Nes IF (2002) Ribosomally synthesized antibacterial peptides in Gram positive bacteria. Curr Drug Targets 3:107–122

Drider D, Fimland G, Hechard Y, McMullen LM, Prevost H (2006) The continuing story of class IIa bacteriocins. Microbiol Mol Biol Rev 70:564–582

Duitman EH, Hamoen LW, Rembold M, Venema G, Seitz H, Saenger W, Bernhard F, Reinhardt R, Schmidt M, Ullrich C, Stein T, Leenders F, Vater J (1999) The mycosubtilin synthetase of *Bacillus subtilis* ATCC6633: a multifunctional hybrid between a peptide synthetase, an amino transferase, and a fatty acid synthase. Proc Natl Acad Sci U S A 96:13294–13299

Eijsink VG, Skeie M, Middelhoven PH, Brurberg MB, Nes IF (1998) Comparative studies of class IIa bacteriocins of lactic acid bacteria. Appl Environ Microbiol 64:3275–3281

Eijsink VG, Axelsson L, Diep DB, Havarstein LS, Holo H, Nes IF (2002) Production of class II bacteriocins by lactic acid bacteria; an example of biological warfare and communication. Antonie Van Leeuwenhoek 81:639–654

Ennahar S, Sashihara T, Sonomoto K, Ishizaki A (2000) Class IIa bacteriocins: biosynthesis, structure and activity. FEMS Microbiol Rev 24:85–106

Fimland G, Johnsen L, Axelsson L, Brurberg MB, Nes IF, Eijsink VG, Nissen-Meyer J (2000) A C-terminal disulfide bridge in pediocin-like bacteriocins renders bacteriocin activity less temperature dependent and is a major determinant of the antimicrobial spectrum. J Bacteriol 182:2643–2648

Fimland N, Rogne P, Fimland G, Nissen-Meyer J, Kristiansen PE (2008) Three-dimensional structure of the two peptides that constitute the two-peptide bacteriocin plantaricin EF. Biochim Biophys Acta 1784:1711–1719

Flynn S, van Sinderen D, Thornton GM, Holo H, Nes IF, Collins JK (2002) Characterization of the genetic locus responsible for the production of ABP-118, a novel bacteriocin produced by the probiotic bacterium Lactobacillus *salivarius* subsp. *salivarius* UCC118. Microbiology 148:973–984

Fontecave M, Atta M, Mulliez E (2004) S-adenosylmethionine: nothing goes to waste. Trends Biochem Sci 29:243–249

Franz CM, Belkum MJ, Holzapfel WH, Abriouel H, Galvez A (2007) Diversity of enterococcal bacteriocins and their groupings in a new classification scheme. FEMS Microbiol Rev 31:293–310

Fregeau Gallagher NL, Sailer M, Niemczura WP, Nakashima TT, Stiles ME, Vederas JC (1997) Three-dimensional structure of leucocin A in trifluoroethanol and dodecylphosphocholine micelles: spatial location of residues critical for biological activity in type IIa bacteriocins from lactic acid bacteria. Biochemistry 36:15062–15072

Ganzle MG, Weber S, Hammes WP (1999) Effect of ecological factors on the inhibitory spectrum and activity of bacteriocins. Int J Food Microbiol 46:207–217

Garneau S, Martin NI, Vederas JC (2002) Two-peptide bacteriocins produced by lactic acid bacteria. Biochimie 84:577–592

Garneau S, Ference CA, van Belkum MJ, Stiles ME, Vederas JC (2003) Purification and characterization of brochocin A and brochocin B(10-43), a functional fragment generated by heterologous expression in *Carnobacterium piscicola*. Appl Environ Microbiol 69:1352–1358

Geis A, Singh J, Teuber M (1983) Potential of lactic streptococci to produce bacteriocin. Appl Environ Microbiol 45:205–211

Gong X, Martin-Visscher LA, Nahirney D, Vederas JC, Duszyk M (2009) The circular bacteriocin, carnocyclin A, forms anion-selective channels in lipid bilayers. Biochim Biophys Acta 1788:1797–1803

Goto Y, Li B, Claesen J, Shi Y, Bibb MJ, van der Donk WA (2010) Discovery of unique lanthionine synthetases reveals new mechanistic and evolutionary insights. PLoS Biol 8:e1000339

Haugen HS, Fimland G, Nissen-Meyer J, Kristiansen PE (2005) Three-dimensional structure in lipid micelles of the pediocin-like antimicrobial peptide curvacin A. Biochemistry 44:16149–16157

Heng NCK, Wescombe PA, Burton JP, Jack RW, Tagg JR (2007) The diversity of bacteriocins in Gram-positive bacteria. In: Riley MA, Chavan MA (eds) Bacteriocins: ecology and evolution. Springer, Berlin

Herranz C, Chen Y, Chung HJ, Cintas LM, Hernandez PE, Montville TJ, Chikindas ML (2001a) Enterocin P selectively dissipates the membrane potential of *Enterococcus faecium* T136. Appl Environ Microbiol 67:1689–1692

Herranz C, Cintas LM, Hernandez PE, Moll GN, Driessen AJ (2001b) Enterocin P causes potassium ion efflux from *Enterococcus faecium* T136 cells. Antimicrob Agents Chemother 45:901–904

Hickey RM, Twomey DP, Ross RP, Hill C (2003) Production of enterolysin A by a raw milk enterococcal isolate exhibiting multiple virulence factors. Microbiology 149:655–664

Jack RW, Tagg JR, Ray B (1995) Bacteriocins of Gram-positive bacteria. Microbiol Rev 59:171–200

Jacob F, Lwoff A, Siminovitch A, Wollman E (1953) Definition de quelque termes relatifs a la lysogenie. Ann Inst Pasteur (Paris) 84:222–224

Jimenez MA, Barrachi-Saccilotto AC, Valdivia E, Maqueda M, Rico M (2005) Design, NMR characterization and activity of a 21-residue peptide fragment of bacteriocin AS-48 containing its putative membrane interacting region. J Pept Sci 11:29–36

Jimenez-Diaz R, Ruiz-Barba JL, Cathcart DP, Holo H, Nes IF, Sletten KH, Warner PJ (1995) Purification and partial amino acid sequence of plantaricin S, a bacteriocin produced by *Lactobacillus plantarum* LPCO10, the activity of which depends on the complementary action of two peptides. Appl Environ Microbiol 61:4459–4463

Joerger MC, Klaenhammer TR (1986) Characterization and purification of helveticin J and evidence for a chromosomally determined bacteriocin produced by *Lactobacillus helveticus* 481. J Bacteriol 167:439–446

Johnsen L, Fimland G, Nissen-Meyer J (2005) The C-terminal domain of pediocin-like antimicrobial peptides (class IIa bacteriocins) is involved in specific recognition of the C-terminal part of cognate immunity proteins and in determining the antimicrobial spectrum. J Biol Chem 280:9243–9250

Jung G (1991) Lantibiotics: a survey. In Nisin and novel lantibiotics G Jung and HG Sahl Editors ESCOM Science Publishers, BV Leiden

Kalmokoff ML, Cyr TD, Hefford MA, Whitford MF, Teather RM (2003) Butyrivibriocin AR10, a new cyclic bacteriocin produced by the ruminal anaerobe *Butyrivibrio fibrisolvens* AR10: characterization of the gene and peptide. Can J Microbiol 49:763–773

Kaur K, Andrew LC, Wishart DS, Vederas JC (2004) Dynamic relationships among type IIa bacteriocins: temperature effects on antimicrobial activity and on structure of the C-terminal amphipathic alpha helix as a receptor-binding region. Biochemistry 43:9009–9020

Kawai Y, Saitoh B, Takahashi O, Kitazawa H, Saito T, Nakajima H, Itoh T (2000) Primary amino acid and DNA sequences of gassericin T, a lactacin F-family bacteriocin produced by *Lactobacillus gasseri* SBT2055. Biosci Biotechnol Biochem 64:2201–2208

Kawai Y, Kemperman R, Kok J, Saito T (2004) The circular bacteriocins gassericin A and circularin A. Curr Protein Pept Sci 5:393–398

Kawulka K, Sprules T, McKay RT, Mercier P, Diaper CM, Zuber P, Vederas JC (2003) Structure of subtilosin A, an antimicrobial peptide from *Bacillus subtilis* with unusual posttranslational modifications linking cysteine sulfurs to alpha-carbons of phenylalanine and threonine. J Am Chem Soc 125:4726–4727

Kemperman R, Kuipers A, Karsens H, Nauta A, Kuipers O, Kok J (2003) Identification and characterization of two novel clostridial bacteriocins, circularin A and closticin 574. Appl Environ Microbiol 69:1589–1597

Klaenhammer TR (1993) Genetics of bacteriocins produced by lactic acid bacteria. FEMS Microbiol Rev 12:39–85

Kodani S, Hudson ME, Durrant MC, Buttner MJ, Nodwell JR, Willey JM (2004) The SapB morphogen is a lantibiotic-like peptide derived from the product of the developmental gene ramS in *Streptomyces coelicolor*. Proc Natl Acad Sci U S A 101:11448–11453

Kodani S, Lodato MA, Durrant MC, Picart F, Willey JM (2005) SapT, a lanthionine-containing peptide involved in aerial hyphae formation in the streptomycetes. Mol Microbiol 58:1368–1380

Kozak W, Bardowski J, Dobrzanski WT (1978) Lactostrepcins – acid bacteriocins produced by lactic streptococci. J Dairy Res 45:247–257

Lawton EM, Cotter PD, Hill C, Ross RP (2007) Identification of a novel two-peptide lantibiotic, haloduracin, produced by the alkaliphile *Bacillus halodurans* C-125. FEMS Microbiol Lett 267:64–71

Leer RJ, van der Vossen JM, van Giezen M, van Noort JM, Pouwels PH (1995) Genetic analysis of acidocin B, a novel bacteriocin produced by *Lactobacillus acidophilus*. Microbiology 141(Pt 7):1629–1635

Maqueda M, Sanchez-Hidalgo M, Fernandez M, Montalban-Lopez M, Valdivia E, Martinez-Bueno M (2008) Genetic features of circular bacteriocins produced by Gram-positive bacteria. FEMS Microbiol Rev 32:2–22

Martin-Visscher LA, van Belkum MJ, Garneau-Tsodikova S, Whittal RM, Zheng J, McMullen LM, Vederas JC (2008) Isolation and characterization of carnocyclin A, a novel circular bacteriocin produced by *Carnobacterium maltaaromaticum* UAL 307. Appl Environ Microbiol 74:4756–4763

Martin-Visscher LA, Gong X, Duszyk M, Vederas JC (2009) The three-dimensional structure of carnocyclin A reveals that many circular bacteriocins share a common structural motif. J Biol Chem 284:28674–28681

Marx R, Stein T, Entian KD, Glaser SJ (2001) Structure of the *Bacillus subtilis* peptide antibiotic subtilosin A determined by 1H-NMR and matrix assisted laser desorption/ionization time-of-flight mass spectrometry. J Protein Chem 20:501–506

McCormick JK, Poon A, Sailer M, Gao Y, Roy KL, McMullen LM, Vederas JC, Stiles ME, Van Belkum MJ (1998) Genetic characterization and heterologous expression of brochocin-C, an antibotulinal, two-peptide bacteriocin produced by *Brochothrix campestris* ATCC 43754. Appl Environ Microbiol 64:4757–4766

Meindl K, Schmiederer T, Schneider K, Reicke A, Butz D, Keller S, Guhring H, Vertesy L, Wink J, Hoffmann H, Bronstrup M, Sheldrick GM, Sussmuth RD (2010) Labyrinthopeptins: a new class of carbacyclic lantibiotics. Angew Chem Int Ed Engl 49:1151–1154

Mogi T, Kita K (2009) Gramicidin S and polymyxins: the revival of cationic cyclic peptide antibiotics. Cell Mol Life Sci 66:3821–3826

Morgan SM, O'Connor PM, Cotter PD, Ross RP, Hill C (2005) Sequential actions of the two component peptides of the lantibiotic lacticin 3147 explain its antimicrobial activity at nanomolar concentrations. Antimicrob Agents Chemother 49:2606–2611

Muriana PM, Klaenhammer TR (1991) Purification and partial characterization of lactacin F, a bacteriocin produced by *Lactobacillus acidophilus* 11088. Appl Environ Microbiol 57:114–121

Nes IF, Diep DB, Havarstein LS, Brurberg MB, Eijsink V, Holo H (1996) Biosynthesis of bacteriocins in lactic acid bacteria. Antonie Van Leeuwenhoek 70:113–128

Nes IF, Holo H, Fimland G, Hauge HH, Nissen-Meyer J (2002) Unmodified peptide-bacteriocins (Class II) produced by lactic acid bacteria. In: Dutton CJ, Haxell MA, McArthur HAI, Wax RG (eds) Peptide antibiotics: discovery, modes of action and applications. Marcel Dekker, New York, pp 81–115

Nissen-Meyer J, Holo H, Havarstein LS, Sletten K, Nes IF (1992) A novel lactococcal bacteriocin whose activity depends on the complementary action of two peptides. J Bacteriol 174:5686–5692

Nissen-Meyer J, Rogne P, Oppegard C, Haugen HS, Kristiansen PE (2009) Structure-function relationships of the non-lanthionine-containing peptide (class II) bacteriocins produced by gram-positive bacteria. Curr Pharm Biotechnol 10:19–37

Oman TJ, van der Donk WA (2009) Insights into the mode of action of the two-peptide lantibiotic haloduracin. ACS Chem Biol 4:865–874

Oppegard C, Rogne P, Emanuelsen L, Kristiansen PE, Fimland G, Nissen-Meyer J (2007) The two-peptide class II bacteriocins: structure, production, and mode of action. J Mol Microbiol Biotechnol 13:210–219

O'Shea EF, Gardiner GE, O'Connor PM, Mills S, Ross RP, Hill C (2009) Characterization of enterocin- and salivaricin-producing lactic acid bacteria from the mammalian gastrointestinal tract. FEMS Microbiol Lett 291:24–34

Pag U, Sahl HG (2002) Multiple activities in lantibiotics–models for the design of novel antibiotics? Curr Pharm Des 8:815–833

Papagianni M, Anastasiadou S (2009) Pediocins: the bacteriocins of Pediococci. Sources, production, properties and applications. Microb Cell Fact 8:3

Pasteur L, Joubert JF (1877) Charbon et septicemie. C R Soc Biol Paris 85:101–115

Piper C, Cotter PD, Ross RP, Hill C (2009) Discovery of medically significant lantibiotics. Curr Drug Discov Technol 6:1–18

Rea MC, Sit CS, Clayton E, O'Connor PM, Whittal RM, Zheng J, Vederas JC, Ross RP, Hill C (2010) Thuricin CD, a postranslationally modified bacteriocin with a narrow spectrum of activity against *Clostridium difficile*. Proc Natl Acad Sci U S A 107:9352–9357

Riley MA, Wertz JE (2002) Bacteriocins: evolution, ecology, and application. Annu Rev Microbiol 56:117–137

Rogers LA, Whittier EO (1928) Limiting factors in lactic fermentation. J Bacteriol 16:211–214

Rogne P, Fimland G, Nissen-Meyer J, Kristiansen PE (2008) Three-dimensional structure of the two peptides that constitute the two-peptide bacteriocin lactococcin G. Biochim Biophys Acta 1784:543–554

Rogne P, Haugen C, Fimland G, Nissen-Meyer J, Kristiansen PE (2009) Three-dimensional structure of the two-peptide bacteriocin plantaricin JK. Peptides 30:1613–1621

Rojo-Bezares B, Saenz Y, Navarro L, Zarazaga M, Ruiz-Larrea F, Torres C (2007) Coculture-inducible bacteriocin activity of *Lactobacillus plantarum* strain J23 isolated from grape must. Food Microbiol 24:482–491

Sawa N, Zendo T, Kiyofuji J, Fujita K, Himeno K, Nakayama J, Sonomoto K (2009) Identification and characterization of lactocyclicin Q, a novel cyclic bacteriocin produced by *Lactococcus* sp. strain QU 12. Appl Environ Microbiol 75:1552–1558

Schindler CA, Schuhardt VT (1964) Lysostaphin: A New Bacteriolytic Agent for the Staphylococcus. Proc Natl Acad Sci U S A 51:414–421

Senes A, Engel DE, DeGrado WF (2004) Folding of helical membrane proteins: the role of polar, GxxxG-like and proline motifs. Curr Opin Struct Biol 14:465–479

Simmonds RS, Simpson WJ, Tagg JR (1997) Cloning and sequence analysis of zooA, a *Streptococcus zooepidemicus* gene encoding a bacteriocin-like inhibitory substance having a domain structure similar to that of lysostaphin. Gene 189:255–261

Simon L, Fremaux C, Cenatiempo Y, Berjeaud JM (2002) Sakacin g, a new type of antilisterial bacteriocin. Appl Environ Microbiol 68:6416–6420

Sprules T, Kawulka KE, Gibbs AC, Wishart DS, Vederas JC (2004) NMR solution structure of the precursor for carnobacteriocin B2, an antimicrobial peptide from *Carnobacterium piscicola*. Eur J Biochem 271:1748–1756

Tagg JR, Dajani AS, Wannamaker LW (1976) Bacteriocins of Gram-positive bacteria. Bacteriol Rev 40:722–756

Tahara T, Oshimura M, Umezawa C, Kanatani K (1996) Isolation, partial characterization, and mode of action of Acidocin J1132, a two-component bacteriocin produced by *Lactobacillus acidophilus* JCM 1132. Appl Environ Microbiol 62:892–897

Twomey D, Ross RP, Ryan M, Meaney B, Hill C (2002) Lantibiotics produced by lactic acid bacteria: structure, function and applications. Antonie Van Leeuwenhoek 82:165–185

Ueda K, Oinuma K, Ikeda G, Hosono K, Ohnishi Y, Horinouchi S, Beppu T (2002) AmfS, an extracellular peptidic morphogen in *Streptomyces griseus*. J Bacteriol 184:1488–1492

Uteng M, Hauge HH, Markwick PR, Fimland G, Mantzilas D, Nissen-Meyer J, Muhle-Goll C (2003) Three-dimensional structure in lipid micelles of the pediocin-like antimicrobial peptide sakacin P and a sakacin P variant that is structurally stabilized by an inserted C-terminal disulfide bridge. Biochemistry 42:11417–11426

Valdes-Stauber N, Scherer S (1994) Isolation and characterization of Linocin M18, a bacteriocin produced by *Brevibacterium linens*. Appl Environ Microbiol 60:3809–3814

Vaughan A, Eijsink VG, O'Sullivan TF, O'Hanlon K, van Sinderen D (2001) An analysis of bacteriocins produced by lactic acid bacteria isolated from malted barley. J Appl Microbiol 91:131–138

Wiedemann I, Bottiger T, Bonelli RR, Wiese A, Hagge SO, Gutsmann T, Seydel U, Deegan L, Hill C, Ross P, Sahl HG (2006) The mode of action of the lantibiotic lacticin 3147 – a complex mechanism involving specific interaction of two peptides and the cell wall precursor lipid II. Mol Microbiol 61:285–296

Willey JM, van der Donk WA (2007) Lantibiotics: peptides of diverse structure and function. Annu Rev Microbiol 61:477–501

Wirawan RE, Swanson KM, Kleffmann T, Jack RW, Tagg JR (2007) Uberolysin: a novel cyclic bacteriocin produced by *Streptococcus uberis*. Microbiology 153:1619–1630

Zendo T, Koga S, Shigeri Y, Nakayama J, Sonomoto K (2006) Lactococcin Q, a novel two-peptide bacteriocin produced by *Lactococcus lactis* QU 4. Appl Environ Microbiol 72:3383–3389

Zheng G, Hehn R, Zuber P (2000) Mutational analysis of the sbo-alb locus of *Bacillus subtilis*: identification of genes required for subtilosin production and immunity. J Bacteriol 182:3266–3273

Chapter 4
Bacteriocins from Gram-Negative Bacteria: A Classification?

Sylvie Rebuffat

Abstract Bacteria produce an arsenal of toxic peptides and proteins, which are termed bacteriocins and play a role in mediating the dynamics of microbial populations and communities. Bacteriocins from Gram-negative bacteria arise mainly from Enterobacteriaceae. They assemble into two main families: high molecular mass modular proteins (30–80 kDa) termed colicins and low molecular mass peptides (between 1 and 10 kDa) termed microcins. The production of colicins is mediated by the SOS response regulon, which plays a role in the response of many bacteria to DNA damages. Microcins are highly stable hydrophobic peptides that are produced under stress conditions, particularly nutrient depletion. Colicins and microcins are found essentially in *Escherichia coli*, but several other Gram-negative species also produce bacteriocin-like substances. This chapter presents the basis of a classification of colicins and microcins.

Introduction

To understanding the diversity and complexity of nature and its components at the different levels of organization, scientists need classification. Since the ancient times, Greeks and Latins tried to classify plants. Since then, classification of organisms has been for many centuries one of the main concerns for scientists. A classification is an orderly arrangement of organisms or objects in a hierarchical series. In this way, biologists reconstruct the pattern of events that have led to the distribution and diversity of life in the course of evolution. The basis of the contemporary phylogenetic classification of organisms results from the double influence of Carl von Linné (1707–1778) and Charles Darwin (1809–1882). It was further modified, according to

S. Rebuffat (✉)
Laboratory of Communication Molecules and Adaptation of Microorganisms, Muséum National d'Histoire Naturelle - CNRS, UMR 7245 CNRS-MNHN, CP 54, 57 rue Cuvier, Paris 75005, France
e-mail: rebuffat@mnhn.fr

D. Drider and S. Rebuffat (eds.), *Prokaryotic Antimicrobial Peptides: From Genes to Applications*, DOI 10.1007/978-1-4419-7692-5_4,

the revolutionary concepts introduced in 1950, by the entomologist Willy Hennig, who founded the *phylogenetic systematics* or *cladistics*. Phylogenetic classifications are often so hardly complex that they are subject to controversies and thus to changes, as fast as novel species are discovered. But apart *natural classification* of species, plants, animals and microorganisms, which is constructed in connection with evolutionary relationships, deciphering the Tree of Life, *artificial classifications* may be structured for convenience, using readily identifiable characters that are not necessarily indicative of phylogenetic relationships. Classifications rely on comparative analyses of characters, functions, properties, molecules and so on that allow bringing out similarities or differences at the organism, cellular or molecular level. Adding more parameters generally allows improving classifications. Indeed, molecules are also classified: they gather according generally to their molecular masses, functionalities, chemical or biological properties as well.

The production of molecules able to inhibit the growth of microorganisms is probably the most widespread defence strategy developed in nature. From mammals to bacteria, antimicrobial peptides contribute to the strategies that allow organisms to fight against pathogens (innate immunity) or to develop themselves at the detriment of others for the conquest of a given niche (competitions). The term bacteriocin was introduced at first in the 1950s by François Jacob (Jacob et al. 1953). It was further used by Tagg et al. (1976) and Klaenhammer (1988), successively to define proteinaceous compounds of bacterial origin being lethal to bacteria that were different from the producing strain. The sensible bacterial species are generally closely related to the producer organism or occupy the same ecological niche. It is worth noting that nowadays, in the literature, the term bacteriocin is very often restricted to the ribosomally synthesized antibacterial peptides of Gram-positive bacteria, principally from lactic acid bacteria. However, the first bacteriocin to be described had been identified in 1925 from the Gram-negative bacterium *Escherichia coli* (Gratia 1925). It inhibited the growth of another *E. coli* strain. This pioneering work has inaugurated the very numerous studies that were further conducted on these antimicrobial (poly)peptides or proteins.

Actually, bacteriocins can be considered today as antimicrobial (poly)peptides or proteins that are ribosomally synthesized by many bacteria, have generally a narrow spectrum of antibacterial activity, and against which the producing strains are protected by a dedicated immunity system. They may undergo or not posttranslational modifications that endow them with very diverse structures and mechanisms of action. It has been estimated that 30–99% of bacteria and archaea very probably are able to produce at least one bacteriocin (Klaenhammer 1988; Riley 1998). Moreover, many strains have been shown to produce two or more bacteriocins (Gordon and O'Brien 2006). Indeed, bacteriocins have been largely found in *E. coli* and some other enterobacteria, as well as in lactic acid bacteria. Therefore, for both historical and economical reasons, colicins from enterobacteria and bacteriocins from lactic acid bacteria have been more extensively studied than other bacteriocins, leading to the very impressive literature on these antimicrobial peptides and proteins. Parallel to colicins, which are large proteins, lower molecular mass bacteriocins have been identified in Enterobacteriaceae (Asensio et al. 1976;

Baquero and Moreno 1984). The name microcin was coined by Asensio to differentiate these lower molecular mass and protease-resistant bacteriocins produced by enteric bacteria from colicins (Asensio et al. 1976). This was the starting point for numerous studies leading to the identification of 14 microcins, and the analysis and characterization of the mechanisms involved in the production and antibacterial activity of seven of them (for reviews refer to Destoumieux-Garzón et al. 2002; Duquesne et al. 2007a; Severinov et al. 2007). It is striking to note that the first bacteriocin to be described, colicin V (ColV) (Gratia 1925), is produced by *E. coli* and is indeed a microcin, based on several parameters including its low molecular mass, noninducible production and dedicated export system (Fredericq et al. 1949; Duquesne et al. 2007a).

Although colicins and microcins are the most studied bacterial antimicrobial (poly)peptides/proteins, *E. coli* is not the single species to produce bacteriocins for killing neighbouring bacteria. A limited number of bacteriocins have been characterized in other Gram-negative organisms. Some of them are related to colicins, showing a similar domain organization, such as pyocins and lumicins in *Pseudomonas* species (Jacob 1954) and *Photorhabdus luminescens* (Sharma et al. 2002), respectively, pesticins in *Yersinia pestis* (Ferber and Brubaker 1979) or klebicins in *Klebsiella pneumoniae* (Chibber and Vadehra 1986) (see below). It is worth noting that *K. pneumoniae* is also the producer of a microcin (de Lorenzo 1984), namely microcin E492. Others, sharing part of the characteristics of lower molecular mass bacteriocins, are frequently termed bacteriocin-like inhibitory substances (BLIS), according to the recommendations of Ray and colleagues (Jack et al. 1995). Such BLIS have been also characterized in archaea. Archaea possess a remarkable variety of cell envelope structure and chemical, which make them having heterogeneous response to Gram staining. However, they are rather considered as being closer to Gram-positive than Gram-negative bacteria (Gupta 1998). For this reason, bacteriocins from archaea will not be considered here.

Identified Gram-negative bacteria producers of BLIS include mainly *Vibrio* (Farkas-Himsley and Seyfried 1962), *Myxococcus* (Munoz et al. 1984) and *Aeromonas* (Messi et al. 2003). A few bacteriocins produced by the genus *Vibrio* have been identified, namely vibriocins isolated from *V. cholerae* (Wahaba 1965) and *V. harveyi* (Mc Call and Sizemore 1979) and the bacteriocins from *V.* sp strain NM10 (Sugita et al. 1997), *V. vulnificus* (Shehane and Sizemore 2002) and *V. mediterranei* (Carraturo et al. 2006). Among these BLIS, a very restricted number has been well identified. In majority, they have been only partly purified and/or characterized. Therefore, BLIS are not prone to classification.

But what criterions have to be considered to elaborate a classification for bacteriocins, given the heterogeneity of bacterial sources, of peptide structures and of mechanisms of action? In the case of Gram-positive bacteria, the high number of bacteriocins isolated, purified and fully characterized makes the answer easier, even if elaborating the classification remains complex. The classification into groups and subgroups relies essentially (1) on the primary structures, including the size of the (poly)peptide and the presence or absence of consensus sequences and of posttranslational modifications, and (2) on the mechanisms of action and bacterial target specificity (for more details on this classification, the reader is referred to Chap. 3 by Paul Cotter). Contrary to antimicrobial peptides from eucaryotes, where the three-dimensional structures acquired by the peptides are one of the important criterions considered for their classification, this aspect is not taken into account for bacteriocins, presumably as a lower number of three-dimensional structures has been determined. In Gram-negative bacteria, except for colicins where the classification is clear enough, a classification based on similar criterions is made more difficult, given the great structural heterogeneity inside the very restricted group of bacteriocins formed by microcins. Finally, it is worth noting that, similar to all classifications, the classification of bacteriocins requires constant revision, as it continuously evolves according to the knowledge that accumulates rapidly, as soon as novel bacteriocins or mechanisms of action are brought to light.

This introductive chapter to bacteriocins from Gram-negative bacteria presents their classification, dealing with colicins, colicin-like bacteriocins and microcins, successively. The less-studied bacteriocins, which are described in this chapter, will not be spoken of later in the book, while for a more extensive description of colicins and microcins, readers are referred to the corresponding chapters, which provide comprehensive overviews of these classes of bacteriocins.

The Colicin Classification

Colicins are the most studied bacteriocins from Gram-negative bacteria (for reviews see Konisky 1982; Braun et al. 1994; Braun et al. 2002; Duché et al. 1995; Cascales et al. 2007). They are high-molecular mass (30–80 kDa) bactericidal proteins (Table 4.1 and Fig. 4.1) produced by many *E. coli* strains in periods of stress. Colicins are produced by *E. coli* strains harbouring one colicinogenic plasmid. The production of colicins is lethal for the producing cells, as a consequence of the concomitant production of a lysis protein, coexpressed with the colicin. The typical colicin operon contains one to three genes: a gene encoding the colicin (*cxa* for colicin X activity), a gene often located downstream from the structural gene that encodes the immunity protein (*cxi* for colicin X immunity or *imX*) and a gene encoding the lysis protein (or bacteriocin release protein, BRP), whose product permits the release of the colicin into the external medium (for reviews, the reader is referred to Van der Wal et al. 1995; Braun et al. 2002). Regulation of colicin gene clusters is mediated by the SOS response regulon, which plays a role in the response of bacteria to DNA damages (Walker 1995). Since 1963, it was demonstrated that the various identified colicins had different modes of action (Nomura 1963). The classification elaborated for colicins relies on both their killing mechanisms and the uptake machineries they use (for reviews, the reader is referred to Braun 1995, Cascales et al. 2007). Indeed, colicins kill target cells through three main mechanisms (Cascales et al. 2007; refer to Chap. 14 by Miklos de Zamarockzy) (Table 4.1) either (1) by making voltage-dependent channels in the inner membrane of the target bacteria, (2) by a nuclease action in the cytoplasm or (3) by degrading

Table 4.1 Main characteristics of group A and B colicins and colicin-like bacteriocins[a]

Name of the bacteriocin Producer	Active bacteriocin Number of residues, gene, 3D structure (PDB)	Receptor	Translocation system	Killing mechanism	Immunity protein Number of residues, gene, 3D structure (PDB)
Group A					
Colicin A *Citrobacter freundii* *Escherichia coli*	592, *caa*, 1COL[b]	BtuB	OmpF TolA, B, Q, R	Pore-forming	178, *cai*
Colicin K *E. coli*	548, *cka*	Tsx	OmpF, OmpA TolA, B, Q, R	Pore-forming	196, *cki*
Colicin N *E. coli*	387, *cna*, 1A87	OmpF	OmpF TolA, Q, R	Pore-forming	131, *cni*
Colicin U *E. coli*	618, *cua*	OmpA	OmpF, LPS TolA, B, Q, R	Pore-forming	174, *cui*
Colicin S4 *E. coli*	499, *csa*	OmpW	OmpF, TolA, B, Q, R	Pore-forming	179, *csi*
Colicin E1 *E. coli*	522, *cea*	BtuB	TolC TolA, Q, R	Pore-forming	113, *ceiA*
Colicin E2 *E. coli*	581, *ceaB*	BtuB	OmpF, TolA, B, Q, R	DNase	86, *ceiB*
Colicin E3 *E. coli*	551, *ceaC*, 1JCH[c]	BtuB	OmpF, TolA, B, Q, R	rRNase	84, *ceiC*, 1JCH[c]
Colicin E4 *E. coli*	177, *cea4*	BtuB	OmpF, TolA, B, Q, R	rRNase	85, *cei4*
Colicin E5 *E. coli*	unknown, *cea5*	BtuB	OmpF, TolA, B, Q, R	tRNase	83, *cei5*
Colicin E6 *E. coli*	551, *cea6*	BtuB	OmpF, TolA, B, Q, R	rRNase	85, *cei6*
Colicin E7 *E. coli*	576, *cea7*, 1M08[d]	BtuB	OmpF, TolA, B, Q, R	DNase	87, *cei7*, 1AYI

(continued)

Table 4.1 (continued)

Name of the bacteriocin Producer	Active bacteriocin Number of residues, gene, 3D structure (PDB)	Receptor	Translocation system	Killing mechanism	Immunity protein Number of residues, gene, 3D structure (PDB)
Colicin E8 *E. coli*	unknown, *cea8*	BtuB	OmpF, TolA, B, Q, R	DNase	85, *cei8*
Colicin E9 *E. coli*	582, *cea9*, 1FSJ	BtuB	OmpF, TolA, B, Q, R	DNase	86, *cei9*, 1IMQ
Cloacin DF13 *Enterobacter cloacae*	561, *ccl*	IutA	TolA, Q, R	rRNase	85, *cim*
Pyocin AP41 *Pseudomonas aeruginosa*	776	(?)	TolA, B, Q, R	DNase	–
Alveicins A *Hafnia alvei*	408, *aat*	(?)	TolA, B, Q, R	Pore-forming	*aai*
Alveicin B *H. alvei*	358, *abt*	(?)	TolA, B, Q, R	Pore-forming	*abi*
Marcescin 28b *Serratia marcescens*	–	Omp4[e]/OmpA[f]	OmpF TolA, B, Q, R	Pore-forming	–
Group B					
Colicin B *E. coli*	510, *cba*, 1RH1	FepA	FepA (?) TonB, ExbB, D	Pore-forming	175, *cbi*
Colicin Ia *E. coli*	626, *cia*, 1CH	Cir	Cir, TonB, ExbB, D	Pore-forming	111, *imm*
Colicin Ib *E. coli*	626, *ciaB*	Cir	Cir (?) TonB, ExbB, D	Pore-forming	115
Colicin 5 *E. coli*	490, *cfa*	Tsx	TolC, TonB, ExbB, D	Pore-forming	43, *cfl*
Colicin 10 *E. coli*	490, *cta*	Tsx	TolC, TonB, ExbB, D	Pore-forming	43, *ckl*
Colicin D *E. coli*	697, *cda*	FepA	FepA (?) TonB, ExbB, D	tRNase (Arg tRNA)	87, *cdi*

Colicin M *E. coli*	271, *cma*	FhuA	FhuA (?) TonB, ExbB, D	Degradation of peptidoglycan	117, *cmi*
Pesticin *E. coli*	357, *pst*	FyuA	FyuA (?) TonB, ExbB, D	Muramidase	141, *pim*
Pyocin S1 *Pseudomonas aeruginosa*	618, *pyoS1A*	ferripyoverdine receptor	(?) TonB, ExbB, D	DNase	87, *pyoS1I*
Pyocin S2 *P. aeruginosa*	690, *pyoS2A*	ferripyoverdine receptor	(?) TonB, ExbB, D	DNase	87, *pyoS2I*
Pyocin S3 *P. aerugoinosa*	768, *pyoS3A*	ferripyoverdine receptor	TonB, ExbB, D	DNase	154, *pyoS3I*
Undetermined group					
Klebicin C *K. pneumoniae* *K. oxytoca*	619, *kca*	(?)	non-TonB	rRNase	84, *kci*
Klebicin D *K. pneumoniae*	716, *kda*	(?)	(?)	tRNase	88, *kdi*
Pyocin S4[g] *P. aeruginosa*	764	(?)	(?)	tRNase	–
Pyocin S5[g] *P. aeruginosa*	498	(?)	(?)	Pore-forming	–
Carocin S1 *Herwinia carotovora*	361, *caroS1K*	(?)	(?)	DNase (?)	129, *caroS1I*

[a] References are given in the main text
[b] C-terminal killing domain
[c] Colicin E3 + immunity protein
[d] Nuclease domain
[e] In *S. marsescens*
[f] In *E. coli*
[g] Predicted by genome sequence in silico analysis

Name of the bacteriocin	**Three-dimensional structure**
Colicin A Group A (PDB: 1COL)	
Colicin N Group A (PDB: 1A87)	
Colicin B Group B (PDB: 1RH1)	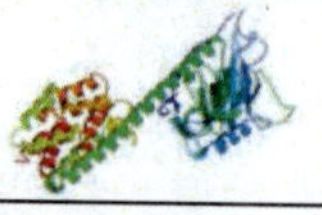
Colicin Ia Group B (PDB: 1CH)	

Fig. 4.1 Three-dimensional structures of pore-forming colicins of groups A and B

the peptidoglycan (Barreteau et al. 2010). The activity of colicins requires a recognition step that uses a number of receptors normally involved in the uptake of essential nutrients, such as vitamin B12 (cobalamin), siderophore-bound iron or nucleosides, which have been parasitized by colicins, helping them entering more efficiently in target bacteria. The receptors that have been hijacked by colicins are principally the siderophore receptors FhuA (hydroxamate siderophores), FepA, Cir and Fiu (catecholate siderophores) (Ferguson and Deisenhofer 2002), the cobalamine receptor BtuB (Taylor et al. 1998) and the nucleoside receptor Tsx (Bremer et al. 1990). Porins, the membrane proteins that form aqueous channels in the outer membrane and control the passive diffusion of specific small metabolites (sugars, phosphates, amino acids and so on) into the bacterial cells (Nikaido and Vaara 1985; Nikaido 2003), are also used by some colicins (Table 4.1). In many cases, the colicin translocation step remains poorly understood since it is questionable if the receptor is used also as a translocator, or the translocation step does need a second protein partner. For example colicin Ia uses Cir for both recognition and translocation (Jakes and Finkelstein 2009), while colicin A uses BtuB for the recognition step and OmpF for translocation (Lazzaroni et al. 2002) (for a detailed description of this aspect, the reader is referred to Chap. 14 by Miklos de Zamarockzy). The receptors used for the recognition step are each coupled to one of two protein machineries, namely the Ton (Krewulak and Vogel 2008) and Tol membrane systems (Lloubès et al. 2001). These two machineries are anchored at the inner membrane and provide energy to the receptors, using the proton-motive force. The Ton system is composed of three inner membrane proteins TonB, ExbB and ExbD, while the Tol system contains several proteins of similar topology TolA, TolQ,

TolR, and a periplasmic protein TolB and an outer membrane-anchored lipoprotein Pal, which is very probably not required for colicin import (Cascales et al. 2007). Colicins of group A require the Tol proteins or a subset of these proteins (e.g. TolA/TolQ for colicin E1 and TolA/TolQ/TolR for colicin N), while all colicins of group B require TonB/ExbB/ExbD (Table 4.1). Moreover, colicins of group A are encoded by small plasmids and are released into the culture medium, while colicins of group B are encoded by large plasmids and are not released into the medium. Group A comprises colicins A, E1 to E9, K, N, U (Cascales et al. 2007), and S4 (Pilsl et al. 1999; Arnold et al. 2009), which are translocated by the Tol machinery; group B comprises colicins B, D, Ia, M, 5 and 10, which use the Ton system (Cascales et al. 2007) (Table 4.1). In addition, colicin-like bacteriocins, which have been characterized in non-*E. coli* strains, have been shown (or are supposed) to split also into the two groups A and B: cloacin DF13 (Oudega and de Graaf 1976; Thomas and Valvano 1993), klebicins (Chavan et al. 2005), pyocin AP41 (Michel-Briand and Baysse 2002), marcescin 28b (Guasch et al. 1995a, b; Enfedaque et al. 1996) and alveicins (Wertz and Riley 2004) fit with group A and pesticins (Rakin et al. 1996) and pyocins S1 to S5 (Michel-Briand and Baysse 2002) with group B (Table 4.1).

Most colicins and colicin-like bacteriocins are organized into three protein functional domains, each corresponding to one step of the mode of action: (1) the central domain is involved in binding to the receptor (R-domain for Receptor), and thus recognizes and adheres to specific regions on the surface of target cells; (2) the N-terminal domain is responsible for translocation (T-domain for Translocation) and enables entry of the bacteriocin into the target cells and (3) the C-terminal domain, which is the active region, is the killing domain (C-domain, for Cytotoxicity) (Braun et al. 1994; Cascales et al. 2007). Inside each of the two groups, A and B, previously defined, colicins and colicin-like bacteriocins may be assembled according to their killing mechanisms, pore-forming, nuclease or peptidoglycan degradation activities.

The Microcin Classification

Microcins are the lower molecular mass bacteriocins produced by Gram-negative bacteria, mostly *E. coli*, in stress conditions (Baquero and Moreno 1984; Duquesne et al. 2007a), and particularly, in poor nutrient conditions. They are peptides ranging from 1 to 10 kDa that are generally resistant to proteases, extreme pHs and temperatures. They are encoded by gene clusters carried by plasmids or in some cases by the chromosome. The gene clusters involved in the production of microcins include a variable number of genes, but show a conserved organization: open reading frames encode the precursor of the microcin, secretion proteins and self immunity factors and, in some cases, posttranslational modification enzymes. In contrast with the bacteriocins from Gram-positive bacteria, where identification of a high number of representatives sharing common structural features and mechanisms of action allows identifying classes and subclasses, and with colicins that share characteristics related to their uptake and mode of cell killing that have been exploited for their classification, classifying microcins appears as a difficult task. This is related to the restricted number of representatives identified until now and the high diversity in structures and mechanisms of action they exhibit (Table 4.2). They also arise from various biosynthetic pathways, including or not the acquisition of different complex posttranslational modifications, and display various mechanisms of uptake. In the 2000s, Pons and colleagues proposed the first microcin classification into two classes, depending on the presence or absence of a posttranslational modification (Gaillard-Gendron et al. 2000; Pons et al. 2002a). However in 2004, while microcin E492 was described as unmodified until then (Pons et al. 2002b), the finding of a modified form of microcin E492 carrying a siderophore at the C terminus (Thomas et al. 2004) prompted our group to propose a novel classification (Duquesne et al. 2007a). This classification takes into account the three following criteria: (1) the presence, nature and localization of posttranslational modifications, (2) the gene cluster organization and (3) the sequences of the leader peptides, therefore assembling microcins into two classes (Table 4.2 and Fig. 4.2).

Class I gathers peptides with a molecular mass below 5 kDa, which have supported extensive backbone posttranslational modifications, namely microcins B17, C7-C51 and J25 (Duquesne et al. 2007b; Severinov et al. 2007). Microcin B17 contains four thiazole and four oxazole rings that result from an unusual posttranslational modification of six glycines, four serines and four cysteines in the 39–66 region of the 69-amino acid precursor of microcin B17. Microcin C7-C51 is an *N*-formylated heptapeptide carrying a C-terminal modified nucleotide: a C-terminal aspartic acid is covalently linked to a phosphoramidate group, itself substituted with both an adenosine moiety and a propylamine chain (Fig. 4.2). Microcin J25 is a lasso peptides, which means that it adopts the typical compact lasso structure consisting of an N-terminal ring formed by a lactam bond between the N-terminal extremity (Gly/Cys) and the side chain of an acidic residue (Asp/Glu) in position 8 or 9, where the C-terminal tail is irreversibly threaded (Rebuffat et al. 2004) (Fig. 4.2). Class II includes higher molecular mass peptides (in the 5–10 kDa range) and is itself further subdivided into two subclasses, IIa and IIb. Class IIa contains plasmid-encoded peptides without posttranslational modification and forming possibly disulfide bonds (MccL, MccV and Mcc24) (Fig. 4.2). Class IIb gathers those chromosome-encoded linear microcins that carry a C-terminal siderophore posttranslational modification (microcins E492, M, H47 and presumably I47 and G47 that have been hypothesized through genome analyses), (Poey et al. 2006; Vassiliadis et al. 2010), (Fig. 4.2). It is to note that microcin 24 does not fit perfectly with the criteria previously defined for any of the classes. This microcin has neither been isolated nor biochemically characterized, but based on its precursor sequence predicted from the gene, it contains neither disulfide bond nor posttranslational modification. Nevertheless, it was incorporated into class IIb to which it was more tightly related when considering its gene cluster organization (Duquesne et al. 2007a). For a more detailed description of the microcins from classes I and II, the reader is referred to Chaps. 15 and 16 by Konstantin Severinov et al. and Vassiliadis et al. 2010, respectively.

Table 4.2 Main characteristics of class I and II microcins[a]

Name of the microcin	Active microcin Number of residues of the precursor of the mature peptide, gene, microcin 3D structure (PDB)	Post-translational modification Name of the genes involved	Receptor	Translocation system	Killing mechanism
Class I					
B17	69, 43, *mcbA*	Thiazole and oxazole rings *mcbB, mcbC, mcbD*	OmpF	OmpF (?) SbmA	DNA gyrase inhibition
C7-C51	7, 7, *cea*	Modified nucleotide *mccB, mccC, mccE*	OmpF	OmpF (?) YejA, B, E, F	Cleavage in the target cell into a modified aspartyl adenylate that inhibits Asp-tRNA synthetase
J25	58, 21, *mcjA*, 1Q71	Lasso structure *mcjB, mcjC*	FhuA	Fhua (?) TonB, ExbB, D, SbmA	RNA polymerase inhibition Mitochondrial proteins and lipids damages
Class II					
Class IIa					
L	105, 90, *mclC*	2 disulfide bonds	Cir	Cir TonB, ExbB, D SdaC	Membrane permeability modification (?)
V	103, 88, *cvaC*	1 disulfide bond	Cir	Cir (?) TonB, ExbB, D	Membrane permeability modification
24	90, 73, *mtfS*	no disulfide bond	(?)	(?)	Mannose permease (ManYZ) targeting (?)
Class IIb					
E492	103, 84, *mceA*	Siderophore anchored at the C terminus *mceC, mceD, mceI, mceJ*	FepA, Cir, Fiu	FepA, Cir, Fiu (?) TonB, ExbB, D	Inner membrane channels ManYZ targeting (mannose permease)

(continued)

Table 4.2 (continued)

Name of the microcin	Active microcin Number of residues of the precursor of the mature peptide, gene, microcin 3D structure (PDB)	Post-translational modification Name of the genes involved	Receptor	Translocation system	Killing mechanism
M	92, 77, *mcmA*	Siderophore anchored at the C-terminus *mcmL, mcmK*	FepA, Cir, Fiu	FepA, Cir, Fiu (?) TonB, ExbB, D	(?)
H47	75, 60, *mchB*	Siderophore anchored at the C-terminus *mchA, mchS1, mchD* *mchC*	FepA, Cir, Fiu	FepA, Cir, Fiu (?) TonB, ExbB, D	F_0F_1 ATP synthetase
I47[b]	77, 62, *mchS2*	Siderophore anchored at the C-terminus	FepA, Cir, Fiu	FepA, Cir, Fiu (?) TonB, ExbB, D	(?)
G492[b]	89, 74, *mceL*	Siderophore anchored at the C-terminus	FepA, Cir, Fiu (?)	FepA, Cir, Fiu (?) TonB, ExbB, D	(?)

[a]References are given in the main text
[b]Predicted by genome sequence in silico analysis

Name of the bacteriocin	Structure
Microcin J25 Class I (1) primary structure; (2) three-dimensional structure (PDB: 1Q71)	(1) G G A G H V P E Y F V G I G T P I S F Y G (2) 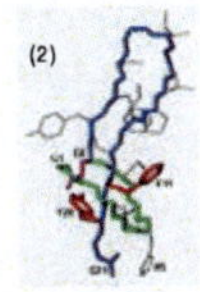
Microcin C7-C51 Class I (1) native microcin; (2) toxic entity generated in target cells	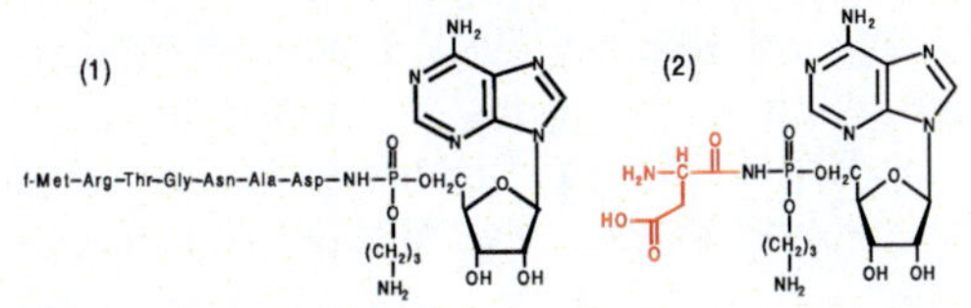
Microcin V Class IIa	ASGRDIAMAI GTLSGQFVAG GIGAAAGGVA GGAIYDYAST HKPNPAMSPS GLGGTIKQKP EGIPSEAWNY AAGRLCNWSP NNLSDVCL
Microcin E492 Class IIb	GETDPNTQLLNDLGNNMAWGAALGAPG GLGSAALGAAGGALQTVGQGLIDHGPV NVPIPVLIGPSWNGSGSGYNSATSSSGSGS 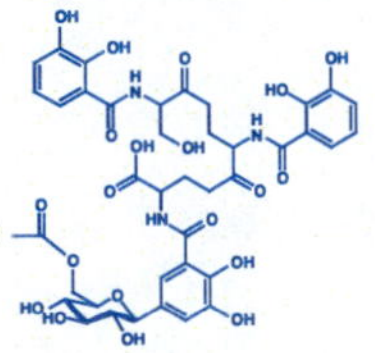

Fig. 4.2 Structures of microcins of classes I and II

Similar to colicins, microcins have receptor-mediated mechanisms of action. This is reflected in their minimal inhibitory concentrations, which are in the nanomolar range, while they are in the micromolar range for antimicrobial peptides of eukaryotic origin, which act through a direct interaction with phospholipid membrane bilayers. To improve their intake into sensible bacteria, microcins hijack receptors involved in the uptake of essential nutrients, such as the iron siderophore receptors (FhuA, FepA, Cir and Fiu) or the porin OmpF (Duquesne et al. 2007a) (Table 4.2). These receptors, which are also exploited by bacteriophages, antibiotics and bacterial toxins, are thus critical for bacteria for which they represent an Achille's hill. In general, translocation of microcins requires the TonB–ExbB–ExbD complex (Duquesne et al. 2007a), which uses the proton-motive force from the cytoplasmic membrane for energy transduction to the outer membrane and its receptors. However, microcin B17 uses the outer-membrane porin OmpF and the protein SbmA, and microcin C7–C51 requires OmpF and the inner-membrane ABC-transporter YeJ to be actively transported through the inner membrane (refer to Chap. 15 by Konstantin Severinov et al.). Microcins exhibit very heterogeneous killing mechanisms (Table 4.2). Class I microcins inhibit vital bacterial enzymes. Microcins B17 and J25 inhibit DNA gyrase

(Heddle et al. 2001) and RNA polymerase (Mukhopadhyay et al. 2004, Adelman et al. 2004), respectively. Microcin C7–C51 needs a preliminary cleavage inside the target bacteria to generate the toxic entity, which is a mime of aspartyl adenylate that inhibits aspartyl tRNA synthetase, thus blocking protein synthesis at the translation step (Metlitskaya et al. 2006). More complex and subtle mechanisms appear to be developed in certain cases, as microcin J25 has been shown to target mitochondria and the respiratory chain (Niklison-Chirou et al. 2010) in addition to its inhibitory effect of RNA polymerase. Class II microcins rather target the inner membrane or their components. Microcin H47 targets the F_0 proton channel of ATP synthetase (Rodríguez and Laviña 2003). Microcins E492 (Lagos et al. 1993; Bieler et al. 2006) and V (Yang and Konisky 1984) both permeabilize the inner membrane. However, microcin E492 not only forms channels in the inner membrane (Lagos et al. 1993) but also requires the ManYZ inner-membrane components of the mannose permease, which is involved in the active uptake of mannose and related hexoses, to exert its bactericidal activity (Bieler et al. 2006). Microcins can also require proteins at the inner membrane for activity, such as SdaC, which is involved in serine uptake, for microcin L (Gérard et al. 2005) or SbmA for microcins B17 (Yorgey et al. 1994) and J25 (De Cristobal et al. 2006). The exact role of these proteins is presumably helping microcins passing through the inner membrane.

Conclusions

In contrary to the classification of bacteriocins from Gram-positive bacteria, which mainly relies on the structural features, the classification of colicins is based on functional criteria that are the recognition machineries and the killing mechanisms. In the case of microcins, due to the high heterogeneity of this restricted class of bacteriocins as regards the structural characteristics, the complexity of the genetic systems and the killing mechanisms developed, the classification has to take into account these different criteria to assemble microcins sharing common characters. For bacteriocins produced by Gram-positive and Gram-negative bacteria, from both commensal and environmental origin, a classification helps pointing more easily similarities or differences in the strategies they developed to make bacteriocinogenic strains better adapted to the natural conditions in a given niche. No doubt that genomic advance will contribute to the rapid identification of novel bacteriocins (refer to Chap. 5 by Oscar Kuipers and colleagues), which will allow describing novel mechanisms at different levels in bacteriocin production and activity, i.e. synthesis of the bacteriocins, immunity of the producing bacteria, penetrating capacity inside target bacteria and killing mechanisms. This will possibly change the classification and also modify and improve our use of bacteriocins as models for deciphering novel, clever and sophisticated strategies of antibacterial activity and of resistance to antimicrobials, thus providing tools for various applications in the environment and for the development of novel antibiotics.

References

Adelman K, Yuzenkova J, La Porta A, Zenkin N, Lee J, Lis JT, Borukhov S, Wang MD, Severinov K (2004) Molecular mechanism of transcription inhibition by peptide antibiotic Microcin J25. Mol Cell 14:753–762

Arnold T, Zeth K, Linke D (2009) Structure and function of colicin S4, a colicin with a duplicated receptor-binding domain. J Biol Chem 284:6403–6413

Asensio C, Pérez-Diaz JC, Martinez MC, Baquero F (1976) A new family of low molecular weight antibiotics from enterobacteria. Biochem Biophys Res Commun 69:7–14

Baquero F, Moreno F (1984) The microcins. FEMS Microbiol Lett 23:117–124

Barreteau H, Bouhss A, Gérard F, Duché D, Boussaid B, Blanot D, Lloubes R, Mengin-Lecreulx D, Touzé T (2010) Deciphering the catalytic domain of colicin M, a peptidoglycan lipid II degrading enzyme. J Biol Chem 285:12378–12389

Bieler S, Silva F, Soto C, Belin D (2006) Bactericidal activity of both secreted and nonsecreted microcin E492 requires the mannose permease. J Bacteriol 188:7049–7061

Braun V, Patzer SI, Hantke K (2002) Ton-dependent colicins and microcins: modular design and evolution. Biochimie 84:365–380, Review

Braun V, Pilsl H, Gross P (1994) Colicins: structures, modes of action, transfer through membranes and evolution. Arch Microbiol 161:199–206, Review

Bremer E, Middendorf A, Martinussen J, Valentin-Hansen P (1990) Analysis of the tsx gene, which encodes a nucleoside-specific channel-forming protein (Tsx) in the outer membrane of *Escherichia coli*. Gene 96:59–65

Carraturo A, Raieta K, Ottaviani D, Russo GL (2006) Inhibition of *Vibrio parahaemolyticus* by a bacteriocin-like inhibitory substance (BLIS) produced by *Vibrio mediterranei* 1. J Appl Microbiol 101:234–241

Cascales E, Buchanan SK, Duché D, Kleanthous C, Lloubes R, Postle K, Riley M, Slatin S, Cavard D (2007) Colicin biology. Microbiol Mol Biol Rev 71:158–229, Review

Chavan M, Rafi H, Wertz J, Goldstone C, Riley MA (2005) Phage associated bacteriocins reveal a novel mechanism for bacteriocin diversification in *Klebsiella*. J Biol Chem 284:6403–6413

Chibber S, Vadehra DV (1986) Purification and characterization of bacteriocin from *Klebsiella pneumoniae* 158. J Gen Microbiol 132:1051–1054

De Cristobal RE, Solbiati JO, Zenoff AM, Vincent PA, Salomon RA, Yuzenkova J (2006) Microcin J25 uptake: His5 of the MccJ25 lariat ring is involved in interaction with the inner membrane MccJ25 transporter protein SbmA. J Bacteriol 188:3324–3328

Destoumieux-Garzón D, Peduzzi J, Rebuffat S (2002) Focus on modified microcins: structural features and mechanisms of action. Biochimie 84:511–519, Review

Duché D, Letellier L, Géli V, Bénédetti H, Baty D (1995) Quantification of group A colicin import sites. J Bacteriol 177:4935–4939

Duquesne S, Destoumieux-Garzón D, Peduzzi J, Rebuffat S (2007a) Microcins, gene-encoded antibacterial peptides from enterobacteria. Nat Prod Rep 24:708–734, Review

Duquesne S, Petit V, Peduzzi J, Rebuffat S (2007b) Structural and functional diversity of microcins, gene-encoded antibacterial peptides from enterobacteria. J Mol Microbiol Biotechnol 13:200–209, Review

Enfedaque J, Ferrer S, Guasch JF, Tomás J, Regué M (1996) Bacteriocin 28b from *Serratia marcescens* N28b: identification of *Escherichia coli* surface components involved in bacteriocin binding and translocation. Can J Microbiol 42:19–26

Farkas-Himsley H, Seyfried PL (1962) Lethal biosynthesis of a new antibacterial principle: vibriocin. Nature 193:1193–1194

Ferber DM, Brubaker RR (1979) Mode of action of pesticin: *N*-acetylglusaminidase activity. J Bacteriol 139:495–501

Ferguson AD, Deisenhofer J (2002) TonB dependent receptors structural perspectives. Biochim Biophys Acta 1565:318–332

Fredericq P, Joiris E, Betz-Barreau M, Gratia A (1949) Recherche des germes producteurs de colicines dans les selles de malades atteints de fièvre paratyphoïde B. CR Soc Biol 143:556–559

Gaillard-Gendron S, Vignon D, Cottenceau G, Graber M, Zorn A, van Dorsselaer A, Pons A-M (2000) Isolation, purification and partial amino acid sequence of a highly hydrophobic new microcin named microcin L produced by *Escherichia coli*. FEMS Microbiol Lett 193:95–98, Erratum in: FEMS Microbiol Lett 2001, 199:151

Gérard F, Pradel N, Wu LF (2005) Bactericidal activity of colicin V is mediated by an inner membrane protein, SdaC, of *Escherichia coli*. J Bacteriol 187:1945–1950

Gordon DM, O'Brien CL (2006) Bacteriocin diversity and the frequency of multiple bacteriocin production in *Escherichia coli*. Microbiology 152:3239–3244

Gratia A (1925) Sur un remarquable exemple d'antagonisme entre deux souches de colibacille. CR Soc Biol 93:1041–1042

Guasch JF, Enfedaque J, Ferrer S, Gargallo D, Regué M (1995a) Bacteriocin 28b, a chromosomally encoded bacteriocin produced by most *Serratia marcescens* biotypes. Res Microbiol 146:477–483

Guasch JF, Ferrer S, Enfedaque J, Viejo MB, Regué M (1995b) A 17 kDa outer-membrane protein (Omp4) from *Serratia marcescens* confers partial resistance to bacteriocin 28b when expressed in *Escherichia coli*. Microbiology 141:2535–2542

Gupta RS (1998) What are archaebacteria: life's third domain or monoderm prokaryotes related to gram-positive bacteria? A new proposal for the classification of prokaryotic organisms. Mol Microbiol 29:695–707

Heddle JG, Blance SJ, Zamble DB, Hollfelder F, Miller DA, Wentzell LM, Walsh CT, Maxwell A (2001) The antibiotic microcin B17 is a DNA gyrase poison: characterisation of the mode of inhibition. J Mol Biol 307:1223–1234

Jack RW, Tagg JR, Ray B (1995) Bacteriocins of Gram positive bacteria. Microbiol Rev 59:171–200

Jacob F, Lwoff A, Siminovitch A, Wollman E (1953) Définition de quelque termes relatifs a la lysogénie. Ann Inst Pasteur (Paris) 84:222–224

Jacob F (1954) Biosynthèse induite et mode d'action d'une pyocine, antibiotique de *Pseudomonas pyocyanea*. Ann Inst Pasteur (Paris) 86:149–160

Jakes KS, Finkelstein A (2009) The colicin Ia receptor Cir is also the translocator for colicin Ia. Mol Microbiol 75:567–578

Klaenhammer TR (1988) Bacteriocins of lactic acid bacteria. Biochimie 70:337–349

Konisky J (1982) Colicins and other bacteriocins with established modes of action. Annu Rev Microbiol 36:125–144, Review

Krewulak KD, Vogel HJ (2008) Structural biology of bacterial iron uptake. Biochim Biophys Acta 1778:1781–1804

Lloubès R, Cascales E, Walburger A, Bouveret E, Lazdunski C, Bernadac A, Journet L (2001) The Tol-Pal proteins of the *Escherichia coli* cell envelope: an energized system required for outer membrane integrity? Res Microbiol 152:523–529, Review

Lagos R, Wilkens M, Vergara C, Cecchi X, Monasterio O (1993) Microcin E492 forms ion channels in phospholipid bilayer membrane. FEBS Lett 321:145–148

Lazzaroni JC, Dubuisson JF, Vianney A (2002) The Tol proteins of *Escherichia coli* and their involvement in the translocation of group A colicins. Biochimie 84:391–397

de Lorenzo V (1984) Isolation and characterization of microcin E492 from *Klebsiella pneumoniae*. Arch Microbiol 139:72–75

Mc Call JO, Sizemore RK (1979) Description of a bacteriocinogenic plasmid in *Beneckea harveyi*. Appl Environ Microbiol 38:974–979

Messi P, Guerrieri E, Bondi M (2003) Bacteriocin-like substance (BLS) production in *Aeromonas hydrophila* water isolates. FEMS Microbiol Lett 220:121–125

Metlitskaya A, Kazakov T, Kommer A, Pavlova O, Praetorius-Ibba M, Ibba M, Krasheninnikov I, Kolb V, Khmel I, Severinov K (2006) Aspartyl-tRNA synthetase is the target of peptide nucleotide antibiotic Microcin C. J Biol Chem 281:18033–18042

Michel-Briand Y, Baysse C (2002) The pyocins of *Pseudomonas aeruginosa*. Biochimie 84:499–510, Review

Mukhopadhyay J, Sineva E, Knight J, Levy RM, Ebright RH (2004) Antibacterial peptide microcin J25 inhibits transcription by binding within and obstructing the RNA polymerase secondary channel. Mol Cell 14:739–751

Munoz J, Arias JM, Montoya E (1984) Production and properties of a bacteriocin from *Myxococcus coralloides* D. J Appl Bacteriol 57:69–74

Nikaido H (2003) Molecular basis of bacterial outer membrane permeability revisited. Microbiol Mol Biol Rev 67:593–656, Review

Nikaido H, Vaara M (1985) Molecular basis of bacterial outer membrane permeability. Microbiol Rev 49:1–32, Review

Niklison-Chirou MV, Dupuy F, Pena LB, Gallego SM, Barreiro-Arcos ML, Avila C, Torres-Bugeau C, Arcuri BE, Bellomio A, Minahk C, Morero RD (2010) Microcin J25 triggers cytochrome c release through irreversible damage of mitochondrial proteins and lipids. Int J Biochem Cell Biol 42:273–281

Nomura M (1963) Mode of action of colicins. Cold Spring Harbour Symp Quant Biol 28:315–324

Oudega B, de Graaf FK (1976) Enzymatic properties of cloacin DF13 and kinetics of ribosome inactivation. Biochim Biophys Acta 425:296–304

Pilsl H, Smajs D, Braun V (1999) Characterization of colicin S4 and its receptor OmpW, a minor protein of the *Escherichia coli* outer membrane. J Bacteriol 181:3578–3581

Poey ME, Azpiroz MF, Laviña M (2006) Comparative analysis of chromosome-encoded microcins. Antimicrob Agents Chemother 50:1411–1418

Pons A-M, Lanneluc G, Cottenceau G, Sablé S (2002a) New developments in non-post translationally modified microcins. Biochimie 84:531–537

Pons AM, Zorn N, Vignon D, Delalande F, Van Dorsselaer A, Cottenceau G (2002b) Microcin E492 is an unmodified peptide related in structure to colicin V. Antimicrob Agents Chemother 46:229–230

Rakin A, Boolgakowa E, Heesemann J (1996) Structural and functional organization of the *Yersinia pestis* bacteriocin pesticin gene cluster. Microbiology 142:3415–3424

Rebuffat S, Blond A, Destoumieux-Garzón D, Goulard C, Peduzzi J (2004) Microcin J25, from the macrocyclic to the lasso structure: implications for biosynthetic, evolutionary and biotechnological perspectives. Curr Protein Pept Sci 5:383–391, Review

Riley MA (1998) Molecular mechanisms of bacteriocins evolution. Annu Rev Genet 32:255–278

Rodríguez E, Laviña M (2003) The proton channel is the minimal structure of ATP synthase necessary and sufficient for microcin H47 antibiotic action. Antimicrob Agents Chemother 47:181–187

Severinov K, Semenova E, Kazakov A, Kazakov T, Gelfand MS (2007) Low-molecular-weight post-translationally modified microcins. Mol Microbiol 65:1380–1394, Review. Erratum in: Mol Microbiol 66:277

Sharma S, Waterfield N, Bowen D, Rocheleau T, Holland L, James R, French-Constant R (2002) The lumicins: novel bacteriocins from *Photorhabdus luminescens* with similarity to the uropathogenic-specific protein (USP) from uropathogenic *Escherichia coli*. FEMS Microbiol Lett 214:241–249

Shehane SD, Sizemore RK (2002) Isolation and preliminary characterization of bacteriocins produced by *Vibrio vulnificus*. J Appl Microbiol 92:322–328

Sugita H, Matsuo N, Hirose Y, Iwato M, Deguchi Y (1997) *Vibrio* sp. strain NM10, isolated from the intestine of a Japanese coastal fish, has an inhibitory effect against *Pasteurella piscicida*. Appl Environ Microbiol 63:4986–4989

Tagg JR, Dajani AS, Wannamaker LW (1976) Bacteriocins of gram-positive bacteria. Bacteriol Rev 40:722–756

Taylor R, Burgner JW, Clifton J, Cramer WA (1998) Purification and characterization of monomeric *Escherichia coli* vitamin B12 receptor with high affinity for colicin E3. J Biol Chem 273:31113–31118

Thomas X, Destoumieux-Garzón D, Peduzzi J, Afonso C, Blond A, Birlirakis N, Goulard C, Dubost L, Thai R, Tabet JC, Rebuffat S (2004) Siderophore peptide, a new type of post-translationally modified antibacterial peptide with potent activity. J Biol Chem 279:28233–28242

Thomas JA, Valvano MA (1993) Role of tol genes in cloacin DF13 susceptibility of *Escherichia coli* K-12 strains expressing the cloacin DF13-aerobactin receptor IutA. J Bacteriol 175:548–552

Van der Wal FJ, Luirink J, Oudega B (1995) Bacteriocin release proteins: mode of action, structure and biotechnological applications. FEMS Microbiol Rev 17:381–399

Vassiliadis G, Destoumieux-Garzón D, Lombard C, Rebuffat S, Peduzzi J (2010) Siderophore microcins form the first family of structure-related antimicrobial peptides from Enterobacteriaceae: isolation and characterization of microcins M and H47. Antimicrob Agents Chemother 54:288–297

Wahaba AH (1965) Vibriocin production in the cholera and El Tor vibrios. Bull World Health Organ 33:661–664

Walker GC (1995) SOS-regulated proteins in translesion DNA synthesis and mutagenesis. Trends Biochem Sci 20:416–420, Review

Wertz JE, Riley MA (2004) Chimeric nature of two plasmids of *Hafnia alvei* encoding the bacteriocins alveicins A and B. J Bacteriol 186:1598–1605

Yang CC, Konisky J (1984) Colicin V-treated *Escherichia coli* does not generate membrane potential. J Bacteriol 158:757–759

Yorgey P, Lee J, Kordel J, Vivas E, Warner P, Jebaratnam D, Kolter R (1994) Posttranslational modifications in microcin B17 define an additional class of DNA gyrase inhibitor. Proc Natl Acad Sci USA 91:4519–4523

Part III
Isolation, Purification, and Production of Antimicrobial Peptides

Chapter 5
Genome Exploitation and Bioinformatics Tools

Anne de Jong, Auke J. van Heel, and Oscar P. Kuipers

Abstract Bioinformatic tools can greatly improve the efficiency of bacteriocin screening efforts by limiting the amount of strains. Different classes of bacteriocins can be detected in genomes by looking at different features. Finding small bacteriocins can be especially challenging due to low homology and because small open reading frames (ORFs) are often omitted from annotations. In this chapter, several bioinformatic tools/strategies to identify bacteriocins in genomes are discussed.

Introduction

Screening for the production of new bacteriocins in bacteria is a challenging endeavour due to the huge screening space (number of potential producer strains × number of potential expression conditions). Limiting the number of potential producer strains by using predictions from genomic data can therefore be a valuable step in the search for and exploitation of new bacteriocins. The availability of more and more genomic data makes this a good starting point for the discovery of novel bacteriocins. Bacteriocins can be mined directly from genomic data based on homology with either known bacteriocins, bacteriocin motifs, or, even more important, bacteriocin biosynthesis genes. Until now, only a few bacteriocins have been discovered by genome mining. Although it is possible to screen for homology to known bacteriocins, this approach is limited to larger bacteriocins and subgroups of bacteriocins, which show high sequence homology. Screening proteins based on conserved motifs is a good alternative, especially for small bacteriocins, but will not discover all, due to the poor conservation of amino acid sequences among bacteriocins.

O.P. Kuipers (✉)
Department of Molecular Genetics, University of Groningen, Groningen Biomolecular Sciences and Biotechnology Institute, PO BOX 11103, 9700 CC, Groningen, The Netherlands
e-mail: O.P.Kuipers@rug.nl

D. Drider and S. Rebuffat (eds.), *Prokaryotic Antimicrobial Peptides: From Genes to Applications*, DOI 10.1007/978-1-4419-7692-5_5,

To enhance the reliability of the predictions and to predict the class to which the bacteriocin belongs, it is of vital importance that the genomic context of a bacteriocin is taken into account. Genomic context of bacteriocins is class dependent and consists of genes encoding modification, regulation, leader processing, transport and immunity proteins. The current understanding of bacteriocins provides a set of general (class dependent) and more specific (subclass dependent) features which can be exploited to explore the vast amount of genomic data to find bacteriocins.

Bioinformatic Tools

Homology Searches in Bacteriocin Databases

The proper analysis tool to determine whether a protein is a bacteriocin strongly depends on the size of the bacteriocin. For proteins larger than 200 amino acids, an alignment algorithm like BLAST is sufficient to determine if a protein is a bacteriocin or not, because large bacteriocins are more conserved in sequence than small ones. In these cases, no specialised bacteriocin database is needed besides the standard nr-database available at the NCBI BLAST server. For blasting smaller proteins (<200 aa), the NCBI nr-database is not suitable and a dedicated database is needed. Two bacteriocins and one more general database are freely accessible for this purpose: (1) the BACTIBASE (Hammami et al. 2007) database holds most known bacteriocins and offers tools like BLAST and a Hidden Markov Model (HMM) for classification of bacteriocins, (2) BAGEL2 (de Jong et al. 2010) which offers, next to automated screening of genomes for bacteriocins with class prediction, an extensive database that allows BLAST searches, (3) PIRSF contains a complete database of proteins but allows good filtering for keywords and therefore can be useful to specifically screen for bacteriocins. The advantage of these bacteriocin-specific databases is that higher cutoff *E*-values for BLAST can be used without increasing the background. The drawback of these dedicated databases is that interesting novel bacteriocins without or with low homology to known ones will not be found using BLAST. The discovery of the two-component lantibiotic haloduracin by McClerren et al. (2006) is an example of a discovery solely based on sequence homology.

Mining for Conserved Protein Domains

In genome annotation projects, functions of proteins are routinely assigned using conserved protein sequence patterns. Many databases that include intelligent search and/or mining tools are available via Web interfaces nowadays (Table 5.1). For mining bacteriocins, some annotation servers contain none or only one bacteriocin

Table 5.1 Online databases suitable for bacteriocin analysis using conserved protein domains

Database	# Bacteriocin domains	URL
PROSITE	1	http://www.expasy.ch/prosite
PFAM	11	http://pfam.sanger.ac.uk
PRINTS	8	http://www.bioinf.manchester.ac.uk/dbbrowser/PRINTS/index.php
TIGRFAM	8	http://www.jcvi.org
PIRSF	NA	http://pir.georgetown.edu
ProDom	NA	http://prodom.prabi.fr

peptide-specific pattern, where others contain up to 11 conserved bacteriocin domains (Table 5.1). The ExPASy proteomics server enables to screen for bacteriocin class IIa family signatures, but does not contain signatures for other types of bacteriocins. TIGRFAM, PRINTS and PFAM contain more conserved protein domains based on bacteriocin alignments. Because these databases do not completely overlap, 14 unique protein patterns could be deduced from them.

Screening Genomic Context

The methods mentioned above allow to screen for bacteriocins that have either homology with known bacteriocins or that at least contain a conserved motif within the bacteriocin. To mine for novel bacteriocins with low homology to known ones, the genomic context of small open reading frames (ORFs) can be screened for genes involved in synthesis, regulation, transport, processing or immunity of bacteriocins. From the literature, it is known that these context genes are conserved between a broad range of species. Especially genes encoding the modification enzymes of lantibiotics have been successfully used to screen genomic data and identify new bacteriocins. Begley et al. (2009) describe the discovery of lichenicidin by LanM (modification enzyme) mining.

Bacteriocin Mining Tool: BAGEL2

The BAGEL2 (de Jong et al. 2010) Webserver, which is the follow-up of BAGEL (de Jong et al. 2006), is still the only fully automated tool available that allows genome mining for bacteriocins, and several publications raised in the literature have used BAGEL successfully (Knoll et al. 2008; Navarro et al. 2008; Holtsmark et al. 2008). This Web-based tool is specialised in mining genomes for bacteriocins using a combined approach of homology search in a bacteriocin database and bacteriocin motif screening together with screening the genome context.

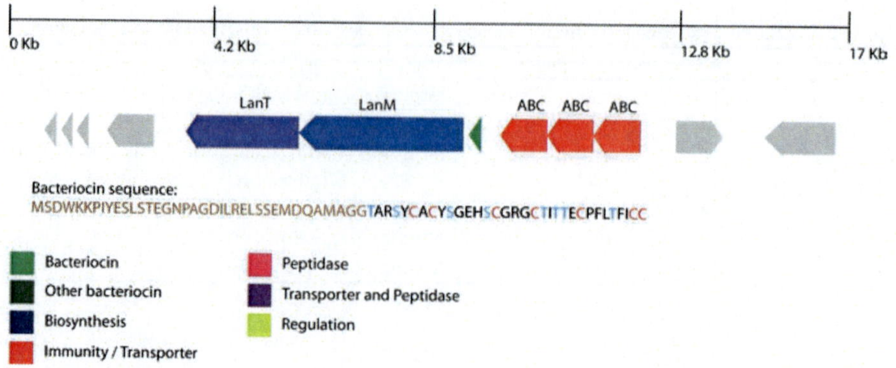

Fig. 5.1 Graphics from the BAGEL2 report of putative bacteriocin SPN23F_12701 of *Streptococcus pneumoniae* ATCC 700669. The leader sequence is indicated in *green* and amino acids involved in dehydration and/or lanthionine ring formation in *red* and *blue*

BAGEL2 does not need annotation of any protein coded by the genome because that is done "on the fly" during the bacteriocin mining process. If the resource is DNA, then an ORF calling is needed to deduce the proteins from the DNA. For this purpose, a good algorithm is available on the BAGEL2 Website. A typical result of a putative bacteriocin and its context genes is shown in Fig. 5.1 where BAGEL2 could annotate three context genes as LanT, LanM and an ABC-transporter. Furthermore, detailed information of the putative bacteriocin is given, including a scoring table showing the most important features that contributed to the positive hit.

Prediction of Small ORFs

Small ORFs are regularly omitted in the annotation process of bacterial genomes when they have low homology to existing proteins. Especially, the small proteins of class I and class II bacteriocins are frequently absent in annotated genomes. For ORF prediction in prokaryotic genomes, the following tools can be used: Glimmer (Delcher et al. 2007) and Prodigal (Table 5.2). Normally, these programs give a FastA formatted file as end product, but for genome context mining the organisation of the genome is also needed, therefore these files must be converted to a GenBank or EMBL format. Fully automated ORF calling systems which give a genbank or EMBL format as output can be found on http://bagel2.molgenrug.nl and http://bioinformatics.biol.rug.nl/websoftware/orf/orf_start.php. Artemis is a popular program for annotation of bacterial genomes and can export results in GenBank format, but is very laborious for a high number of samples and therefore not suitable for high-throughput approach. Alternatively, the genome data can be submitted at NCBI or at the Doe Joint Genome Institute where the data will be processed upon request.

Table 5.2 Web links to programs and analysis servers

Program/server	URL
BLAST	http://blast.ncbi.nlm.nih.gov/
BACTIBASE	http://bactibase.pfba-lab-tun.org
BAGEL2	http://bagel2.molgenrug.nl
PIRSF	http://pir.georgetown.edu
GILMMER	http://www.cbcb.umd.edu/software/glimmer
Zcurve	http://tubic.tju.edu.cn/Zcurve_B
Artimis	http://www.sanger.ac.uk/resources/Software/Artemis
NBCI Annotation	http://www.ncbi.nlm.nih.gov/Genbank/genomesubmit_annotation.html
JGI Annotation	http://merced.jgi-psf.org/cgi-bin/img_er_submit/main.cgi
NCBI Genome Projects	http://www.ncbi.nlm.nih.gov/genomeprj
FIVA	http://bioinformatics.biol.rug.nl/standalone/fiva
MOTIFATOR	http://bioinformatics.biol.rug.nl/standalone/motifator
DISCLOSE	http://bioinformatics.biol.rug.nl/standalone/disclose

High-Throughput Data Mining

The number of completely sequenced genomes is increasing fast, but currently more and more genome sequences remain unfinished, especially in meta-genomics projects, although they are undoubtedly very interesting data sources for bacteriocin mining. The NCBI Entrez Genome Project database consists of complete and incomplete (in progress) genome sequencing projects. Currently (January 2010), 3,215 prokaryotes genome projects are available from this database. High-throughput screening in these kinds of sequence data is possible with the command line driven version of BAGEL2 (available on request at http://bagel2.molgen.rug.nl).

Transcriptome Analysis

To get further insight in the expression and transcriptional regulation of putative bacteriocins and their accompanying immunity, regulation and synthesis of genes, analysis of transcriptome data derived from DNA-microarray or deep-sequencing experiments can be performed using the tools FIVA, MOTIFATOR and DISCLOSE. FIVA (Blom et al. 2007) is used to discover differential expression of functional classes between two experiments. Therefore, a functional class should first be defined for the bacteriocin and its context genes. MOTIFATOR (Blom et al. 2009) predicts *cis*-regulatory motifs using the gene expression profiles of functional classes. Furthermore, it creates an interactive visualisation of the results. For analysing multiple experiments, DISCLOSE (Blom et al. 2008) is used which is also used for the discovery of de novo DNA motifs involved in the transcriptional regulation. Using these tools will speed up validation of expression of bacteriocin gene clusters and identify possible motifs as targets for regulation. Moreover, they will provide possible links between bacteriocin production and co-expression of other genes, enabling the reconstruction of gene regulatory networks to which the bacteriocins belong.

Conclusion

Prediction of bacteriocins and the (sub)class it belongs becomes more and more reliable due to the increasing number of class-specific protein motifs and, very importantly, by taking the genomic context into account. The Webserver BAGEL2 applies decision rules for combinations of bacteriocin protein patterns and the function of context genes, which further enhance the prediction of the bacteriocin (sub) class. Although mining for novel bacteriocins is a challenging task, modern bioinformatic tools allow discovery of thousands of new bacteriocin candidates in (un) finished genomes and in meta-genomics data.

References

Begley M, Cotter PD, Hill C, Ross RP (2009) Identification of a novel two-peptide lantibiotic, lichenicidin, following rational genome mining for LanM proteins. Appl Environ Microbiol 75:5451–5460

Blom EJ, Bosman DWJ, Van Hijum SAFT, Breitling R, Tijsma L, Silvis R, Roerdink JBTM, Kuipers OP (2007) FIVA: Functional Information Viewer and Analyzer extracting biological knowledge from transcriptome data of prokaryotes. Bioinformatics 23:1161–1163

Blom EJ, van Hijum SA, Hofstede KJ, Silvis R, Roerdink JB, Kuipers OP (2008) DISCLOSE: DISsection of CLusters Obtained by SEries of transcriptome data using functional annotations and putative transcription factor binding sites. BMC Bioinform 9:535

Blom EJ, Roerdink JBTM, Kuipers OP, Van Hijum SAFT (2009) MOTIFATOR: detection and characterization of regulatory motifs using prokaryote transcriptome data. Bioinformatics 25:550–551

de Jong A, van Hijum SAFT, Bijlsma JJE, Kok J, Kuipers OP (2006) BAGEL: a web-based bacteriocin genome mining tool. Nucleic Acids Res 34:W273–W279

de Jong A, van Heel AJ, Kok J, Kuipers OP (2010) BAGEL2: mining for bacteriocins in genomic data. Nucleic Acids Res 38:W647–W651

Delcher AL, Bratke KA, Powers EC, Salzberg SL (2007) Identifying bacterial genes and endosymbiont DNA with Glimmer. Bioinformatics 23:673–679

Hammami R, Zouhir A, Hamida B, Fliss I (2007) BACTIBASE: a new web-accessible database for bacteriocin characterization. BMC Microbiol 7:89

Holtsmark I, Eijsink VH, Brurberg M (2008) Bacteriocins from plant pathogenic bacteria. FEMS Microbiol Lett 280:1–7

Knoll C, Divol B, du Toit M (2008) Genetic screening of lactic acid bacteria of oenological origin for bacteriocin-encoding genes. Food Microbiol 25:983–991

McClerren AL, Cooper LE, Quan C, Thomas PM, Kelleher NL, van der Donk WA (2006) Discovery and in vitro biosynthesis of haloduracin, a two-component lantibiotic. Proc Natl Acad Sci U S A 103:17423–17428

Navarro L, Rojo-Bezares B, Sáenz Y, Díez L, Zarazaga M, Ruiz-Larrea F, Torres C (2008) Comparative study of the pln locus of the quorum-sensing regulated bacteriocin-producing *L. plantarum* J51 strain. Int J Food Microbiol 128:390–394

Chapter 6
Design and Engineering Strategies for Synthetic Antimicrobial Peptides

Alessandro Tossi

Abstract Thousands of antimicrobial peptides (AMPs) of prokaryotic, fungal, plant, or animal origin have been identified, and their potential as lead compounds for the design of novel therapeutic agents in the treatment of infection, for stimulating the immune system, or in countering septic shock has been widely recognized. Added to this is their possible use in prophylaxis of infectious diseases for animal or plant protection, for disinfection of surgical instruments or industrial surfaces, and for food preservation among other commercially important applications. Since the early eighties, AMPs have been subject to a vast number of studies aimed at understanding what determines their potency and spectrum of activities against bacterial or fungal pathogens, and at maximizing these while limiting cytotoxic activities toward host cells. Much research has also been directed toward understanding specific mechanisms of action underlying the antimicrobial activity and selectivity, to be able to redesign the peptides for optimal performance. A central theme in the mode of action of many AMPs is their dynamic interaction with biological membranes, which involves various properties of these peptides such as, among others, surface hydrophobicity and polarity, charge, structure, and induced conformational variations. These features are often intimately interconnected so that engineering peptides to independently adjust any one property in particular is not an easy task. However, solid-phase peptide synthesis allows the use of a large repertoire of nonproteinogenic amino acids that can be used in the rational design of peptides to finely tune structural and physicochemical properties and precisely probe structure–function relationships.

A. Tossi (✉)
Department of Life Sciences, University of Trieste, Via Giorgieri 1, Trieste 34127, Italy
e-mail: atossi@units.it

D. Drider and S. Rebuffat (eds.), *Prokaryotic Antimicrobial Peptides: From Genes to Applications*, DOI 10.1007/978-1-4419-7692-5_6,

Introduction

Antimicrobial peptides (AMPs) are ubiquitous host defense molecules and an impressive body of literature covers their isolation, purification, classification, and biology in both prokaryotic and eukaryotic organisms, as well as their modes of action in vivo, in vitro, and in model systems, and their redesign for development as potential therapeutic agents. This chapter limits itself to considering some facets of the design and engineering of AMPs used for probing structure–activity relationships and optimizing specific characteristics. For a more extensive understanding of these essential components of immunity, readers are referred to the many reviews (for examples see Hancock 1997; Zasloff 2002; Tossi and Sandri 2002; Brogden 2005; Hancock and Sahl 2006) as well as other chapters of this book, which provide brief but comprehensive overviews on historical, biological, and functional aspects of AMPs.

Endogenous AMPs (also known as host defense peptides or HDPs) have an ancient origin, possibly preceding that of eukaryotic organisms, and so have a very long evolutionary history. Furthermore, they are produced by epithelial cells or phagocytes that are at the interface between a producing host organism and the constantly varying microbial biota in its environment, so their genes are subject to strong adaptive pressures. In fact, a recurrent theme in the evolution of the gene-encoded HDPs appears to be repeated duplication events followed by rapid diversification (Hughes 1999; Vanhoye et al. 2003; Maxwell et al. 2003; Patil et al. 2004; Tennessen 2005). Time and hypervariation have resulted in an extraordinary molecular diversity, so that classification according to sequence or structure is only possible on rather broad terms (see Chaps. 1 and 3), and AMP/HDPs with novel structural features are continuously discovered. Added to this is a disconcerting functional multiplicity; many HDPs seem to have developed the capacity to interact concurrently with different components of target microorganisms, initiating diverse inactivation pathways (Brogden 2005; Hancock and Sahl 2006). Several HDPs also interact with and variously attract and/or stimulate host immune cells, thus participating in other innate or adaptive responses by the host and indirectly contributing to pathogen inactivation (Oppenheim et al. 2007). Any attempt at the rational redesign or engineering of AMPs needs to take this complexity into account.

Notwithstanding the diversity of gene-encoded AMPs, two broad structural classes stand out for their abundance and prevalence throughout the eukaryotic kingdom: helical peptides and defensins. The former group comprises linear, cationic peptides that usually undergo a transition to an amphipathic, helical conformation on contact with biological membranes. This favors insertion of the so-formed hydrophobic face of the helix into anionic microbial membranes, initiating processes that eventually lead to channel formation or membrane lysis, as described by the Huang–Matsuzaki–Shai model (Matsuzaki 1998; Yang et al. 2001; Shai 2002; Huang 2006), or inhibition of membrane-associated metabolic processes (Pag et al. 2008). Peptides presenting this simple but very effective scaffold are found in organisms ranging from bacteria to humans (Tossi et al. 2000;

Tossi and Sandri 2002). Defensins have a more complex scaffold consisting of a small, triple-stranded β-sheet platform onto which a short helix is often grafted, stabilized by the presence of three or four disulfide bonds. Their mechanism of action is less clear but involves interaction with and likely partial insertion into biological membranes as an initial step in their mechanisms of action (Aerts et al. 2008; Sass et al., 2008). Defensins of different types are produced by fungi, plants, and vertebrate and invertebrate animals (Antcheva et al. 2006; Wong et al. 2007) and possibly have a prokaryotic origin (Zhu 2007). Both the simpler helical scaffold and more complex defensin platform are present in peptides from widely diverging organisms, ranging from plants to vertebrate and invertebrate animals, and therefore represent cases either of convergent evolution to or of long-term conservation of structural elements that are likely to be important for antimicrobial activity (Tossi et al. 2000; Yount and Yeaman 2004; Yeaman and Yount 2007).

This chapter briefly describes some of the design methods that have been used to study the mode of action of helical or defensin AMPs from vertebrate and invertebrate animals and how both natural and nonproteinogenic amino acid building blocks can be used in the rational design of peptides based on these scaffolds.

Rational Design Methods

Four general approaches have been used to design AMPs and guide SAR studies: (1) sequence modification of naturally occurring HDPs so as to produce congeners, fragments, or hybrids of various types; (2) minimalist approaches based on the maintenance of specific structural features (e.g., the amphipathic helix); (3) generation of peptide libraries or arrays; and (4) sequence–template-guided approaches. These methods have most often been applied to the simpler helical type peptides but are also commonly extended to other types, such as beta-hairpin peptides, proline-rich peptides, or defensins. The ultimate aim is generally to obtain molecules that are easier and more cost effective to produce, are more stable and/or bioavailable, and display increased antimicrobial activity accompanied by decreased toxicity toward host cells, in view of potential therapeutic applications. These approaches can be combined in several permutations, and applied in iterative cycles, resulting in a form of "directed evolution" of the AMPs toward an optimal activity against particular pathogens.

Peptide Congeners, Fragments, and Hybrids

In this approach, naturally occurring AMP sequences are modified by single or multiple replacements, deletions, or additions of one or more residues, carried out simultaneously or in sequence. Alternatively, fragments are generated from a natural peptide either by truncating at the N- or C-termini or by excising internal regions predicted to be important for activity. A further popular variation on this theme is that of assembling chimeric peptides from the active segments of different natural AMPs (see also Chap. 6). A good example of sequential mutations leading to a structure with improved activity is pexiganan (Lamb and Wiseman 1998; Gottler and Ramamoorthy 2009), a variant of magainin that has come close to approval for clinical use in the treatment of infections associated with diabetic foot ulcers. In common with many other examples, these variations resulted in a considerably increased cationicity and improved amphipathicity. Similarly, directed mutations of the fungal defensin, plectasin (Mygind et al. 2005) that also increased positive charge and modulated surface hydrophobicity, have resulted in an analogue currently under development as an antibiotic.

Studies on peptide fragments have also led to shortened AMPs with unaltered or sometimes even improved properties, with respect to the parent peptides, both for helical and defensin-type HDPs (e.g. Braff et al. 2005; Varkey et al. 2006; Krishnakumari et al. 2006; Antcheva et al. 2009, among many others). However, it should be stressed that variations such as alterations in the sequence or truncation may result in loss of structure-stabilizing factors, such as disulfide bonds or internal salt bridges, which can severely alter the structure of peptides. This may cause them to switch to different modes of membrane interaction, thus also altering their mechanisms of action toward microbial or host cells (Morgera et al. 2008a; Tomasinsig et al. 2008, 2009; Antcheva et al. 2009). Design of hybrid peptides has been extensively applied to the study of helical peptides of insect and amphibian origin (e.g., cecropins, melittin, magainin, and PGLa; see Chap. 6 for a comprehensive report).

Minimalist Approaches and AMP mimics

Most often, minimalist AMPs are designed de novo based purely on broad structural requirement, such as a compliance with the amphipathic α-helical scaffold, while limiting the types of residues used to one or the other of the basic amino acids lysine and arginine, and one or two of the hydrophobic residues alanine, leucine/isoleucine, phenylalanine, or tryptophan. This can result in potent antimicrobial agents but ignores finer structural attributes selected for by evolution, leading to reduced selectivity for microbial with respect to host cells. Taken to an extreme, these approaches also decrease the peptidic nature of the molecule, to the extent of doing away with the peptide backbone altogether, in attempts to improve bioavailability. Numerous minimalist peptide mimics have been described, including peptoids (in which the side chain is placed on an N-substituted glycine, Chongsiriwatana et al. 2008), oligo-acyl-lysines (OAKs, Rotem and Mor 2009), beta-peptides (β-amino acid oligomers with cationic, amphiphilic helical structures, Porter et al. 2002), and ceragenins (cholic-acid based scaffolding with appended amine groups, Lai et al. 2008) among several others.

AMP Combinatorial Libraries and Arrays

Synthetic or biological combinatorial libraries are powerful tools that can help in rapidly obtaining optimized classes of active compounds, especially when combined with directed evolution (Castro et al. 2006). In the past, solid-state chemical synthesis methods have somewhat limited the size of both the peptides that could be synthesized and the libraries themselves. However, powerful new methodologies such as spot synthesis are making these approaches more accessible (Winkler et al. 2009; Hilpert et al. 2005). Furthermore, the creation of a library can be made simpler by limiting the amino acid types that are used to build the peptides (converging with the minimalist approach) or by acting combinatorially only at certain positions in natural peptides (converging with the sequence modification methods) (Blondelle and Houghten 1996). As an alternative to chemical syntheses, biosynthetic peptide libraries can be obtained by making use of the phage-display technology (Pini et al. 2005) or methods such as the "Suicide Expression System" (SES). The latter method involves involves the controlled expression of systematically mutated AMP genes in susceptible bacteria (Kristensen and Yaver 2004), and by identifying those that result in bacterial self-inactivation under the control of an inducer.

Sequence Templates (an Evolutionary Approach)

Sequence templates can be obtained by comparing structurally homologous stretches from a large collection of naturally occurring HDPs and abstracting conserved patterns in terms of residue type (e.g., charged, polar, hydrophobic, etc.). Depending on the origin of the peptides used to create each collection, one can probe for structural aspects that have been selected for by evolution at different levels. Comparing helical peptides from insects, amphibians, and mammals, for example, allowed defining important generic scaffold characteristics that have arisen by convergent evolution or have been conserved over hundreds of millions of years (Zelezetsky and Tossi 2006). Considering just orthologues of the helical cathelicidin in primates instead allowed to better understand specific structural characteristics of the scaffold that have arisen over tens of millions of years (Zelezetsky et al. 2006). On an intermediate level, comparing many paralogues from different vertebrate animals allowed to determine a minimal generic beta-defensin scaffold (Antcheva et al. 2009).

The rationale in developing templates is not to determine significant sequence homologies or positional conservation of specific amino acids in the natural peptides that are compared (which is in any case likely to be poor due to the fact that evolution of HDPs in animals is often accelerated). It is rather to identify underlying patterns in the distribution of different types of residues (charged, neutral polar, structure determining, or hydrophobic), and the maintenance of physico-chemical parameters that may be correlated to potency and specificity (Zelezetsky and Tossi 2006). Using templates as a guide can then significantly reduce the number of variants that need to be synthesized in order to obtain useful structure/activity information for a given HDP, with respect to the conventional "sequence modification" approach. Furthermore, it ensures that evolutionarily conserved, and therefore functionally important, sequence patterns are maintained. Once developed, templates are also compatible with minimalist and high-throughput library approaches. The following section provides an overview of how templates can be rationally filled with appropriate amino acid residues, so as to design or engineer peptides with a broad-spectrum antimicrobial activity and to probe the structural factors that influence their antimicrobial potency and selectivity.

Amino Acid Building Blocks for Peptide Engineering

Parameters Influencing AMP Activity

Helical HDPs and defensins have different scaffolds, which display some common characteristics (e.g., a net cationicity balanced by the presence of hydrophobic residues resulting in amphipathic structures) and other quite different parameters (e.g., the nature of their amphipathic structures), affecting activity.

The helical amphipathic structure segregates polar/charged residues on one face of the helix cylinder (the polar sector) and hydrophobic residues on the other (the hydrophobic sector), which can then be inserted into biological membranes. Structural and physicochemical parameters that influence the folding/insertion process include size, residue arrangement, propensity for helical structuring, net charge, global hydrophobicity, amphipathicity, angle subtended by each face on a helical wheel projection, and the side-chain depths of each face (Tossi et al. 2000; Giangaspero et al. 2001; Zelezetsky and Tossi 2006). These parameters are intimately related so that modifications aimed at altering one can result in significant changes to one or more of the others as well. For example, increasing the number of hydrophobic residues necessarily increases the size of the hydrophobic sector, affecting amphipathicity, and similarly for the number of residues in the polar sector. For shorter peptides (<20 residues), increasing the positive charge can bring cationic side chains next to each other when the helix forms (especially if separated by three or four residues in the primary sequence). The resulting repulsion reduces the propensity for helix formation.

Defensins have considerably more complex structures, so that while it is possible to quantitatively assess such parameters as overall charge and global hydrophobicity, other parameters such as amphipathicity are not as well defined or easily controllable as in helical peptides. In any case, sequence templates, by providing the possibility of generating simplified model peptides with more defined structural and physicochemical characteristics, have been quite useful in helping to explain the general mode of action of both types of peptides, and in helping to explain the behavior of natural HDPs. For example, they have helped explain differential interactions with the microbial cell wall under different conditions, the effect of interactions with medium or serum components on biological activities, and how antimicrobial or host-cell modulating activities are affected by the molecular evolution of peptide orthologues in different species (Zelezetsky et al. 2005a; Zelezetsky and Tossi 2006; Tomasinsig et al. 2008; Antcheva et al. 2009). The wide repertoire of nonproteinogenic amino acids (in principle over 400 alongside the 20 natural ones) combined with specific sequence templates provides a powerful tool for these studies. Their use requires an accurate assessment of side-chain hydrophobicities. However, the published hydrophobicity index scales include only natural amino acids. A consensus hydrophobicity scale that includes some nonproteinogenic amino acids and appropriate analysis software has thus been developed ad hoc and is available on the Trieste University Web site at http://www.bbcm.units.it/~tossi/antimic.html. It allows quantitative estimates of global per residue hydrophobicity, and global or relative amphipathicity for sequence stretches of homogeneous conformation.

Charge and Hydrophobicity

These parameters are modulated by balancing the content of amino acid residues with charged, neutral polar and aliphatic/aromatic side chains in the peptide. Cationicity is widely recognized as being important for the activity of HDPs and is normally provided by the presence of Lys and/or Arg residues. Arg, however, undergoes added H-bonding and π interactions that need to be taken into account. For a helical scaffold, the polar/charged residues Glu, Gln, and Orn (ornithine) and the aliphatic residue Nle (norleucine) are a good choice to start with for probing charge effects (see Table 6.1). Their side chains are of similar size and so ensure that the polar and hydrophobic sectors of the helix have similar depths and that the anionic Glu, neutral Gln, and cationic Orn have similar hydrophobicity index values. For an 18-residue helix, it is thus possible, by interchanging them, to vary charge from +1 to +6 without significantly affecting the global hydrophobicity, amphipathicity, or helix-forming propensity of the sequence (Zelezetsky et al. 2005a; Zelezetsky and Tossi 2006). Further increasing the cationicity can, however, lead to conformational destabilization due to excessive charge density in the polar sector. Decreasing cationicity by introducing anionic residues can instead facilitate helix formation due to intramolecular salt-bridge formation.

The global hydrophobicity is normally modulated by altering the residues in the hydrophobic face as well as by altering its width on the helix cylinder (i.e., the angle subtended by the hydrophobic sector on a helical wheel projection). In a helical template derived from a combination of invertebrate and vertebrate animal peptides, interface positions between polar and hydrophobic faces were undefined, allowing to modulate face widths by inserting either polar or hydrophobic residues. Interestingly, glycine is often present at the interface. Overall hydrophobicity is also decreased by switching from aromatic to long aliphatic to short aliphatic residues. Use of long-chain α-branched amino acids such as Dpg (dipropylglycine) or Deg (diethylglycine) (Table 6.1) increases the density of aliphatic side chains in the

Table 6.1 Examples of amino acid building blocks for assembling antimicrobial peptides using solid-phase peptide synthesis

Name	Altered property	Structure
Glutamine (Gln) Glutamic acid (Glu) Ornithine (Orn)	Charge	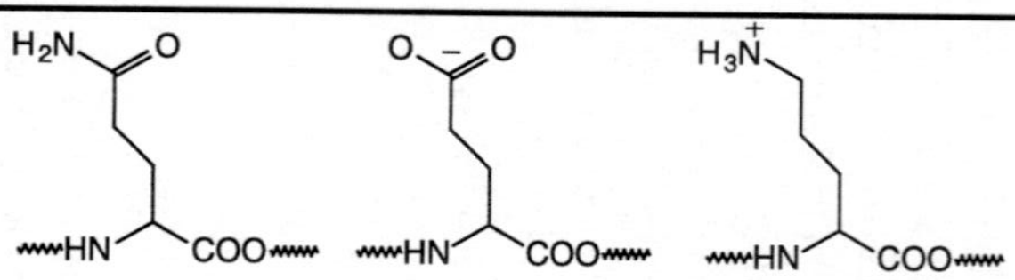 **Gln** ($q = 0$, $H_i = -6.0$) **Glu** ($q = -1$, $H_i = -8.3$) **Orn** ($q = +1$, $H_i = -9.0$)
Norleucine (Nle) Norvaline (Nva) Aminobutyric acid (Abu) Alanine (Ala) Dipropylglycine (Dpg) Diethylglycine (Deg)	Hydrophobicity (hydrophobic sector depth)	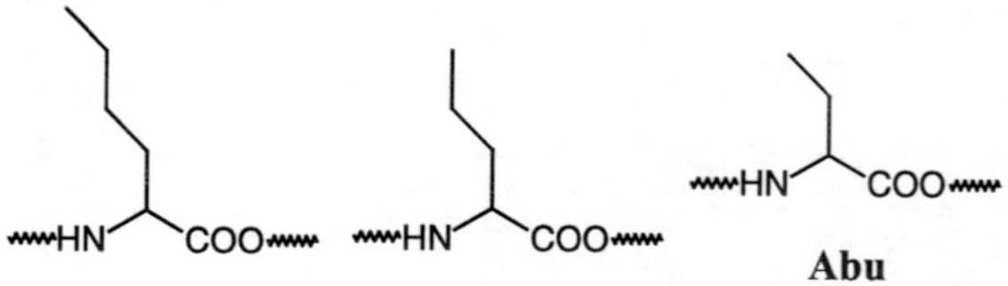 **Nle** ($q = 0$, $H_i = +9.5$) **Nva** ($q = 0$, $H_i = +5.3$) **Abu** ($q = 0$, $H_i = +1.7$) 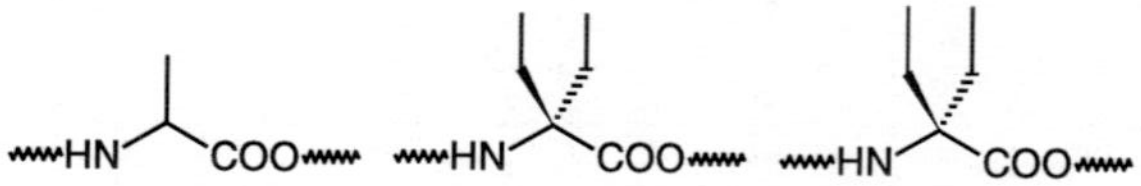 **Ala** ($q = 0$, $H_i = -1.1$) **Dpg** ($q = 0$, $H_i = +13.7$) **Deg** ($q = 0$, $H_i = +6.0$)

Alpha-amino-isobutyric acid[1] (Aib) 1-amino-1-cylopentanecarboxylic acid Proline (Pro) Glycine (Gly) D-Enantiomers (D-AA)	Conformational stability	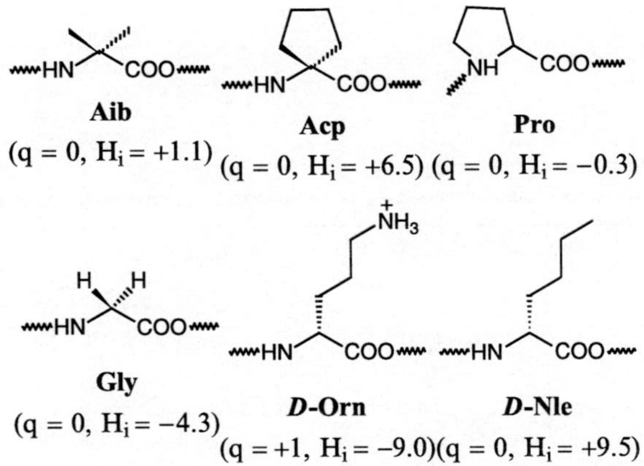
Lysine (Lys) Ornithine (Orn) Diaminobutyric acid (Dab)	Polar sector depth	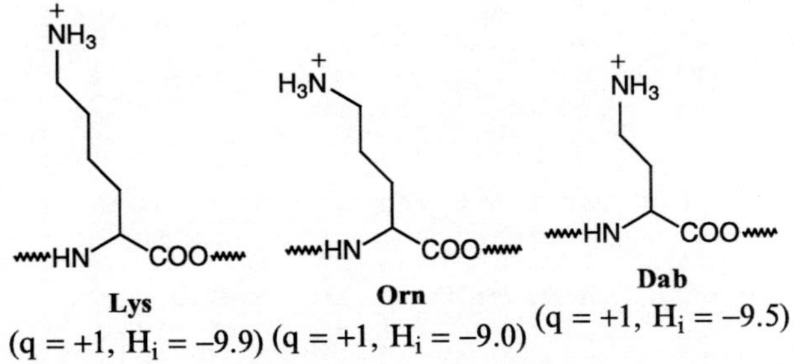

(continued)

Table 6.1 (continued)

Name	Altered property	Structure
Diaminopropionic acid (Dap) Homoserine (Hse) Serine (Ser)		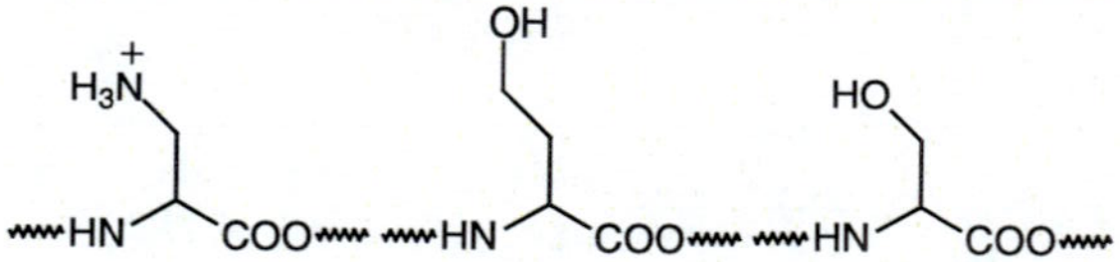 **Dap** (q = +1, H_i = –9.3) **Hse** (q = 0, H_i = –3.5) **Ser** (q = 0, H_i = –2.4)
Cyanophenylalanine[2] (CN-Phe) Cyanoalanine[2] (CN-Ala)	Fluorescence/IR probe	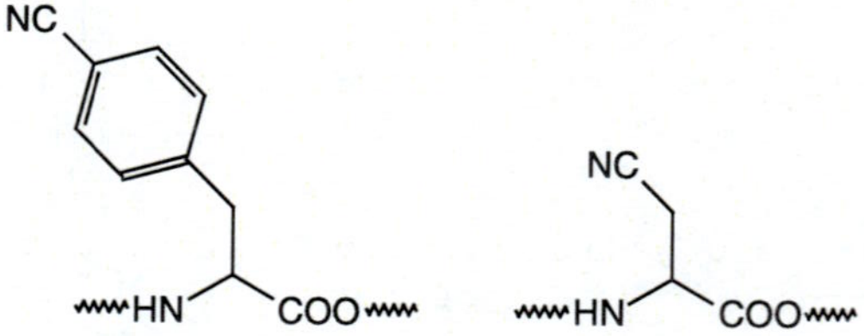 **CN-Phe** **CN-Ala** λ_{ex} = 240 nm, λ_{em} = 300 nm, IR 2228 cm^{-1} (H_20) → 2238 cm^{-1} (org)

q=charge; H_i=residue side-chain hydrophobicity index according to an extended consensus scale developed at Trieste University (see www.bbcm.units.it/~HydroMcalc/Hydromcalc.html)

[1]Toniolo et al. 2001

[2]Tucker et al. 2004; Tang et al. 2009

hydrophobic face, markedly increasing global hydrophobicity. Unlike the short-chain alpha-branched residue aminoisobutiric acid (Aib, see below), these residues do not affect helix stability (Zelezetsky et al. 2005a).

For defensins, the more complex scaffold makes quantification of parameters such as amphipathicity and surface depths less meaningful, and use of nonproteinogenic amino acids is thus less significant. A template derived from avian and mammalian β-defensins in principle allows variation of charge up to +11, as observed in some natural peptides such as human hBD3 (Antcheva et al. 2009), with about half the positions being open to hydrophopic residues, for control of hydrophobicity.

Surface Properties and Amphipathicity

For the helical scaffold, surface properties can be altered by either varying the size of the polar and hydrophobic faces (angle subtended on a helical wheel) or varying the depth of each face by controlling the size of residue side chains. As mentioned above, the respective size of the two faces can be altered by interchanging Gly with polar or hydrophobic residues at their interface. Switching between nonproteinogenic residues such as Nle (norleucine), Nva (norvaline), and Abu (aminobutyric acid) (Table 6.1) allows for a gradual decrease of hydrophobic face depth. Similarly, it is possible to vary the polar sector depth by switching between Lys, Orn, Dab (diaminobutyric acid), and Dap (diaminopropionic acid) (Table 6.1). This variation is useful in probing the relevance of the so-called snorkel or periscope effect from the polar sector when the helical peptide binds parallel to the membrane surface and then sinks with its hydrophobic sector into the membrane lipid bilayer (Monne et al. 1998; Zelezetsky and Tossi 2006).

Amphipathicity can be modulated by simply interchanging polar with hydrophobic residues so as to decrease the homogeneity of the respective faces. This can be taken to an extreme in completely scrambled peptides. As residue content is unaltered, global hydrophobicity and charge are not affected by this. By redistributing residues in the sequence it is however possible to inadvertently favor other types of regular amphipathic conformations (such as extended β-sheet-like or other types of helical conformations), which may show some form of antimicrobial activity. A simple computational tool at Trieste University (http://www.bbcm.units.it/~tossi/antimic.html) allows simple visualizations of residue distribution to check for this.

The dynamic interaction of helical AMPs with membranes requires the formation of an amphipathic conformation, and this occurs equally well in *all*-D enantiomers (composed only of D-amino acids), which form left-handed helices with similar surfaces to the right-handed ones. For several naturally derived or artificial AMPs, these enantiomers in fact display comparable antimicrobial properties in vitro, and these are likely to be improved in vivo, as they are less prone to be sequestered by stereospecific interactions with serum components or removal by proteolysis, thus displaying improved bioavailability. Possibly for the same reasons, however, all-*D* peptides have been found to be somewhat more cytotoxic than the natural enantiomers (Pacor et al. 2002; Tomasinsig et al. 2009).

Conformational Stability

For helical AMPs, this active conformation normally occurs only on interaction with biological membranes, while they are disordered in aqueous solution. Both a propensity for helix formation and the presence of an amphipathic residue arrangement are required for the peptide to insert into the lipid bilayer. It is possible to modulate the propensity for helix formation in various different ways. The appropriate placement of anionic and cationic residues at positions *i*+3 or *i*+4 from each other results in helix-stabilizing salt bridges. If the number of intramolecular side-chain attractions exceeds that of repulsions, helix formation is strongly favored. However, as each salt bridge comes at the cost of a positive charge, only longer peptides can accommodate several bridges while maintaining a high global cationicity. This trait may have been selected for in the evolution of some primate cathelicidin. Thus, the human LL-37 (37 residues, charge +6) and its orangutan orthologue (charge +4) have an excess of attractions and form more-or-less stable helices even in aqueous solution, while rhesus RL-37 (charge +10), with an excess of repulsion, behaves in the canonical manner and only forms a helix at the membrane surface, being randomly coiled in aqueous solution (Zelezetsky et al. 2006). These features were found to modulate their interactions with both microbial and host cells (Morgera et al. 2008b; Tomasinsig et al. 2008, 2009).

For shorter peptides, it is possible to stabilize the helical conformation considerably using specific nonproteinogenic, α-branched amino acids, due to the Thorpe–Ingold effect (Toniolo et al. 2001). Aib (aminoisobutyric acid) is well known in this respect and indeed is found in natural, nonribosomally synthesized helical antimicrobial peptaibols such as alamethicin (Duclohier 2007). Acp (1-amino-1-cylopentanecarboxylic acid) (Table 6.1), with a cyclic α-branched side chain, is particularly effective, and a few residues interspersed in the hydrophobic sector are sufficient to ensure helix formation in aqueous solution, even in the absence of membranes, while its hydrophobicity is comparable to that of aliphatic residues commonly present in this sector (Zelezetsky et al. 2005a). It is interesting to note that short, stable helical peptides of this type show an antimicrobial and cytotoxic behavior in some ways similarly to those of stable helical HDPs such as LL-37, thus helping to explain some aspects of their mode of action (Zelezetsky et al. 2005a; Zelezetsky et al. 2006).

Helix destabilization is sometimes useful in probing functional aspects of AMPs. This is achieved by introducing residues such as proline, preferably in the polar sector of the helix, or by introducing D-enantiomers of amino acids already present. Substitution of two successive residues with their enantiomers significantly reduces the helix-forming propensity (Krause et al. 1995; Zelezetsky et al. 2005a).

It is interesting to note that diastereomers of helical AMPs have been found to selectively maintain antimicrobial activity toward some bacteria (Zelezetsky et al. 2005a), with significantly reduced cytotoxicity toward host cells and improved bioavailability (Oren et al. 2002). Introduction of Gly residues may render the helix more flexible and also sometimes correlates with a better selectivity for bacterial with respect to host cells (Zelezetsky et al. 2005b).

Peptide Capping, Cyclization, Linearization, and Oligomerization

Peptide capping can render peptides more stable, protecting them from exopeptidases. This can be effected either by acetylating the N-terminus and/or amidating the C-terminus (which respectively also results in a decrease or increase of cationicity) or by introducing a pyroglutamic acid at the N-terminus, as present in some natural AMPs, including β-defensins (Antcheva et al. 2009). Cyclization, either by introducing a disulfide bridge between N- and C-terminal cysteines or by covalent linkage of the termini, has also been used to modulate AMP properties and to increase peptide stability and bioavailability. Interestingly, a substantial maintenance of the helical conformation was observed on end-to-end cyclization of natural helical peptides or disulfide bridge cyclization of an artificial one, although effects on antimicrobial potency and cytotoxicity were quite variable (Krishnakumari et al. 1999; Unger et al. 2001). End-to-end cyclization of defensins improved antimicrobial activity by reducing salt sensitivity (Yu et al. 2000).

Many studies have been carried out to test the effects of the linearization of defensins, by removing disulfide bonds, regarding activities toward both microbial and host cells (Wu et al. 2003; Kluver et al. 2005; Antcheva et al. 2009). These studies often report that antimicrobial activity is either unaffected or actually improved. It is, however, highly unlikely that a structural scaffold that has been maintained throughout the evolution of these peptides could be so easily dispensable for their biological functions. Rather, the scaffold supports sequences displaying a certain balance of cationic and hydrophobic residues necessary for the peptides' activity. This balance remains in the linearized versions, although residues are likely arranged in distinctly different structures, allowing different modes of membrane interaction and resulting in different mechanisms of bacterial inactivation (Antcheva et al. 2009).

Both helical AMPs and defensins may oligomerize at the membrane surface. Covalent dimerization can be useful to probe the relevance of these processes to biological functions. For helical peptides, covalent dimerization has been effected by placing cysteine residues at the N- or C-termini, so as to produce head-to-head (parallel) or head-to-tail (antiparallel) homo- or heterodimers (Hara et al. 2001a, b), and used to probe the effect of dimerization on membrane pore formation and stability. An alternative means of producing parallel dimers is that of carrying out solid-phase synthesis starting from an unprotected lysine residue, so that identical helices grow from both the α- and ε-amine groups. This technique has also been used to form multiply branched dendrimers with interesting antimicrobial properties (Pini et al. 2005) that are being developed as potential novel antibiotics.

For defensins, loss of one Cys residue has been reported to result in covalent intermolecular dimer formation in mammalian β-defensins and results in somewhat altered antimicrobial activities (Circo et al. 2002; Campopiano et al. 2004). Starting from a β-defensin sequence template, covalent dimerization of this type was useful in probing the relevance of noncovalent dimerization to biological activities in primate β-defensins 2 and 3 (Antcheva et al. 2009).

Peptide Labeling with Spectroscopic Probes

For membrane-active AMPs, the possibility of following membrane interactions via spectroscopic techniques is very useful. This is most often carried out by using fluorescence or infrared techniques, based on internal or introduced spectrophores. In the first case, the presence of a tryptophan residue is highly desirable, as its fluorescence is very sensitive to environmental transitions (e.g., from bulk solution to the membrane lipid bilayer) and also to inter- or intramolecular quenching from quencher groups specifically placed in the external aqueous environment or lipid acyl chains. However, some care should be taken if tryptophan residues are not naturally present and must be introduced, as they are known to favor partitioning at the membrane surface interface and may thus affect the mechanism of action. Tryptophan also has interesting CD properties useful for probing the rigidity of its environment (Agashe et al. 1995). An alternative is to replace naturally present Phe or Tyr residues with the nonproteinogenic residue CN-Phe (cyanophenylalanine, see Table 6.1) (Tucker et al. 2004; Tang et al. 2009). This has useful emission properties, as its fluorescence intensity is markedly reduced in organic environments such as the lipid bilayer, and is also strongly quenched by Cl- ions present in bulk solution. It can also be used as an IR probe, as its IR spectrum changes significantly on passing from an aqueous to a membrane environment. Both these effects have been useful in better understanding the interaction of human β-defensin 3 with model membranes (Morgera et al. 2008a). CN-Ala (cyanoalanine) can also be used as a nonaromatic IR probe.

Introduced fluorophores are often based on fluorescein isothiocyanate (FITC), BODIPY, or the Alexa dyes (Mattiuzzo et al. 2007; Scocchi et al. 2008; Benincasa et al. 2009; Morgera et al. 2009). These can be covalently conjugated to α-(N-terminal) or ε-amine groups (lysines), or to the SH groups of cysteines. Care should be taken when linking to naturally present amine groups, as formation of the amide bond removes a positive charge, which can be important for antimicrobial activity (addition of fluorophores such FITC, which are anionic, further decreases cationicity). For this reason, it is sometimes useful to introduce extra N- or C-terminal cysteine or lysine residues specifically for linking of the fluoropores. In this respect, anecdotal accounts indicate that introduction of the fluorophore at the C-terminus is less likely to affect the mode of action of AMPs than that at the

N-terminus. Combined with extracellular quenchers, fluorescently labelled AMPs are also very useful for following peptide internalization into bacterial cells (Benincasa et al. 2009).

Conclusions

AMP sequence templates, in combination with the use of rationally selected non-proteinogenic amino acids, are powerful tools for obtaining structure/activity data for different classes of AMPs, and for designing peptides with optimized antimicrobial properties. Together with other design methodologies, they have allowed a better understanding of how AMPs function and what structural and physicochemical features are important for their activities. All these methodologies should, however, be applied with caution. HDPs can have multiple modes of action against bacteria, and many are multifunctional, also modulating the activities of host immune cells. Even quite slight alterations in sequence can cause significant alterations in structure, which rather than modulating a particular functional feature, may cause a switch to a different mode of action than that being probed. It is therefore essential to subject newly designed analogues of any given AMP or HDP to an extensive series of structural and functional assessments, using diverse biochemical, biophysical, and microbiological assays, before any firm conclusions as to the relevance of structural variations on the biological functions can be made.

Acknowledgments This work was supported by an FVG regional grant under L33/2008 and the EU project PIAP-GA-2008-218191.

References

Aerts AM, François IE, Cammue BP, Thevissen K (2008) The mode of antifungal action of plant, insect and human defensins. Cell Mol Life Sci 65:2069–2079

Agashe VR, Shastry MC, Udgaonkar JB (1995) Initial hydrophobic collapse in the folding of barstar. Nature 377:754–757

Antcheva N, Zelezetsky I, Tossi A (2006) Cationic antimicrobial peptides: the defensins. In: Kastin A (ed) Handbook of biologically active peptides. Elsevier, Amsterdam, pp 55–66

Antcheva N, Morgera F, Creatti L, Vaccari L, Pag U, Pacor S, Shai Y, Sahl HG, Tossi A (2009) Artificial beta-defensin based on a minimal defensin template. Biochem J 421:435–447

Benincasa M, Pacor S, Gennaro R, Scocchi M (2009) Rapid and reliable detection of antimicrobial peptide penetration into gram-negative bacteria based on fluorescence quenching. Antimicrob Agents Chemother 53:3501–354

Blondelle SE, Houghten RA (1996) Novel antimicrobial compounds identified using synthetic combinatorial library technology. Trends Biotechnol 14:60–65

Braff MH, Hawkins MA, Di Nardo A, Lopez-Garcia B, Howell MD, Wong C, Lin K, Streib JE, Dorschner R, Leung DY, Gallo RL (2005) Structure–function relationships among human cathelicidin peptides: dissociation of antimicrobial properties from host immunostimulatory activities. J Immunol 174:4271–4278

Brogden KA (2005) Antimicrobial peptides: pore formers or metabolic inhibitors in bacteria. Nat Rev Microbiol 3:238–250

Campopiano DJ, Clarke DJ, Polfer NC, Barran PE, Langley RJ, Govan JR, Maxwell A, Dorin JR (2004) Structure-activity relationships in defensin dimers: a novel link between beta-defensin tertiary structure and antimicrobial activity. J Biol Chem 279:48671–4869

Castro MS, Cilli EM, Fontes W (2006) Combinatorial synthesis and directed evolution applied to the production of alpha-helix forming antimicrobial peptides analogues. Curr Protein Pept Sci 7:473–478

Chongsiriwatana NP, Patch JA, Czyzewski AM, Dohm MT, Ivankin A, Gidalevitz D, Zuckermann RN, Barron AE (2008) Peptoids that mimic the structure, function, and mechanism of helical antimicrobial peptides. Proc Natl Acad Sci 105:2794–9279

Circo R, Skerlavaj B, Gennaro R, Amoroso A, Zanetti M (2002) Structural and functional characterization of hBD-1(Ser35), a peptide deduced from a DEFB1 polymorphism. Biochem Biophys Res Commun 293:586–592

Duclohier H (2007) Peptaibiotics and peptaibols: an alternative to classical antibiotics? Chem Biodivers 4:1023–1026

Giangaspero A, Sandri L, Tossi A (2001) Amphipathic a helical antimicrobial peptides – A systematic study of the effects of structural and physical properties on biological activity. Eur J Biochem 268: 5589–5600

Gottler L, Ramamoorthy A (2009) A structure, membrane orientation, mechanism, and function of pexiganan – A highly potent antimicrobial peptide designed from magainin. Biochim Biophys Acta 1788:1680–1686

Hancock REW (1997) Peptide antibiotics. Lancet 349:418–422

Hancock RE, Sahl HG (2006) Antimicrobial and host-defense peptides as new anti-infective therapeutic strategies. Nat Biotechnol 24:1551–1557

Hara T, Kodama H, Kondo M, Wakamatsu K, Takeda A, Tachi T, Matsuzaki K (2001a) Effects of peptide dimerization on pore formation: Antiparallel disulfide-dimerized magainin 2 analogue. Biopolymers 58:437–446

Hara T, Mitani Y, Tanaka K, Uematsu N, Takakura A, Tachi T, Kodama H, Kondo M, Mori H, Otaka A, Nobutaka F, Matsuzaki K (2001b) Heterodimer formation between the antimicrobial peptides magainin 2 and PGLa in lipid bilayers: a cross-linking study. Biochemistry 40:12395–12399

Hilpert K, Volkmer-Engert R, Walter T, Hancock RE (2005) High-throughput generation of small antibacterial peptides with improved activity. Nat Biotechnol 23:1008–1012

Huang HW (2006) Molecular mechanism of antimicrobial peptides: the origin of cooperativity. Biochim Biophys Acta 1758:1292–1302

Hughes AL (1999) Evolutionary diversification of the mammalian defensins. Cell Mol Life Sci 56:94–103

Kluver E, Schulz-Maronde S, Scheid S, Meyer B, Forssmann W-G, Adermann K (2005) Structure–activity relation of human β-defensin 3: influence of disulfide bonds and cysteine substitution on antimicrobial activity and cytotoxicity. Biochemistry 44:9804–9816

Krause E, Beyermann M, Dathe M, Rothemund S, Bienert M (1995) Location of an amphipathic alpha-helix in peptides using reversed-phase HPLC retention behavior of D-amino acid analogs. Anal Chem 67:252–258

Krishnakumari V, Sharadadevi A, Sitaram N, Nagaraj R (1999) Consequences of introducing a disulfide bond into an antibacterial and hemolytic peptide. J Pept Res 54:528–535

Krishnakumari V, Singh S, Nagaraj R (2006) Antibacterial activities of synthetic peptides corresponding to the carboxy-terminal region of human b-defensins 1–3. Peptides 27:2607–2613

Kristensen H-H, Yaver D (2004) Antimicrobial peptides. Innov Pharm Technol (2004) 77–82

Lai XZ, Feng Y, Pollard J, Chin JN, Rybak MJ, Bucki R, Epand RF, Epand RM, Savage PB (2008) Ceragenins: cholic acid-based mimics of antimicrobial peptides. Acc Chem Res 41:1233–1240

Lamb HM, Wiseman LR (1998) Pexiganan acetate. Drugs 56:1047–1052

Mattiuzzo M, Bandiera A, Gennaro R, Benincasa M, Pacor S, Antcheva N, Scocchi M (2007) Role of the *Escherichia coli* SbmA in the antimicrobial activity of proline-rich peptides. Mol Microbiol 66:151–163

Matsuzaki K (1998) Magainins as paradigm for the mode of action of pore forming polypeptides. Biochim Biophys Acta 1376:391–400

Maxwell AI, Morrison GM, Dorin JR (2003) Rapid sequence divergence in mammalian beta-defensins by adaptive evolution. Mol Immunol 40:413–421

Monne M, Nilsson I, Johansson M, Elmhed N, von Heijne G (1998) Positively and negatively charged residues have different effects on the position in the membrane of a model transmembrane helix. J Mol Biol 284:1177–1183

Morgera F, Antcheva N, Pacor S, Quaroni L, Berti F, Vaccari L, Tossi A (2008a) Structuring and interactions of human β-defensins 2 and 3 with model membranes. J Pep Sci 14:518–523

Morgera F, Vaccari L, Antcheva N, Scaini D, Pacor S, Tossi A (2008b) Primate cathelicidin orthologues display different structures and membrane interactions. Biochem J 417:727–735

Morgera F, Vaccari L, Creatti L, Antcheva N, Tossi A (2009) The study of biomolecular-membrane interactions using surface plasmon resonance spectroscopy. In: Castanho M (ed) Membrane active peptides: methods and results on structure and function. IUL Publishers (International University Line), La Jolla, CA, USA

Mygind PH, Fischer RL, Schnorr KM, Hansen MT, Sönksen CP, Ludvigsen S, Raventós D, Buskov S, Christensen B, De Maria L, Taboureau O, Yaver D, Elvig-Jørgensen SG, Sørensen MV, Christensen BE, Kjaerulff S, Frimodt-Moller N, Lehrer RI, Zasloff M, Kristensen HH (2005) Plectasin is a peptide antibiotic with therapeutic potential from a saprophytic fungus. Nature 437:975–980

Oren Z, Ramesh J, Avrahami D, Suryaprakash N, Shai Y, Jelinek R (2002) Structures and mode of membrane interaction of a short alpha helical lytic peptide and its diastereomer determined by NMR, FTIR, and fluorescence spectroscopy. Eur J Biochem 269:3869–3880

Oppenheim JJ, Tewary P, de la Rosa G, Yang D (2007) Alarmins initiate host defense. Adv Exp Med Biol 601:185–194

Pacor S, Giangaspero A, Bacac M, Sava G, Tossi A (2002) Analysis of the cytotoxicity of synthetic antimicrobial peptides on mouse leucocytes: implications for systemic use. J Antimicrob Chemother 50:339–348

Pag U, Oedenkoven M, Sass V, Shai Y, Shamova O, Antcheva N, Tossi A, Sahl HG (2008) Analysis of in vitro activities and modes of action of synthetic antimicrobial peptides derived from an alpha-helical 'sequence template'. J Antimicrob Chemother 61:341–352

Patil A, Hughes AL, Zhang G (2004) Rapid evolution and diversification of mammalian alpha-defensins as revealed by comparative analysis of rodent and primate genes. Physiol Genomics 20:1–11

Pini A, Giuliani A, Falciani C, Runci Y, Ricci C, Lelli B, Malossi M, Neri P, Rossolini GM, Bracci L (2005) Antimicrobial activity of novel dendrimeric peptides obtained by phage display selection and rational modification. Antimicrob Agents Chemother 49:2665–2672

Porter EA, Weisblum B, Gellman SH (2002) Mimicry of host-defense peptides by unnatural oligomers: antimicrobial beta-peptides. J Am Chem Soc 124:7324–7330

Rotem S, Mor A (2009) Antimicrobial peptide mimics for improved therapeutic properties. Biochim Biophys Acta 1788:1582–1592

Sass V, Pag U, Tossi A, Bierbaum G, Sahl HG (2008) Mode of action of human beta-defensin 3 against Staphylococcus aureus and transcriptional analysis of responses to defensin challenge. Int J Med Microbiol 298:619–633

Scocchi M, Mattiuzzo M, Benincasa M, Antcheva N, Tossi A, Gennaro R (2008) Investigating the mode of action of proline-rich antimicrobial peptides using a genetic approach: a tool to identify new bacterial targets amenable to the design of novel antibiotics. Methods Mol Biol 494:161–176

Shai Y (2002) Mode of action of membrane active antimicrobial peptides. Biopolymers 66:236–248

Tang J, Yin H, Qiu J, Tucker MJ, DeGrado WF, Gai F (2009) Using two fluorescent probes to dissect the binding, insertion, and dimerization kinetics of a model membrane peptide. J Am Chem Soc 131:3816–3817

Toniolo C, Crisma M, Formaggio F, Peggion C (2001) Control of peptide conformation by the Thorpe–Ingold effect (C alpha-tetrasubstitution). Biopolymers 60:396–419

Tossi A, Sandri L, Giangaspero A (2000) Amphipathic, alpha-helical antimicrobial peptides. Biopolymers 55:4–30

Tossi A, Sandri L (2002) Molecular diversity in gene-encoded, cationic antimicrobial polypeptides. Curr Pharm Des 8:743–761

Tennessen JA (2005) Molecular evolution of animal antimicrobial peptides: widespread moderate positive selection. J Mol Evol 61:445–455

Tomasinsig L, Pizzirani C, Skerlavaj B, Pellegatti P, Gulinelli S, Tossi A, Di Virgilio F, Zanetti M (2008) The human cathelicidin LL-37 modulates the activities of the P2X7 receptor in a structure-dependent manner. J Biol Chem 283:30471–30481

Tomasinsig L, Morgera F, Antcheva N, Pacor S, Skerlavaj B, Zanetti M, Tossi A (2009) Structure dependence of biological activities for primate cathelicidins. J Pept Sci 15:576–582

Tucker MJ, Getahun Z, Nanda V, DeGrado WF, Gai F (2004) A new method for determining the local environment and orientation of individual side chains of membrane-binding peptides. J Am Chem Soc 126:5078–5079

Unger T, Oren Z, Shai Y (2001) The effect of cyclization of magainin 2 and melittin analogues on structure, function, and model membrane interactions: implication to their mode of action. Biochemistry 40:6388–6397

Vanhoye D, Bruston F, Nicolas P, Amiche M (2003) Antimicrobial peptides of hylid and ranin frogs originated from a 150-million-year-old ancestral precursor with a conserved signal peptides but hypermutable antimicrobial domain. Eur J Biochem 270:2068–2081

Varkey J, Singh S, Nagaraj R (2006) Antibacterial activity of linear peptides spanning the carboxy-terminal beta-sheet domain of arthropod defensins. Peptides 27:2614–2623

Winkler DF, Hilpert K, Brandt O, Hancock RE (2009) Synthesis of peptide arrays using SPOT-technology and the CelluSpots-method. Methods Mol Biol 570:157–174

Wong JH, Xia L, Ng TB (2007) A review of defensins of diverse origins. Curr Protein Pept Sci 8:446–459

Wu Z, Hoover DM, Yang D et al (2003) Engineering disulfide bridges to dissect antimicrobial and chemotactic activities of human beta-defensin 3. Proc Natl Acad Sci USA 100:8880–8885

Yang L, Harroun TA, Weiss TM, Ding L, Huang HW (2001) Barrel-stave model or toroidal model? A case study on melittin pores. Biophys J 81:1475–1485

Yount NY, Yeaman MR (2004) Multidimensional signatures in antimicrobial peptides. Proc Natl Acad Sci USA 101:7363–7368

Yeaman MR, Yount NY (2007) Unifying themes in host defence effector polypeptides. Nat Rev Microbiol 5:727–740

Yu Q, Lehrer RI, Tam JP (2000) Engineered salt-insensitive alpha-defensins with end-to-end circularized structures. J Biol Chem 275:3943–3949

Zasloff M (2002) Antimicrobial peptides of multicellular organisms. Nature 415:389–395

Zelezetsky I, Pacor S, Pag U, Papo N, Shai Y, Sahl H-G, Tossi A (2005a) Controlled alteration of the shape and conformational stability of α-helical cell-lytic peptides: effect on mode of action and cell specificity. Biochem J 390:177–188

Zelezetsky I, Pag U, Sahl HG, Tossi A (2005b) Tuning the biological properties of amphipathic alpha-helical antimicrobial peptides: rational use of minimal amino acid substitutions. Peptides 26:2368–2376

Zelezetsky I, Tossi A (2006) α-Helical antimicrobial peptides: using a sequence template to guide structure–activity relationship studies. Biochim Biophys Acta 1758:1436–1449

Zelezetsky I, Pontillo A, Puzzi L, Antcheva N, Segat L, Pacor S, Crovella S, Tossi A (2006) Evolution of the primate cathelicidin. Correlation between structural variations and antimicrobial activity. J Biol Chem 281:19861–19871

Zhu S (2007) Evidence for myxobacterial origin of eukaryotic defensins. Immunogenetics 59:949–954

Chapter 7
Purification Techniques of Bacteriocins from Lactic Acid Bacteria and Other Gram-Positive Bacteria

Lucila Saavedra and Fernando Sesma

Abstract The search for new antimicrobial peptides produced by lactic acid bacteria and other Gram-positive microorganisms has become an interesting field of research in the past decades.

The fact that bacteriocins are active against numerous foodborne and human pathogens, are produced by generally regarded as safe (GRAS) microorganisms, and are readily degraded by proteolytic host systems makes them attractive candidates for biotechnological applications. However, before suggesting or choosing a new bacteriocin for future technology developments, it is necessary to elucidate its biochemical structure and its mode of action, which may be carried out once the bacteriocin is purified to homogeneity.

This chapter focuses on describing the main strategies used for the purification of numerous bacteriocins.

Introduction

The production of antimicrobial peptides (AMPs) is widely distributed among all living organisms where they play an essential role as part of the innate immune defenses. After the initial discovery of AMPs in insects and amphibians, many more have subsequently been identified and isolated. Probably, the best characterized AMPs are those derived from insects (cecropins), amphibians (magainins), and humans (α- and β-defensins) (Zasloff 2002; Hancock and Sahl 2006).

Regardless of the origin of the AMPs, they share distinguishing features such as being short (between 10 and 50 amino acids) and cationic (charge of generally +2 to +9), imparted by the presence of multiple lysine and arginine and with a

F. Sesma (✉)
Centro de Referencia para Lactobacilos (CERELA-CONICET), Chacabuco 145.
San Miguel de Tucumán, Tucumán, Argentina
e-mail: fsesma@cerela.org.ar

D. Drider and S. Rebuffat (eds.), *Prokaryotic Antimicrobial Peptides: From Genes to Applications*, DOI 10.1007/978-1-4419-7692-5_7,

substantial portion (≥30% or more) of hydrophobic residues. However, their primary sequence and structure may vary considerably among the groups (Hancock and Sahl 2006).

In bacteria, AMPs are called bacteriocins including those produced by Gram-positive and Gram-negative bacteria (Riley and Wertz 2002). Among the Gram-positive group of microorganisms, lactic acid bacteria (LAB) are particularly prolific in bacteriocin production (Klaenhammer 1988).

In the last 20 years, many studies have focused on bacteriocins from LAB due to their potential biotechnological applications. Many of them have shown both in vitro and in vivo to be effective against foodborne human pathogens and spoilage microorganisms (Fimland et al. 2005; Guinane et al. 2005; Lawton et al. 2007; Castellano et al. 2008). In food, antibiotics are not allowed, so the utilization of these natural antagonistic additives with preservative or antimicrobial properties has become a trademark approach in food safety and preservation. The in situ or ex situ production of bacteriocins constitutes the two alternative directions for food biocontrol: (1) the use of bacteriocin-producing LAB or (2) the direct addition of bacteriocin preparations, either synthetic or purified from the culture supernatant of the producer strains. Such applications could be greatly facilitated with the development of efficient procedures for detection, quantification, and purification of bacteriocins (Martinez et al. 2000). In this regard, only nisin, a posttranslationally modified lantibiotic produced by certain strains of *Lactococcus lactis*, is presently licensed for food use in a partially purified form.

Current drawbacks for the extensive use of this kind of peptides are the high cost of a large-scale production mainly attributed to (a) the complex culture media used for growing LABs and (b) the nature of bacteriocins, which entails a difficult and expensive purification process. A general procedure is not yet available.

Strategies for Purification of LAB and Other Gram-Positive Bacteriocins

Peptide bacteriocins are classified on the basis of structural and functional characteristics. The Class I group is called "lantibiotics" because of the distinctive posttranslational modifications, whereas Class II group are those bacteriocins lacking such modifications. In the classification suggested by Cotter et al. (2005), the Class II bacteriocin group presents four subclasses, although another classification system has been proposed (Heng et al. 2007).

Besides their nature, several approaches for the analytical purification to homogeneity of LAB bacteriocins have been published through the years (Table 7.1) (Carolissen-Mackay et al. 1997; Suarez et al. 1997; Parente and Ricciardi 1999; Guyonnet et al. 2000; Uteng et al. 2002; Berjeaud and Cenatiempo 2004; Saavedra et al. 2004a).

The classical strategy includes the following: (a) concentration of the peptide present in the culture supernatant by salt precipitation (Muriana and

Table 7.1 Purification strategies of selected bacteriocins produced by LAB and other Gram-positive bacteria

Lantibiotics (Class I)	Purification strategy	Reference
Bovicin HJ50	Solvent extraction (isopropanol), acetone concentration, HIC, Gel filtration	Xiao et al. (2004)
Epidermin and gallidermin	Chloroform extraction, RP-HPLC	Bonelli et al. (2006)
Epicidin 280 and Pep5	HIC, cation-exchange (CM-sephadex C-25), RP-HPLC	Heidrich et al. (1998)
Epilancin 15X	FPLC, cation-exchange, HID, RP-HPLC	Ekkelenkamp et al. (2005)
Lacticin 481	Ammonium sulfate precipitation Gel filtration (Ultragel ACA202), C_{18} RP-HPLC	Piard et al. (1992)
Lacticin 3147 (2p)	Amberlite XAD-16, RP-HPLC, RP-FPLC	McAuliffe et al. (1998)
Mersacidin	Serdolit PAD-I resin, RP-HPLC	Hsu et al. (2003)
Michiganin	Ammonium sulfate precipitation, cation-exchange (SP-sepharose), RP-FPLC	Holtsmark et al. (2006)
Mutacin H-29B	Cell pellet extraction (ethanol–HCL), C_{18} Sep-Pak cartridges, RP-HPLC	Nicolas et al. (2006)
Nukacin ISK1	Amberlite XAD-16, cation-exchange (SP-Sepharose), RP-HPLC	Aso et al. (2005)
Nisin A	Single-step immunoaffinity chromatography	Suarez et al. (1997)
Nisin A	Trypsin digestion at pH 5, HPLC	Coughlin and Crabb (2002)
Nisin A (and derivatives) Patent	Cell pellet extraction (isopropanol), SPE C_{18} RP-HPLC	Cotter and Hill (2009) WO/2009/135945
Nisin Q	Amberlite XAD-16, ion-exchange (SP-sepharose) RP-HPLC	Zendo et al. (2003)
Nisin Z	Single-step immunoaffinity (magnetic beads coated with antinisin Z monoclonal antibodies)	Prioult et al. (2000)
Salivaricin A2	XAD-2 resin, DEAE-Sephadex, CM-Sephadex FPLC Superose 12HR 10/30, RP-HPLC	Ross et al. (1993)
Subtilin	Supernatant concentrated with Tosohaas Toyopearl Butyl-650 resin concentration, C_{18} RP-HPLC	Parisot et al. (2008)

(continued)

Table 7.1 (continued)

Class IIa bacteriocins	Purification strategy	Reference
Bifidocin B	Absorption in diatomite calcium silicate, cation-exchange (CM-cellulose)	Yildirim et al. (1999)
Carnobacteriocin A	Amberlite XAD-8, Gel filtration (Sephadex LH-60) RP-HPLC	Worobo et al. (1994)
Enterocin CRL35	Ammonium sulfate precipitation – Gel filtration Biogel P-6 – cation-exchange chromatography – HPLC	Farias et al. (1996)
Enterocin P	Ammonium sulfate precipitation, Gel filtration cation-exchange (Q-sepharose), HIC, RP-FPLC	Cintas et al. (1997)
Leucocin A	Ammonium sulfate precipitation, Amberlite XAD-2, Gel filtration (Sephadex G25), RP-HPLC	Hastings et al. (1991)
Mundticin KS	Ammonium sulfate precipitation, cation-exchange (SP-Toyopearl), C_{18} SPE cartridge	Kawamoto et al. (2002)
Pediocin PA-1	Amberlite XAD-8, SPE C_{18} column, RP-HPLC	Kaur et al. (2004)
Piscicolin 126	Ammonium sulfate precipitation, cation-exchange (CM-Sepharose, RP-HPLC)	Jack et al. (1996)
Piscicocin CS526	Ammonium sulfate precipitation, Gel filtration (Sephadex G50), cation-exchange (SP-Sepharose), C_{18} SPE cartridge, RP-HPLC	Yamazaki et al. (2005)
Plantaricin 423	Rotavap concentration, ultrafiltration, RP-HPLC	van Reenen et al. (1998)
Mesentericin Y105 Sakacin A Sakacin P Enterocin A Pediocin PA-1 Divercin V41	Cation-exchange (CM-cellulose), SPE C_{18} cartridge, RP-HPLC	Guyonnet et al. (2000)

Class IIb bacteriocins	Purification strategy	Reference
ABP-118	Ammonium sulfate precipitation, Amberlite XAD-16, Ion-exchange (SP-Sepharose), RP-FPLC	Flynn et al. (2002)
Brochocin C	Butanol extraction – Acetone precipitation Gel filtration (Sephadex G-50)	McCormick et al. (1998)
Enterocin C	Ammonium sulfate precipitation, cation-exchange SP-Sepharose, HIC, RP-FPLC	Maldonado-Barragan et al. (2009)

(continued)

Table 7.1 (continued)

Class IIb bacteriocins	Purification strategy	Reference
Enterocin 1071	Ammonium sulfate precipitation, cation-exchange SP-Sepharose	Balla et al. (2000)
Lactacin F	Ultrafiltration, ammonium sulfate precipitation Gel filtration (Sephacryl S-300), RP-HPLC	Muriana and Klaenhammer (1991)
Lactocin 705	Adsorption–desorption; RP-HPLC	Palacios et al. (1999)
Lactococcin G	Ammonium sulfate precipitation, cation-exchange SP-Sepharose, HIC, RP-FPLC	Nissen-Meyer et al. (1992)
Lactococcin Q	Acetone precipitation, cation-exchange SP-Sepharose, RP-HPLC	Zendo et al. (2006)
Mutacin IV	Chloroform extraction, HIC (Source 15 RPC)	Qi et al. (2001)
Plantaricin S	Ammonium sulfate precipitation SP-Sepharose, HIC, RP-FPLC	Jimenez-Diaz et al. (1995)
Plantaricin E/F and J/K	Ammonium sulfate precipitation, cation-exchange SP-Sepharose, HIC, RP-FPLC	Anderssen et al. (1998)
Plantaricin NC8	Ammonium sulfate precipitation, cation-exchange SP-Sepharose, HIC, RP-FPLC	Maldonado et al. (2003)
Thermophilin 13	TCA precipitation, HIC (Source 15 PHE resin), RP chromatography (Resource RP column)	Marciset et al. (1997)

Circular bacteriocins	Purification strategy	Reference
AS-48	Cation-exchange CM-Sephadex, Gel filtration, (Biogel P-6), RP-chromatography	Galvez et al. (1989)
Butyrivibriocin AR10	Ammonium sulfate precipitation, ultrafiltration RP-chromatography	Kalmokoff and Teather (1997)
Carnocyclin A	Amberlite XAD-16 resin SPE C_{18} cartridge, RP-HPLC	Martin-Visscher et al. (2008)
Circularin A	Ammonium sulfate precipitation butanol extraction, RP-HPLC	Kemperman et al. (2003)
Gassericin A and reutericin 6	HIC, RP-chromatography (LiChroprep RP-8 resin)	Kawai et al. (2004)
Subtilosin A	Anion-exchange (High Q), butanol extraction, gel filtration (Sephadex LH-20)	Zheng and Slavik (1999)
Uberolysin	Ammonium sulfate precipitation, cation-exchange (CM Macroprep), Gel filtration (Superdex 75) RP-HPLC	Wirawan et al. (2007)
Lactocyclicin Q	Adsorption–desorption, cation-exchange (SP-Sepharose), HIC, RP-HPLC	Sawa et al. (2009)

Klaenhammer 1991) or acid extraction (Yang et al. 1992), (b) ion-exchange chromatography, (c) hydrophobic interaction, and (d) gel filtration and/or reversed-phase chromatography (RP-HPLC).

Since bacteriocins are secreted into the culture medium during bacterial growth and considering the relatively low specific production of these peptides, a first necessary step is the concentration of the cell-free culture supernatant (CS). Therefore, sequential precipitation or adsorption steps are routinely performed. Salt precipitation with ammonium sulfate is generally used. Although most bacteriocins display a reduced activity at high salt concentrations, ammonium sulfate as concentrated as 80% saturation does not interfere with the antimicrobial activity. In addition, vacuum concentration and extraction with organic solvents (acetone, ethanol, chloroform) were also reported (Holck et al. 1994, 1996; Coventry et al. 1996; Burianek and Yousef 2000; Taylor et al. 2007).

On the other hand, LABs are fastidious microorganisms that require rich and complex culture medium (LAPTg broth, MRS broth, Tryptic Soy broth, M17G broth) for growth, which often contain considerable amounts of small peptides (3,000–6,000 Da) in the range of most bacteriocins [10–30 $g\,l^{-1}$ compared to a bacteriocin concentration of 10–100 mg l^{-1} (Parente and Ricciardi 1999)]. Because of the presence of these contaminants, freeze-drying or any other direct removal of water is not suitable. An appropriate alternative would be the use of a culture medium that allows the maximum production of the peptides with minimum interference with the purification scheme. Concerning this issue, Vera Pingitore et al. (2009) have recently optimized the production of salivaricin CRL1328 in a chemically defined medium lacking interference peptides, which facilitates its purification. Among medium constituents that interfere not only with bacteriocin purification but also with bacteriocin production is Tween 80. One example is gassericin A where Tween 80 from the culture medium has been replaced by oleic acid, which resulted in a 4,500-fold increase in specific activity (Kawai et al. 1994).

An interesting shorter and inexpensive alternative to concentrate LAB bacteriocins was described by Yang et al. (1992), in which bacteriocins such as pediocin AcH, nisin, sakacin A, and leuconocin Lcm1 were adsorbed onto producer bacteria at pH 5–7 and desorbed later by lowering the pH. This method is based on the fact that most bacteriocins have a specific range of pH where they are completely adsorbed onto cell surfaces. Therefore, after an overnight incubation, cultures are heated to inactivate the bacteria. Then, pH is adjusted to assure complete adsorption of bacteriocins followed by a number of washes in order to get rid of culture medium contaminants. Finally, the peptides are released by dropping the pH with strong acids and using 50 mM sodium dodecyl sulfate (SDS). Daba et al. (1994) also purified pediocin PA-1 by acid extraction (at pH 2) and a final RP-HPLC, using a C_{18} column maintained at 39°C throughout the separation. The peptide was released with acetonitrile–0.1% trifluoroacetic acid (TFA) gradient. Following Yang scheme, other adsorbents have been reported to be effective. For instance, nisin, pediocin PO2, brevicin 286, and piscicolin 126, were extracted from fermentation media by adsorption onto Micro-Cel (a food-grade diatomite calcium silicate anticaking agent) and subsequent desorption with 1% sodium deoxycholate and 1%

SDS (Coventry et al. 1996). These authors also showed that this protocol resulted in bacteriocin preparations free of proteinaceous contaminant substances. Another adsorption/desorption technique was implemented by Janes et al. (1998) where nisin, pediocin RS2, leucocin BC2, lactocin GI3, and enterocin CS1 were bound to rice hull ash or silicic acid. This methodology raised a relatively purer sample, although an additional dialysis step must be performed with a resulting decreased recovery of the peptide. In addition, several lantibiotics have also been extracted from CS using different resins (Amberlite XAD 16; XAD-2) or unpolar resins such as Serdolit PAD-I (Ross et al. 1993; McAuliffe et al. 1998; Zendo et al. 2003; Aso et al. 2008). In general, this seems to be a cleaner method that gives a purer final sample with less contaminants. However, it is not suitable for all bacteriocins; for instance, the yield of two-component bacteriocins is often low and inappropriate for large-scale purification (Nissen-Meyer et al. 1992; Anderssen et al. 1998).

Since the concentration steps only reduce the working volume and do not provide a high degree of purity, several subsequent chromatographic steps are still required. Taking advantages of the cationic and hydrophobic nature of bacteriocins, different types of column of chromatography are used. As a first choice, a cation-exchange column [SP (sulfopropyl)-Sepharose or CM (carboxymethyl)-Sephadex] is used in most of the procedures (Table 7.1), and the release of the peptides is often achieved using a linear NaCl gradient. Finally, active fractions are pooled and concentrated by RP-HPLC or RP-FPLC. The most common column packings are silica or polymeric supports to which straight chain of hydrocarbons ranging from C_4 to C_{18} are bonded. Then, bacteriocins are eluted using a gradient of water-miscible organic solvents such as methanol, isopropanol, or acetonitrile (Liu and Hansen 1990; Floriano et al. 1998; Palacios et al. 1999; Berjeaud and Cenatiempo 2004; Saavedra et al. 2004a). This is usually the last step of the preparative purification. This final step is always included, not only to check the purity of the samples but also to eliminate the last contaminants.

This general protocol, though it is the most widely used, is time consuming, and the final recovery of peptides is generally low, which makes it unsuitable for large-scale purification for biotechnological applications. To circumvent these problems and improve the recovery of bacteriocins, various reports offering advantages in the proceeding have been published. Callewaert and de Vuyst (1999) reported the isolation of amylovorins L471 [two active antimicrobial peptides (4,800 and 5,800 Da)] directly from the crude fermentation broth of *Lactobacillus amylovorus* DCE 471 through expanded bed adsorption using a strong cation-exchanger; however, this methodology did not allow a good recovery of the sample. Later on, the same authors proposed a novel three-step method for the purification of one of the peptides where the concentration of the sample was followed by a chloroform/methanol extraction/precipitation step and finally ending in a single separation by RP-FPLC (Callewaert et al. 1999; De Vuyst et al. 1996;). Guyonnet et al. (2000) also reported a three-step method for the purification of class IIa bacteriocins or anti-*Listeria* peptides. Basically, the authors modified two steps of the previous published protocol (Biet et al. 1998) consisting of the elimination of color contaminants from the culture medium by modifying the extent and temperature of sterilization; moreover, the ammonium sulfate precipitation was replaced by cation-exchange chromatography. Although the purification yield varied between 10 and 60%, they claimed that the purified bacteriocins appeared to be at least 95% pure. Uteng et al. (2002) shortened the protocol even more when they published a two-step purification protocol for the same type of peptides. The first step consists in the direct application of the bacterial culture on a SP-Sepharose Fast Flow cation-exchange column (Amersham Pharmacia Biotech); in these conditions, the cationic bacteriocins are eluted with 1 M NaCl. In the second step, the active fraction issued from the cation-exchanger is applied at a high flow rate on a low-pressure Resource reversed-phase column (Amersham Pharmacia Biotech) and eluted with a propanol gradient. The final recovery of the sample is significantly increased from 10 to 20% using the original methodology to 90–100% with the new and improved protocol (Uteng et al. 2002).

In addition to the schemes described above, immunoaffinity chromatography represents a relatively novel and specific strategy that allows a rapid and clean way to purify bacteriocins. For instance, antinisin A monoclonal antibodies were coupled to a HiTrap N-hydroxysuccinimide-activated column for a successfully one-step purification process (Suarez et al. 1997). In the same trend, specific antibodies against enterocin B, nizin Z, pediocin PA-1, and enterocin P have also been generated and used for the purification and detection of these bacteriocins (Rose et al. 2001; Prioult et al. 2000; Gutierrez et al. 2004; Naghmouchi et al. 2008). Although specific antibodies against Lacticin RM, enterocin A, and an heterologously produced pediocin PA-1 were also reported, they were mainly used for the development of detection and quantification methods rather than the purification of these bacteriocins (Keren et al. 2004; Martinez et al. 2000). Although the yield obtained using this technique appears to be variable among the bacteriocins tested, it seems to be an interesting alternative that needs further improvements to achieve a good recovery of the sample.

Detection of Purified Bacteriocins

The detection or visualization of the antimicrobial peptides during different purification stages represents an important part of the whole process. The most widely used methodologies are the radial diffusion assay and the spot-on-lawn technique where aliquots from different stages of the process are tested against a target cell usually grown on an agar base culture medium. The difference between both the methods is that in the first one, samples from the stages are loaded into small sample wells (e.g., 3 mm diameter), while in the second, small aliquots are spotted onto a lawn of the sensitive cell (Steinberg and Lehrer 1997). The inhibitory activity of active fractions is visualized as “inhibition halos,” i.e., clear zones where no growth of target cell is evident. Also, microdilution assay is also reported, although the procedure is time consuming and requires approximately ten times more sample.

A modified SDS-PAGE is another very useful technique to examine and estimate the molecular mass of partially or completely purified bacteriocins (Schagger and von Jagow 1987). The inconvenience is that small hydrophobic peptides tend to diffuse out of the polyacrylamide gel during staining, which makes it difficult to visualize them after electrophoresis. To overcome this problem, Bhunia et al. (1988) described a method where stained/destained gel was overlaid with a lawn of the target cell showing a clear zone of inhibition around the lowermost protein band, indicating that this band contained the antimicrobial activity. This technique (gel overlay assay or bioassay) has shown to be very useful for direct detection of antimicrobial peptides (Fig. 7.1). However, depending on the yield of the purification process, several laboratory staining protocols [Coomasie Blue, Silver staining, SYPRO-Ruby (Molecular Probes-Invitrogen), Lumitein] have also been reported for SDS-PAGE of bacteriocins (Onda et al. 2003; Ball and Karuso 2007; Vera Pingitore et al. 2007, 2009; Nock et al. 2008). In addition, new fluorescent dyes offer the advantages of staining and cutting off the bands from the gels, which allow downstream analysis such as mass spectrometry or sequencing. Alternatively, a direct transfer to a PVDF membrane is also possible, which can be followed by the standard procedure.

As a complementary approach for the detection and quantification of bacteriocins, the generation of specific antibodies and the development of immunochemical methods had also proved to be effective, sensitive, and specific.

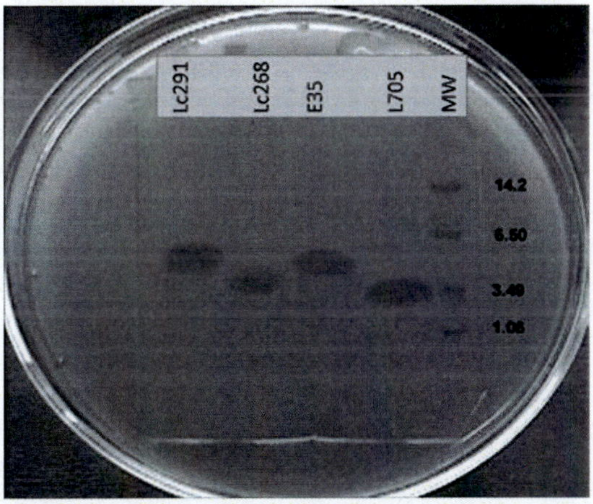

Fig. 7.1 Detection of antimicrobial activities of different bacteriocins after SDS-PAGE by using a bioassay. Lc291 and Lc268 correspond to bacteriocins produced by *Lactococcus lactis* strains (Palacios et al. unpublished); E35 is enterocin CRL35 (4.3 kDa; Saavedra et al. 2004b) and L705 is a two-peptide bacteriocin (L705alpha, 3.5 kDa and L705beta 3.3 kDa, Cuozzo et al. 2003) produced by *Lactococcus curvatus* CRL705; MW, Low-molecular-mass marker (Sigma). (Source: Palacios, 2000. Used with permission of the author)

Conclusion

The purification process of bacteriocins from LAB and related Gram-positive bacteria is often cumbersome due to a combination of (1) small amounts of bacteriocin produced and (2) the interference of contaminant hydrophobic peptides present in the complex culture media that "mimic" the chromatographic behavior of these antimicrobial compounds. This has led to a vast number of publications of protocols offering new or different alternatives to overcome the problems. The improvement of these techniques, along with the development of special and low-cost fermentation media, will have a huge impact not only on basic research but also on the industries due to the tremendous potential of LAB bacteriocins in food and medical fields.

References

Anderssen EL, Diep DB, Nes IF, Eijsink VG, Nissen-Meyer J (1998) Antagonistic activity of *Lactobacillus plantarum* C11: two new two-peptide bacteriocins, plantaricins EF and JK, and the induction factor plantaricin A. Appl Environ Microbiol 64:2269–2272

Aso Y, Okuda K, Nagao J, Kamenasa Y, Phuong NTB, Koga H, Shioya K, Sashihara T, Nakayama J, Sonomoto K (2005) A novel type of immunity protein, NukH, for the lantibiotic nukacin ISK-1 produced by *Staphylococcus warneri* ISK-1. Biosci Biotechnol Biochem 69:1403–1410

Aso Y, Takeda A, Sato M, Takahashi T, Yamamoto T, Yoshikiyo K (2008) Characterization of lactic acid bacteria coexisting with a nisin Z producer in Tsuda-turnip pickles. Curr Microbiol 57:89–94

Ball MS, Karuso P (2007) Mass spectral compatibility of four proteomics stains. J Proteome Res 6:4313–4320

Balla E, Dicks LM, Du Toit M, Van Der Merwe MJ, Holzapfel WH (2000) Characterization and cloning of the genes encoding enterocin 1071A and enterocin 1071B, two antimicrobial peptides produced by *Enterococcus faecalis* BFE 1071. Appl Environ Microbiol 66:1298–1304

Berjeaud JM, Cenatiempo Y (2004) Purification of antilisterial bacteriocins. Methods Mol Biol 268:225–233

Bhunia AK, Johnson MC, Ray B (1988) Purification, characterization and antimicrobial spectrum of a bacteriocin produced by *Pediococcus acidilactici*. J Appl Bacteriol 65:261–268

Biet F, Berjeaud JM, Worobo RW, Cenatiempo Y, Fremaux C (1998) Heterologous expression of the bacteriocin mesentericin Y105 using the dedicated transport system and the general secretion pathway. Microbiology 144(Pt 10):2845–2854

Bonelli RR, Schneider T, Sahl HG, Wiedemann I (2006) Insights into in vivo activities of lantibiotics from gallidermin and epidermin mode-of-action studies. Antimicrob Agents Chemother 50:1449–1457

Burianek LL, Yousef AE (2000) Solvent extraction of bacteriocins from liquid cultures. Lett Appl Microbiol 31:193–197

Callewaert R, De Vuyst L (1999) Expanded bed adsorption as a unique unit operation for the isolation of bacteriocins from fermentation media. Bioseparation 8:159–168

Callewaert R, Holo H, Devreese B, Van Beeumen J, Nes I, De Vuyst L (1999) Characterization and production of amylovorin L471, a bacteriocin purified from *Lactobacillus amylovorus* DCE 471 by a novel three-step method. Microbiology 145:2559–2568

Carolissen-Mackay V, Arendse G, Hastings JW (1997) Purification of bacteriocins of lactic acid bacteria: problems and pointers. Int J Food Microbiol 34:1–16

Castellano P, Belfiore C, Fadda S, Vignolo G (2008) A review of bacteriocinogenic lactic acid bacteria used as bioprotective cultures in fresh meat produced in Argentina. Meat Sci 79:483–499

Cintas LM, Casaus P, Havarstein LS, Hernandez PE, Nes IF (1997) Biochemical and genetic characterization of enterocin P, a novel sec-dependent bacteriocin from *Enterococcus faecium* P13 with a broad antimicrobial spectrum. Appl Environ Microbiol 63:4321–4330

Cotter PD, Hill C (2009) Patent WO/2009/135945.

Cotter PD, Hill C, Ross RP (2005) Bacteriocins: developing innate immunity for food. Nat Rev Microbiol 3:777–788

Coughlin RT, Crabb JH (2002) Patent US6794181.

Coventry MJ, Gordon JB, Alexander M, Hickey MW, Wan J (1996) A food-grade process for isolation and partial purification of bacteriocins of lactic acid bacteria that uses diatomite calcium silicate. Appl Environ Microbiol 62:1764–1769

Cuozzo SA, Castellano P, Sesma F, Vignolo G, Raya RR (2003) Differential roles of the two-component peptides of lactocin 705 in antimicrobial activity. Current Microbiol 46:180–183

Daba H, Lacroix C, Huang J, Simard RE, Lemieux L (1994) Simple method of purification and sequencing of a bacteriocin produced by *Pediococcus acidilactici* UL5. J Appl Bacteriol 77:682–688

De Vuyst L, Callewaert R, Pot B (1996) Characterization of the antagonistic activity of *Lactobacillus amylovorus* DCE 471 and large scale isolation of its bacteriocin amylovorin L471. Syst Appl Microbiol 19:9–20

Ekkelenkamp MB, Hanssen M, Hsu S-TD, de Jong A, Milatovic D, Verhoef J, van Nuland NAJ (2005) Isolation and structural characterization of epilancin 15X, a novel lantibiotic from a clinical strain of *Staphylococcus epidermidis*. FEBS Lett 579:1917–1922

Farias ME, Farias RN, de Ruiz Holgado AP, Sesma F (1996) Purification and N-terminal amino acid sequence of Enterocin CRL35, a 'pediocin-like' bacteriocin produced by *Enterococcus faecium* CRL35. Lett Appl Microbiol 22:417–419

Fimland G, Johnsen L, Dalhus B, Nissen-Meyer J (2005) Pediocin-like antimicrobial peptides (class IIa bacteriocins) and their immunity proteins: biosynthesis, structure, and mode of action. J Pept Sci 11:688–696

Floriano B, Ruiz-Barba JL, Jimenez-Diaz R (1998) Purification and genetic characterization of enterocin I from *Enterococcus faecium* 6T1a, a novel antilisterial plasmid-encoded bacteriocin which does not belong to the pediocin family of bacteriocins. Appl Environ Microbiol 64:4883–4890

Flynn S, van Sinderen D, Thornton GM, Holo H, Nes IF, Collins JK (2002) Characterization of the genetic locus responsible for the production of ABP-118, a novel bacteriocin produced by the probiotic bacterium *Lactobacillus salivarius* subsp. *salivarius* UCC118. Microbiology 148:973–984

Galvez A, Gimenez-Gallego G, Maqueda M, Valdivia E (1989) Purification and amino acid composition of peptide antibiotic AS-48 produced by *Streptococcus* (Enterococcus) *faecalis* subsp. *liquefaciens* S-48. Antimicrob Agents Chemother 33:437–441

Guinane CM, Cotter PD, Hill C, Ross RP (2005) Microbial solutions to microbial problems; lactococcal bacteriocins for the control of undesirable biota in food. J Appl Microbiol 98:1316–1325

Gutierrez J, Criado R, Citti R, Martin M, Herranz C, Fernandez MF, Cintas LM, Hernandez PE (2004) Performance and applications of polyclonal antipeptide antibodies specific for the enterococcal bacteriocin enterocin P. J Agric Food Chem 52:2247–2255

Guyonnet D, Fremaux C, Cenatiempo Y, Berjeaud JM (2000) Method for rapid purification of class IIa bacteriocins and comparison of their activities. Appl Environ Microbiol 66:1744–1748

Hancock RE, Sahl HG (2006) Antimicrobial and host-defense peptides as new anti-infective therapeutic strategies. Nat Biotechnol 24:1551–1557

Hastings JW, Sailer M, Johnson K, Roy KL, Vederas JC, Stiles ME (1991) Characterization of leucocin A-UAL 187 and cloning of the bacteriocin gene from *Leuconostoc gelidum*. J Bacteriol 173:7491–7500

Heidrich C, Pag U, Josten M, Metzger J, Jack RW, Bierbaum G, Jung G, Sahl HG (1998) Isolation, characterization, and heterologous expression of the novel lantibiotic epicidin 280 and analysis of its biosynthetic gene cluster. Appl Environ Microbiol 64:3140–3146

Heng NCK, Wescombe P, Burton JP, Jack R, John R (2007) The diversity of bacteriocins in gram-positive bacteria. Springer, Heidelberg

Holck AL, Axelsson L, Huhne K, Krockel L (1994) Purification and cloning of sakacin 674, a bacteriocin from *Lactobacillus sake* Lb674. FEMS Microbiol Lett 115:143–149

Holck A, Axelsson L, Schillinger U (1996) Divergicin 750, a novel bacteriocin produced by *Carnobacterium divergens* 750. FEMS Microbiol Lett 136:163–168

Holtsmark I, Mantzilas D, Eijsink VG, Brurberg MB (2006) Purification, characterization, and gene sequence of michiganin A, an actagardine-like lantibiotic produced by the tomato pathogen *Clavibacter michiganensis* subsp. *michiganensis*. Appl Environ Microbiol 72:5814–5821

Hsu S-TD, Breukink E, Bierbaum G, Sahl HG, de Kruijff B, Kaptein R, van Nuland NAJ, Bonvin AMJJ (2003) NMR study of mersacidin and lipid II interaction in dodecylphosphocholine micelles. Conformational changes are a key to antimicrobial activity. J Biol Chem 278:13110–13117

Jack RW, Wan J, Gordon J, Harmark K, Davidson BE, Hillier AJ, Wettenhall RE, Hickey MW, Coventry MJ (1996) Characterization of the chemical and antimicrobial properties of piscicolin 126, a bacteriocin produced by *Carnobacterium piscicola* JG126. Appl Environ Microbiol 62:2897–2903

Janes ME, Nannapaneni R, Proctor A, Johnson MG (1998) Rice hull ash and silicic acid as adsorbents for concentration of bacteriocins. Appl Environ Microbiol 64:4403–4409

Jimenez-Diaz R, Ruiz-Barba JL, Cathcart DP, Holo H, Nes IF, Sletten KH, Warner PJ (1995) Purification and partial amino acid sequence of plantaricin S, a bacteriocin produced by *Lactobacillus plantarum* LPCO10, the activity of which depends on the complementary action of two peptides. Appl Environ Microbiol 61:4459–4463

Kalmokoff ML, Teather RM (1997) Isolation and characterization of a bacteriocin (Butyrivibriocin AR10) from the ruminal anaerobe *Butyrivibrio fibrisolvens* AR10: evidence in support of the widespread occurrence of bacteriocin-like activity among ruminal isolates of *B. fibrisolvens*. Appl Environ Microbiol 63:394–402

Kaur K, Andrew LC, Wishart DS, Vederas JC (2004) Dynamic relationships among type IIa bacteriocins: temperature effects on antimicrobial activity and on structure of the C-terminal amphipathic alpha helix as a receptor-binding region. Biochemistry 43:9009–9020

Kawai Y, Saito T, Toba T, Samant SK, Itoh T (1994) Isolation and characterization of a highly hydrophobic new bacteriocin (gassericin A) from *Lactobacillus gasseri* LA39. Biosci Biotechnol Biochem 58:1218–1221

Kawai Y, Kemperman R, Kok J, Saito T (2004) The circular bacteriocins gassericin A and circularin A. Curr Protein Pept Sci 5:393–398

Kawamoto S, Shima J, Sato R, Eguchi T, Ohmomo S, Shibato J, Horikoshi N, Takeshita K, Sameshima T (2002) Biochemical and genetic characterization of mundticin KS, an antilisterial peptide produced by *Enterococcus mundtii* NFRI 7393. Appl Environ Microbiol 68:3830–3840

Kemperman R, Kuipers A, Karsens H, Nauta A, Kuipers O, Kok J (2003) Identification and characterization of two novel clostridial bacteriocins, circularin A and closticin 574. Appl Environ Microbiol 69:1589–1597

Keren T, Yarmus M, Halevy G, Shapira R (2004) Immunodetection of the bacteriocin lacticin RM: analysis of the influence of temperature and Tween 80 on its expression and activity. Appl Environ Microbiol 70:2098–2104

Klaenhammer TR (1988) Bacteriocins of lactic acid bacteria. Biochimie 70:337–349

Lawton EM, Ross RP, Hill C, Cotter PD (2007) Two-peptide lantibiotics: a medical perspective. Mini Rev Med Chem 7:1236–1247

Liu W, Hansen JN (1990) Some chemical and physical properties of nisin, a small-protein antibiotic produced by *Lactococcus lactis*. Appl Environ Microbiol 56:2551–2558

Maldonado A, Ruiz-Barba JL, Jimenez-Diaz R (2003) Purification and genetic characterization of plantaricin NC8, a novel coculture-inducible two-peptide bacteriocin from *Lactobacillus plantarum* NC8. Appl Environ Microbiol 69:383–389

Maldonado-Barragan A, Caballero-Guerrero B, Jimenez E, Jimenez-Diaz R, Ruiz-Barba JL, Rodriguez JM (2009) Enterocin C, a class IIb bacteriocin produced by *E. faecalis* C901, a strain isolated from human colostrum. Int J Food Microbiol 133:105–112

Marciset O, Jeronimus-Stratingh MC, Mollet B, Poolman B (1997) Thermophilin 13, a nontypical antilisterial poration complex bacteriocin, that functions without a receptor. J Biol Chem 272:14277–14284

Martinez JM, Kok J, Sanders JW, Hernandez PE (2000) Heterologous coproduction of enterocin A and pediocin PA-1 by *Lactococcus lactis*: detection by specific peptide-directed antibodies. Appl Environ Microbiol 66:3543–3549

Martin-Visscher LA, van Belkum MJ, Garneau-Tsodikova S, Whittal RM, Zheng J, McMullen LM, Vederas JC (2008) Isolation and characterization of carnocyclin a, a novel circular bacteriocin produced by *Carnobacterium maltaromaticum* UAL307. Appl Environ Microbiol 74:4756–4763

McAuliffe O, Ryan MP, Ross RP, Hill C, Breeuwer P, Abee T (1998) Lacticin 3147, a broad-spectrum bacteriocin which selectively dissipates the membrane potential. Appl Environ Microbiol 64:439–445

McCormick JK, Poon A, Sailer M, Gao Y, Roy KL, McMullen LM, Vederas JC, Stiles ME, Van Belkum MJ (1998) Genetic characterization and heterologous expression of brochocin-C, an antibotulinal, two-peptide bacteriocin produced by *Brochothrix campestris* ATCC 43754. Appl Environ Microbiol 64:4757–4766

Muriana PM, Klaenhammer TR (1991) Purification and partial characterization of lactacin F, a bacteriocin produced by *Lactobacillus acidophilus* 11088. Appl Environ Microbiol 57:114–121

Naghmouchi K, Drider D, Fliss I (2008) Purification of pediocin PA-1 by immunoaffinity chromatography. J AOAC Int 91:828–832

Nicolas G, Morency H, LaPointe G, Lavoie MC (2006) Mutacin H-29B is identical to mutacin II (J-T8). BMC Microbiol 6:36

Nissen-Meyer J, Holo H, Havarstein LS, Sletten K, Nes IF (1992) A novel lactococcal bacteriocin whose activity depends on the complementary action of two peptides. J Bacteriol 174:5686–5692

Nock CM, Ball MS, White IR, Skehel JM, Bill L, Karuso P (2008) Mass spectrometric compatibility of Deep Purple and SYPRO Ruby total protein stains for high-throughput proteomics using large-format two-dimensional gel electrophoresis. Rapid Commun Mass Spectrom 22:881–886

Onda T, Yanagida F, Tsuji M, Shinohara T, Yokotsuka K (2003) Production and purification of a bacteriocin peptide produced by *Lactococcus* sp. strain GM005, isolated from Miso-paste. Int J Food Microbiol 87:153–159

Palacios J (2000) Estudio químico y biotecnológico de bacteriocinas producidas por bacterias lácticas. Tesis Doctoral. Universidad Nacional de Tucumán.

Palacios J, Vignolo G, Farias ME, de Ruiz Holgado AP, Oliver G, Sesma F (1999) Purification and amino acid sequence of lactocin 705, a bacteriocin produced by *Lactobacillus casei* CRL 705. Microbiol Res 154:199–204

Parente E, Ricciardi A (1999) Production, recovery and purification of bacteriocins from lactic acid bacteria. Appl Microbiol Biotechnol 52:628–638

Parisot J, Carey S, Breukink E, Chan WC, Narbad A, Bonev B (2008) Molecular mechanism of target recognition by subtilin, a class I lanthionine antibiotic. Antimicrob Agents Chemother 52:612–618

Piard JC, Muriana PM, Desmazeaud MJ, Klaenhammer TR (1992) Purification and partial characterization of lacticin 481, a lanthionine-containing bacteriocin produced by *Lactococcus lactis* subsp. *lactis* CNRZ 481. Appl Environ Microbiol 58:279–284

Prioult G, Turcotte C, Labarre L, Lacroix C, Fliss I (2000) Rapid purification of nisin Z using specific monoclonal antibody-coated magnetic beads. Int Dairy J 10(9):627–633

Qi F, Chen P, Caufield PW (2001) The group I strain of *Streptococcus mutans*, UA140, produces both the lantibiotic mutacin I and a nonlantibiotic bacteriocin, mutacin IV. Appl Environ Microbiol 67:15–21

Riley MA, Wertz JE (2002) Bacteriocins: evolution, ecology, and application. Ann Rev Microbiol 56:117–137

Rose NL, Sporns P, McMullen LM (2001) Production of antibodies against enterocin B for immunological detection and purification purposes. Food Agric Immunol 13:225–239

Ross KF, Ronson CW, Tagg JR (1993) Isolation and characterization of the lantibiotic salivaricin A and its structural gene salA from *Streptococcus salivarius* 20P3. Appl Environ Microbiol 59:2014–2021

Saavedra L, Castellano P, Sesma F (2004a) Purification of bacteriocins produced by lactic acid bacteria. Methods Mol Biol 268:331–336

Saavedra L, Minahk C, de Ruiz HAP, Sesma F (2004b) Enhancement of the enterocin CRL35 activity by a synthetic peptide derived from the NH2-terminal sequence. Antimic Agents Chemother 48(7):2778–2781

Sawa N, Zendo T, Kiyofuji J, Fujita K, Himeno K, Nakayama J, Sonomoto K (2009) Identification and characterization of lactocyclicin Q, a novel cyclic bacteriocin produced by *Lactococcus* sp. strain QU 12. Appl Environ Microbiol 75:1552–1558

Schagger H, von Jagow G (1987) Tricine-sodium dodecyl sulfate-polyacrylamide gel electrophoresis for the separation of proteins in the range from 1 to 100 kDa. Anal Biochem 166:368–379

Steinberg D, Lehrer R (1997) Designer assays for antimicrobial peptides. In: Shafer WM (ed) Antibacterial peptides protocol. Humana, Totowa, NJ

Suarez AM, Azcona JI, Rodriguez JM, Sanz B, Hernandez PE (1997) One-step purification of nisin A by immunoaffinity chromatography. Appl Environ Microbiol 63:4990–4992

Taylor TM, Davidson PM, Zhong Q (2007) Extraction of nisin from a 2.5% commercial nisin product using methanol and ethanol solutions. J Food Prot 70:1272–1276

Uteng M, Hauge HH, Brondz I, Nissen-Meyer J, Fimland G (2002) Rapid two-step procedure for large-scale purification of pediocin-like bacteriocins and other cationic antimicrobial peptides from complex culture medium. Appl Environ Microbiol 68:952–956

van Reenen CA, Dicks LM, Chikindas ML (1998) Isolation, purification and partial characterization of plantaricin 423, a bacteriocin produced by *Lactobacillus plantarum*. J Appl Microbiol 84:1131–1137

Vera Pingitore E, Salvucci E, Sesma F, Nader ME (2007) Different strategies for purification of antimicrobial peptides from lactic acid bacteria. In: Mendez-Vilas A (ed) Communicating current research and educational topics and trends in applied microbiology. Formatex, Badajoz, pp 557–568

Vera Pingitore E, Hebert EM, Sesma F, Nader-Macias ME (2009) Influence of vitamins and osmolites on growth and bacteriocin production by *Lactobacillus salivarius* CRL 1328 in a chemically defined medium. Can J Microbiol 55:304–310

Wirawan RE, Swanson KM, Kleffmann T, Jack RW, Tagg JR (2007) Uberolysin: a novel cyclic bacteriocin produced by *Streptococcus uberis*. Microbiology 153:1619–1630

Worobo RW, Henkel T, Sailer M, Roy KL, Vederas JC, Stiles ME (1994) Characteristics and genetic determinant of a hydrophobic peptide bacteriocin, carnobacteriocin A, produced by *Carnobacterium piscicola* LV17A. Microbiology 140(Pt 3):517–526

Xiao H, Chen X, Chen M, Tang S, Zhao X, Huan L (2004) Bovicin HJ50, a novel lantibiotic produced by *Streptococcus bovis* HJ50. Microbiology 150:103–108

Yamazaki K, Suzuki M, Kawai Y, Inoue N, Montville TJ (2005) Purification and characterization of a novel class IIa bacteriocin, piscicocin CS526, from surimi-associated *Carnobacterium piscicola* CS526. Appl Environ Microbiol 71:554–557

Yang R, Johnson MC, Ray B (1992) Novel method to extract large amounts of bacteriocins from lactic acid bacteria. Appl Environ Microbiol 58:3355–3359

Yildirim Z, Winters DK, Johnson MG (1999) Purification, amino acid sequence and mode of action of bifidocin B produced by *Bifidobacterium bifidum* NCFB 1454. J Appl Microbiol 86:45–54

Zasloff M (2002) Antimicrobial peptides of multicellular organisms. Nature 415:389–395

Zendo T, Fukao M, Ueda K, Higuchi T, Nakayama J, Sonomoto K (2003) Identification of the lantibiotic nisin Q, a new natural nisin variant produced by *Lactococcus lactis* 61-14 isolated from a river in Japan. Biosci Biotechnol Biochem 67:1616–1619

Zendo T, Koga S, Shigeri Y, Nakayama J, Sonomoto K (2006) Lactococcin Q, a novel two-peptide bacteriocin produced by *Lactococcus lactis* QU 4. Appl Environ Microbiol 72:3383–3389

Zheng G, Slavik MF (1999) Isolation, partial purification and characterization of a bacteriocin produced by a newly isolated *Bacillus* subtilis strain. Lett Appl Microbiol 28:363–367

Chapter 8
Natural and Heterologous Production of Bacteriocins

Luis M. Cintas, Carmen Herranz, and Pablo E. Hernández

Abstract Bacteriocins are ribosomally synthesized antimicrobial peptides produced by bacteria, and their use as natural and nontoxic food preservatives has been the source of considerable interest for the research community. In addition, bacteriocins have been investigated for their potential use in human and veterinary applications and in the animal production field. In the native bacterial strain, most bacteriocins are synthesized as biologically inactive precursors, with N-terminal extensions, that are cleaved concomitantly during export of the bacteriocin by dedicated ABC transporters, or the general secretory pathway (GSP) or Sec-dependent pathway. However, a few bacteriocins are synthesized without an N-terminal extension, and others are circularized through a head-to-tail peptide bond, complicating the elucidation of their processing and transport across the cytoplasmic membrane. The high cost of synthetic bacteriocin synthesis and their low yields from many natural producers recommends the exploration of recombinant microbial systems for the heterologous production of bacteriocins. Other advantages of such systems include production of bacteriocins in safer hosts, increased bacteriocin production, control of bacteriocin gene expression, production of food ingredients with antimicrobial activity, construction of multibacteriocinogenic strains with a wider antagonistic spectrum, a better adaptation of the selected hosts to food environments, and providing antagonistic properties to lactic acid bacteria (LAB) used as starter, protective, or probiotic cultures. The recombinant production of bacteriocins mostly relies on the use of expression vectors that replicate in Gram-negative bacteria, Gram-positive bacteria, and yeasts, whereas the production of bacteriocins in heterologous LAB hosts may be essentially based on the expression of native biosynthetic genes, by exchanging or replacing leader peptides and/or dedicated processing and secretion systems (ABC transporters), or by fusion of mature bacteriocins to signal peptides that act as secretion signals.

P.E. Hernández (✉)
Departamento de Nutrición, Bromatología y Tecnología de los Alimentos, Facultad de Veterinaria, Universidad Complutense de Madrid (UCM) 28040 Madrid, Spain
e-mail: ehernan@vet.ucm.es

D. Drider and S. Rebuffat (eds.), *Prokaryotic Antimicrobial Peptides: From Genes to Applications*, DOI 10.1007/978-1-4419-7692-5_8,

Introduction

Compounds with antimicrobial activity are produced by a wide range of mammals, birds, insects, plants, and microorganisms, while ribosomally synthesized antimicrobial peptides produced by bacteria are generally referred to as bacteriocins (Klaenhammer 1993; Nes et al. 1996; Eijsink et al. 2002). Bacteriocin production is common in bacteria and mostly in lactic acid bacteria (LAB), and since bacteriocins produced by LAB display antimicrobial activity against a broad range of Gram-positive bacteria and, to a lesser extent, Gram-negative bacteria including spoilage and foodborne pathogenic microorganisms, they are attracting considerable interest for their potential use as natural and nontoxic food preservatives (Cintas et al. 2001; Cleveland et al. 2001; Cotter et al. 2005; Deegan et al. 2006; Gálvez et al. 2007). However, bacteriocins may also be useful in human and veterinary applications. As the emergence of bacterial antibiotic resistance is an increasing problem in medical treatments and animal production, there is a pressing need for novel or alternative sources of antimicrobial agents. Potential applications for bacteriocins include the treatment of local and systemic bacterial and viral infections of humans and food-producing animals (Wachsman et al. 1999, 2003; O'Connor et al. 2006; Dezwaan et al. 2007; Klostermann et al. 2008; Knoetze et al. 2008; Kang et al. 2009). The advantages of bacteriocins in clinical applications include their broad-spectrum antagonistic activity against bacteria and their low propensity for development of resistance. Hence, the complementary use of antibiotics together with bacteriocins can be an alternative approach for preventing the emergence of resistant bacterial pathogens (Sit and Vederas 2008). However, concomitant use of bacteriocins as therapeutic agents and food preservatives should be avoided in the animal production and food field. As the mechanisms of target cell recognition, producer cell self-protection (immunity), and bacteriocin resistance of sensitive bacterial cells become better understood in the bacteriocin research field, such advances would also permit the design of efficient bacteriocins against unrecognized sensitive bacteria and to circumvent bacterial resistance (Diep et al. 2007). LAB not only add to food quality and safety, but may also exert direct positive effects on human and animal health (Parvez et al. 2006). Moreover, bacteriocin-producing isolates may also be useful as probiotics with anti-infective effects in humans and animals (Corr et al. 2007; Gillor et al. 2008; Shin et al. 2008). Accordingly, the knowledge, evaluation, and improvement of the natural and heterologous production of bacteriocins in microbial hosts hold a considerable fundamental and applied interest.

Natural Production of Bacteriocins

Bacteriocins are ribosomally synthesized peptides produced and exported by both Gram-negative and Gram-positive bacteria for the apparent purpose of destroying their competitors. They generally act at the bacterial membrane by combining pleiotropic membrane activities with targeting of specific molecules and are extremely potent, often exhibiting their antagonistic effects at nanomolar concentrations. Bacteriocins are also naturally present in foods and are nontoxic to mammals (Cotter et al. 2005; Sit and Vederas 2008). Although classification schemes for bacteriocins are made to provide a common nomenclature, they are still subject to revision. While the classification scheme devised by Klaenhammer (1993) comprised four bacteriocin classes, recent proposals recommend classification of bacteriocins into two major groups: the class I lantibiotics, containing posttranslationally modified amino acids, and the class II nonlantibiotics, containing nonmodified amino acids (Cotter et al. 2005; Nes et al. 2007). Although there is also a lack of consensus on how to subdivide class I and II bacteriocins into subclasses, the class II bacteriocins have been categorized into several subgroups: the subclass IIa (pediocin-like bacteriocins containing the N-terminal conserved motif YGNGVxC), the subclass IIb (bacteriocins whose full activity is dependent on the complementary action of two different peptides), the subclass IIc (leaderless bacteriocins), the subclass IId (circular bacteriocins), and the subclass IIe (bacteriocins that do not belong to any of the above groups) (Fimland et al. 2005; Drider et al. 2006; Franz et al. 2007; Nes et al. 2007). However, some researchers maintain that a class III division should be retained for heat-sensitive, large bacteriocins, with a further division into subclass IIIa (bacteriolysins) and IIIb (nonlytic proteins) (Heng and Tagg 2006). Furthermore, recommendations are also made for the circular, postranslationally modified bacteriocins, to be upgraded to a new class IV bacteriocins (Kemperman et al. 2003; Maqueda et al. 2004; Heng and Tagg 2006; Maqueda et al. 2008). The circular bacteriocins constitute a unique family of active proteins in which the N- and C-terminal ends are linked to form a circular backbone. The term "circular" has been adopted to distinguish this group of gene-encoded proteins from the classic "cyclic" peptides produced by microorganisms via multiple steps of enzymatic synthesis (Maqueda et al. 2008). Nine class IV bacteriocins have been reported to date and these can be further divided into two major groups according to their primary structures, biochemical characteristics, and genetic arrangements (Kawai et al. 2009).

Most bacteriocins produced by LAB are synthesized as biologically inactive precursors or propeptides containing an N-terminal extension, whereas the mature peptides are often cationic, amphiphilic, membrane-permeabilizing molecules. The N-terminal extensions of most lantibiotics and nonlantibiotics are of the so-called double-glycine type (leader peptide) and are cleaved concomitantly with export across the cytoplasmic membrane by dedicated ATP-binding cassette transporters (ABC transporters) and their accessory proteins (Håvarstein et al. 1995; Venema et al. 1995). However, some class II bacteriocins, such as acidocin B, divergicin A, bacteriocin 31, enterocin P, lactococcin 972, listeriocin 743A, propionicin T1, enterolysin A, and hiracin JM79/bacteriocin T8 (Sánchez et al. 2007a), contain N-terminal extensions of the so-called Sec type (signal peptide), which are proteolytically cleaved concomitantly with bacteriocin externalization by the general secretory pathway (GSP) or Sec-dependent pathway. They contain canonical Sec signal peptides consisting of a positively charged N-terminus, a hydrophobic core, and a defined cleavage site that is removed by a specific signal peptidase during translocation (van Wely et al. 2001; Herranz and Driessen 2005).

However, a few bacteriocins described to date are synthesized without an N-terminal extension, including enterocin L50 (EntL50A and EntL50B), enterocin Q (EntQ), enterocin EJ97, aureocin A70, and aureocin A53 (Cintas et al. 1998; Sánchez-Hidalgo et al. 2003; Cintas et al. 2000; Criado et al. 2006a). The production of two bacteriocins, designated LsbA and LsbB, synthesized with and without an N-terminal extension, respectively, was shown to be mediated by LmrB a multidrug resistance (MDR) transporter protein involved in the secretion and immunity of both antimicrobial peptides (Gajic et al. 2003). The genetic characterization of the EntQ production and immunity revealed that EntqB is involved in an ABC transporter-mediated secretion, while EntqC confers immunity to the bacteriocin (Criado et al. 2006a). For the circular bacteriocins such as enterocin AS-48, butyrivibriocin AR10, circularin A, gassericin A and reutericin 6, subtilosin A, uberolysin, carnocyclin A, and lactocyclin Q, the expression of structural genes must be combined with the activity of proteins involved in their maturation (cleavage/circularization) and secretion outside the cells, as well as with the production of multifaceted immunity mechanisms essential to ensure bacterial self-protection (Kemperman et al. 2003; Maqueda et al. 2008; Ito et al. 2009; Martin-Visscher et al. 2009). The operons coding for production of several of the circular bacteriocins have been elucidated, enabling the identification of putative immunity proteins and accesory proteins involved in the production and export of these peptides. However, the function of the leader peptide that directs the processing and secretion of the mature bacteriocin, and the details of how cyclization of the secreted bacteriocin occurs remain unknown (Martin-Visscher et al. 2008).

The production of most lantibiotics and class II bacteriocins synthesized with a leader peptide relies on a well-conserved genetic organization including the following four genes often organized in one or two operon-like structures in gene clusters: (1) the structural gene encoding the probacteriocin, (2) a gene encoding the immunity protein, (3) a gene encoding a dedicated ATP-binding cassette (ABC transporter) required for processing and transport of the bacteriocin, and (4) a gene encoding an accessory protein required for proper bacteriocin externalization (Nes et al. 1996; Drider et al. 2006). However, bacteriocins synthesized with a signal peptide show a genetic organization of only two colinear genes, the structural gene and the putative immunity gene (Cintas et al. 1997; Drider et al. 2006). The leaderless peptide bacteriocins are produced as single or multiple homologous peptides, each encoded by individual genes localized in tandem repeats (Cintas et al. 1998; Criado et al. 2006a; Nes et al. 2007). Gene clusters involved in the production of, and immunity to, circular bacteriocins include structural genes encoding the probacteriocin, putative biosynthetic and processing genes, putative immunity genes, putative ABC-type transporters, and regulatory genes (Maqueda et al. 2008).

In most cases, bacteriocin production appears to be regulated, and the production of many bacteriocins including lantibiotics, class II and circular bacteriocins, seems to involve a quorum-sensing mode of regulation mediated by a secreted peptide pheromone (induction peptide; IP), a histidine protein kinase (HPK), and a response regulator (RR) (Diep et al. 1995; Kuipers et al. 1995; Diep et al. 1996). The genes of these regulatory determinants are normally organized within the same operon and have a low basal expression at low cell density. This regulatory operon can be activated by the presence of specific bacteria, by environmental modifications, or by a critical threshold concentration of the secreted IP when the cell density reaches a certain level (hence the term "quorum sensing"). The IP signal activates the gene activator RR via a series of phosphorylation reactions mediated by the sensor HPK. Subsequently, the phosphorylated RR binds to regulated promoters and activates a defined set of genes including its own operon, which eventually triggers a burst in bacteriocin production (Diep et al. 2009). Furthermore, components of well-characterized regulatory systems involved in quorum-sensing bacteriocin production have been exploited in the construction of a number of gene expression vectors for protein expression (Mierau and Kleerebezem 2005; Diep et al. 2009). In addition to the cell density-dependent regulation, the biosynthesis of many different bacteriocins is regulated in response to temperature, medium composition, ionic strength, and pH (Drider et al. 2006; Criado et al. 2006b). Generally, bacteriocin production by LAB is a growth-associated process ceasing at the end of the exponential phase. However, good cell growth does not necessarily result in high bacteriocin production. Many times, the biosynthesis of bacteriocins is stimulated by stress conditions leading to lower growth rates and cell yields but to higher bacteriocin activity (Neysens et al. 2003).

Heterologous Production of Bacteriocins

One factor limiting the use of bacteriocins in commercial applications is their high cost of production due to the low and, in many cases, controlled production rates by the producer strains. Bacteriocins can also be produced by chemical synthesis and/or in vitro translation systems. However, a problem with both is the requirement to form intramolecular disulfide bonds to produce a biologically active peptide. The high cost of synthetic peptide synthesis and the low yields from natural producers drive the exploration of recombinant microbial systems for the production of antimicrobial peptides. However, the production of peptides and proteins by heterologous hosts depends on many factors such as the expression vector, the host strain, and the use of signal sequences for alternative protein secretion pathways, in the sense that the choice of each component must confer the successful expression of the gene of interest.

Nevertheless, all efforts toward production of bacteriocins in heterologous hosts should merit recognition, since their production and functional expression would lead to the production of bacteriocins in safer hosts, increased bacteriocin production, control of bacteriocin gene expression, production of food ingredients with antimicrobial activity, construction of multibacteriocinogenic strains with a wider antagonistic spectrum, a better adaptation of selected hosts to food environments, and providing antagonistic properties to LAB used as starter, protective, or probiotic

Table 8.1 Expected outcomes for production of bacteriocins in heterologous hosts

(a) Increase bacteriocin production
(b) Production of bacteriocins in safer hosts
(c) Control of bacteriocin gene expression
(d) Production of food ingredients with antimicrobial activity
(e) Construction of multibacteriocinogenic strains with a wider antagonistic spectrum
(f) Better adaptation of selected hosts to food environments
(g) Provide antagonistic properties to LAB used as starter, protective, or probiotic cultures

cultures (Table 8.1). It should also be recognized that enterococci are LAB that are members of the gut microbiota of humans and animals and part of the food microbiota, and their ability to produce antimicrobial peptides, referred to as enterocins, may be useful to control undesirable microorganisms in food and feed, in human medicine, and in veterinary applications (Franz et al. 2003; Foulquié-Moreno et al. 2006; Sánchez et al. 2007b). However, because enterocins may be produced by enterococcal species carrying antibiotic-resistance genes and/or genes coding for potential virulence factors, it would be desirable to express the bacteriocins from an alternative and safer bacterium. For this reason, the production of enterocins in other microbial hosts is growing rapidly (Gutiérrez et al. 2005a; Gutiérrez et al. 2006). The emergence of multiple antibiotic-resistant enterococci as agents of nosocomial disease and the presence of virulence factors among food isolates require a careful safety evaluation of isolates intended for potential biotechnological use. Enterococcal bacteriocins produced by heterologous hosts or added as cell-free preparations may still be atractive for application in food preservation or in other potential medical and veterinary applications (Gutiérrez et al. 2006; Sánchez et al. 2008).

Heterologous Production of Bacteriocins by Gram-Negative Bacteria

Among prokaryotic expression systems, the highest protein production levels are probably obtained using *E. coli* as the host strain, although the most commonly used production strategies are intracellular (in the periplasm or cytoplasm) and involve expensive and often problematic downstream purification processes (Morello et al. 2008). The use of Gram-negative bacteria for heterologous production of bacteriocins may also present obstacles due to major differences in codon usage, potential toxicity of the recombinant protein, stability of the mRNA transcripts, incomplete processing of signal sequences, variable secretion efficiency, formation of inclusion bodies in the cytosol and periplasm when using strong promoters, and incorrect formation of disulfide bonds (Choi and Lee 2004).

Nevertheless, a number of bacteriocins have been expressed in Gram-negative hosts. When the pediocin PA-1 operon (*ped*), from *P. acidilactici* PAC1.0, producer of pediocin PA-1 (PedA-1), was expressed in both *E. coli* and a sensitive LAB strain, under control of its own promoter, this permitted the production of active PedA-1 and the functional analysis of the *ped* operon (Venema et al. 1995). The mature form of PedA-1/AcH could also be produced and secreted in *E. coli*, without coexpression of its own processing and transport genes (*pedB* and *pedC*), if fused to a secretory maltose-binding protein (MBP) released into the culture medium when expressed in a periplasmic leaky host (Miller et al. 1998). Fusing alkaline phosphatase to mature divergicin A, produced by *Cb. divergens* LV13, resulted in secretion of this enzyme in the periplasmic space and the supernatant of an *E. coli* host (Worobo et al. 1995). Fusion of the signal peptide of divergicin A to mature mesentericin Y105 behind an inducible lac promoter (P*lac*) led to mesentericin Y105 intracellular production in transformed *E. coli* DH5α cells (Biet et al. 1998). Piscicolin 126, a bacteriocin produced by *Cb. piscicola* JG126, was cloned into an *E. coli* expression system and expressed as a thioredoxin–piscicolin 126 cytoplasmic fusion protein. Purified recombinant piscicolin 126 was obtained after cyanogen bromide (CNBr) cleavage of the fusion protein followed by reverse-phase high-performance liquid chromatography (RP-HPLC) (Gibbs et al. 2004). A synthetic gene (*dvnRV41*) encoding the mature divercin V41 (DvnV41), a bacteriocin produced by *Cb. divergens* V41, was fused as a translational intracytoplasmic fusion protein with thioredoxin and overexpressed with the pET-32b vector under control of the T7 RNA polymerase promoter in the *E. coli* Origami (DE3) (pLysS/pCR03) host strain. The fusion protein TRX-$(His)_6$-DvnRV41 accumulated in the cytoplasm in soluble form and was further cleaved by enzymatic hydrolysis or acid cleavage to release pure active DvnRV41. Such an expression and purification system permitted the production of 23 mg l^{-1} of DvnRV41 during cultivation of the producer strain in shake flasks (Richard et al. 2004). Biomass and bacteriocin yields (30 and 74 mg ml^{-1} of pure DvnRV41) were both increased from *E. coli* Origami (DE3) (pLysS/pCR03) cultures by varying the inducer concentration and the use of controlled batch and fed-batch cultures (Yildirim et al. 2007). The expression of enterolysin A (EnlA), a heat-labile class III bacteriocin from *E. faecalis* II/1 has also been achieved in *E. coli*. The expression of the *enlA* structural gene led to the synthesis of a functionally active His-tagged EnlA (Nigutova et al. 2008).

The cloning and production of enterocin P (EntP), a Sec-dependent bacteriocin produced by *E. faecium* P13 (Cintas et al. 1997), was also evaluated in *E. coli*. When mature EntP (*entP*) was cloned in the vector pETBlue-1 under control of the inducible T7 *lac* promoter to obtain plasmid pJG01, the transformation of this plasmid in the *E. coli* Tuner (DE3)pLacI host permitted the production and quantification of EntP in the supernatant (SN), cellular soluble protein fraction (CSF), and inclusion bodies (IB) of the *E. coli* Tuner (DE3)pLacI (pJG01) cells. Production of EntP in the supernatants of the *E. coli* producers was very low, but biologically active EntP was recovered at a high efficiency by immunoaffinity chromatography using specific polyclonal anti-EntP antibodies (Gutiérrez et al. 2005a). Cloning of *entP* into the expression vector pCM80, under control of the methanol dehydrogenase promoter (P*mxaF*) to generate plasmid pS25, and its transformation into the methylotrophic bacteria *Methylobacterium extorquens* ATCC 55366 also permitted the extracellular production of EntP, but with a low production, and an antimicrobial activity much lower than that deduced from its presence in the supernatant (Gutiérrez et al. 2005b). The low production of EntP by *E. coli* and *M. extorquens* as compared to *E. faecium* may respond to many factors such as that the N-terminal signal peptides and other components of the Sec-dependent system may differ among these bacteria, to differences in codon usage, to the stability and translational efficiency of the mRNA, to differences in secretion patterns among these bacteria, and to the rapid degradation of the secreted EntP by the host cell proteases. Nevertheless, the successful production of EntP by *E. coli* and *M. extorquens* might permit the heterologous production in these hosts of other bacteriocins, peptides, or proteins by fusing their structural genes to the nucleotide sequence of the EntP signal peptide (Gutiérrez et al. 2005b).

The expression of thioredoxin–bacteriocin chimeric proteins in *E. coli* has resulted in the production of intracellular soluble proteins that require chemical, enzymatic, or acid hydrolysis to release the recombinant fused protein (Gibbs et al. 2004; Richard et al. 2004; Yildirim et al. 2007). However, of major interest would be the development of systems for the cloning and expression of self-cleaving forms of secretable mature bacteriocins. Such a system has been developed based on the construction of an expression vector pSuV1 containing an inducible T7 RNA polymerase promoter, the pectate lyase secretion signal (*pelB*), the fusion partner open reading frame of the intein (VMA intein) and chitin-binding domain (CBD), and a multicloning site (MCS) for cloning of the bacteriocins of interest (Ingham et al. 2005). Codon-optimized genes encoding the active mature region of the bacteriocins BacR1, DvnV41, EntP, PedA-1, and piscicolin 126 were inserted into pSuV1 in frame with the PelB secretion signal to facilitate export of the mature protein, whereas the VMA intein-CBD of the fusion partner permitted a column-based adsorption of the fusion protein. Mature bacteriocins were released from the bound fusion following cleavage under reducing conditions in a dithiothreitol-reduced environment. The production levels of around 1 mg ml^{-1} for all bacteriocins are comparatively lower, as high, or higher than those previously reported for related bacteriocins using other heterologous production systems (Ingham et al. 2005; Yildirim et al. 2007). An optimization of the production and purification processes of the carnobacteriocins Cbn BM1 and Cbn B2 from *Cb. maltaromaticum* CP5, by their heterologous expression in *E. coli*, has also been described (Jasniewski et al. 2008). The genes encoding mature bacteriocins were cloned into an pET32a expression system to produce a thioredoxin fusion protein that also contained a His tag for affinity purification and an enzymatic or chemical site for cleavage with enterokinase or CNBr. Recombinant *E. coli* BL21 (DE3) hosts were cultivated following a fed-batch fermentation process, while the overexpression of the fusion protein was improved by replacing the inducer by lactose. The fusion proteins were purified by thermal coagulation followed by affinity chromatography; the thioredoxin fusion protein was removed by CNBr instead of enterokinase, and the carnobacteriocins were recovered by RP-HPLC. These conditions permitted production of up to 320 mg of pure bacteriocin per liter of culture, which is up to tenfold higher than what has been described for other *E. coli* heterologous expression systems

(Jasniewski et al. 2008). However, for the successful production of bacteriocins in heterologous recombinant hosts, a high production of the bacteriocin is not the only important aspect. Obtaining of a peptide with an antimicrobial activity and an specific antimicrobial activity, similar or higher to that of the bacteriocin produced by the natural producer strain, is also essential.

Heterologous Production of Bacteriocins by Lactic Acid Bacteria

The production of bacteriocins by Gram-positive LAB may not only permit the production of antimicrobial peptides in microbial hosts with a generally recognized as safe (GRAS) or a qualified presumption of safety (QPS) status (EFSA 2007), but also accomplish many of the objectives described in Table 8.1. The production of bacteriocins in heterologous LAB strains likely to be used as starters or as producers of food ingredients with antimicrobial activity is desirable, as is the development of bacteriocin-producing heterologous LAB strains with a view of the potential uses of the bacteriocins in food, medical, and veterinary applications. Bacteriocins produced by LAB strains would be secreted into the medium, thus facilitating their recovery and purification. While bacteriocins with high antimicrobial activity are naturally produced by LAB of food origin, interest for their production in heterologous LAB adapted to specific food substrates is growing (Buyong et al. 1998; Coffey et al. 1998; Rodríguez et al. 2005; Reviriego et al. 2007; Liu et al. 2008).

The production of bacteriocins in heterologous LAB requires the use of expression vectors that in addition to an origin of replication of a wide host range may contain a multiple cloning site (MCS) flanked by gene expression signals preceeded by a recognized ribosome-binding site (RBS), and a gene product that confers a resistance marker. The recognition during the mid-nineties of the existence of regulated strong promoters driving lantibiotic production in lactococci, as well as the production of nonmodified bacteriocins in other LAB, were landmarks in bacteriocin research. Furthermore, these findings opened up new avenues for development of gene expression systems for LAB that would allow gene expression to be both highly efficient and strongly regulated (Mierau and Kleerebezem 2005; Diep et al. 2009). Kuipers et al. (1995) initially described the autoregulatory mechanism for production of the lactococcal bacteriocin nisin, and this marked the birth of one of the most successful and widely used Gram-positive gene expression systems, subsequently dubbed as NICE (NIsin-Controlled gene Expression) (de Ruyter et al. 1996; Mierau and Kleerebezem 2005). Numerous expression vectors have been constructed for translational and transcriptional fusions, and for the intracellular production or secretion of the gene product. Plasmid pNZ8048 is the most commonly used expression vector for translational fusions, and NZ9000 (with the *nisK* and *nisR* genes integrated in its chromosome) is the most commonly used *L. lactis* host strain. Using a dual plasmid system, the NICE system has been evaluated for their usefulness in different LAB hosts, and single vectors that carry both the *nisRK* genes and the *nisA* promoter (P*nisA*) have also been developed, simplifying the use of single and suitable expression vectors among different heterologous LAB hosts (Mierau and Kleerebezem 2005).

A system for the efficient inducible heterologous production of bacteriocins was also described based on a versatile two-plasmid system for expression of class II bacteriocins in a nonbacteriocin-producing *Lb. sakei* strain (Axelsson et al. 1998). The first plasmid (pSAK20) contained the genes necessary for transcriptional activation of the sakacin A promoter as well as for processing and export of prebacteriocins with so-called double-glycine leader peptides. The second plasmid (a pLPV111 derivative) contained the structural and immunity genes of the bacteriocin of interest fused to the sakacin A promoter. Using this system, efficient heterologous expression of sakacin P, PedA-1, and piscicolin 61 was obtained at levels equal to or higher than those obtained with the corresponding wild-type producer strain (Axelsson et al. 1998). Versatile expression vectors based on the sakacin A (*sap* regulon) (pSIP300 series) and the sakacin P (*spp* regulon) (pSIP400 series) have also been developed for expression studies of several homologous and heterologous genes of interest (reporter genes) in *Lb. plantarum* NC8 and *Lb. sakei* Lb790 (Sørvig et al. 2005).

Another novel expression system for use in *L. lactis* hosts has been developed by combining the use of the P_{Zn}*zitR* expression system and the SP_{Exp4} signal peptide into an *htrA* mutant *L. lactis* strain (Llull and Poquet 2004; Morello et al. 2008). Proteases are known to be involved in inactivation of native proteins, processing of propeptides, and proteolysis of secreted recombinant proteins, and a *L. lactis* strain deficient in both its major proteases, intracellular (ClpP) and extracellular (HtrA), has been shown as a useful host for the high-level production of stable heterologous proteins (Cortes-Perez et al. 2006). For *L. lactis* as a host, the inactivation of the HtrA protease or complementation of the Sec machinery with the *B. subtilis* SecDF accessory protein results in an increase in heterologous protein yields (Morello et al. 2008). Other results indicate that suppression of the acid tolerance response (ATR) in wild-type *L. lactis* may serve as a better strategy than the use of a *L. lactis htrA* mutant for production of secreted recombinant proteins (Sriraman and Jayaraman 2008). A short synthetic propeptide can be fused between a SP and the mature protein to improve the efficiency of heterologous secretion in *L. lactis*, as shown for several proteins (Morello et al. 2008). However, their usefulness for secretion of bacteriocins from expression vectors remains to be proven.

The examples detailed above highlight the importance of promoter choice in lactococci and lactobacilli gene expression systems. Inducible expression systems are often preferable when the aim is to overproduce a desired protein at high levels. However, when a steady-state gene expression is required, inducible systems are less suitable, and constitutive promoters could be used as an alternative for such applications. A large effort has been performed to identify constitutive promoters functional in lactococci and lactobacilli and a corresponding range of these promoters, including synthetic promoters, have been identified (Van der Vossen et al. 1987; Rud et al. 2006). Nevertheless, inducible expression systems usually yield expression levels higher than those obtained with the best constitutive promoters (Gutiérrez et al. 2006; Diep et al. 2009).

Table 8.2 Production of bacteriocins in heterologous lactic acid bacteria (LAB) hosts

(a)	Expression of native biosynthetic genes
(b)	Exchange or replacement of leader peptides and/or dedicated processing and secretion systems (ABC transporters)
(c)	Fusion of mature bacteriocins to signal peptides (SP) that act as secretion signals

The use of genetically modified strains in food-related applications may demand cloning vectors devoid of antibiotic resistance genes. Food-grade cloning systems utilizing different selection methods to obtain recombinant strains have been developed for a number of LAB. Preferred methods are complementation of auxotrophic phenotypes for carbohydrate or DNA metabolism, and the use of dominant selection methods, such as resistance to bacteriocins or metal ions for the positive selection of transformants (Takala and Saris 2002; Liu et al. 2005; Sridhar et al. 2006). Several food-grade vectors for integration of the genes of interest by gene-replacement strategies or single crossover integration in the LAB genomes have also been described, including vectors that either do not replicate at all in the LAB hosts or are conditionally nonreplicative (Leenhouts et al. 1998; Henrich et al. 2002; Simoes-Barbosa et al. 2004).

Another important consideration for bacteriocin gene expression is the likely location of the expressed bacteriocin peptide. Production of secreted recombinant proteins provides advantages over systems that culminate in cytosolic production. A number of putative signal peptides (SP) from different LAB have been selected for their ability to drive secretion of various model proteins (Brockmeier et al. 2006; Mathiesen et al. 2008). However, despite tremendous efforts in understanding secretion in Gram-positive bacteria, it is generally difficult to predict how efficient protein secretion can be genetically engineered (Brockmeier et al. 2006; Diep et al. 2009). Nevertheless, production of bacteriocins in heterologous LAB hosts may be essentially based on (a) the expression of native biosynthetic genes, (b) exchanging or replacing leader peptides and/or dedicated processing and secretion systems (ABC transporters), or (c) fusion of mature bacteriocins to signal peptides that act as secretion signals (Table 8.2) (Rodríguez et al. 2002; Gutiérrez et al. 2006).

Heterologous Production of Bacteriocins Based on Expression of Native Biosynthetic Genes

Cloning of the genes involved in the synthesis, processing, and secretion of bacteriocins in plasmid vectors that replicate in heterologous hosts constitutes the easiest and simplest procedure for the production of bacteriocins. Following this approach, bacteriocins from different classes and subclasses could be produced by different LAB hosts, usually at concentrations below that of the natural producer (Rodríguez et al. 2002). However, production of PedA-1 was increased in *L. lactis* LL108 by increasing the copy number of the pMC117-encoded *ped* operon (Chikindas et al. 1995).

Later on, *L. lactis* MM210, a strain used in Cheddar cheese manufacture, was selected as a host for pMC117. In experimental cheeses, the counts of *L. monocytogenes* decreased in the presence of the lactococcal starter culture (Buyong et al. 1998). In a second example, the cloning of the structural (*entA*), immunity (*entiA*), and processing and transport genes (*entTD*) of the bacteriocin enterocin A (EntA), produced by *E. faecium* DPC1146, into the expression vector pENT02 containing the lactococcal P_{32} promoter permitted the production of EntA by different *E. faecalis* and *L. lactis* hosts, although plasmid and phenotypic instability was observed (O'Keeffe et al. 1999). The cloning of the complete operon (10,769 bp) of the bacteriocin ABP-118, produced by *Lb. salivarius* UCC118, in the expression vector pSF05, a derivative of pNZ8048 containing the P*nisA* inducible promoter, permits the production and functional expression of the bacteriocin in the supernatants of transformed *Lb. plantarum*, *L. lactis*, and *B. cereus* hosts containing a second plasmid with the functional *nisRK* genes (Flynn et al. 2002).

Lacticin 3147 is a broad-spectrum, two-peptide lantibiotic whose genetic determinants are located on two divergent operons found naturally in the lactococcal plasmid pMRC01. Surprisingly, cloning of the two operons in a high-copy-number vector (pOM02) did not lead to a higher production of lacticin 3147 than that observed with the same construct into a low-copy-number plasmid pMR01. However, the antimicrobial activity of the *L. lactis* MG1363 (pMRC01, pOM02) derivative was much higher than that of the strains containing either plasmid alone. After introducing different combinations of lacticin 3147 genes into *L. lactis* MG1363 (pMRC01), it was determined that while the provision of additional copies of genes encoding the biosynthetic/production machinery and the regulator LtnR is a requirement for high-level overproduction, the presence of additional copies of the structural genes is not (Cotter et al. 2006). When enterocin AS-48, a thermostable circular bacteriocin produced by *E. faecalis* S-48 encoded by the *as-48* gene cluster, ribosomally synthesized as a propeptide and posttranslationally modified by a head-to-tail peptide bond, was cloned in the expression vector pAM401, the successfully transformed *E. faecalis* JH2-2 (pAM401-81) strain produced enterocin AS-48 as efficiently as the native S-48 strain. However, the full expression of enterocin AS-48 derived from cloning of the *as-48* gene cluster in different expression vectors transformed in various enterococcal, lactococcal, and lactobacilli hosts was not possible in organisms not belonging to the genus *Enterococcus*, because only a partial expression of its immunity was obtained (Fernández et al. 2007).

Heterologous Production of Bacteriocins by Exchanging or Replacing Leader Peptides and/or Dedicated Processing and Secretion Systems (ABC Transporters)

Since most bacteriocins are synthesized with N-terminal extensions that determine their processing and secretion, many research efforts have been directed toward production of bacteriocins in heterologous LAB hosts through the exchange of leader peptides and/or dedicated processing and secretion systems of the dedicated transport system (DTS), or by replacement of leader peptides by signal peptides recognized by the general secretory pathway (GSP) or Sec-dependent pathway. The recognition that leader peptides of the double-glycine type (Gly-Gly at positions 1 and 2 of the leader peptide) are cleaved during the export of the bacteriocin and function as export signals for ABC transporters constitutes the basis for the exchange or replacement of the leader peptides for secretion of the antimicrobial peptides by heterologous hosts (Håvarstein et al. 1995; Van Belkum et al. 1997).

Based on the homology among the N-terminal extensions of the bacteriocin lactacin F (LacF), produced by *Lb. johnsonii* VPI 11088, and the carnobacteriocins A, BM1, and BM2, produced by *Cb. piscicola* LV17, Allison et al. (1995) noticed the capacity of transformed cells of *Cb. piscicola* with *lacF* to produce LacF. This observation implied that the machinery for the translocation and processing of the carnobacteriocins was also able to facilitate the secretion of LacF. Since this report, several further reports have described the production of bacteriocins using heterologous ABC transporters, although heterologous production was not as high as that measured in the homologous hosts with their own dedicated transport machinery. However, it remained unclear whether this was due to low expression of the genes in a heterologous background or because the precursors are not efficiently recognized and/or processed by the heterologous export proteins (Van Belkum et al. 1997; Rodríguez et al. 2002). When the leader peptides of leucocin A, lactococcin A, and colicin V were fused to mature divergicin A (DvnA), a bacteriocin secreted via the general secretory pathway, and introduced via expression vectors into *Lc. gelidum*, *L. lactis*, and *E. coli* strains carrying the dedicated export and processing systems for leucocin A, lactococcin A, and colicin V, respectively, the highest efficiency of DvnA secretion was observed when the leader peptides were homologous to the host's transport proteins. However, in some cases, the leader peptides did not direct DvnA secretion, suggesting that some leader peptides are poorly recognized or not recognized at all by the proteolytic domain of heterologous ABC transporters (Worobo et al. 1995; Van Belkum et al. 1997). As described previously, a procedure for the inducible heterologous production of bacteriocins was also developed based on a versatile two-plasmid system for expression of class II bacteriocins in a non-bacteriocin-producing *Lb. sakei* strain. Efficient expression of sakacin P, PedA-1, and piscicolin 61 was obtained at levels equal to, or slightly higher than, those obtained with the corresponding wild-type producer strain (Axelsson et al. 1998).

Horn et al. (1998) also determined that pediocin PA-1 (PedA-1), produced by *P. acidilactici* 347, and lactococcin A (LcnA), produced by *L. lactis* subsp. *lactis diacetylactis* WM4, were likely candidates for translocation and processing of PedA-1 via the replacement of leader peptides and the use of heterologous translocators. With this aim, these researchers designed a vector system containing an in-frame fusion of the lactococin A (*lcnA*) promoter, the *lcnA* leader sequence, and the mature pediocin PA-1 (*pedA*) for transformation into *L. lactis* IL1403 harboring chromosomal genes analogous (*lcnC'*, *lcnD'*) to those encoding the LcnA translocation and processing system. The resulting *L. lactis* strains secreted PedA-1 with an antimicrobial activity of, approximately, 25% of that displayed by *P. acidilactici* 347. This heterologous production system was enhanced by the introduction in *L. lactis* of the LcnA-dedicated translocation genes, *lcnC* and *lncD*, and by increasing the gene dosage of the cloned genes using multicopy plasmids. The yield of LnCD-mediated translocation of PedA-1 was also influenced by the choice of *L. lactis* host strain. In the same study, *L. lactis* strains with the ability to coproduce nisin A (NisA) and PedA-1 were constructed by introducing the recombinant plasmid for production of PedA-1 into a NisA-producing strain (Horn et al. 1999). However, when the enterocin A (EntA) and/or PedA-1 structural genes, containing their own leader peptides, were introduced in plasmid vectors transformed in *L. lactis* IL1403, a very low-level production and/or coproduction of both bacteriocins was achieved. This reduction in bacteriocin production may be due to the low copy number of the chromosomal *lcnC'* and *lcnD'* analogs in *L. lactis* IL1403, and/or to inefficient recognition of the resulting gene products by the LcnA translocation apparatus (Martínez et al. 2000). The production of PedA-1 was further enhanced by placing the chimeric genes encoding the fusion leader of lactococcin A-propediocin PA-1 or procolicin V under control of the inducible *nisA* promoter (P*nisA*) and the LcnA-dedicated secretion genes (*lcnCD*) into *L. lactis* strains expressing the two component regulator NisRK, leading to the nisin-controlled production of PedA-1 and colicin V in nisin- and nonnisin-producing *L. lactis* strains (Horn et al. 2004).

The design of LAB as coproducers of bacteriocins with activity for Gram-positive (NisA) and Gram-negative (colicin V) bacteria is of great interest, as it is the capacity of the recombinant plasmids responsible for the production of PedA-1 to transform lactococcal strains of dairy origin for their use as starter culture adjuncts in cheese making (Rodríguez et al. 2005). The introduction of the plasmid(s) encoding the fusion between the LcnA leader sequence and propediocin PA-1, the *lcnCD* genes of the LcnA secretion machinery, and the PedA-1 immunity gene (*pedB*) into nisin- and nonnisin-producing lactococcal hosts, has also led to an increase in the activity of PedA-1 compared with the equivalent *pedB*-devoid systems. In addition, a nisin-producing strain with the ability to produce PedA-1 with an antimicrobial activity equivalent to that of *P. acidilactici* 347 has been obtained (Arqués et al. 2008).

Heterologous Production of Bacteriocins by Fusion to Signal Peptides that Act as Secretion Signals

The recognition, also during the mid-nineties, that a number of bacteriocins were synthesized with N-terminal extensions of the so-called Sec type and used the GSP or Sec-dependent pathway for translocation and processing prompted interest for their production in homologous and heterologous LAB hosts. The Sec system is a universally conserved protein export system that translocates unfolded proteins across the cell membrane via a protein-conducting pore formed by the SecYEG complex and a molecular motor, the ATPase SecA. Secretory proteins are equipped with an N-terminal signal peptide (SP) that functions as a target and recognition signal for signal peptidases that remove the signal peptide from the translocated protein, resulting in the release of the mature protein or peptide (Fekkes and Driessen 1999). Divergicin A (DvnA), produced by *Cb. divergens* LV13, was the first bacteriocin reported to be synthesized with a SP to access the GSP. Production of DvnA in other *Carnobacterium* spp. and *L. lactis* hosts was determined after their transformation with expression vectors containing only the two genes associated with DvnA production (*dvnA*) and immunity (*dviA*) (Worobo et al. 1995).

In-frame fusions of the SP of DvnA (SP_{DvnA}) with the mature sequence of carnobacteriocin B2 (McCormick et al. 1996), colicin V (McCormick et al. 1999), and enterocin B (Franz et al. 1999) permitted the production of these bacteriocins in the supernatants of different LAB hosts, including *L. lactis*. The production of mesentericin Y105 (MesY), produced by *Lc. mesenteroides* Y105 via a dedicated transport system (DTS), was also evaluated using two expression vectors. In one of them, the mature *mesY* and its putative immunity gene (*mesI*) were fused to the SP_{DvnA} under control of a constitutive *lac* promoter (pFBYC07), whereas plasmid pFBYC04 contained the entire mesentericin Y105 gene cluster (*mesYICDE*). The production of MesY by a nonbacteriocin-producing *Lc. mesenteroides* strain transformed with pFBYC07 was approximately fourfold lower than the production of the same bacteriocin by *Lc. mesenteroides* Y105, while the level of MesY produced by the same strain transformed with pFBYC04 was similar to that obtained with the natural producer strain (Biet et al. 1998). Although it remains unclear why most bacteriocins have a dedicated secretion and processing system if provided with an adequate SP (McCormick et al. 1996), it would seem likely that bacteriocins would employ a dedicated export machinery because they drive a more efficient process (Biet et al. 1998). The bacteriocin PedA-1, produced by *P. acidilactici*, has also been expressed in the heterologous host *Bifidobacterium longum* MG1 and secreted from the transformed strain by fusing the bacteriocin gene to the SP of the bifidobacterial α-amylase. The recombinant *B. longum* MG1 derivative, secreting PedA-1, killed *L. monocytogenes* cells in a coculture (Moon et al. 2005).

Enterocin P, a Sec-dependent bacteriocin produced by *E. faecium* P13, was also evaluated for its production and functional expression in *L. lactis* (Herranz and Driessen 2005; Gutiérrez et al. 2006). The EntP structural gene (*entP*) with or without the EntP immunity gene (*entiP*) was cloned in (1) plasmid pMG36c under the control of a constitutive promoter, (2) plasmid pNG8048e under the control of the inducible P*nisA* promoter, and (3) the integration vector pINT29 (Fig. 8.1), and the introduction of the recombinant vectors in *L. lactis* resulted in the production of biologically active EntP in the supernatants of *L. lactis* IL1403 and *L. lactis* NZ9000, and the coproduction of NisA and EntP in *L. lactis* DPC5598 (Fig. 8.2). The presence of *entP* was the minimum requirement for the production of biologically active EntP, whereas the production of EntP, detected by specific polyclonal anti-EntP antibodies and a noncompetitive indirect enzyme-linked immunosorbent assay (NCI-ELISA), by the recombinant *L. lactis* strains depended on expression vector, the host strain, and the presence of the *entiP* gene in the constructs of the recombinant strains. The highest amount of EntP was produced with derivatives containing *entP* and *entiP* for both *L. lactis* IL1403 and *L. lactis* NZ9000. These derivatives produced up to 5- to 6-fold more EntP than *E. faecium* P13 (Gutiérrez et al. 2006).

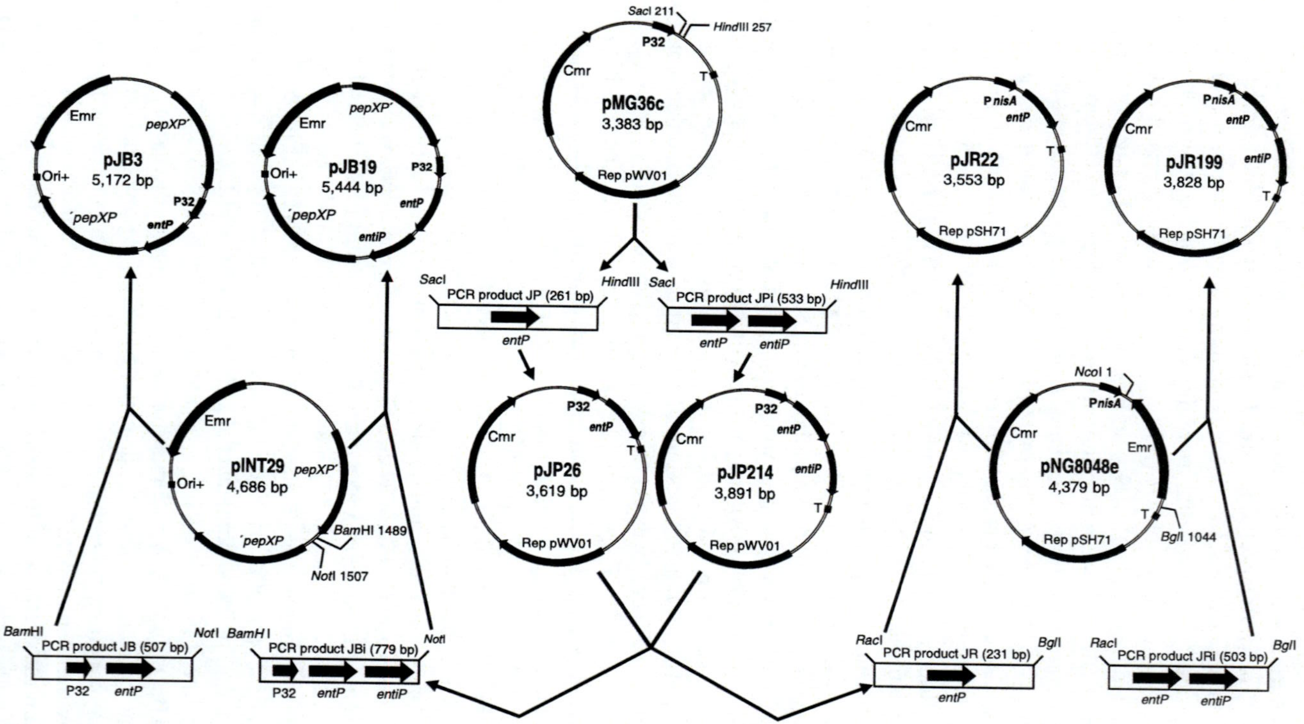

Fig. 8.1 Construction of the recombinant expression vectors for production of enterocin P in *L. lactis*. Sizes of plasmids are given in base pairs. Only relevant restriction enzymes are given. P_{32}, constitutive lactococcal promoter; P*nisA*, inducible nisin A promoter; T, transcription terminator; Rep, lactococcal replicons pWV01 and pSH71; Ori^{+}, pWV01 origin of replication; *pepXP'* and *'pepXP*, 5' and 3' sequence of *pepX* gene, respectively; *entP*, enterocin P structural gene; *entiP*, enterocin P immunity gene; Cmr, choramphenicol resistance; Emr, erythromycin resistance. Reproduced from Gutiérrez et al. (2006) with kind permission of Springer Science and Business Media

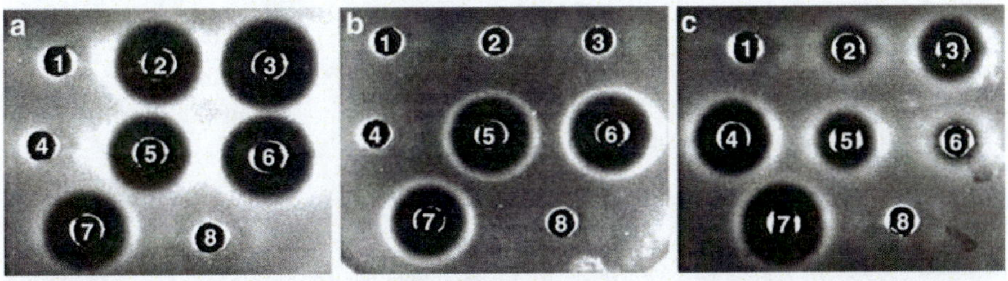

Fig. 8.2 Production of enterocin P by *L. lactis* as determined by the agar well diffusion test with *E. faecium* T136 as the indicator strain. (**a**) Supernatants of lactococcal strains carrying constitutive promoter constructs: (1) *L. lactis* IL1403 (pMG36c), (2) *L. lactis* IL1403 (pJP26), (3) *L. lactis* IL1403 (pJP214), (4) *L. lactis* NZ9000 (pMG36c), (5) *L. lactis* NZ9000 (pJP26), and (6) *L. lactis NZ9000* (pJP214). (**b**) Supernatants of *L. lactis* NZ9000 carrying inducible promoter constructs: (1, 4) pNG8048e, (2, 5) pJR22, and (3, 6) pJR199, before (1, 2, 3) and after (4, 5, 6) induction of the cultures with nisin. (**c**) Supernatants of *L. lactis* NZ9000 cells transformed with the integrative vectors: (1) pINT29, (2) pJB3, (3) pJB19, EntP from *E. faecium* P13 at 8 mg ml^{-1} (4), 2 mg ml^{-1} (5), and 0.5 mg ml^{-1} (6). Supernatants of *E. faecium* P13 (7) and *E. faecium* T136 (8) were used as controls. Reproduced from Gutiérrez et al. (2006) with kind permission of Springer Science and Business Media

Interestingly enough, the production of EntP in the supernatants of most of the recombinant *L. lactis* strains was, in most of the cases, comparatively higher than production of a number of bacteriocins in other LAB hosts using dedicated ABC transport systems (Chikindas et al. 1995; van Belkum et al. 1997; Axelsson et al. 1998; Biet et al. 1998; Horn et al. 1998, 1999; Horn et al. 2004; Arqués et al. 2008) or the Sec-dependent system (Worobo et al. 1995; Biet et al. 1998; McCormick et al. 1999; Moon et al. 2005), suggesting that whereas production of EntP in *E. coli* hosts was rather low (Gutiérrez et al. 2005a), the EntP signal peptide (SP_{EntP}) of *E. faecium* efficiently directs the processing and secretion of EntP in *L. lactis* (Gutiérrez et al. 2006). Accordingly, fusions between the SP_{EntP} and other mature bacteriocins may allow *L. lactis* to secrete bacteriocins in the absence of specific immunity and secretion proteins, as well as to serve as a model system for the secretion of numerous peptides or proteins of interest in the food industry by *L. lactis* and other LAB (Gutiérrez et al. 2006).

It has also been reported that the dehydration and cyclization of the lantibiotic nisin propeptide still occur when the nisin leader is preceeded by the signal peptide of Usp45 (SP_{Usp45}) and that the Sec system of *L. lactis* efficiently secretes dehydrated therapeutic peptides (Kuipers et al. 2006). Cloning of the synthetic *dvnRV41* gene with the codon usage for *E. coli*, encoding mature DvnV41, fused in frame to the SP_{Usp45} and under control of the P*nisA* promoter in the pSEC-ET vector, originated the recombinant plasmid pSEC:DvnRV41. The induction with NisA of the transformed *L. lactis* NZ9000 (pSEC:DvnRV41) cells permitted secretion of DvnRV41 into the supernatant. However, although the levels of DvnRV41 produced by the *L. lactis* cells were similar to those produced by the natural host, *Cb. divergens* V41, their antimicrobial and specific antimicrobial activities were lower than

Table 8.3 Factors that may affect the production and antimicrobial activity of bacteriocins in heterologous microbial hosts

Regulatory responses to a secretion stress due to overproduction of bacteriocins
Differences in components of the Sec-dependent translocon and Sec machinery
Characteristics of the signal peptide (SP)
Characteristics of the mature bacteriocin
Efficiency of recognition of the SP by signal peptidases
Protein quality control network involving folding factors and housekeeping proteases
Host cell physiology and genera and species of the heterologous LAB hosts
Copy number of the plasmid expression vector
Characteristics of the promoter that drives gene expression
Level of transcription of the two-component regulatory systems
Differences in codon usage, stability, and translational efficiency of the produced mRNA
Proper or improper bacteriocin disulfide bond formation (DSB)
Efficiency of posttranslational processing of probacteriocins to active bacteriocins
Intramolecular folding of bacteriocins that maintain the propeptide in a secretion-incompetent conformation
Spontaneous chemical modification of amino acid residues
Bacteriocin oxidation events
Conformational modifications of bacteriocins to less active extracellular forms
Differences in protein-folding efficiency and turnover
Variable bacteriocin production in multibacteriocinogenic hosts
Bacteriocin self-aggregation
Degradation of bacteriocins by host-cell proteases, intracellular and extracellular
Bacteriocin retention within the cytoplasmic membrane
Interactions of the hydrophobic bacteriocins with the producer cell membrane
Adsorption of bacteriocins to the producer cells or to growth media components

expected from this level of bacteriocin. Since the antagonistic activity of DvnRV41 was not affected by the ClpP and HtrA proteases of the producing *L. lactis* hosts, several other factors such as those described in Table 8.3 could explain the differences in the antimicrobial activity between the natural DvnV41 and the DvnRV41 produced by *L. lactis* (Bermúdez-Humarán et al. 2007).

Hiracin JM79/bacteriocin T8, another Sec-dependent bacteriocin produced by *E. hirae* DCH5 (Sánchez et al. 2007a), was also evaluated for its production and expression in other LAB hosts. Two recombinant plasmids, derivatives of either pMG36c or pNZ8048c and containing the HirJM structural gene (*hirJM79*) with or without the HirJM79 immunity gene (*hiriJM79*), permitted the production of biologically active HirJM79 in the supernatants of *L. lactis* IL1403, *L. lactis* NZ9000, *Lb. sakei* Lb790, and *E. faecalis* JH2-2; the coproduction of HirJM79 and NisA in *L. lactis* DPC5598; and the coproduction of HirJM79 and EntP by *E. faecium* L50/14-2. All recombinant LAB evaluated produced larger quantities of HirJM79 than *E. hirae* DCH5, although the antimicrobial activities of most transformants were lower than that predicted from their production of HirJM79. It is possible to speculate that disulfide bond formation (DSB), a universally conserved mechanism for stabilizing extracytoplasmic proteins, is not performed adequately by lactococci, that the bacteriocin could undergo conformational modifications rendering less active extracellular HirJM79 forms, or that some of the factors cited in Table 8.3 are responsible for the lower antimicrobial activity of HirJM79 by the recombinant LAB hosts (Sánchez et al. 2008).

To evaluate the potential of the signal peptide of EntP (SP_{EntP}) to direct the synthesis, processing, and secretion of mature bacteriocins in other LAB, chimeric genes encoding the Sp_{entP} fused to mature PedA-1 (*pedA*), with or without its immunity gene (*pedB*), were cloned into the expression vector pMG36c to generate recombinant plasmids that after transformation into competent *L. lactis* IL1403, *L. lactis* NZ9000, and *L. lactis* DPC5598 cells permitted the detection and quantification of PedA-1 and the coproduction of NisA and PedA-1 in supernatants of the producer hosts. Although only a 261-bp fragment is required for the production and functional expression of PedA-1 in *L. lactis*, its production was less efficient (47–55%), and its antimicrobial and specific antimicrobial activities (12–44%) were lower than its production and antagonistic activity by the *P. acidilactici* PLBH9 natural PedA-1 producer (Martín et al. 2007a). However, chimeras of the SP_{entP} fused to mature enterocin A (*entA*), with and without its immunity gene (*entiA*), cloned in the same vector as cited above permitted the production and functional expression of EntA in *L. lactis* IL1403 and *L. lactis* NZ9000, and the coproduction of NisA and EntA by *L. lactis* DPC5598. Mature EntA (*entA*) was the minimum requirement for the synthesis, processing, and secretion of biologically active EntA by *L. lactis* in the absence of specific induction, immunity, and secretion proteins. The production of EntA by the recombinant *L. lactis* hosts was larger (1.5 to 2.4-fold), although the antimicrobial activity and antimicrobial specific activity of the EntA produced by *L. lactis* was lower (44–58%) than expected (Martín et al. 2007b). It is hypothesized that some of the factors cited in Table 8.3 are responsible for the lower production and antimicrobial activity of PedA-1 by *L. lactis*. The N-terminus of mature PedA-1 may also affect its translocation across the cytoplasmic membrane of *L. lactis* hosts, as the presence of basic amino acids has been shown to reduce the secretion of peptides, whereas acidic or neutral residues may improve its secretion (Le Loir et al. 2001).

Heterologous Production of Bacteriocins by Yeasts

Yeasts have not been fully exploited as alternative hosts for production of bacteriocins, although they provide attractive expression platforms given the ease of genetic manipulation and the option of a single fermentation design. Whereas the initial yeast system for heterologous protein production was based on the yeast *Saccharomyces cerevisiae*, recent alternative yeast production systems include, among others, the methylotrophic species *Pichia pastoris*, *Hansenula polymorpha* (*Pichia angusta*), and *Klyveromyces lactis*, and the dimorphic species *Arxula adeninivorans* and *Yarrowia lipolytica* (Gellissen et al. 2005; Böer et al. 2007a). As no single yeast system is likely to be optimal for production and expression of all proteins, the development and use of a wide-range vector system may permit the assessment of several yeast expression platforms in parallel for criteria such as efficient processing or secretion (Steinborn et al. 2006; Böer et al. 2007b).

The structural *pedA* gene, coding for mature PedA-1, was first expressed in *S. cerevisiae* Y294 under control of the alcohol dehydrogenase I gene promoter and terminator, whereas secretion of the peptide was directed by the yeast mating pheromone α-factor 1 secretion signal (MFα1s). A difference in PedA-1 activity levels was observed among transformants grown on solid or liquid media, which was ascribed to the possibility that the recombinant PedA-1 molecules become stuck within the yeast plasma membrane or remained cell wall associated (Schoeman et al. 1999). The bacteriocin plantaricin 423 was also cloned into the multicopy yeast shuttle vector YepPla423 and heterologously produced by *S. cerevisiae* L5366h (YepPla423). Clear zones of inhibition were observed around colonies of the yeast transformant when loaded with the indicator strains, but no inhibitory activity could be detected in the supernatant of the recombinant yeast strain without concentration (Van Reenen et al. 2002).

Expression of the *pedA* gene was also evaluated after its cloning in the pPICZαA expression vector using *P. pastoris* KM71H (Mut^S) as the host strain. Although a significant concentration of extracellular PedA-1 was confirmed through antibody specific assays, the bacteriocin showed no biological activity. Strong evidence indicated that some "collagen-like" material was tightly associated, most probably via covalent binding, with the recombinant bacteriocin (Beaulieau et al. 2005).

However, when the *entP* gene encoding mature EntP was cloned into pPICZαA in frame to the MFα1s without the Glu-Ala spacers adjacent to the Kex2 protease cleavage site, giving plasmid pJC31, and the recombinant vector was linearized and transformed into *P. pastoris* X-33 competent cells, the *P. pastoris* X-33t1 derivative showed potent direct antimicrobial activity. Moreover, maximum production of EntP by *P. pastoris* X-33t1 grown in the complex medium BMMY was 3.7-fold higher, its antimicrobial activity 16-fold higher, and its specific antimicrobial activity 4.3-fold higher, than production of EntP by the natural producer *E. faecium* P13 (Gutiérrez et al 2005c). Similarly, the heterologous production and functional expression of HirJM79, produced by *E. hirae* DCH5, has also been achieved by the *P. pastoris* X-33TH host transformed with pPICH5, a derivative of pPICZαA. However, whereas the production of HirJM79 by *P. pastoris* X-33TH, grown in the BMMY medium, was higher than that by *E. hirae* DCH5 (Fig. 8.3), its antimicrobial activity and specific antimicrobial activity were lower than that expected from their production of HirJM79 (Sánchez et al. 2008).

The structural genes of the two leaderless peptides of enterocin L50 (EntL50A and EntL50B) from *E. faecium* L50 have also been cloned, separately (*entL50A* or *entL50B*) and together (*entL50AB*), into a segregationally stable expression and secretion vector for *S. cerevisiae*, named pYABD01, under control of the galactose inducible promoter P_{gal1}. The generation of recombinant *S. cerevisiae* strains heterologously expressing and secreting biologically active EntL50A and EntL50B demonstrates the suitability of the MFα1s-containing vector pYABD01 to direct processing and secretion of these antimicrobial peptides through the *S. cerevisiae*

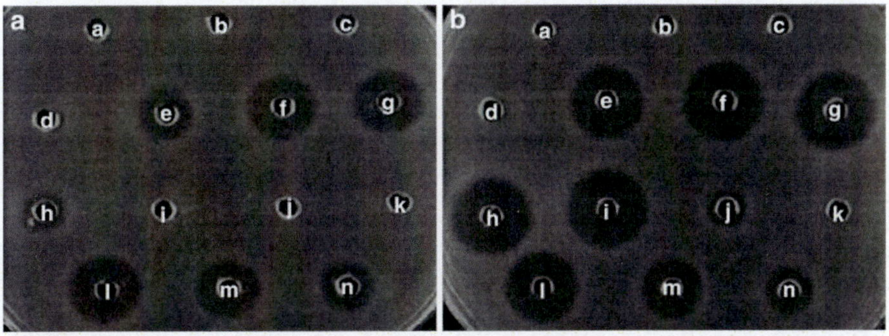

Fig. 8.3 Production of hiracin JM79 by *P. pastoris* X-33TH as determined by the agar well diffusion test with *E. faecium* T136 as the indicator strain. Supernatants of *P. pastoris* X-33TH, grown in (**a**) the minimal medium BMM or (**b**) the complex medium BMMY, after 0 (d), 4 (e), 6 (f), 8 (g), 10 (h), 12 (i), and 24 h (j) of incubation. The supernatants of *P. pastoris* X-33 cultures grown in BMM or BMMY for 0 (a), 6 (b), and 12 h (c), the supernatants of *E. hirae* DCH5, not diluted (l), fourfold diluted (m), and 16-fold diluted (n), and the supernatant of *E. faecium* T136 (k), were all used as negative controls. Reproduced from Sánchez et al. (2008) with permission from the American Society for Microbiology (ASM)

Sec system. However, despite promising results, further research is needed to optimize the enterocin L50 production, increase bacteriocin stability under oxidative stress, and achieve paired production of EntL50A and EntL50B by a single yeast strain (Basanta et al. 2009).

Conclusions and Perspectives

If the use of bacteriocins as natural antimicrobial agents in food, veterinary, and medical applications is ever to meet the high expectations of the research community, then high-level production of active bacteriocin is essential. Similarly, all efforts to improve their production in heterologous microbial hosts are welcome. However, bacteriocin production levels and their biological activity vary, mostly in an unpredictable manner, depending on the expression vector, the gene of interest, and the host strain. Apparently, the effectiveness of the expression systems depend on subtle properties that are not straightforward to predict as they are often case specific. Accordingly, further research efforts should be performed for a more efficient genetically engineered bacteriocin production, secretion, and stabilization of the bacteriocin molecule.

A deeper understanding and overcoming of the factors that may affect the production and antimicrobial activity of mature bacteriocins fused to SP would surely improve the production and antagonistic activity of the bacteriocins produced in heterologous LAB. It is difficult to ascertain secretion efficiencies and each combination of host strain and target bacteriocin may need individual optimization.

Novel computer tools should also be developed to identify the complex interaction patterns between SP and the mature part of the secretion target bacteriocin. In fact, selecting an optimal SP for each individual bacteriocin of interest may be an optimal approach.

The development of expression vectors with novel inducible or constitutive promoters, improved peptide tags, and/or SP with a more efficient recognition by signal peptidases, and the use and design of synthetic genes with an adapted codon usage could improve not only the stability and translational efficiency of the produced mRNA but also the production of single and multiple bacteriocins, or the production of hybrid bacteriocins with improved properties. The use of food-grade self-cloning expression/secretion vectors for the overproduction of bacteriocins would also help to overcome the legislation over the use of genetically modified organisms for certain biotechnological applications. This implies further the development of plasmid-based systems using food-grade genes for positive selection of transformants and novel systems for the integration of the genes of interest into the host chromosome, and that antibiotic selection markers should be avoided.

Furthermore, the improvement of the genetic tools for the overproduction of bacteriocins by LAB hosts may also be valuable for the design and evaluation of other LAB as potential cell factories for the production of many other bioactive peptides (i.e., immunomodulatory, antihypertensive, hypocholesterolemic, etc.) of biotechnological interest. Whereas the production of bacteriocins by yeasts has mostly relied on the use of *S. cerevisiae* and *P. pastoris* as the heterologous producers, the recent development of novel yeast production hosts and wide host-range vector systems awaits the assessment of such expression platforms for the efficient production of bacteriocins in other yeasts.

Acknowledgments Most of the work of our group referred to in this manuscript was supported by grants BIOT-CT910263 and BIOT-CT943055 from the Commission of the European Communities (CEC), by grant 07/0026/00 from the Comunidad de Madrid (CAM), by grant PR248/02-1168 from the Fundacion Danone-UCM, by grant PR41/06-15051 from the Banco Santander-UCM, by grants GR85/06, CCG07/AGR-2546, and GR58/08 from the UCM, by grants S-0505/AGR/0265 and S-2009/AGR-1489 from the CAM, and by grants ALI91-0255, ALI94-1026, ALI97-0559, AGL2000-0707, AGL2003-01508, AGL2006-01042, and AGL2009-08348 from the Ministerio de Educación y Ciencia (MEC). We also acknowledge the research results generated by J.M. Rodríguez, O.J. Sobrino, W.L. Moreira, P. Casaus, M.I. Martínez, J.M. Martínez, J. Gutiérrez, R. Citti, M. Martín, R. Criado, B. Gómez-Sala, A. Basanta, J. Sánchez, J. Borrero, L. Gútiez, J.J. Jiménez, C. Campanero, and E. Muñoz.

References

Allison GE, Worobo RW, Stiles ME, Klaenhammer TR (1995) Heterologous expression of the lactacin F peptides by *Carnobacterium piscicola* LV17. Appl Environ Microbiol 61:1371–1377

Arqués JL, Rodríguez JM, Gasson MJ, Horn N (2008) Immunity gene *pedB* enhances production of pediocin PA-1 in naturally resistant *Lactococcus lactis* strains. J Dairy Sci 91:2591–2594

Axelsson L, Katla T, Bjørnslett M, Eijsink VGH, Holck A (1998) A system for heterologous expression of bacteriocins in *Lactobacillus sake*. FEMS Microbiol Lett 168:137–143

Basanta A, Herranz C, Gutiérrez J, Criado R, Hernández PE, Cintas LM (2009) Development of bacteriocinogenic strains of *Saccharomyces cerevisae* heterologously expressing and secreting the leaderless enterocin L50 peptides L50A and L50B from *Enterococcus faecium* L50. Appl Environ Microbiol 75:2382–2392

Beaulieau L, Groleau D, Miguez CB, Jetté JF, Aomari H, Subirade M (2005) Production of pediocin PA-1 in the methylotrophic yeast *Pichia pastoris* reveals unexpected inhibition of its biological activity due to the presence of collagen-like material. Prot Exp Purif 43:11–125

Bermúdez-Humarán LG, Rihakova J, Langella P, Demerova K, Nazef L, Prévost H, Drider D (2007) Antimicrobial activity of divercin RV41 produced and secreted by *Lactococcus lactis*. J Mol Microbiol Biotechnol 13:259–263

Biet F, Berjeaud JM, Worobo RW, Cenatiempo Y, Fremaux C (1998) Heterologous expression of the bacteriocin mesentericin Y105 using the dedicated transport system and the general secretory pathway. Microbiology 144:2845–2854

Böer E, Steinborn G, Kunze G, Gellissen G (2007a) Yeast expression platforms. Appl Microbiol Biotechnol 77:513–523

Böer E, Steinborn G, Matros A, Hans-Peter M, Gellissen G, Kunze G (2007b) Production of interleukin-6 in *Arxula adenninivorans*, *Hansenula polymorpha* and *Saccharomyces cerevisae* by applying the wide-range yeast vector (CoMed™) system to simultaneous comparative assessment. FEMS Yeast Res 7:1181–1187

Brockmeier U, Caspers M, Freudl R, Jockwer A, Noll T, Eggert T (2006) Systematic screening of all signal peptides from *Bacillus subtilis*: a powerful strategy in optimizing heterologous protein secretion in Gram-positive bacteria. J Mol Biol 362:393–402

Buyong N, Kok J, Luchansky JB (1998) Use of a genetically enhanced, pediocin-producing starter culture, *Lactococcus lactis* subsp. *lactis* NM217, to control *Listeria monocytogenes* in cheddar cheese. Appl Environ Microbiol 64:4842–4845

Chikindas ML, Venema K, Ledeboer AM, Venema G, Kok J (1995) Expression of lactococcin A and pediocin PA-1 in heterologous hosts. Lett Appl Microbiol 21:183–189

Choi JH, Lee SY (2004) Secretory and extracellular production of recombinant proteins using *Escherichia coli*. Appl Microbiol Biotechnol 64:625–635

Cintas LM, Casaus P, Håvarstein LS, Hernández PE, Nes IF (1997) Biochemical and genetic characterization of enterocin P, a novel *sec*-dependent bacteriocin from *Enterococcus faecium* P13 with a broad antimicrobial spectrum. Appl Environ Microbiol 63:4321–4330

Cintas LM, Casaus P, Holo H, Hernández PE, Nes IF, Håvarstein LS (1998) Enterocins L50A and L50B, two novel bacteriocins from *Enterococcus faecium* L50, are related to staphylococcal hemolysins. J Bacteriol 180:1988–1994

Cintas LM, Casaus P, Herranz C, Håvarstein LS, Holo H, Hernández PE, Nes IF (2000) Biochemical and genetic evidence that *Enterococcus faecium* L50 produces enterocins L50A and L50B, the *sec*-dependent enterocin P, and a novel bacteriocin secreted without an N-terminal extension termed enterocin Q. J Bacteriol 182:6806–6814

Cintas LM, Casaus P, Herranz C, Nes IF, Hernández PE (2001) Bacteriocins of lactic acid bacteria. Food Sci Technol Int 7:281–305

Cleveland J, Montville TJ, Nes IF, Chikindas ML (2001) Bacteriocins: safe, natural antimicrobials for food preservation. Int J Food Microbiol 71:1–20

Coffey A, Ryan M, Ross RP, Hill C, Arendt E, Schwarz G (1998) Use of a broad-host-range bacteriocin-producing *Lactococcus lactis* transconjugant as an alternative starter for salami manufacture. Int J Food Microbiol 43:231–235

Corr SC, Li Y, Riedel CU, O'Toole PW, Hill C, Gahan CGM (2007) Bacteriocin production as a mechanism for the antiinfective activity of *Lactobacillus salivarius* UCC118. Proc Natl Acad Sci U S A 104:7617–7621

Cortes-Perez NG, Poquet I, Oliveira M, Gratadoux JJ, Madsen SM, Miyoshi A, Cortier G, Azevedo P, Langella P, Bermúdez-Humarán LG (2006) Construction and characterization of a *Lactococcus lactis* strain defficient in intracellular ClpP and extracellular HtrA proteases. Microbiology 152:2611–2618

Cotter PD, Hill C, Ross RP (2005) Bacteriocins: developing innate immunity for food. Nat Rev Microbiol 3:777–788

Cotter PD, Draper LA, Lawton EM, McAuliffe O, Hill C, Ross RP (2006) Overproduction of wild-type and bioengineered derivatives of the lantibiotic lacticin 3147. Appl Environ Microbiol 72:4492–4496

Criado, R, Diep DB, Aakra Å, Gutiérrez J, Nes IF, Hernández PE, Cintas LM (2006a) Complete sequence of the enterocin Q-encoding plasmid pCIZ2 from the multiple bacteriocin producer *Enterococcus faecium* L50 and genetic characterization of enterocin Q production and immunity. J Bacteriol 72:6653–6666

Criado R, Gutiérrez J, Martín M, Herranz C, Hernández PE, Cintas LM (2006b) Immunochemical characterization of temperature-regulated production of enterocin L50 (EntL50A and EntL50B), enterocin P, and enterocin Q by *Enterococcus faecium* L50. Appl Environ Microbiol 72:7634–7643

de Ruyter PGGA, Kuipers OP, de Vos WM (1996) Controlled gene expression systems for *Lactococcus lactis* using the food-grade inducer nisin. Appl Environ Microbiol 62: 3662–3667

Deegan LH, Cotter PD, Hill C, Ross P (2006) Bacteriocins: biological tools for bio-preservation and shelf-life extension. Int Dairy J 16:1058–1071

Dezwaan DC, Mequio MJ, Littell JS, Allen JP, Rossbach S, Pybus V (2007) Purification and characterization of enterocin 62-6, a two peptide bacteriocin produced by a vaginal strain of *Enterococcus faecium*: potential significance in bacterial vaginosis. Microb Ecol Health Dis 19:241–250

Diep DB, Håvarstein LS, Nes IF (1995) A bacteriocin-like peptide induces bacteriocin synthesis in *Lactobacillus plantarum* C11. Mol Microbiol 18:631–639

Diep DB, Håvarstein LS, Nes IF (1996) Characterization of the locus responsible for the bacteriocin production in *Lactobacillus plantarum* C11. J Bacteriol 178:4472–4483

Diep DB, Skaugen M, Salehian Z, Nes IF (2007) Common mechanisms of target cell recognition and immunity to class II bacteriocins. Proc Natl Acad Sci U S A 104:2384–2389

Diep DB, Mathiesen G, Eijsink VG, Nes IF (2009) Use of lactobacilli and their pheromone-based regulatory mechanism in gene expression and drug delivery. Curr Pharm Biotechnol 10:62–73

Drider D, Fimland G, Héchard Y, McMullen LM, Prevost H (2006) The continuing history of class IIa bacteriocins. Microbiol Mol Biol Rev 70:564–582

Eijsink VGH, Axelsson D, Diep DB, Havårstein LS, Holo H, Nes IF (2002) Production of class II bacteriocins by lactic acid bacteria: an example of biological warfare and communication. Antonie van Leeuwenhoek 81:369–654

European Food Safety Authority (EFSA) (2007) Opinion of the Scientific Committee on a request from EFSA on the introduction of a Qualified Presumption of Safety (QPS) approach for assessment of selected microorganisms referred to EFSA. EFSA J 587:1–16

Fekkes P, Driessen AJ (1999) Protein targeting to the bacterial cytoplasmic membrane. Microbiol Mol Biol Rev 63:161–173

Fernández M, Martínz-Bueno M, Martín MC, Valdivia E, Maqueda M (2007) Heterologous expression of enterocin AS-48 in several strains of lactic acid bacteria. J Appl Microbiol 102:1350–1361

Fimland G, Johnsen L, Dalhus B, Nissen-Meyer J (2005) Pediocin-like antimicrobial peptides (class IIa bacteriocins) and their immunity proteins: biosynthesis, structure, and mode of action. J Pept Sci 11:688–696

Flynn S, van Synderen D, Thornton GM, Holo H, Nes IF, Collins JK (2002) Characterization of the genetic locus responsible for the production of ABP-118, a novel bacteriocin produced by the probiotic bacterium *Lactobacillus salivarius* subsp. *salivarius* UCC118. Microbiology 148:973–984

Foulquié-Moreno MR, Sarantinopoulos P, Tsakalidou E, de Vuyst L (2006) The role and application of enterococci in food and health. Int J Food Microbiol 106:1–24

Franz CMAP, Worobo RW, Quadri LEN, Schillinger U, Holzapfel WM, Vederas JC, Stiles ME (1999) Atypical genetic locus associated with constitutive production of enterocin B by *Enterococcus faecium* BFE 900. Appl Environ Microbiol 65:2170–2178

Franz CMAP, Stiles ME, Schleifer KH, Holzapfel WH (2003) Enterococci in foods – a conundrum for food safety. Int J Food Microbiol 88:105–122

Franz CMAP, van Belkum MJ, Holzapfel WH, Abriouel H, Gálvez A (2007) Diversity of enterococcal bacteriocins and their grouping in a new classification scheme. FEMS Microbiol Rev 31:293–310

Gajic O, Buist G, Kojic M, Topisirovic L, Kuipers OP, Kok J (2003) Novel mechanism of bacteriocin scretion and immunity carried out by lactococcal multidrug resistance proteins. J Biol Chem 278:34291–34298

Gálvez A, Abriouel H, Lucas-López R, Omar NB (2007) Bacteriocin-based strategies for food biopreservation. Int J Food Microbiol 120:51–70

Gellissen G, Kunze G, Gaillardin C, Cregg JM, Berardi E, Veenhuis M, van der Klei I (2005) New yeast expression platforms based on methylotrophic *Hansenula polymorpha* and *Pichia pastoris* and on dimorphic *Arxula adeninivorans* and *Yarrowia lipolytica*: a comparison. FEMS Yeast Res 5:1079–1096

Gibbs GM, Davidson BE, Hillier AJ (2004) Novel expression system for large-scale production and purification of recombinant class IIa bacteriocins and its application to piscicolin 126. Appl Environ Microbiol 70:3292–3297

Gillor O, Etzion A, Riley MA (2008) The dual role of bacteriocins as anti- and probiotics. Appl Microbiol Biotechnol 81:591–606

Gutiérrez J, Criado R, Citti R, Martín M, Herranz C, Nes IF, Cintas LM, Hernández PE (2005a) Cloning, production and functional expression of enterocin P, a Sec-dependent bacteriocin produced by *Enterococcus faecium* P13, in *Escherichia coli*. Int J Food Microbiol 103:239–250

Gutiérrez J, Bourque D, Criado R, Choi YJ, Cintas LM, Hernández PE, Míguez CB (2005b) Heterologous extracellular production of enterocin P from *Enterococcus faecium* P13 in the methylotrophic bacterium *Methylobacterium extorquens*. FEMS Microbiol Lett 248:125–131

Gutiérrez J, Criado R, Martín M, Herranz C, Cintas LM, Hernández PE (2005c) Production of enterocin P, an antilisterial pediocin-like bacteriocin from *Enterococcus faecium* P13, in *Pichia pastoris*. Antimicrob Agents Chemother 49:3004–3008

Gutiérrez J, Larsen R, Cintas LM, Kok J, Hernández PE (2006) High-level heterologous production and functional expression of the Sec-dependent bacteriocin from *Enterococcus faecium* P13, in *Lactococcus lactis*. Appl Microbiol Biotechnol 72:41–51

Håvarstein LS, Diep DB, Nes IF (1995) A family of bacteriocin ABC transporters carry out proteolytic processing of their substrates concomitantly with export. Mol Microbiol 16:229–240

Heng NCK, Tagg JR (2006) What's in a name? Class distinction for bacteriocins. Nat Rev Microbiol 4, doi:10.1038/nrmicro1273-c1.

Henrich B, Klein JR, Weber B, Delorme C, Renault P, Wegmann U (2002) Food-grade delivery system for controlled gene expression in *Lactococcus lactis*. Appl Environ Microbiol 68:5429–5436

Herranz C, Driessen AJM (2005) *Sec*-mediated secretion of enterocin P by *Lactococcus lactis*. Appl Environ Microbiol 71:1959–1963

Horn N, Martínez MI, Martínez JM, Hernández PE, Gasson MJ, Rodríguez JM, Dodd H (1998) Production of pediocin PA-1 by lactococcus lactis using the lactococcin A secretory apparatus. Appl Environ Microbiol 64:818–823

Horn N, Martínez MI, Martínez JM, Hernández PE, Gasson MJ, Rodríguez JM, Dodd H (1999) Enhanced production of pediocin PA-1 and coproduction of nisin and pediocin PA-1 by *Lactococcus lactis*. Appl Environ Microbiol 65:4443–4450

Horn N, Fernández A, Dodd HM, Gasson MJ, Rodríguez JM (2004) Nisin-controlled production of pediocin PA-1 and colicin V in nisin- and non- nisin-producing *Lactococcus lactis* strains. Appl Environ Microbiol 70:5030–5032

Ingham AB, Sproat KW, Tizard MLV, Moore RJ (2005) A versatile system for the expression of nonmodified bacteriocins in *Escherichia coli*. J Appl Microbiol 98:676–683

Ito Y, Kawai Y, Arakawa K, Honme Y, Sasaki T, Saito T (2009) Conjugative plasmid of *Lactobacillus gasseri* LA39 that carries genes for peoduction of an immunity to the circular bacteriocin gassericin A. Appl Environ Microbiol 75:6340–6351

Jasniewski J, Cailliez-Grimal C, Gelhaye E, Revol-Junelles AM (2008) Optimization of the production and purification processes of carnobacteriocins Cbn BM1 and Cbn B2 from *Carnobacterium maltaromaticum* CP5 by heterologous expression in *Escherichia coli*. J Microbiol Methods 73:41–48

Kang BS, Seo J-G, Lee G-S, Kim J-H et al (2009) Antimicrobial activity of enterocins from *Enterococcus faecalis* SL-5 against *Propionibacterium acnes*, the causative agent in acne vulgaris, and its therapeutic effect. J Microbiol 47:101–109

Kawai Y, Kusnadi J, Kemperman R, Kok J, Ito Y, Endo M, Arakawa K, Uchida H, Nishimura J, Kitazawa H, Saito T (2009) DNA sequencing and homologous expression of a small peptide conferring immunity to gassericin A, a circular bacteriocin produced by *Lactobacillus gasseri* LA39. Appl Environ Microbiol 75:1324–1330

Kemperman R, Jonker M, Nauta A, Kuipers OP, Kok J (2003) Functional analysis of the gene cluster involved in production of the bacteriocin circularin A by *Clostridium beijirinckii* ATCC 25752. Appl Environ Microbiol 69:5839–5848

Klaenhammer TR (1993) Genetics of bacteriocins produced by lactic acid bacteria. FEMS Microbiol Rev 12:39–85

Klostermann K, Crispie F, Flynn J, Ross RP, Hill C, Meaney W (2008) Intramamary infusion of a live culture of *Lactococcus lactis* for treatment of bovine mastitis: comparaison with antibiotic treatment in field trials. J Dairy Res 75:365–373

Knoetze H, Todorov SD, Dicks LMT (2008) A class IIa peptide from *Enterococcus mundtii* inhibits bacteria associated with otitis media. Int J Antimicrob Agents 31:228–234

Kuipers OP, Beerthuyzen MM, de Ruyter PG, Luesink EJ, de Vos WM (1995) Autoregulation of nisin biosynthesis in *Lactococcus lactis* by signal transduction. J Biol Chem 270:27299–27304

Kuipers A, Wierenga J, Rink R, Kluskens LD, Driessen AJM, Kuipers OP, Moll GN (2006) Sec-mediated transport of posttranslationally dehydrated peptides in *Lactococcus lactis*. Appl Environ Microbiol 72:7626–7633

Le Loir Y, Nouialle S, Commissaire L, Bretign L, Gruss A, Langella P (2001) Signal peptide and propeptide optimization for heterologous protein secretion in *Lactococcus lactis*. Appl Environ Microbiol 67:4119–4127

Leenhouts K, Bolhuis A, Venema G, Kok J (1998) Construction of a food-grade multiple-copy integration system for *Lactococcus lactis*. Appl Microbiol Biotechnol 49:417–423

Liu CQ, Su P, Khunajakr N, Deng YM, Sumual S, Kim WS, Tandianus JE, Dunn NW (2005) Development of food-grade cloning and expression vectors for *Lactococcus lactis*. J Appl Microbiol 98:127–135

Liu L, O'Conner P, Cotter PD, Hill C, Ross RP (2008) Controlling *Listeria monocytogenes* in Cottage cheese through heterologous production of enterocin A by *Lactococcus lactis*. J Appl Microbiol 104:1059–1066

Llull D, Poquet I (2004) New expression system tightly controlled by zinc availability in *Lactococcus lactis*. Appl Environ Microbiol 70:5398–5406

Maqueda M, Gálvez A, Sánchez-Barrena MJ, González C, Albert A, Rico M, Valdivia E (2004) Peptide AS-48: Prototype of a new class of cyclin bacteriocins. Curr Prot Pept Sci 5:399–416

Maqueda M, Sánchez-Hidalgo M, Fernández M, Montalbán-López M, Valdivia E, Martínez-Bueno M (2008) Genetic features of circular bacteriocins produced by Gram-positive bacteria. FEMS Microbiol Rev 32:2–22

Martín M, Gutiérrez J, Criado R, Herranz C, Cintas LM, Hernández PE (2007a) Chimeras of mature pediocin PA-1 fused to the signal peptide of enterocin P permits the cloning, production, and expression of pediocin PA-1 in *Lactococcus lactis*. J Food Prot 70:2792–2798

Martín M, Gutiérrez J, Criado R, Herranz C, Cintas LM, Hernández PE (2007b) Cloning, production and expression of the bacteriocin enterocin A produced by *Enterococcus faecium* PLBC21 in *Lactococcus lactis*. Appl Microbiol Biotechnol 76:667–675

Martínez JM, Kok J, Sanders W, Hernández PE (2000) Heterologous co-production of enterocin A and pediocin PA-1 by *Lactococcus lactis*: detection by specific peptide-directed antibiotics. Appl Environ Microbiol 66:3543–3549

Martin-Visscher LA, van Belkum MJ, Garneau-Tsodikova S, Whittal RM, Zheng J, McMullen L, Vederas JC (2008) Isolation and characterization of carnocyclin A, a novel circular bacteriocin produced by *Carnobacterium maltaromaticum* UAL307. Appl Environ Microbiol 74:4756–4763

Martin-Visscher LA, Gong X, Duszyk M, Vederas JC (2009) The 3D structure of carnocyclin A reveals that many circular bacteriocins share a common structural motif. J Biol Chem 284:28674–28681

Mathiesen G, Sveen A, Piard JC, Axelsson L, Eijsink VGH (2008) Heterologous protein secretion by *Lactobacillus plantarum* using homologous signal peptides. J Appl Microbiol 105:215–226

McCormick JK, Worobo RW, Stiles ME (1996) Expression of the antimicrobial peptide carnobacteriocin B2 by a signal peptide-dependent general secretory pathway. Appl Environ Microbiol 62:4095–4099

McCormick JK, Klaenhammer TR, Stiles ME (1999) Colicin V can be produced by lactic acid bacteria. Lett Appl Microbiol 29:37–41

Mierau I, Kleerebezem M (2005) 10 years of the nisin-controlled gene expression system (NICE) in *Lactococcus lactis*. Appl Microbiol Biotechnol 68:705–717

Miller KW, Schamber R, Chen Y, Ray B (1998) Production of an active chimeric pediocin AcH in *Escherichia coli* in the absence of processing and secretion genes from the *Pediococcus* pap operon. Appl Environ Microbiol 64:14–20

Moon GS, Pyun YR, Park MS, Ji GE, Kim WJ (2005) Secretion of recombinant pediocin PA-1 by *Bifidobacterium longum*, using the signal sequence for bifidobacterial α-amylase. Appl Environ Microbiol 71:5630–5632

Morello E, Bermúdez-Humarán LG, Llull D, Solé V, Miraglio N, Langella P, Poquet I (2008) *Lactococcus lactis*, an efficient cell factory for recombinant protein production and secretion. J Mol Microbiol Biotechnol 14:48–8

Nes IF, Diep DB, Håvarstein LS, Brurberg MB, Eijsink V, Holo H (1996) Biosynthesis of bacteriocins in lactic acid bacteria. Antonie Van Leeuwenhoek 70:113–128

Nes IF, Diep DB, Holo H (2007) Bacteriocin diversity in *Streptococcus* and *Enterococcus*. J Bacteriol 189:1189–1198

Neysens P, Messens W, Gevers D, Swings J, de Vuyst L (2003) Biphasic kinetics of growth and bacteriocin production with *Lactobacillus amylovorus* DCE 471 occur under stress conditions. Microbiology 149:1073–1082

Nigutova K, Serencova L, Piknova M, Javorsky P, Pristas P (2008) Heterologous expression of functionally active enterolysin A, a class III bacteriocin from *Enterococcus faecalis*, in *Escherichia coli*. Prot Exp Purif 60:20–24

O'Connor EB, O'Riordan B, Morgan SM, O'Mullane DM, Ross RP, Hill C (2006) A lacticin 3147 enriched food ingredient reduces *Streptococcus mutans* isolated from the human oral cavity in saliva. J Appl Microbiol 100:1251–1260

O'Keeffe T, Hill C, Ross RP (1999) Chracterization and heterologous expression of the genes encoding enterocin A production, immunity, and regulation in *Enterococcus faecium* DPC1146. Appl Environ Microbiol 65:1506–1515

Parvez S, Malik KA, Ah Kang S, Kim HY (2006) Probiotics and their fermented food products are benefitial for health. J Appl Microbiol 100:1171–1185

Reviriego C, Fernández L, Rodríguez JM (2007) A food-grade system for production of pediocin PA-1 in nisin-producing and non-nisin-producing *Lactococcus lactis* strains: application to inhibit *Listeria* growth in a cheese model system. J Food Prot 70:2512–2517

Richard C, Drider D, Elmorjani K, Marion D, Prévost H (2004) Heterologous expression and purification of active divercin V41, a class IIa bacteriocin encoded by a synthetic gene in *Escherichia coli*. J Bacteriol 186:4276–4284

Rodríguez JM, Martínez MI, Horn N, Dodd HM (2002) Heterologous production of bacteriocins by lactic acid bacteria. Int J Food Microbiol 80:101–116

Rodríguez E, Calzada J, Arqués JL, Rodríguez JM, Nuñez M, Medina M (2005) Antimicrobial activity of pediocin-producing *Lactococcus lactis* on *Listeria monocytogenes*, *Staphylococcus aureus* and *Escherichia coli* O157:H7 in cheese. Int Dairy J 15:51–57

Rud I, Jensen PR, Naterstad K, Axelsson L (2006) A synthetic promoter library for constitutive gene expression in *Lactobacillus plantarum*. Microbiology 152:1011–1019

Sánchez J, Diep DB, Herranz C, Nes IF, Cintas LM, Hernández PE (2007a) Amino acid and nucleotide sequence, adjacent genes, and heterologous expression of hiracin JM79, a *sec*-dependent bacteriocin produced by *Enterococcus hirae* DCH5, isolated from Mallard ducks (*Anas platyrhynchos*). FEMS Microbiol Lett 270:227–236

Sánchez J, Basanta A, Gómez-Sala B, Herranz C, Cintas LM, Hernández PE (2007b) Antimicrobial and safety aspects, and biotechnological potential of bacteriocinogenic enterococci isolated from Mallard ducks (*Anas platyrhynchos*). Int J Food Microbiol 117:295–305

Sánchez J, Borrero J, Gómez-Sala B, Basanta A, Herranz C, Cintas LM, Hernández PE (2008) Cloning and heterologous production of hiracin JM79, a Sec-dependent bacteriocin produced by *Enterococcus hirae* DCH5, in lactic acid bacteria and *Pichia pastoris*. Appl Envriron Microbiol 74:2471–2479

Sánchez-Hidalgo M, Maqueda M, Gálvez A, Abriouel H, Valdivia E, Martínez-Bueno M (2003) The genes coding enterocin EJ97 production by *Enterococcus faecalis* EJ97 are located on a conjugative plasmid. Appl Environ Microbiol 69:1633–1641

Schoeman H, Vivier MA, Du Toit M, Dicks LMT, Pretorius IS (1999) The development of bactericidal yeast strains by expressing the *Pediococcus acidilactici* pediocin gene (*pedA*) in *Saccharomyces cerevisae*. Yeast 15:647–656

Shin MS, Han SK, Ji AR, Kim KS, Lee WK (2008) Isolation and characterization of bacteriocin-producing bacteria from the gastrointestinal tract of broiler chickens for probiotic use. J Appl Microbiol 105:2203–2212

Simoes-Barbosa A, Abreu H, Silva Neto A, Gruss A, Langella P (2004) A food-grade delivery system for *Lactococcus lactis* and evaluation of inducible gene expression. Appl Microbiol Biotechnol 65:61–67

Sit CS, Vederas JC (2008) Approaches to the discovery of new antibacterial agents based on bacteriocins. Biochem Cell Biol 86:116–123

Sørvig E, Mathiesen G, Naterstad K, Eijsink VGH, Axelsson L (2005) High-level inducible expression in *Lactobacillus sakei* and *Lactobacillus plantarum* using versatile expression vectors. Microbiology 151:2439–2449

Sridhar VR, Smaianov VV, Steele JL (2006) Construction and evaluation of food-grade vectors for *Lactococcus lactis* using aspartate aminotransferase and α-galactocidase as selectable markers. J Appl Microbiol 101:161–171

Sriraman K, Jayaraman G (2008) HtrA is essential for efficient secretion of recombinant proteins by *Lactococcus lactis*. Appl Environ Microbiol 74:7442–7446

Steinborn G, Böer E, Scholz A, Tag K, Kunze G, Gellissen G (2006) Application of a wide-range yeast vector (CoMed™) system to recombinant protein production in dimorphic *Arxula adeninivorans*, methylotrophic *Hansenula polymorpha* and other yeasts. Microbial Cell Factories 5:33

Takala TM, Saris PEJ (2002) A food-grade cloning vector for lactic acid bacteria based on the nisin immunity gene *nisI*. Appl Microbiol Biotechnol 59:467–471

Van Belkum MJ, Worobo RW, Stiles ME (1997) Double-glycine-type leader peptides direct secretion of bacteriocins by ABC transporters: colicin V secretion in *Lactococcus lactis*. Mol Microbiol 23:1293–1301

Van der Vossen JMBM, van der Lelie D, Venema G (1987) Isolation and characterization of *Streptococcus cremoris* Wg2-specific promoters. Appl Environ Microbiol 53:2452–2457

Van Reenen CA, Chikindas ML, Van Zyl WH, Dicks LMT (2002) Characterization and heterologous expression of a class IIa bacteriocin, plantaricin 423 from *Lactobacillus plantarum* 423, in *Saccharomyces cerevisae*. Int J Food Microbiol 81:29–40

Van Wely KHM, Swaving J, Freudl R, Driessen AJM (2001) Translocation of proteins across the cell envelope of Gram-positive bacteria. FEMS Microbiol Rev 25:437–454

Venema K, Kok J, Marugg JD, Toonen A, Ledeboer AM, Venema G, Chikindas ML (1995) Functional analysis of the pediocin operon of *Pediococcus acidilactici* PAC1.0: PedB is the immunity protein and PedD is the precursor processing enzyme. Mol Microbiol 17:515–522

Wachsman MB, Farías ME, Takeda E, Sesma F, de Ruíz Holgado AP, Torres RA, Coto CE (1999) Antiviral activity of enterocin CRL35 against herpesviruses. Int J Antimicrob Agents 12:293–299

Wachsman MB, Castilla V, de Ruiz Holgado AP, de Torres RA, Sesma F, Coto CE (2003) Enterocin CRL35 inhibits late stages of HSV-1 and HSV-2 replication in vitro. Antiviral Res 58:17–24

Worobo RW, van Belkum MJ, Sailer M, Roy KL, Vederas JC, Stiles ME (1995) A signal peptide secretion-dependent bacteriocin from *Carnobacterium divergens*. J Bacteriol 177:3143–3149

Yildirim S, Konrad D, Calvez S, Drider D, Prevost H, Lacroix C (2007) Production of recombinant bacteriocin divercin V41 by high cell density *Escherichia coli* batch and fed-batch cultures. Appl Microbiol Biotechnol 77:525–553

Part IV
Genetics, Biosynthesis, Structure, and Mode of Action of AMP from Gram-Positive Bacteria

Chapter 9
Genetics, Biosynthesis, Structure, and Mode of Action of Lantibiotics

Anneke Kuipers, Rick Rink, and Gert N. Moll

Abstract Lantibiotics are lanthionine-containing peptide antibiotics. They are characterized by having *meso*-lanthionine(s) and/or β-methyllanthionine(s) or both. These intramolecular monosulfide cross-links render the peptide resistant against breakdown by peptidases. Moreover, in several cases, the (methyl)lanthionines are essential for interaction with the so-called docking molecule lipid II. The best known lantibiotic, nisin, highly effectively inhibits growth of target cells via two mechanisms: (1) abduction of the cell wall precursor lipid II from the septum and (2) formation of pores composed of lipid II and nisin. (Methyl)lanthionines result from two enzyme-catalyzed posttranslational modifications: dehydration of serines/threonines and coupling of the resulting dehydro amino acids to cysteines. Besides the localization of the thioether bridges and dehydro amino acids in the lantibiotics, also the three-dimensional structure of some lantibiotics has been resolved by NMR. Genes encoding proteins involved in the biosynthesis of lantibiotics are present in clusters and may comprise combinations of the following genes in varying order: a structural gene that encodes a leader peptide and the lantibiotic propeptide, modification enzyme(s), a transporter responsible for the export of the lantibiotic and in some cases for cleavage of the leader peptide, a leader peptidase, a so-called immunity protein involved in self-protection of the host cell, components of a transporter also involved in self-protection, and two components of an autoinduction system.

Introduction

The name lantibiotics was introduced more than two decades ago (Schnell et al. 1988). Lantibiotics are ribosomally produced dehydroresidue- and (methyl)lanthionine-containing peptides (Fig. 9.1). Lanthionines are thioether-bridged amino acids

---

G.N. Moll (✉)
BiOMaDe Technology Foundation, Nijenborgh 4, 9747 Groningen, The Netherlands
e-mail: Moll@biomade.nl

D. Drider and S. Rebuffat (eds.), *Prokaryotic Antimicrobial Peptides: From Genes to Applications*, DOI 10.1007/978-1-4419-7692-5_9,

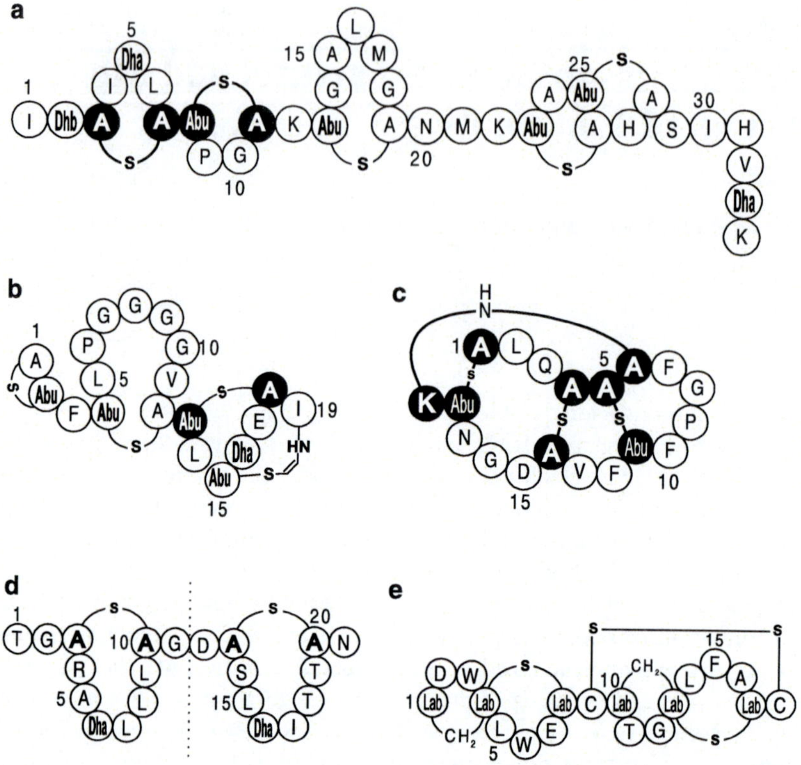

Fig. 9.1 Examples of lantibiotics. (**a**) nisin, (**b**) mersacidin, (**c**) duramycin, (**d**) sapB and (**e**) labyrinthopeptin A2. (Methyl)lanthionines and lysinoalanine in black are conserved within structural groups termed A, B, and C distinguished by Rink et al. (2005). The vertical dotted line indicates local structural symmetry in this morphogenetic lantibiotic (D). "Lab" in Figure E is labionine

(Fig. 9.2a). They are predominantly produced by Gram-positive bacteria, and those with antibiotic activity are principally effective against Gram-positive bacteria. Besides *meso*-lanthionine (Ala-S-Ala) and β-methyllanthionine, several other post-translational modifications may occur in lantibiotics (Table 9.1).

Nisin was the first lantibiotic described in literature (Rogers and Whittier 1928) and is the most studied lantibiotic. It is produced by different *Lactococcus lactis* strains. Already in 1969, nisin was approved for use as a food preservative (Delves-Broughton 2005). Nisin has a broad activity spectrum against Gram-positive bacteria, including strains of *Staphylococcus*, *Streptococcus*, *Micrococcus*, *Lactobacillus*, *Bacillus*, *Listeria*, and *Clostridium*, (Thomas et al. 2000), and has antimicrobial activity in the nanomolar range (de Vos et al. 1993). Despite its world wide application as a food additive, virtually no resistance against nisin has occurred. Owing to their stability, high activity, and virtual absence of resistance development, lantibiotics are promising candidates for biomedical application. Their features make the

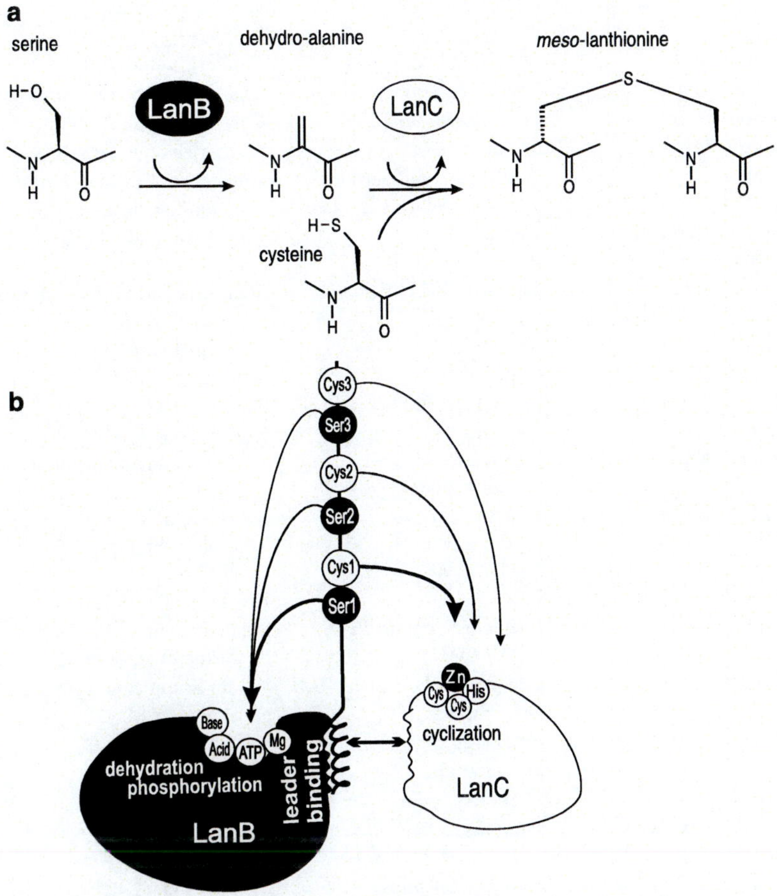

Fig. 9.2 (**a**) Enzyme-catalyzed dehydration and cyclization. (**b**) Hypothetical distributive mechanism of lantibiotic modification. A leader peptide binds to the modification enzyme, thus bringing the modifiable residues in the vicinity of the catalytic site. Due to the different distances to the catalytic site, residues 1, 2, and 3 will most frequently be modified in the order 1, 2, and 3, but all other orders may also occur. Dehydratase and cylase activities may alternate (Kuipers et al. 2008) due to vicinity of the leader-binding sites and or due to vicinity of the catalytic sites

search for new lantibiotic variants that are further improved by genetic engineering an exciting and relevant approach in the battle against multiple-antibiotic-resistant pathogens (Kuipers et al. 1992; Rink et al. 2007b; Field et al. 2008).

Nisin, subtilin, epidermin, Pep5, and some similar lantibiotics were first designated as type A lantibiotics, which are rod-shaped, flexible with an elongated structure, and mainly act by forming pores in the bacterial membrane (Jung 1991). Type B lantibiotics (e.g., cinnamycin, duramycin, and ancovenin) were discerned as having a higher degree of cyclization resulting in structures that are more globular and being devoid of pore-forming activity. To date, more than 60 different lantibiotics

Table 9.1 Residues occurring within lantibiotics

Residue	Lantibiotic	Reference
meso-Lanthionine	Most lantibiotics	Chatterjee et al. (2005b)
β-Methyllanthionine	Most lantibiotics	Chatterjee et al. (2005b)
2,3-Dehydroalanine	Most lantibiotics	Chatterjee et al. (2005b)
(Z)-2,3-Dehydrobutyrine	Most lantibiotics	Chatterjee et al. (2005b)
S-(Z)-Aminovinyl-D-cysteine	Epidermin, gallidermin, cypemycin	Allgaier et al. (1986), Kellner et al. (1988), Minami et al. (1994)
S-Aminovinyl-D-3-methylcysteine	Mersacidin	Chatterjee et al. (1992)
D-Alanine	Lactocin S, lactocin 3147	Skaugen et al. (1994); Martin et al. (2004)
Alloisoleucine	Cypemycin	Minami et al. (1994)
Dimethylalanine	Cypemycin	Minami et al. (1994)
Amide	Nisin(1-32)amide	Chan et al. (1989)
Erythro-3-hydroxy-L-aspartic acid	Cinnamycin, duramycin, duramycin B, duramycin C	Kessler et al. (1988); Naruse et al. (1989); Fredenhagen et al. (1991)
(2S, 8S)-Lysinoalanine	Cinnamycin, duramycin, duramycin B, duramycin C	Kessler et al. (1988); Naruse et al. (1989); Fredenhagen et al. (1991)
2-Oxobutyryl	Pep5	Kellner et al. (1989)
2-Oxopropionyl	Lactocin S, epilancin K7 (3-31)	Skaugen et al. (1994); van de Kamp et al. (1995a)
Hydroxypropionyl	Epilancin K7	van de Kamp et al. (1995b)
Acetyl	Paenibacillin	He et al. (2008)
Sulfoxide	Actagardine	Boakes et al. (2009)
disulfide	Thermophilin 1277	Kabuki et al. (2009)
5-Chloro-trypthopan	Microbisporicin	Castiglione et al. (2008)
Mono-/bis-hydroxylated proline	Microbisporicin	Castiglione et al. (2008)
Labionine	Labyrinthopeptins	Meindl et al. (2010)

have been discovered (Bierbaum and Sahl 2009), and several different posttranslational modifications have been described (Willey and van der Donk 2007; Table 1). After the finding of many new lantibiotics, the clarity of the old type A and type B classification diminished, and two (Oman and van der Donk 2010) or three (Pag and Sahl 2002; Willey and van der Donk 2007) groups were proposed on the basis of novel criteria. The three-group classification placed all lantibiotics in one of three classes initially on the basis of the biosynthesis machinery used for maturation of the peptide (class I and class II) or the absence of antibiotic activity (class III). Moreover, in class III lantibiotics, LanM maturation enzymes might act via a mechanism distinct from that of class II LanM enzymes.

Several detailed reviews on all lantibiotics exist (Chatterjee et al. 2005b; Bierbaum and Sahl 2009; Asaduzzaman and Sonomoto 2009) and also reviews on clinical applications of lantibiotics (Smith and Hillman 2008), on subgroups of lantibiotics (Lawton et al. 2007; Dufour et al. 2007; Willey et al. 2006), on a single lantibiotic (Lubelski et al. 2008), and on the application of lantibiotic enzymes for modifying nonlantibiotic peptides (Moll et al. 2010). In this chapter, we focus on the genetics, synthesis, structure, and mode of action of lantibiotics including the most recent developments.

Genetics and Biosynthesis

The genes involved in lantibiotic synthesis are arranged in clusters. These gene clusters can be organized on a transposon (nisin), on the chromosome (subtilin), or on a plasmid (epidermin). Genes on these clusters have been designated the generic locus symbol *lan* (de Vos et al. 1991). In 1993, the gene cluster involved in the nisin biosynthesis was unraveled (van der Meer et al. 1993, Kuipers et al. 1993a). Besides the gene products required for the biosynthesis of the peptides, proteins that are needed for the processing (LanP), translocation (LanT), self-protection/immunity (LanI, LanEFG), and regulation (LanRK) are also encoded. Per type, many of these proteins encoded on the different gene clusters show amino acid homology, which indicates that indeed they have similar functions (Siezen et al. 1996; Kuipers et al. 1993a; Qiao et al. 1996).

The lantibiotics nisin, epidermin, and Pep5 belong to the class I lantibiotics. In class I lantibiotics, the prepeptide LanA is modified by two distinct enzymes, LanB and LanC. The lantibiotic prepeptide contains a leader sequence that is thought to be necessary for targeting the propeptide part to the separate modifying, processing, and translocating enzymes. LanB dehydrates the serines and threonines in the propeptide part of LanA, and LanC couples these dehydrated residues regio- and stereoselectively to cysteines to form respectively *meso*lanthionines and β-methyllanthionines. After translocation of the modified peptide via an ABC transporter LanT, the leader part is, in most class I lantibiotic systems, removed by a protease LanP, releasing the active lantibiotic (Willey and van der Donk 2007).

In class II lantibiotics, e.g., lacticin 481, mersacidin, and actagardin, only one enzyme is responsible for dehydration and cyclization of the propeptide LanA. These bifunctional LanM enzymes as well as LanB enzymes are both composed of about 1000 amino acids and share no sequence homology. The N-terminal parts of LanM enzymes are responsible for the dehydration reaction, but surprisingly, their sequences are about 400 amino acids shorter than those of LanB dehydratases, and they have no similarity to LanB enzymes. LanC enzymes are composed of around 400 amino acids. The C-terminal part of LanM enzymes has low sequence homology with the LanC enzymes (Siezen et al. 1996), including three zinc- coordinating amino acids (Patton and van der Donk 2005). Knockouts of one of these zinc ligands completely abolished the cyclase activity of NisC or LctM (Li et al. 2006, Paul et al. 2007). Another dissimilarity to class I lantibiotics is the dual functionality of LanT. Before translocation of the modified peptide, the peptide is intracellularly processed by the conserved N-terminal protease part of LanT (Pag and Sahl 2002, Willey and van der Donk 2007). Class II lantibiotics also comprises the two-component lantibiotics, e.g., lacticin 3147 (Ryan et al. 1996) and haloduracin (McClerren et al. 2006). In the case of lacticin 3147, the two prepeptides LtnAα and LtnAβ are each separately modified by the two corresponding enzymes, respectively, LtnM1 and LtnM2. After modification, both peptides are processed and translocated by one LtnT enzyme. The gene cluster of lacticin 3147 contains a gene coding for an additional posttranslational modification enzyme, LtnJ. This enzyme converts some dehydroalanines in the prepeptides Ltnα and Ltnβ to D-alanines (Ryan et al. 1999).

The third class of lantibiotics was initially discerned as (methyl)lanthionine-containing peptides often devoid of antimicrobial activity. Instead, they would have other –for instance, morphogenetic– features that may be beneficial to the producing cells. The three lantibiotics that were first found for this group were SapB (Kodani et al. 2004; Fig. 9.1d), SapT (Kodani et al. 2005), and AmfS (Ueda et al. 2002). SapB and SapT are believed to be biosurfactants that may have a positive effect on the surface of aerial hyphae of the producer strains. In addition, SapT has antimicrobial activity against *Bacillus cereus* (Kodani et al. 2005). Interestingly, RamC, the presumed SapB modification enzyme has an N-terminal domain that resembles a Ser/Thr kinase and a central dimerization domain (Kodani et al. 2004). Furthermore, the enzymes involved in the biosynthesis of SapB and AmfS share homology with the C-terminal part of other LanM enzymes except for the zinc ligands, which are not conserved.

Three recent and exciting publications further characterized class III lantibiotics. Labyrinthopeptins were discovered by Meindl and coworkers by following the activity of labyrinthopeptin A2 against *Herpex simplex* virus. In addition, labyrinthopeptin A2 has an excellent efficacy against neuropathic pain in an in vivo mouse model. Labyrinthopeptins belong to class III lantibiotics and contain labionine, a carbacyclic triamino acid (Meindl et al. 2010; Fig. 9.1e). Subsequently, the in vitro reconstitution of the prelabyrinthopeptin A2 biosynthesis was demonstrated, which required guanosine triphosphate for the phosphorylation and dehydratation reaction of serines (Müller et al. 2010). A paper from the group of van der Donk demonstrated that a class III enzyme LanL comprises a kinase domain, which phosphorylates Ser/Thr, a phosphoSer/Thr lyase domain, and a cyclase domain comprising a zinc finger. It was proposed that LanL enzymes have evolved from stand-alone protein Ser/Thr kinases, phospho-Ser/Thr lyases, and enzymes catalyzing thiol alkylation. The name lantipeptides was suggested for compounds that by structure and biosynthesis are related to lantibiotics but that are not known to possess antimicrobial activity (Goto et al. 2010).

Engineering of Lantibiotics

With the elucidation of gene clusters involved in the biosynthetic pathways of lantibiotics, genetic engineering of lantibiotics became the next challenge. Much more studies have been performed on the engineering of the lantibiotics than on the mutagenesis of their leader peptides. The existence of natural variants among lantibiotics (e.g., nisin

A/nisin Z, nisin Q (Fukao et al. 2008), nisin U (Wirawan et al. 2006), epidermin/gallidermin) and the high homology between certain lantibiotics (i.e., nisin/subtilin/lantibiotic 97518 (Maffioli et al. 2009), mutacin II/lacticin 481) suggest that the identity of amino acids present at certain locations is flexible. Indeed, by site-directed mutagenesis of the structural genes and the development of expression systems, many lantibiotic variants were designed and produced in vivo (Cotter et al. 2005).

In 1992, the first nisin mutants were reported (Kuipers et al. 1992) followed by many other nisin mutants (reviewed by: Kuipers et al. 1996; Lubelski et al. 2008). Interestingly, a T2S mutant had increased activity (Kuipers et al. 1996) and some nisin hinge region mutants had antimicrobial activity against Gram-negative species (Yuan et al. 2004). By altering the charge of the nisin lantibiotic, solubility could be improved (Yuan et al. 2004). Randomization of the codons of the amino acids in nisin's ring A and ring B yielded a large number of mutants (Rink et al. 2007b). Nisin ring A mutants I4K/S5F/L6I and I4K/L6I showed enhanced activity against some target strains (Rink et al. 2007b), as did mutations M21V, N20P, and K22T in the hinge region (Field et al. 2008). Ring A mutants were obtained with enhanced activity against some strains, mutants that were not recognized by the self-protection systems, whereas opening of ring B caused loss of antimicrobial activity while the induction capacity remained intact (Rink et al. 2007b). Early successful mutagenesis studies also concerned subtilin (Liu and Hansen 1992) and gallidermin (Ottenwälder et al. 1995).

The first novel thioether bridge was introduced in Pep5. By the substitution A19C, a β- methyllanthionine was introduced in the peptide, which was formed between the Dhb on position 16 and the introduced cysteine at position 19. This mutant exhibited increased proteolytic stability against chymotrypsin and Lys-C. However, the novel thioether bridge had a negative effect on the antimicrobial activity of Pep 5 (Bierbaum et al. 1996). Also, in the class II lantibiotics, comprising mutacin II (Chen et al. 1998), mersacidin (Szekat et al. 2003), cinnamycin (Widdick et al. 2003), and actagardine (Boakes et al. 2009), new variants were made by site-directed mutagenesis. A systematic mutant analysis by alanine scanning of the two-peptide lantibiotic lacticin 3147 revealed the areas within the peptide that are amenable to changes and areas that are essential for the production. None of the mutants displayed an antimicrobial activity higher than that of the wild-type producer (Cotter et al. 2006). Succesful methods and strategies have been developed to engineer new lantibiotic variants and analyze libraries (Cortés et al. 2009).

By random mutagenesis and NNK scanning of nukacin ISK-1, a series of nukacin ISK-1 variants was generated to identify the positional importance of individual residues responsible for antimicrobial activity (Islam et al. 2009). Furthermore, by random mutagenesis of mersacidin, 80 mutants that produced mature mersacidin at good levels were made, and novel variants were obtained with improved overall bioactivity, such as F3W (Appleyard et al. 2009).

Novel lantibiotics with enhanced antimicrobial activity may be lethal for the producer itself. Many *pre*lantibiotics are inactive; apparently, the presence of the N-terminal leader peptide keeps these prelantibiotics inactive. To avoid lethality of engineered lantibiotics, a production system can be used without leader peptidase

(Rink et al. 2007b). After production, the leader can be removed. Another approach is using an in vitro modification system. The lantibiotics lacticin 481 and the two peptide lantibiotic haloduracin were both modified successfully by incubation of the precursor peptide with the LanM enzymes in vitro (Xie et al. 2004; McClerren et al. 2006). Furthermore, NisC could successfully cyclize dehydrated prenisin (Li et al. 2006).

Lantibiotic chimeras of nisin and subtilin have been made (Chakicherla and Hansen 1995). By replacing the propeptide-encoding sequence of one lantibiotic by the propeptide-encoding sequence of another lantibiotic, lantibiotics have been produced by the biosynthesis machinery of a system within the same class (Kuipers et al 1993b; Patton et al. 2008), and also of another class. Class II pneumococcins have been dehydrated, cyclized, and exported by Class I enzymes from the nisin system (Majchrzykiewicz et al. 2010). Overall, the biosynthetic systems used for the biosynthesis of lantibiotics seem to have a remarkable flexibility.

Mechanistic Aspects of the Biosynthesis of Lantibiotics

Lantibiotic genes code for prepeptides, which are composed of a leader peptide and a modifiable propeptide. The leader peptide appears to have a leading role in the posttranslational modification processes and in export (Patton et al. 2008). It induces the LanB/LanM-catalyzed dehydration of serines and threonines, the LanC/LanM-catalyzed cyclization, and the LanT-mediated export. However, it was demonstrated that presenting the leader in trans, not attached to the substrate, still leads to modification of the structural peptide (Levengood et al. 2007; Patton et al. 2008). Some, but only strongly reduced, LctM activity has also been demonstrated in the absence of the leader peptide. Since LanB, LanC, LanM, and LanT can be active in the absence of other lantibiotic enzymes, it is likely that each enzyme itself has a leader peptide-binding site. Possible roles of the leader could thus be stabilizing of an active conformation of the enzyme (Patton et al 2008) and bringing the modifiable substrate propeptide in the vicinity of the active center of the modification enzymes (Fig. 9.2b).

Alignments of leader peptides suggest at least two groups (Plat et al. 2010). One group containing an FNLD sequence, mainly occurring in Class I, LanB and LanC-modified prelantibiotics (Chatterjee et al. 2005b; Table 9.2). Mutagenesis of residues in this box affected nisin biosynthesis (van der Meer et al. 1994) and Pep5 production (Neis et al. 1997). Another group with an EVxxxEL sequence occur in Class II, LanM-modified prelantibiotics (Chatterjee et al. 2005b; Table 9.2). Some mutations of some leader peptide residues in mutacin II eliminated biosynthesis, whereas other mutations only affected the level of production (Chen et al. 2001). In the leader peptide of lacticin 481, not one of the conserved and not-conserved residues appears essential. Introduction of prolines, however, seems to interfere with the functionality of the leader peptide (Patton et al. 2008). For nukacin ISK-1, the importance of the α-helicity of the leader peptide was demonstrated (Nagao et al. 2009).

Table 9.2 Diversity of lantibiotic leader peptides. In bold: residues (partly) conserved within an alignment (sub)group

Lantibiotic	Leader peptide sequence	References for peptide and/or alignment
Nisin	MSTKD**FNLD**LVSVSKKDSGAS**PR**	Gross and Morell (1971); Chatterjee et al. (2005b)
Mutacin II	MNKLNSNAVVSLN**E**VSDS**EL**DTIL**GG**	Woodruff et al. (1998); Chatterjee et al. (2005b)
Cytolysin LS	MLNKENQENYYSNKLELVGPSFEELSLEE MEAIQGSGDVQAE	Gilmore et al. (1994)
Cinnamycin	MTASILQQSVVDADFRAALLENPAAFGASA AALPTPVEAQDQASLDFWTKDIAATEAFA	Widdick et al. (2003)
Michiganin	MNDILETETPVMVSPRWDMLLDAGEDTSP SVQTQIDAEFRRVVSPYM	Holtsmark et al. (2006)

The leader peptides from two component lantibiotics seem to differ from the aforementioned leader peptides, e.g., plantaricin Wα/Wß (Holo et al. 2001), staphylococcin C55α/ß (Navaratna et al. 1999). The leader peptides from the two-component lantibiotic cytolysin (Gilmore et al. 1994, Table 2) are processed at two cleavage sites in each peptide.

Most lantibiotic leader peptides are composed of about 20–35 amino acids. In contrast, the leader peptides from actagardine (Boakes et al. 2009), michiganin (Holtsmark et al. 2006, Table 9.2), and mersacidin (Bierbaum et al. 1995) are composed of respectively 45, 47, and 48 amino acids, whereas the leader peptide from the globular-shaped lantibiotic cinnamycin (Kessler et al. 1988) is even composed of 59 amino acids (Widdick et al. 2003; Table 9.2). Structures of some leader peptide-dependent lantibiotic modification enzymes may provide further insight into the roles of the leader peptide in the intriguing biosynthesis of lantibiotics. Indeed, the structure of NisC indicates a leader peptide binding site (Li et al. 2006).

Data on the substrate specificity of the lantibiotic modification enzymes have been obtained mainly for the nisin and lacticin 481 modification enzymes. The nisin biosynthesis machinery NisBTC proved to be highly versatile and can be used for the introduction of thioether rings in a broad spectrum of nonlantibiotic peptides (Rink et al. 2005; Kluskens et al. 2005; Rink et al. 2007c). Furthermore, successful dehydration of threonines/serines seems to be influenced by the flanking residues. Hydrophobic flanking residues on one or both sides may favor dehydration, whereas the simultaneous presence of hydrophilic flanking residues on both sides seems to disfavor dehydration (Rink et al. 2007a). Lantibiotic cyclases can catalyze the coupling of dehydroalanines and dehydrobutyrines to cysteines yielding lanthionines and β- methyllanthionines, respectively. However, dehydroalanines are reactive and can, in the absence of cyclase action, also spontaneously form a lanthionine when reacting with a cysteine under mild conditions (Rink et al. 2007c). Such nonenzymatic ring closure can result in a mixture of stereoisomers (Burrage et al. 2000).

Not only are small thioether-bridged peptides synthesized by utilization of NisBTC, but a more complicated substrate peptide with the sequence ITPGC-KATVECKITGPCKATVECK can also be successfully modified to a peptide with four thioether linkages. Also, introduction of intertwined thioether rings is possible, thanks to the stereo- and regiospecificity of NisC (Rink et al. 2007c).

The lacticin 3147 enzymes LtnT and LtnM2 were also successfully used in vivo for the introduction of thioether bridges in nonnatural peptide substrates. Before translocation of the nonnatural substrate by LtnT, the peptide first has to be processed. The LtnA2 leader is intracellularly cleaved off by the same LtnT enzyme. Not all designed peptides could in vivo be produced via the Ltn enzymes. It is not clear whether the processing or the translocation itself limits the production of nonlantibiotic peptides in the case of cells containing LtnT (Kuipers et al. 2008).

Studies of lantibiotic biosynthesis systems for nisin, subtilin, and nukacin ISK1 revealed that the modification enzymes and transporters are arranged in multimeric membrane-associated enzyme complexes (Siegers et al. 1996; Kiesau et al. 1997; Nagao et al. 2005). Additionally, a study suggested the presence of a multimeric enzyme complex and the importance thereof for optimal production of prenisin (van den Berg van Saparoea et al. 2008). Alternatively, the presence of thioether rings might impose a structure, which is transported more efficiently by NisT (Kuipers et al. 2008) and is capable of autoinduction (Kuipers et al. 1995).

The composing enzymes of the nisin synthethase complex are separately functional in the absence of other lantibiotic enzymes (Kuipers et al. 2004, Kuipers et al. 2006). In the absence of other nisin enzymes, NisB-dehydrated peptides were exported via the Sec pathway when the nisin leader was preceded by a Sec signal sequence. Prenisin with or without preceding Sec or Tat signal was intracellularly fully modified by NisB and NisC in the absence of NisT, which precludes the necessity of either NisT or Sec action. In view of the above case, it is difficult to comprehend the proposed enzyme-complex-dependent NisT-driven-modification working model for nisin biosynthesis by Lubelski et al. (2009).

In vitro activity of the biosynthetic enzyme of lacticin 481 (Xie et al. 2004) of both haloduracin modification enzymes (McClerren et al. 2006), and of the cyclase of nisin (Li et al. 2006) has been attained (Li et al. 2009). Although LctM does not display an evident ATP-binding domain, ATP is necessary for functionality of LctM (Chatterjee et al. 2005a) and likely – in view of the kinase domain- ATP or GTP is necessary for RamC, the modification enzyme of SapB (Kodani et al. 2004). With the powerful in vitro thioether modification system, the substrate specificity and several mechanistic aspects of the LctM 481 enzyme were explored. Like NisB, LctM has high substrate promiscuity. LctM can dehydrate a range of nonlantibiotic peptides when attached to the N-terminal leader peptide (Chatterjee et al. 2006; Levengood and van der Donk 2008; Patton et al. 2008). Whereas hydrophilic amino acids, especially negatively charged amino acids, which flank dehydratable substrate residues, disfavor NisB-mediated dehydration, this does not appear to be the case for LctM (Chatterjee et al. 2006). Semisynthetic LctA derivates with nonproteinogenic amino acids like β-amino acids, D-amino acids, and N-alkyl amino acids, derived by expressed protein ligation, are successfully modified by LctM (Zhang and van der Donk 2007; Levengood et al. 2009a; Levengood et al. 2009b).

LctM phosphorylates the serines and threonines of its substrate prior to the dehydration of these residues (Chatterjee et al. 2005a; You and van der Donk 2007). An LctM T405A mutant was not affected in phosphorylation of serines and threonines of its substrate but was hampered in the phosphate-elimination step and thereby lost the ability to dehydrate its substrate. This mutant turned out to be a highly efficient kinase for a broad range of peptide substrates with serines, provided that the lacticin leader was N-terminally present (You et al. 2009). Recently, distributive and a tendency to an apparently directional behavior of the LctM enzyme were revealed (Lee et al. 2009). LctM has a high, though not strict, propensity for an apparent N- to C- directionality. When the leader peptide is present in trans, dehydrations were nondirectional. Also, intermediates were found, which were not completely dehydrated but already contained rings within their N-terminal region. This latter finding suggests that the dehydration and the cyclization activity of LctM can be alternating activities. This alternating feature was also observed for NisB- and NisC activity (Kuipers et al. 2008; Fig. 9.2b). The latter data were supported by data obtained by Lubelski et al., who also suggested that NisB and NisC are acting in an N- to C- direction (Lubelski et al. 2009). On the basis of the above observations, we propose a model depicted in Fig. 9.2b. When NisB and NisC have each a fixed leader peptide-binding site close to the active site, the distances of the residues that have to be modified in the prepeptide part of nisin to the active site are set. A residue that is located close to the leader and to the active site will have a much higher chance to be modified compared to a residue further away. The result will appear as directionality, but is actually the result of different binding constants of the residues determined by sequence and distance.

Lantibiotics and lantibiotic enzymes constitute a fascinatingly rich and diverse field of research. Important new mechanistic insights may follow from the crystallization of lantibiotic enzymes other than the cyclase NisC (Li et al. 2006) and the decarboxylase MrsD (Blaesse et al. 2003). In vitro reconstitution of LanB activity, which has not been attained despite many efforts in several laboratories, will hopefully be realized one day and facilitate further mechanistic studies.

Structure

The (methyl)lanthionines give lantibiotics their unique features such as thermostability and proteolytic resistance, and most (methyl)lanthionines are essential for high antimicrobial activity. The primary structure of lantibiotics is highly variable. Aligning the sequences of unmodified propeptides led to three structural groups with one or more conserved (methyl)lanthionine positions, represented by nisin (Fig. 9.1a), mersacidin (Fig. 9.1b), and duramycin (Fig. 9.1c), and a remaining number of apparently unrelated lantibiotics containing among others the morphogenetic sapB (Fig. 9.1d) (Rink et al. 2005). On the basis of primary lantibiotic structures also other groups have been discerned (Chatterjee et al. 2005b; Twomey et al. 2002).

Structural data have been reported on some lantibiotics that share the (methyl) lanthionine positions of rings A and B of nisin (Fig. 9.1a). The structure of nisin was elucidated by chemical degradation (Gross and Morell 1971) and further studied by nuclear magnetic resonance (NMR) spectroscopy in the presence and absence of membrane-mimicking agents (van de Ven et al. 1991; Lian et al. 1992; van den Hooven et al. 1993, Van den Hooven et al. 1996). The data indicated an overall extended conformation and the presence of two amphipathic screw-shaped domains consisting of the N-terminal A-, B-, and C-rings, and the C-terminal fused rings D and E that are joined by a flexible hinge region. Similarly, for gallidermin, in the presence of the structure-inducing solvent trifluoroethanol, an extended amphiphilic screw-shape structure was observed (Freund et al. 1991a, b). Also, the solution structure of mutacin 1140 measured in acetonitrile–water (80:20) indicated rigidity within the lanthionine rings as well as the flexibility of the C-terminal part (Smith et al. 2003).

NMR studies have also been performed on lantibiotics that share the position of the third ring of mersacidin (Fig. 9.1b). Mutacin II consists of an N-terminal helix formed by residues 1-8 (Novak et al. 1997). CD studies on SA-FF22 in lipid-mimicking conditions indicated a significant change compared to the structure in aqueous environment (Jack et al. 1994). The NMR solution structure of plantaricin C indicates a flexible, positively charged N-terminus and a rigid globular C-terminal domain (Turner et al. 1999). The mersacidin NMR structure in methanol solution is globular and has three domains formed by the thioether ring spanning residues (Prasch et al. 1997; Hsu et al. 2003). The X-ray crystallography structure of mersacidin (Schneider et al. 2000) largely resembled the solution structure. The NMR solution structure of actagardine was determined in a water–acetonitrile mixture. Actagardine has a compact globular structure composed of two domains. The N-terminal domain consists of a single lanthionine ring, while the C-terminal domain is composed of three intertwined methyllanthionine rings. Residues 7-8, 9-12, and 17-19 form a small, three-stranded β-sheet with one antiparallel and two parallel strands (Zimmermann and Jung 1997). Preliminary NMR solution structure data of LtnA1 (Martin et al. 2004) indicate a globular shape resembling mersacidin (Hsu et al. 2003).

Detailed structural studies are also known for highly globular lantibiotics which share the (methyl)lanthionine and lysinoalanine structure of duramycin (Fig. 9.1c). The NMR solution structure of cinnamycin has been determined in DMSO and a water–acetic acid mixture (Kessler et al. 1991). The conformation of cinnamycin changes in the presence of SDS micelles (Kessler et al. 1992). The lipophilic part of cinnamycin changed in the presence of SDS bilayers (Kessler et al. 1992) and in the presence of 1-dodecanoyl-*sn*-glycerophosphoethanolamine (C12-LPE) (Wakamatsu et al. 1990) due to interactions with hydrophobic segments of the lipids.

Taken together, detailed structural information for several lantibiotics has been obtained in solution and in the presence of lipophilic membrane-mimicking agents. Generally, lantibiotics contain a few rigid domains, and their conformation changes when interacting with lipophilic or membrane-mimetic surroundings. Furthermore, mechanistically highly important structures describing the interaction of lantibiotics with docking molecules have been obtained (Hsu et al. 2004).

Modes of Action

Pore Formation

Many studies have been performed on the capacity of several lantibiotics to permeabilize membranes. These studies involved liposomes, proteoliposomes, cell-membrane vesicles, black lipid membranes, lipid monolayers, and intact cells and provided much information on the biophysical aspects of the interaction of lantibiotics with these different membrane models. However, up to three orders of magnitude higher concentrations of nisin were needed for permeabilization of model membranes than for pore formation in intact cells. This difference was explained by an excellent study by Sahl and coworkers who were the first to demonstrate that nisin interacts with lipid II, a precursor in the cell wall synthesis. Nisin activity could be eliminated by adding ramoplanin, a compound interacting with lipid II (Brötz et al. 1998a). Subsequent studies demonstrated that hybrid pores composed of nisin and lipid II with a stoichiometry of 1:2 were formed in the target cell membrane. The presence of lipid II in liposome strongly reduced the concentration of nisin required for pore formation (Breukink et al. 1999). The presence of lipid II in black lipid membranes reduced the required threshold membrane potential from about 100 mV to 5–10 mV, allowed the induction of pores by not only trans-negative but also trans-positive membrane potential, and strongly prolonged the lifetime of the formed nisin pores from milliseconds to seconds. Nisin can also form pores in cells when there is no transmembrane electrical potential at all (Moll et al. 1997). Dissipation of a transmembrane electrical potential by itself is not leading to growth inhibition, but dissipation of the transmembrane pH gradient is sufficient for complete growth inhibition (Moll et al. 1999). The structure of nisin lipid II complexes reveals a pyrophosphate cage formed by ring A (Hsu et al. 2004). Interaction with lipid II, leading to pore formation, has also been suggested for a central part of the LtnA1 peptide from the two-component lantibiotic lacticin 3147 (Cotter et al. 2006). Interaction of LtnA1 with lipid II was experimentally demonstrated, and the role of LtnA2 in pore formation has been elucidated (Wiedemann et al. 2006b). Pep5 and epilancin K7 do not interact with lipid II or lipid I but still have high antibacterial activity. Likely, this high activity might be explained in the future by the interaction with (an) other docking molecule(s).

Antibacterial Activity by Interaction With Lipid II Without Pore Formation

Resistance against vancomycin is increasingly threatening. Hybrids of vancomycin and just an N-terminal nisin fragment, nisin(1-12), which does not span the membrane, could prevent the raise of such resistance. These hybrids simultaneously interact with two different sites of lipid II (Arnusch et al. 2008). Epidermin, which shares the pattern of ring A and ring B with nisin, also interacts with lipid II. Likely, other lantibiotics, e.g., mutacin I, mutacin III, mutacin 1140, mutacin B-Ny266, gallidermin, ericin A, ericin S, subtilin, nisin Z, nisin Q, which all share this ring pattern of ring A and B interact with lipid II. Nisin can interact with lipid I and lipid II. Nisin and epidermin cause an accumulation of lipid I in in vitro peptidoglycan synthesis (Brötz et al. 1998b). Truncated nisin(1-23) variants have significant activity but are unable to dissipate the membrane potential (Rink et al. 2007b). Interaction of epidermin and mutacin 1140 with lipid II has been demonstrated. Epidermin whose activity can be even higher than that of nisin does not form pores, since it is too short to span the membrane; in thin model membranes, it does permeabilize the membrane. Breukink and coworkers demonstrated that nisin is able to dislocate lipid II from the septum, thus inhibiting growth of the cell wall (Hasper et al. 2006). It might be worthwhile to investigate whether this mechanism might be responsible for the activity of truncated nisin(1-23) and short lantibiotics that share the ring pattern of nisin's ring A and ring B. Also, other mechanisms have been described. Plantaricin C forms complexes with lipid I and lipid II and inhibits lipid II synthesis and the addition of the first glycine of the pentapeptide chain of lipid II (Wiedemann et al. 2006a, b). Mersacidin and actagardine, whose structures are entirely different from nisin, also bind to lipid II but do not form pores (Brötz et al. 1995; 1997; 1998b). Instead, they exert in vivo high antibacterial activity by blocking the transglycosylation step in the peptidoglycan synthesis.

Inhibition of the Outgrowth of Spores

Some lantibiotics such as nisin and subtilin inhibit the germination of spores of the species from the genera *Bacillus* and *Clostridium* (Thomas et al. 2002). It was suggested that the mechanism of inhibition resulted from the reaction of thiol groups in proteins of the spores with dehydroalanine in position 5 of subtilin and nisin (Morris et al. 1984). Indeed, replacement of the dehydroalanine in position 5 for an alanine caused loss of the capacity to inhibit the outgrowth of spores. However, nisin ringA mutants with other residues in position 5 even had enhanced activity in inhibiting the outgrowth of spores (Rink et al. 2007b).

Other Activities

Cinnamycin and its natural variant duramycin effectively inhibit phospholipase A2 by binding to its phosphatidylethanolamine substrate (Märki et al. 1991). The stoichiometry of the binding is 1:1, as measured by NMR, and binding appears to be specific for (lyso)phosphatidylethanolamine since no binding was observed with other phospholipids (Hosoda et al. 1996; Wakamatsu et al. 1990). Since in eukaryote cells amino phospholipids are usually localized in the inner layer of the plasma membrane, either the lantibiotic has to translocate across the bilayer or it has to induce lipid flip-flop, or both. Lantibiotic-induced lipid flip-flop has been reported for nisin (Moll et al. 1998), but not yet for cinnamycin. Nisin and Pep5 induce autolysis of Staphylococcal strains. It is thought that these cationic peptides bind to lipoteichoic and teichoic acids and thereby displace and activate N-acetyl-alanine aminidase and N-acetylglucosaminidase, which normally interact with teichoic acids (Bierbaum and Sahl 1985). Ancovenin modulates the activity of angiotensin I-converting enzyme (Kido et al 1983). Antiviral activity and activity against neuropathic pain have been reported for some class III lantibiotics (Meindl et al 2010).

Development of Resistance of Gram-Positive Bacteria Against Lantibiotics

Despite its worldwide application, little resistance has been developed against nisin. Resistance against lantibiotics can be due to among others changes of the cell wall and or the cell membrane and has been reviewed in detail (Chatterjee et al. 2005b). Gravesen and coworkers have proposed that nisin resistance in a *Listeria monocytogenes* strain was due to shielding of lipid II from nisin through its binding to a penicillin-binding protein (Gravesen et al. 2001; Gravesen et al. 2004). Nisin resistance in *M. flavus* appeared to be independent of lipid II levels (Kramer et al. 2004). Studies on a resistant *Lactococcus lactis* strain, which was able to grow at a 75-fold higher nisin concentration than the parent strain, demonstrated that less nisin was able to bind to lipid II in the membranes of the resistant strain. The cell wall of the resistant strain displayed significantly increased thickness at the septum. Comparison of modifications in lipoteichoic acid revealed an increase in D-alanyl esters and galactose as substituents in the resistant strain, resulting in a less negatively charged cell wall. Shielding lipid II and thereby decreasing abduction of lipid II and pore formation appeared to be a major defense mechanism of *L. lactis* against nisin (Kramer et al. 2008).

Lantibiotic Activities Exploited for Clinical Developments

Animal studies demonstrated the in vivo efficacy of several lantibiotics. Human cells do not possess the target of several lantibiotics, i.e., lipid II. Duramycin (also called: Moli1901) is being developed for the treatment of reduced mucociliary clearance in cystic fibrosis. The molecular target of duramycin is the phospholipid phosphatidylethanolamine present in the cellular membrane. Duramycin binds to the polar head of phosphatidylethanolamine and induces changes in intracellular calcium levels, which in turn activate calcium-dependent chloride channels. These alternative calcium-activated chloride channels produce an output of chloride and water. This may compensate for reduced or absent cystic-fibrosis-transmembrane-conductance-regulator/Cl^- channel function in cystic fibrosis patients. A phase II study in cystic fibrosis patients showed that the inhalation of duramycin over 5 days resulted in an improvement in pulmonary function parameters (Grasemann et al. 2007). Phase II clinical studies are also being carried out on duramycin for treating dry eyes (Grasemann et al. 2007). By opening the abovementioned alternate salt channel, duramycin promotes the hydration of epithelial tissue. Patients with dry eye disease may thereby experience increased hydration of the eyes. A product that contains *Streptococus salivarius* that produces salivaricin is marketed for oral care to counteract infections (Boakes and Wadman 2008).

Perspectives

Existing and novel effective lantibiotics are of great interest because of the increase in resistance to multiple antibiotics. The identification of docking molecules and the characterization of the detailed molecular interaction with docking molecules will enable contributions to the development of improved lantibiotics. Lantibiotics may also have therapeutic activities that are entirely different from antibiotic activities such as the anti-inflammatory inhibition of pospholipase A2 by duramycin via binding of the phospholipase substrate, phosphatidylethanolamine. Mechanistic studies on the lantibiotic modification enzymes will allow for their directed engineering and subsequent aimed for application. Thioether bridges may also stabilize a variety of peptides, which are drugs for diverse indications (Kluskens et al. 2005; Kluskens et al. 2009; Kuipers et al. 2009; Levengood and van der Donk 2008; Rink et al. 2010). By stabilization, these therapeutic peptides are less sensitive to proteolytic breakdown and accordingly need less frequent administration and/or in a lower dose. In addition, stabilization may allow oral and pulmonary delivery of short peptides (de Vries et al. 2010). These delivery ways are more patient-friendly than injection. Furthermore, the structural constraint resulting from the introduction of thioether bridges may enhance the receptor specificity and/or the efficacy of the receptor interaction, thus enhancing the therapeutic potential.

References

Allgaier H, Jung G, Werner RG et al (1986) Epidermin: sequencing of a heterodetic tetracyclic 21-peptide amide antibiotic. Eur J Biochem 160:9–22

Appleyard AN, Choi S, Read DM et al (2009) Dissecting structural and functional diversity of the lantibiotic mersacidin. Chem Biol 16:490–498

Arnusch CJ, Bonvin AMJJ, Verel AM et al (2008) The vancomycin-nisin(1-12) hybrid restores activity against vancomycin resistant enterococci. Biochemistry 47:12661–12663

Asaduzzaman SM, Sonomoto K (2009) Lantibiotics: Diverse activities and unique modes of action. J Biosc Bioeng 107:475–487

Bierbaum G, Sahl HG (1985) Induction of autolysis of staphylococci by the basic peptide antibiotics Pep 5 and nisin and their influence on the activity of autolytic enzymes. Arch Microbiol 141:249–254

Bierbaum G, Sahl HG (2009) Lantibiotics: mode of action, biosynthesis and bioengineering. Curr Pharm Biotechnol 10:2–18

Bierbaum G, Brötz H, Koller K-P et al (1995) Cloning, sequencing and production of the lantibiotic mersacidin. FEMS Microbiol Lett 127:121–126

Bierbaum G, Szekat C, Josten M et al (1996) Engineering of a novel thioether bridge and role of modified residues in the lantibiotic Pep5. Appl Environ Microbiol 62:385–392

Blaesse M, Kupke T, Huber R et al (2003) Structure of MrsD, an FAD-binding protein of the HFCD family. Acta Crystallogr D Biol Crystallogr 59:1414–1421

Boakes S, Wadman S (2008) The therapeutic potential of lantibiotics. Innov Pharm Technol 27:22–25

Boakes S, Cortés J, Appleyard AN et al (2009) Organization of the genes encoding the biosynthesis of actagardine and engineering of a variant generation system. Mol Microbiol 72:1126–1136

Breukink E, Wiedemann I, van Kraaij C et al (1999) Use of the cell wall precursor lipid II by a pore-forming peptide antibiotic. Science 286:2361–2364

Brötz H, Bierbaum G, Markus A et al (1995) Mode of action of the lantibiotic mersacidin: inhibition of peptidoglycan biosynthesis via a novel mechanism? Antimicrob Agents Chemother 39:714–719

Brötz H, Bierbaum G, Reynolds P et al (1997) The lantibiotic mersacidin inhibits peptidoglycan biosynthesis at the level of transglycosylation. Eur J Biochem 246:193–197

Brötz H, Josten M, Wiedemann I et al (1998a) Role of lipid-bound peptidoglycan precursors in the formation of pores by nisin, epidermin and other lantibiotics. Mol Microbiol 30:317–327

Brötz H, Bierbaum G, Leopold K et al (1998b) The lantibiotic mersacidin inhibits peptidoglycan synthesis by targeting lipid II. Antimicrob Agents Chemother 42:154–160

Burrage S, Raynham T, Williams G et al (2000) Biomimetic synthesis of lantibiotics. Chemistry 6:1455–1466

Castiglione F, Lazzarini A, Carrano L et al (2008) Determining the structure and mode of action of microbisporicin, a potent lantibiotic active against multiresistant pathogens. Chem Biol 15:22–31

Chakicherla A, Hansen JN (1995) Role of the leader and structural regions of prelantibiotic peptides as assessed by expressing nisin-subtilin chimeras in *Bacillus subtilis* 168, and characterization of their physical, chemical, and antimicrobial properties. J Biol Chem 270:23533–23539

Chan WC, Bycroft BW, Lian L-Y et al (1989) Isolation and characterisation of two degradation products derived from the peptide antibiotic nisin. FEBS Lett 252:29–36

Chatterjee S, Chaterjee S, Lad SJ et al (1992) Mersacidin, a new antibiotic from Bacillus Fermentation, isolation, purification and chemical characterization. J Antibiot 45:832–838

Chatterjee C, Miller LM, Leung YL et al (2005a) Lacticin 481 synthetase phosphorylates its substrate during lantibiotic production. J Am Chem Soc 127:15332–15333

Chatterjee C, Paul M, Xie L et al (2005b) Biosynthesis and mode of action of lantibiotics. Chem Rev 105:633–684

Chatterjee C, Patton GC, Cooper L et al (2006) Engineering dehydro amino acids and thioethers into peptides using lacticin 481 synthetase. Chem Biol 13:1109–1117

Chen P, Novak J, Kirk M et al (1998) Structure-activity study of the lantibiotic mutacin II from *Streptococcus mutans* T8 by a gene replacement strategy. Appl Environ Microbiol 64:2335–2340

Chen P, Qi FX, Novak J et al (2001) Effect of amino acid substitutions in conserved residues in the leader peptide on biosynthesis of the lantibiotic mutacin II. FEMS Microbiol Lett 195:139–144

Cortés J, Appleyard AN, Dawson MJ (2009) Whole-cell generation of lantibiotic Variants. Methods Enzymol 458:559–574

Cotter PD, Hill C, Ross RP (2005) Bacterial lantibiotics: strategies to improve therapeutic potential. Curr Protein Pept Sci 6:61–75

Cotter PD, Deegan LH, Lawton EM et al (2006) Complete alanine scanning of the two-component lantibiotic lacticin 3147: generating a blueprint for rational drug design. Mol Microbiol 62:735–747
de Vos WM, Jung G, Sahl HG (1991) In: Jung G, Sahl HG (eds) Nisin and novel lantibiotics: proceedings of the first international workshopon lantibiotics. Escom Publishers, Leiden, pp 457–463
de Vos WM, Mulders JW, Siezen RJ et al (1993) Properties of nisin Z and distribution of its gene, nisZ, in *Lactococcus lactis*. Appl Environ Microbiol 59:213–218
de Vries L, Reitzema-Klein CE, Arkema-Meter A et al (2010) Oral and pulmonary delivery of thioether-bridged angiotensin-(1-7). Peptides 31:893–898
Delves-Broughton J (2005) Nisin as a food preservative. Food Aust 57:525–527
Dufour A, Hindre T, Haras D et al (2007) The biology of lantibiotics from the lacticin 481 group is coming of age. FEMS Microbiol Rev 31:134–167
Field D, Connor PM, Cotter PD et al (2008) The generation of nisin variants with enhanced activity against specific gram-positive pathogens. Mol Microbiol 69:218–230
Fredenhagen A, Märki F, Fendrich G et al (1991) Duramycin B and C, two new lanthionine-containing antibiotics as inhibitors of phospholipase A2, and structural revision of duramycin and cinnamycin. In: Jung G, Sahl HG (eds) Nisin and novel lantibiotics. Escom. Leiden, pp 131–140
Freund S, Jung G, Gutbrod O et al (1991a) The three-dimensional solution structure of gallidermin determined by NMR-based molecular graphics. In: Jung G, Sahl HG (eds) Nisin and novel lantibiotics, Escom Leiden, pp 91–102
Freund S, Jung G, Gutbrod O et al (1991b) The solution structure of the lantibiotic gallidermin. Biopolymers 31:803–811
Fukao M, Obita T, Yoneyama F et al (2008) Complete covalent structure of nisin Q, new natural nisin variant, containing post-translationally modified amino acids. Biosci Biotechnol Biochem 72:1750–1755
Gilmore MS, Segarra RA, Booth MC et al (1994) Genetic structure of the *Enterococcus faecalis* plasmid pAD1-encoded cytolytic toxin system and its relation to lantibiotic determinants. J Bacteriol 176:7335–7344
Goto Y, Li B, Claesen J et al (2010) Discovery of unique lanthionine synthetases reveals new mechanistic and evolutionary insights. PLoS Biol 8:e1000339
Grasemann H, Stehling F, Brunar H et al (2007) Inhalation of moli1901 in patients with cystic fibrosis. Chest 131:1461–1466
Gravesen A, Sorensen K, Aarestrup FM et al (2001) Spontaneous nisin-resistant *Listeria monocytogenes* mutants with increased expression of a putative penicillin-binding protein and their sensitivity to various antibiotics. Drug Resist 7:127–135
Gravesen A, Kallipolitis B, Holmstrom K et al (2004) pbp2229-mediated nisin resistance mechanism in *Listeria monocytogenes* confers cross-protection to class IIa bacteriocins and affects virulence gene expression. Appl Environ Microbiol 70:1669–1679
Gross E, Morell JL (1971) The structure of nisin. J Am Chem Soc 93:4634–4635
Hasper HE, Kramer NE, Smith JL et al (2006) An alternative bactericidal mechanism of action for lantibiotic peptides that target lipid II. Science 313:1636–1637
He Z, Yuan C, Zhang L, Yousef AE (2008) N-terminal acetylation in paenibacillin, a novel lantibiotic. FEBS Lett 582:2787–2792
Holo H, Jeknic Z, Daeschel M et al (2001) Plantaricin W from *Lactobacillus plantarum* belongs to a new family of two-peptide lantibiotics. Microbiology 147:643–651
Holtsmark I, Mantzilas D, Eijsink VG et al (2006) Purification, characterization, and gene sequence of michiganin A, an actagardine-like lantibiotic produced by the tomato pathogen *Clavibacter michiganensis* subsp. *michiganensis*. Appl Environ Microbiol 72:5814–5821
Hosoda K, Ohya M, Kohno T et al (1996) Structure determination of an immunopotentiator peptide, cinnamycin, complexed with lysophosphatidyl-ethanolamine by 1H-NMR1. J Biochem 119:226–230
Hsu ST, Breukink E, Bierbaum G et al (2003) NMR study of mersacidin and lipid II interaction in dodecylphosphocholine micelles. Conformational changes are a key to antimicrobial activity. J Biol Chem 278:13110–13117

Hsu ST, Breukink E, Tischenko E et al (2004) The nisin-lipid II complex reveals a pyrophosphate cage that provides a blueprint for novel antibiotics. Nat Struct Mol Biol 11:963–967

Islam MR, Shioya K, Nagao J et al (2009) Evaluation of essential and variable residues of nukacin ISK-1 by NNK scanning. Mol Microbiol 72:1438–1447

Jack RW, Carne A, Metzger J et al (1994) Elucidation of the structure of SA-FF22 lanthionine-containing antibacterial peptide produced by *Streptococcus pyogenes* strain FF22. Eur J Biochem 220:455–462

Jung G (1991) Lantibiotics: a survey. In: Jung G, Sahl HG (eds) Nisin and novel lantibiotics: proceedings of the first international workshop on lantibiotics. Escom Publishers, Leiden, pp 1–34

Kabuki T, Uenishi H, Seto Y et al (2009) A unique lantibiotic, thermophilin 1277, containing a disulfide bridge and two thioether bridges. J Appl Microbiol 106:853–862

Kellner R, Jung G, Hörner T et al (1988) Gallidermin: a new lanthionine-containing polypeptide antibiotic. Eur J Biochem 177:53–59

Kellner R, Jung G, Kaletta C et al (1989) Pep5: structure elucidation of a large lantbiotic. Angew Chem Int Ed Eng 28:616–619

Kessler H, Steuernagel S, Will M et al (1988) The structure of the polycyclic nonadecapeptide Ro 09-0198. Helv Chim Acta 71:1924–1929

Kessler H, Seip S, Wein T et al (1991) Structure of cinnamycin (Ro 09-0198) In: Jung G, Sahl HG (eds) Nisin and novel lantibiotics: proceedings of the first international workshopon lantibiotics. Escom Publishers, Leiden, pp 76–90

Kessler H, Mierke DF, Saulitis J et al (1992) The structure of Ro 09-0198 in different environments. Biopolymers 32:427–433

Kido Y, Hamakado T, Yoshida T et al (1983) Isolation and characterization of ancovenin, a new inhibitor of angiotensin I converting enzyme, produced by actinomycetes. J Antibiot (Tokyo) 36:1295–1299

Kiesau P, Eikmanns U, Gutowski-Eckel Z et al (1997) Evidence for a multimeric subtilin synthetase complex. J Bacteriol 179:1475–1481

Kluskens LD, Kuipers A, Rink R et al (2005) Post-translational modification of therapeutic peptides by NisB, the dehydratase of the lantibiotic nisin. Biochemistry 44:12827–12834

Kluskens LD, Nelemans SA, Rink R et al (2009) Angiotensin-(1-7) with thioether bridge: an angiotensin-converting enzyme-resistant, potent angiotensin-(1-7) analog. J Pharmacol Exp Ther 328:849–854

Kodani S, Hudson ME, Durrant MC et al (2004) The SapB morphogen is a lantibiotic-like peptide derived from the product of the developmental gene *ramS in Streptomyces coelicolor*. Proc Natl Acad Sci U S A 101:11448–11453

Kodani S, Lodato MA, Durrant MC et al (2005) SapT, a lanthionine-containing peptide involved in aerial hyphae formation in the streptomycetes. Mol Microbiol 58:1368–1380

Kramer NE, Smid EJ, Kok J et al (2004) Resistance of Gram-positive bacteria to nisin is not determined by lipid II levels. FEMS Microbiol Lett 239:157–161

Kramer NE, Hasper HE, van den Bogaard PT et al (2008) Increased D-alanylation of lipoteichoic acid and a thickened septum are main determinants in the nisin resistance mechanism of *Lactococcus lactis*. Microbiology 154:1755–1762

Kuipers OP, Rollema HS, Yap WM et al (1992) Engineering dehydrated amino acid residues in the antimicrobial peptide nisin. J Biol Chem 267:24340–24346

Kuipers OP, Beerthuyzen MM, Siezen RJ et al (1993a) Characterization of the nisin gene cluster *nisABTCIPR* of *Lactococcus lactis*. Requirement of expression of the nisA and nisI genes for development of immunity. Eur J Biochem 216:281–291

Kuipers OP, Rollema HS, de Vos WM et al (1993b) Biosynthesis and secretion of a precursor of nisin Z by Lactococcus lactis, directed by the leader peptide of the homologous lantibiotic subtilin from Bacillus subtilis. FEBS Lett 330:23–27

Kuipers OP, Beerthuyzen MM, de Ruyter PG et al (1995) Autoregulation of nisin biosynthesis in *Lactococcus lactis* by signal transduction. J Biol Chem 270:27299–27304

Kuipers OP, Bierbaum G, Ottenwälder B et al (1996) Protein engineering of lantibiotics. Antonie Van Leeuwenhoek 69:161–169

Kuipers A, de Boef E, Rink R et al (2004) NisT, the transporter of the lantibiotic nisin, can transport fully modified, dehydrated, and unmodified prenisin and fusions of the leader peptide with non-lantibiotic peptides. J Biol Chem 279:22176–22182

Kuipers A, Wierenga J, Rink R et al (2006) Sec-mediated export of posttranslationally dehydrated peptides in *Lactococcus lactis*. Appl Environ Microbiol 72:7626–7633

Kuipers A, Meijer-Wierenga J, Rink R et al (2008) Mechanistic dissection of the enzyme complexes involved in biosynthesis of lacticin 3147 and nisin. Appl Environ Microbiol 74:6591–6597

Kuipers A, Rink R, Moll GN (2009) Translocation of a thioether-bridged azurin peptide fragment via the sec pathway in *Lactococcus lactis*. Appl Environ Microbiol 75:3800–3802

Lawton EM, Ross RP, Hill C et al (2007) Two-peptide lantibiotics: a medical perspective. Mini Rev Med Chem 7:1236–1247

Lee MV, Ihnken LA, You YO et al (2009) Distributive and directional behavior of lantibiotic synthetases revealed by high-resolution tandem mass spectrometry. J Am Chem Soc 131: 12258–12264

Levengood MR, van der Donk WA (2008) Use of lantibiotic synthetases for the preparation of bioactive constrained peptides. Bioorg Med Chem Lett 18:3025–3028

Levengood MR, Patton GC, van der Donk WA (2007) The leader peptide is not required for post-translational modification by lacticin 481 synthetase. J Am Chem Soc 129:10314–10315

Levengood MR, Kerwood CC, Chatterjee C et al (2009a) Investigation of the substrate specificity of lacticin 481 synthetase by using nonproteinogenic amino acids. Chembiochem 10:911–919

Levengood MR, Knerr PJ, Oman TJ et al (2009b) In vitro mutasynthesis of lantibiotic analogues containing nonproteinogenic amino acids. J Am Chem Soc 131:12024–12025

Li B, Yu JP, Brunzelle JS, Moll GN et al (2006) Structure and mechanism of the lantibiotic cyclase involved in nisin biosynthesis. Science 311:1464–1467

Li B, Cooper LE, van der Donk WA (2009) In vitro studies of lantibiotic biosynthesis. Methods Enzymol 458:533–557

Lian LY, Chan WC, Morley SD et al (1992) Solution structures of nisin A and its two major degradation products determined by NMR. Biochem J 283:413–420

Liu W, Hansen JN (1992) Enhancement of the chemical and antimicrobial properties of subtilin by site-directed mutagenesis. J Biol Chem 267:25078–25085

Lubelski J, Rink R, Khusainov R et al (2008) Biosynthesis, immunity, regulation, mode of action and engineering of the model lantibiotic nisin. Cell Mol Life Sci 65:455–476

Lubelski J, Khusainov R, Kuipers OP (2009) Directionality and coordination of dehydration and ring formation during biosynthesis of the lantibiotic nisin. J Biol Chem 284:25962–25972

Maffioli SI, Potenza D, Vasile F et al (2009) Structure revision of the lantibiotic 97518. J Nat Products 72:605–607

Majchrzykiewicz JA, Lubelski J, Moll GN et al (2010) Production of a class II two-component lantibiotic of *Streptococcus pneumoniae* using the class I nisin synthetic machinery and leader sequence. Antimicrob Agents Chemother 54:1498–1505

Märki F, Hänni E, Fredenhagen A et al (1991) Mode of action of the lanthionine- containing peptide antibiotics duramycin, duramycin B and C, and cinnamycin as indirect inhibitors of phospholipase A2. Biochem Pharmacol 42:2027–2035

Martin NI, Sprules T, Carpenter MR et al (2004) Structural characterization of lacticin 3147, a two-peptide lantibiotic with synergistic activity. Biochemistry 43:3049–3056

McClerren AL, Cooper LE, Quan C et al (2006) Discovery and in vitro biosynthesis of haloduracin, a two-component lantibiotic. Proc Natl Acad Sci U S A 103:17243–17248

Meindl K, Schmiederer T, Schneider K et al (2010) Labyrinthopeptins: a new class of carbacyclic lantibiotics. Angew Chem Int Ed Engl 49:1151–1154

Minami Y, Yoshida KI, Azuma R et al (1994) Structure of cypemycin, a new peptide antibiotic. Tetrahedron Lett 35:80001–80004

Moll GN, Clark J, Chan WC et al (1997) Role of transmembrane pH gradient and membrane binding in nisin pore-formation. J Bacteriol 179:135–140

Moll GN, Konings WN, Driessen AJ (1998) The lantibiotic nisin induces transmembrane movement of a fluorescent phospholipid. J Bacteriol 180:6565–6570

Moll GN, Konings WN, Driessen AJ (1999) Bacteriocins: mechanism of membrane insertion and pore formation. Antonie Van Leeuwenhoek 76:185–198

Moll GN, Kuipers A, Rink R (2010) Microbial engineering of dehydro-aminoacids and lanthionines in nonlantibiotic peptides. Antonie van Leeuwenhoek 97:319–333

Morris SL, Walsh RC, Hansen JN (1984) Identification and characterization of some bacterial membrane sulfhydryl groups which are targets of bacteriostatic and antibiotic action. J Biol Chem 259:13590–13594

Müller WM, Schmiederer T, Ensle P et al (2010) In vitro biosynthesis of the prepeptide of type-III lantibiotic labyrinthopeptin A2 including formation of a C–C bond as a post-translational modification. Angew Chem Int Ed Engl 49:2436–2440

Nagao J, Aso Y, Sashihara T et al (2005) Localization and interaction of the biosynthetic proteins for the lantibiotic, Nukacin ISK-1. Biosci Biotechnol Biochem 69:1341–1347

Nagao J, Morinaga Y, Islam MR et al (2009) Mapping and identification of the region and secondary structure required for the maturation of the nukacin ISK-1 prepeptide. Peptides 30:1412–1420

Naruse N, Tenmyo O, Tomita K et al (1989) Lanthiopeptin, a new peptide antibiotic. Production, isolation and properties of lanthiopeptin. J Antibiot (Tokyo) 42:837–845

Navaratna MADB, Sahl HG, Tagg JR (1999) Identification of genes encoding two-component lantibiotic production in *Staphylococcus aureus* C55 and other phage group II S. aureus strains and demonstration of an association with the exofoliate toxin B gene. Infect Immun 67:4268–4271

Neis S, Bierbaum G, Josten M et al (1997) Effect of leader peptide mutations on biosynthesis of the lantibiotic Pep5. FEMS Microbiol Lett 149:249–255

Novak J, Chen P, Kirk M et al (1997) Structure and activity of the lantibiotic mutacin II. Evidence for two domains? Nat Biotechnol Short Rep 8:87

Oman TJ, van der Donk WA (2010) Follow the leader: the use of leader peptides to guide natural product biosynthesis. Nat Chem Biol 6:9–18

Ottenwälder B, Kupke T, Brecht S et al (1995) Isolation and characterization of genetically engineered gallidermin and epidermin analogs. Appl Environ Microbiol 61:3894–3903

Pag U, Sahl HG (2002) Multiple activities in lantibiotics-models for the design of novel antibiotics? Curr Pharm Des 8:815–833

Patton GC, van der Donk WA (2005) New developments in lantibiotic biosynthesis and mode of action. Curr Opin Microbiol 8:543–551

Patton GC, Paul M, Cooper LE et al (2008) The importance of the leader sequence for directing lanthionine formation in lacticin 481. Biochemistry 47:7342–7351

Paul M, Patton GC, van der Donk WA (2007) Mutants of the zinc ligands of lacticin 481 synthetase retain dehydration activity but have impaired cyclization activity. Biochemistry 46:6268–6276

Plat A, Kluskens LD, Kuipers A et al (2010) Requirements of nisin's engineered leader peptide for inducing modification, export and cleavage. Appl Environ Microbiol, Epub ahead of print

Prasch T, Naumann T, Markert RL et al (1997) Constitution and solution conformation of the antibiotic mersacidin determined by NMR and molecular dynamics. Eur J Biochem 244:501–512

Qiao M, Ye S, Koponen O et al (1996) Regulation of the nisin operons in *Lactococcus lactis* N8. J Appl Bacteriol 80:626–634

Rink R, Kuipers A, de Boef E et al (2005) Lantibiotic structures as guidelines for the design of peptides that can be modified by lantibiotic enzymes. Biochemistry 44:8873–8882

Rink R, Wierenga J, Kuipers A et al (2007a) Production of dehydroamino acid-containing peptides by *Lactococcus lactis*. Appl Environ Microbiol 73:1792–1796

Rink R, Wierenga J, Kuipers A et al (2007b) Dissection and modulation of the four distinct activities of nisin by mutagenesis of rings A and B and by C-terminal truncation. Appl Environ Microbiol 73:5809–5816

Rink R, Kluskens LD, Kuipers A et al (2007c) NisC, the cyclase of the lantibiotic nisin, can catalyze cyclization of designed nonlantibiotic peptides. Biochemistry 46:13179–13189

Rink R, Arkema-Meter A, Baudoin I et al (2010) To protect peptide pharmaceuticals against peptidases. J Pharmacol Toxicol Methods 61:210–218

Rogers LA, Whittier EO (1928) Limiting factors in lactic fermentation. J Bacteriol 16:211–229

Ryan MP, Rea MC, Hill C et al (1996) An application in cheddar cheese manufacture for a strain of *Lactococcus lactis* producing a novel broad-spectrum bacteriocin, lacticin 3147. Appl Environ Microbiol 62:612–619

Ryan MP, Jack RW, Josten M et al (1999) Extensive post-translational modification, including serine to D-alanine conversion, in the two-component lantibiotic, lacticin 3147. J Biol Chem 274:37544–37550

Schneider TR, Kärcher J, Pohl E et al (2000) Ab initio structure determination of the lantibiotic mersacidin. Acta Crystallogr D Biol Crystallogr 56:705–713

Schnell N, Entian KD, Schneider U et al (1988) Prepeptide sequence of epidermin, a ribosomally synthesized antibiotic with four sulphide-rings. Nature 333:276–278

Siegers K, Heinzmann S, Entian KD (1996) Biosynthesis of lantibiotic nisin. Posttranslational modification of its prepeptide occurs at a multimeric membrane-associated lanthionine synthetase complex. J Biol Chem 271:12294–12301

Siezen RJ, Kuipers OP, de Vos WM (1996) Comparison of lantibiotic gene clusters and encoded proteins. Antonie van Leeuwenhoek 69:171–184

Skaugen M, Nissen-Meyer J, Jung S et al (1994) In vivo conversion of L-serine to D-alanine in a ribosomally synthesized polypeptide. J Biol Chem 269:27183–27185

Smith L, Hillman J (2008) Therapeutic potential of type A (I) lantibiotics, a group of cationic peptide antibiotics. Curr Opin Microbiol 11:401–408

Smith L, Zachariah C, Thirumoorthy R et al (2003) Structure and dynamics of the lantibiotic mutacin 1140. Biochemistry 42:10372–10384

Szekat C, Jack RW, Skutlarek D et al (2003) Construction of an expression system for site-directed mutagenesis of the lantibiotic mersacidin. Appl Environ Microbiol 69:3777–3783

Thomas LV, Clarkson MR, Delves-Broughton J (2000) Nisin. In *Natural Food Antimicrobial Systems*, edited by AS. Naidu. CRC Press. Boca Raton, USA, pp. 463–524

Thomas LV, Ingram RE, Bevis HE et al (2002) Effective use of nisin to control Bacillus and Clostridium spoilage of a pasteurized mashed potato product. J Food Prot 65:1580–1585

Turner DL, Brennan L, Meyer HE et al (1999) Solution structure of plantaricin C, a novel lantibiotic. Eur J Biochem 264:833–839

Twomey D, Ross RP, Ryan M et al (2002) Lantibiotics produced by lactic acid bacteria: structure, function and applications. Antonie Van Leeuwenhoek 82:165–185

Ueda K, Oinuma K, Ikeda G et al (2002) AmfS, an extracellular peptidic morphogen in *Streptomyces griseus*. J Bacteriol 184:1488–1492

van de Kamp M, Horstink LM, van den Hooven HW et al (1995a) Sequence analysis by NMR spectroscopy of the peptide lantibiotic epilancin K7 from *Staphylococcus epidermidis* K7. Eur J Biochem 227:757–771

van de Kamp M, van den Hooven HW, Konings RNH et al (1995b) Elucidation of the primary structure of the lantibiotic epilancin K7 from *Staphylococcus epidermis* K7 – cloning of the epilancin-K7-encoding gene and NMR analysis of mature epilancin K7. Eur J Biochem 230:587–600

van de Ven FJM, Van den Hooven HW, Konings RN et al (1991) NMR studies of lantibiotics. The structure of nisin in aqueous solution. Eur J Biochem 202:1181–1188

van den Berg van Saparoea HB, Bakkes PJ, Moll GN et al (2008) Distinct contributions of the nisin biosynthesis enzymes NisB and NisC and transporter NisT to prenisin production by *Lactococcus lactis*. Appl Environ Microbiol 74:5541–5548

van den Hooven HW, Fogolari F, Rollema HS et al (1993) NMR and circular dichroism studies of the lantibiotic nisin in non-aqueous environments. FEBS Lett 319:189–194

van den Hooven HW, Doeland CC, Van de Kamp M et al (1996) Three-dimensional structure of the lantibiotic nisin in the presence of membrane-mimetic micelles of dodecylphosphocholine and of sodium dodecylsulphate. Eur J Biochem 235:382–393 van der Meer JR, Polman J, Beerthuyzen MM et al (1993) Characterization of the *Lactococcus lactis* nisin A operon genes nisP, encoding a subtilisin-like serine protease involved in precursor processing, and nisR, encoding a regulatory protein involved in nisin biosynthesis. J Bacteriol 175:2578–2588

van der Meer JR, Rollema HS, Siezen R et al (1994) Influence of amino acid substitutions in the nisin leader peptide on biosynthesis and secretion of nisin by *Lactococcus lactis*. J Biol Chem 269:3555–3562

Wakamatsu K, Choung SY, Kobayashi T et al (1990) Complex formation of peptide antibiotic Ro09-0198 with lysophosphatidylethanolamine: 1H NMR analyses in dimethyl sulfoxide solution. Biochemistry 29:113–118

Widdick DA, Dodd HM, Barraille P et al (2003) Cloning and engineering of the cinnamycin biosynthetic gene cluster from *Streptomyces cinnamoneus cinnamoneus* DSM 40005. Proc Natl Acad Sci U S A 100:4316–4321

Wiedemann I, Böttiger T, Bonelli RR et al (2006a) Lipid II-based antimicrobial activity of the lantibiotic plantaricin C. Appl Environ Microbiol 72:2809–2814

Wiedemann I, Böttiger T, Bonelli RR et al (2006b) The mode of action of the lantibiotic lacticin 3147 – a complex mechanism involving specific interaction of two peptides and the cell wall precursor lipid II. Mol Microbiol 61:285–296

Willey JM, van der Donk WA (2007) Lantibiotics: peptides of diverse structure and function. Annu Rev Microbiol 61:477–501

Willey JM, Willems A, Kodani S et al (2006) Morphogenetic surfactants and their role in the formation of aerial hyphae in *Streptomyces coelicolor*. Mol Microbiol 59:731–742

Wirawan RE, Klesse NA, Jack RW et al (2006) Molecular and genetic characterization of a novel nisin variant produced by *Streptococcus uberis*. Appl Environ Microbiol 72:1148–1156

Woodruff WA, Novák J, Caufield PW (1998) Sequence analysis of mutA and mutM genes involved in the biosynthesis of the lantibiotic mutacin II in *Streptococcus mutans*. Gene 206:37–43

Xie L, Miller LM, Chatterjee C et al (2004) Lacticin 481: in vitro reconstitution of lantibiotic synthetase activity. Science 303:679–681

You YO, van der Donk WA (2007) Mechanistic investigations of the dehydration reaction of lacticin 481 synthetase using site-directed mutagenesis. Biochemistry 46:5991–6000

You YO, Levengood MR, Ihnken LA et al (2009) Lacticin 481 synthetase as a general serine/threonine kinase. ACS Chem Biol 4:379–385

Yuan J, Zhang ZZ, Chen XZ et al (2004) Site-directed mutagenesis of the hinge region of nisinZ and properties of nisinZ mutants. Appl Microbiol Biotechnol 64:806–815

Zhang X, van der Donk WA (2007) On the substrate specificity of dehydration by lacticin 481 synthetase. J Am Chem Soc 129:2212–2213

Zimmermann N, Jung G (1997) The three-dimensional solution structure of the lantibiotic murein-biosynthesis-inhibitor actagardine determined by NMR. Eur J Biochem 246:809–881

Chapter 10
Class IIa Bacteriocins: Current Knowledge and Perspectives

Yanath Belguesmia, Karim Naghmouchi, Nour-Eddine Chihib, and Djamel Drider

Abstract Lactic acid bacteria (LAB) are known to produce antibacterial peptides and small proteins called bacteriocins, which enable them to compete against other bacteria in the environment. Bacteriocins fall structurally and chemically into three different classes, I, II, and III. Bacteriocins are ribosomally synthesized peptides with antagonism against closely related bacteria. This late observation has evolved because bacteriocins active against Gram-negative bacteria have recently been reported. Members of class IIa bacteriocins, referred to as pediocin-like bacteriocins, are among the most studied bacteriocins. This chapter is aimed at providing an updated review on the biology of class IIa bacteriocins.

Introduction

Class IIa bacteriocins are peptides below 10 kDa composed of a varying number of amino acids, ranging from 37 for sakacin G (Simon et al. 2002) to 58 for mundticin L (Feng et al. 2009). Sequence alignment of class IIa bacteriocins displays two distinct regions referred to as N- and C-terminal regions (Fig. 10.1). While, the N-terminal region is hydrophilic and contains the consensus sequence Y-G-N-G-V-X_1-C-X_2-K/N-X_3-X_4-C (where X is any amino acid), the C-terminal region is variable, hydrophobic, and/or amphiphilic. Other characteristics unifying all members of class IIa bacteriocins are (1) their potent activity against *Listeria* spp., (2) their resistance to elevated temperatures and extreme pHs, and (3) their cystibiotic feature attributed to the presence of at least one disulfide bridge, which is crucial for antibacterial activity.

D. Drider (✉)
Department of Research into Sanitary Risks and Biotechnology of Reproduction, UPSP 5301 DGER, Nantes-Atlantic National College of Veterinary Medicine, Food Science and Engineering (ONIRIS), La Chantrerie Cedex 03, BP 40706, 44307 Nantes, France
e-mail: djamel.drider@oniris-nantes.fr

D. Drider and S. Rebuffat (eds.), *Prokaryotic Antimicrobial Peptides: From Genes to Applications*, DOI 10.1007/978-1-4419-7692-5_10,

Producer strain	bacteriocin	amino acid sequence	Molecular mass (Da)	Reference
Bacillus coagulans I4	coagulin	N_{ter}-KYYGNGVTCGKHSCSVDWGKATTCIINNGAMAWATGGHQGTHKC-C_{ter}	4612	Le Marrec et al. 2000
Pediococcus acidilactici PAC-1. 0	pediocin PA-1	N_{ter}-KYYGNGVTCGKHSCSVDWGKATTCIINNGAMAWATGGHQGNHKC-C_{ter}	4629	Henderson et al. 1992
Enterococcus mundtii NFRI7393	mundticin KS	N_{ter}-KYYGNGVSCNKKGCSVDWGKAIGIIGNNSAANLATGGAAGWKS-C_{ter}	4287	Kawamoto et al, 2002
Enterococcus mundtii CUGF08	mundticin L	N_{ter}-MKKLTSKEMAQVVGGKYYGNGLSCNKKGCSVDWGKAIGIIGNNSAANLATGGAAGWKS-C_{ter}	4301	Feng et al. 2009
Enterococcus mundtii AT06	mundticin	N_{ter}-KYYGNGVSCNKKGCSVDWGKAIGIIGNNSAANLATGGAAGWSK-C_{ter}	4287	Bennik et al. 1998
Enterococcus avium 208	avicin A	N_{ter}-TYYGNGVSCNKKGCSVDWGKAI-C_{ter}	4288	Birri et al. 2010
Carnobacterium piscicola CS526	piscicocin CS526	N_{ter}-KYYGNGLSxNKKGxTVDWGTAIGIIGNNAAANxATGGAAGxNK-C_{ter}	4430	Yamazaki et al. 2005
Carnobacterium piscicola JG26	piscicolin 126	N_{ter}-KYYGNGVSCNKNGCTVDWSKAIGIIGNNAAANLTTGGAAGWNKG-C_{ter}	4416	Jack et al. 1996
Lactobacillus sakei 5	sakacin 5X	N_{ter}-KYYGNGLSCNKSGCSVDWSKAISIIGNNAVANLTTGGAAGWKS-C_{ter}	4083	Vaughan et al. 2001
Listeria innocua 743	listeriocin 743A	N_{ter}-KSYGNGVHCNKKKCWVDWGSAISTIGNNSAANWATGGAAGWKS-C_{ter}	4484	Kalmokoff et al. 2001
Lactobacillus sakei LTH673	sakacin P	N_{ter}-KYYGNGVHCGKHSCTVDWGTAIGNIGNNAAANWATGGNAGWNK-C_{ter}*	4437	Tichaczek et al. 1994
Leuconostoc mesenteroides TA33a	leucocin C	N_{ter}-KYYGNGVHCTKSGCSVNWGEAFSAGVHRLANGGNGFW-C_{ter}	4598	Papathanasopoulus et al. 1998
Carnobacterium divergens M35	divergicin M35	N_{ter}-TKYYGNGVYCNSKKCWVDWGTAQGCIDVVIGQLGGGIPGKGKC-C_{ter}	4518	Tahiri et al. 2004
Carnobacterium divergens V41	divercin V41	N_{ter}-TKYYGNGVYCNSKKCWVDWGQASGCIGQTVVGGWLGGAIPGKC-C_{ter}	4509	Métivier et al. 1998
Escherichia coli* Origami (DE3)*	**divercin RV41**	**N_{ter}-MDPTKYYGNGVYCNSKKCWVDWGQASGCIGQTVVGGWLGGAIPGKC-C_{ter}****	4923	Richard et al. 2004
Enterococcus faecium CTC492/T136	enterocin A	N_{ter}-TTHSGKYYGNGVYCTKNKCTVDWAKATTCIAGMSIGGFLGGAIPGKC-C_{ter}	4829	Aymerich et al. 1996
Lactobacillus plantarum 423	plantaricin 423	N_{ter}-KYYGNGVTCGKHSCSVNWGQAFSCSVSHLANFGHGKC-C_{ter}	3500	Van Reenen et al. 1998
Lactobacillus sakei 2512	sakacin G	N_{ter}-KYYGNGVSCNSHGCSVNWGQAWTCGVNHLANGGHGVC-C_{ter}	3839	Simon et al. 2002
Lactobacillus plantarum C19	plantaricin C19	N_{ter}-KYYGNGLSCSKKGCTVNWGQAFSCGVNRVATAGHGKx-C_{ter}	3845	Atrih et al. 2001
Leuconostoc mesenteroides TA33a	leucocin A	N_{ter}-KYYGNGVHCTKSGCSVNWGEAFSAGVHRLANGGNGFW-C_{ter}*	3930	Papathanasopoulus et al. 1998
Leuconostoc mesenteroides Y105	mesentericin Y105	N_{ter}-KYYGNGVHCTKSGCSVNWGEAASAGIHRLANGGNGFW-C_{ter}	3868	Fleury et al. 1996
Lactococcus lactis MMFII	lactococcin MMFII	N_{ter}-TSYGNGVHCNKSKCWIDVSELETYKAGTVSNPKDILW-C_{ter}	4142	Ferchichi et al. 2001
Lactobacillus curvatus LTH 1171	curvacin A	N_{ter}-ARSYGNGVYCNNKKCWVNRGEATQSIIGGMISGWAGLAGM-C_{ter}*	4309	Tichaczek et al. 1992
Enterococcus faecium P13	enterocin P	N_{ter}-ATRSYGNGVYCNNSKCWVNWGEAKENIAGIVISGWASGLAGMGH-C_{ter}	4993	Cintas et al. 1997
Carnobacterium piscicola LV17B	carnobacteriocin BM1	N_{ter}-AISYGNGVYCNKEKCWVNKAENKQAITGIVIGGWASSLAGMGH-C_{ter}	4524	Quadri et al. 1994
Enterococcus faecium RC714	bacteriocin RC714	N_{ter}-ATYYGNGLYCNKEKCWVDWNQAKGEIGKIIVNGWVNHGPWAP-C_{ter}	5100	Del Campo et al. 2001
Enterococcus faecium T8	bacteriocin T8	N_{ter}-ATYYGNGLYCNKEKCWVDWNQAKGEIGKIIVNGWVNHGPWAPRR-C_{ter}	5008	De Kwaadsteniet et al. 2006
Enterococcus faecalis YI717	bacteriocin 31	N_{ter}-ATYYGNGLYCNKQKCWVDWNKASREIGKIIVNGWVQHGPWAPR-C_{ter}	4780	Tomita et al. 1996
Pediococcus pentosaceus ATCC 25745	penocin A	N_{ter}-KYYGNGVHCGKKTCYVDWGQATASIGKIIVNGWTQHGPWAHR-C_{ter}	4684	Diep et al. 2006
Enterococcus faecalis K-4	enterocin SE-K4	N_{ter}-ATYYGNGVYCNKQKCWVDWSRARSEIIDRGVKAYVNGFTKVLG-C_{ter}	5356	Eguchi et al. 2001
Carnobacterium piscicola LV17B	carnobacteriocin B2	N_{ter}-VNYGNGVSCSKTKCSVNWGQAFQERYTAGINSFVSGVASGAGSIGRRP-C_{ter}*	4969	Quadri et al. 1994
Streptococcus uberis strain E	ubericin A	N_{ter}-KTVNYGNGLYxNQKKxWVNWSETATTIVNNSIMNGLTGGN-C_{ter}	5270	Heng et al. 2007
Enterococcus faecium NRRL B-30746	enterocin E50-52	N_{ter}-TTKNYGNGVCNSVNWCQCGNVWASCNLATGCAAWLCKLA-C_{ter}	3339	Svetoch et al. 2008
Pediococcus pentosaceus 31-1	pentocin 31-1	N_{ter}-VIADYGNGVRxATLL-C_{ter}	5592	Zhang et al. 2009
Lactobacillus salivarius NRRL B-30514	bacteriocin OR-7	N_{ter}-KTYYGTNGVHCTKNSLWGKVRLKNMKYDQNTTYMGRLQDILLGWATGAFGKTFH-C_{ter}	5123	Stern et al. 2006

*Bacteriocins with three dimensional structures yet solved. ** Recombinant divercin RV41

Fig. 10.1 Sequences alignment of class IIa bacteriocins. Alignment was performed with ClustalW2 software, available at http://www.ebi.ac.uk/Tools/clustalw2/index.html

Class IIa bacteriocins were formerly considered as "narrow"-spectrum antibiotics, with antimicrobial activity directed against related strains. Recently, some class IIa bacteriocins, such as bacteriocin OR-7, enterocin E50-52, and enterocin E760, have been shown to be active against both Gram-negative and Gram-positive bacteria, including *Campylobacter jejuni, Yersinia* spp., *Salmonella* spp., *Escherichia coli* O157:H7, *Shigella dysenteriae, Morganella morganii, Staphylococcus aureus,* and *Listeria* spp. (Line et al. 2008; Stern et al. 2006; Svetoch et al. 2008a, b). The activity of class IIa bacteriocins seems to be tightly dependent on the number of disulfide bridges, as evidenced by a principal component analysis, which has revealed a strong correlation between the spectrum of activity and the presence of an extra disulfide bridge (Richard et al. 2006).

From studies on class IIa bacteriocins, it has become clear that an YGNGV (Tyr-Gly-Asn-Gly-Val) pediocin box is a prerequisite for class IIa bacteriocins. However, class IIa bacteriocins with YGNGL motif (replacement of valine by leucine) are reported in the literature; this is the case of bacteriocin 31 (Tomita et al. 1996), plantaricin C19 (Atrih et al. 2001), and sakacin 5x (O'Mahony et al. 2000). The residues leucine and valine have similar hydrophobicities, but leucine does not have the β-methyl side chain of valine, which provides a force restraining the freedom of the φ and ψ torsion angles and may have an effect on the stereo-structure of the bacteriocin (Feng et al. 2009). Therefore, we suggest that the highly conserved pentapeptide motif or "pediocin box" of class IIa bacteriocins should instead be denoted by YGNG[V/L].

The most recently reported repertoire of class IIa bacteriocins constituted 28 peptides. (Nissen-Meyer et al. 2009). Here, we wish to complete that repertoire by adding pentocin 31-1 (Zhang et al. 2009), bacteriocin 0R-7 (Stern et al. 2006), ubericin A (Heng et al. 2007), mundticin L (Feng et al. 2009), enterocin E50-52 (Svetoch et al. 2008a), and avicin A (Birri et al. 2010). Briefly the sequence of pentocin 31-1, although it is only partial, does contain the YGNGV box. Bacteriocin OR7 is composed of 54 aminoacids and does contain the YGTNGV motif and one cysteine residue. Ubericin A is composed of 40 amino acids and has an YGNGL motif at the N-terminal half. Mundticin L is composed of 58 amino acids and exhibits the YGNGL signature. Enterocin E50-52 does contain the pediocin-box sequence YGNGV from positions 5-9 within the 39 amino acids of the peptide (Svetoch et al. 2008a). The main characteristics of these novel class IIa bacteriocins are given in Table 10.1, while their respective sequences are shown in Fig. 10.1.

As stated above, the C-terminal region is less conserved and rather diverse. The sequence alignment of the mature peptides has allowed for the division of class IIa bacteriocins into four subgroups (Nissen-Meyer et al. 2009). There is a flexible hinge at the conserved Asp 17 in peptides of subgroup 1 and also at Asn17/Asp17 in peptides of subgroup 2, 3, and 4. The conserved region within the "pediocin box" directs the nonspecific binding to the cell surface, and the C-terminal domain is involved in target specificity (Johnsen et al. 2005). The three dimensional (3D) structures of several class IIa bacteriocins have been determined, as for leucocin A (Frégeau Gallagher et al. 1997), plantaricin

Table 10.1 Biochemical traits and origins of novel class IIa bacteriocins

Class IIa bacteriocin	Producer strain	Origin	Molecular weight (kDa)	Nucleotide or protein sequence accession number	Source
Bacteriocin OR-7	*Lactobacillus salivarius* NRRL B-30514	Cecum of a commercial broiler	5,123		Stern et al. (2006)
Ubericin A	*Streptococcus uberis* E		5,270	EF203953	Heng et al. (2007)
Pentocin 31-1	*Lactobacillus pentosus* 31-	Traditional Chinese fermented Xuan-Wei ham	5,592		Zhang et al. (2009)
Mundticin L	*Enterococcus mundtii* CUGF08	Alfalfa sprouts		FJ899708.	Feng et al. (2009)
Avicin A	*Enterococcus avium*	Infants	4.285	FJ851402.	Birri et al. (2010)
Enterocin E 50 – 52	*Enterococcus faecium* (NRRL B-30746)	Cecal contents of commercial broilers	3.339	P85148	Svetoch et al. (2008a)

A (Hauge et al. 1998), sakacin P (Uteng et al. 2003), carnobacteriocin B2 (Sprules et al. 2004), curvacin A (Haugen et al. 2005). Circular dichroism (CD) and nuclear magnetic resonance revealed that the above class IIa bacteriocins adopted an unfolded structure in water, while they were able to adopt their structure when just in contact with membrane-mimicking environments (Frégeau Gallagher et al. 1997; Haugen et al. 2005; Watson et al. 2001). As class IIa bacteriocins are constituted of N- and C-terminal domains, one has to understand how each domain is structured. The conserved N-terminal region of about 17 residues forms a three-stranded antiparallel β-sheet-like structure that is stabilized by a conserved disulfide bridge. The hydrophobic C-terminal region usually forms a hairpin-like structure consisting of an amphiphilic α-helix (from about residue 18 to about residue 33) followed by a rather extended C-terminal tail that folds back onto the central α-helix (Fig. 10.2; Nissen-Meyer et al. 2009). Notwithstanding that all of the above class IIa bacteriocins possess only one disulfide bridge. The effect of temperature on the antibacterial activity of class IIa bacteriocins has been largely investigated. At higher temperatures, peptides with a unique disulfide bridge showed partial disruption of the helical structure, while those carrying an additional disulphide bridge at the C-terminal half retained their tertiary structure and α-helix (Kaur et al. 2004). Among class IIa bacteriocins with two disulfide bridges, divercin V41 is likely the most active one and also the studied model. A recent review reported specifically the achievements realized on this peptide during the last decade (Rihakova et al. 2009b).

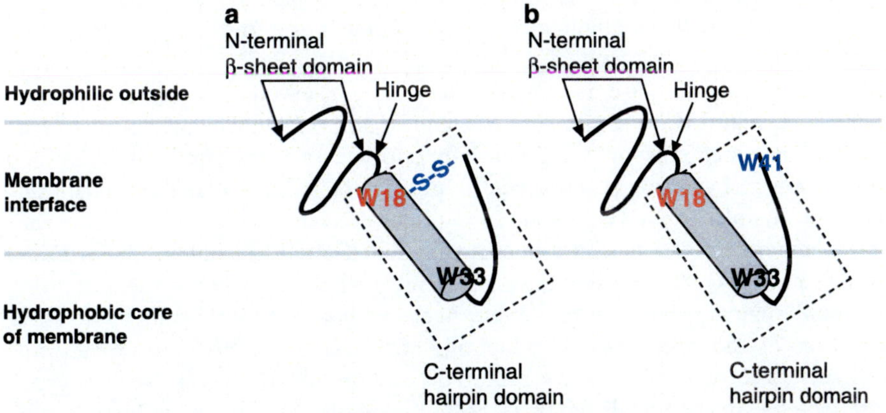

Fig. 10.2 A cartoon depiction of the structure and orientation in membranes of class IIa bacteriocins that belong to subgroups 1 and 2. In class IIa bacteriocins, the C-terminal hairpin structure is stabilized by (**a**) a disulfide bridge (such as in enterocin A, divercin V41,divergicin M35, coagulin, pediocin PA-1, sakacin G, and plantaricin 423), and by (**b**) an interface-localized tryptophan residue near the C-terminal end of the bacteriocin (such as in the other subgroup 1 and 2 bacteriocins). Tryptophan residues that become localized in the membrane–water interface are indicated by either a *red* or a *blue* W, tryptophan residues that become localized in the hydrophobic core of the membrane by a *black* W, and the disulfide bridge by -S-S-

Biosynthesis of Class IIa Bacteriocins

Intercellular communication plays a key role in the regulation of several physiological processes in gram-positive bacteria. Cell–cell communication is often mediated by secreted inducer peptide pheromones (IPs), which upon reaching a threshold concentration in the environment specifically activate a cognate membrane-localized histidine protein kinase (HPK). The process of quorum sensing is regulated by the production and monitoring of quorum sensing molecules (QSMs). When an apparent threshold concentration of QSMs (reflecting high bacterial density) is reached, the bacterial colony in concert 'switches on' the density-dependent trait. Quorum-sensing systems used for the regulation of class IIa bacteriocin production are composed of three gene products and termed three-component regulatory systems (Nes and Eijsink 1999). The three components are (1) the inducer peptide (a peptide pheromone), (2) the transmembrane histidine protein kinase (pheromone receptor), and (3) the cytosolic response regulator (Nes and Eijsink 1999). The inducer peptide is ribosomally synthesized at low levels as a prepeptide, which is cleaved and secreted through the dedicated bacteriocin ABC transporter (Ennahar et al. 2000; Nes and Eijsink 1999). At a certain concentration threshold of the externalized inducer peptide, the transmembrane histidine kinase is activated, leading to the autophosphorylation of a conserved histidine residue at the cytosolic side of the transmembrane protein (Cho et al. 2001; Nes and Eijsink 1999). Plantaricin A inducer peptide was shown to interact extensively with cell membranes, and stereospecific interactions with the histidine kinase are thus necessary to elicit the histidine kinase response (Hauge et al. 1998; Kristiansen et al. 2005). Then, the activated histidine kinase interacts with its cognate response regulator protein through transphosphorylation, and the phosphate group residing on the histidine residue of the activated histidine kinase is transferred to a conserved Asp residue in the response regulator (Cho et al. 2001; Nes and Eijsink 1999). The phosphorylated, and thus activated, response regulator functions as a transcriptional activator, which binds to bacteriocin gene-specific promoters and stimulates transcription (Nes and Eijsink 1999). The response regulator activates the genes encoding the three-component system, and a positive feedback circuit is thus initiated (Nes and Eijsink 1999). At a certain time, essentially all bacteriocin producer cells in the population are believed to secrete bacteriocins, and this result in a rapid activation of the bacteriocin production. More insights into quorum-sensing mechanisms have been recently reported (Straume et al. 2007). Such a gene repression mechanism mediated by truncated regulators was also found in quorum sensing-based bacteriocin systems (in *Lactobacillus sakei* LTH673 and *NC8-pln* in *Lb. plantarum* NC8), suggesting that this mode of repression might represent a common means applied by bacteria to downregulate certain quorum sensing-based pathways.

Mode of Action of Class IIa Bacteriocins

The antibacterial activity of bacteriocins is thought to be mediated by the induction of membrane permeability of the target cell membrane, probably by the formation of ion-selective pores, leading to dissipation of the proton-motive force and depletion of intracellular ATP (Chikindas et al. 1993; Herranz et al. 2001a; Herranz et al. 2001b). The electrostatic interactions become less significant (2–2,000-fold in the antibacterial activity) with bacteriocin variants presenting a modified net positive charge (substitution of positively charged residues by neutral ones) (Miller et al. 1998a).

The N-terminal β-sheet-like domain mediates binding of the class IIa bacteriocin to the target cell surface through electrostatic interactions (Chen et al. 1997a, b; Kazazic et al. 2002). The more hydrophobic C-terminal hairpin-like domain penetrates obliquely into the hydrophobic core of target membranes, mediating membrane leakage. Finally, the modification of the pediocin box was shown to alter and modify the antimicrobial activity of the peptide and also its interactions with liposomes (Fimland et al. 2006). For example, studies carried out on pediocin-PA-1/AcH have revealed the abilities of this peptide to induce leakage of K^+ ions, amino acids, and other low-molecular-weight molecules from sensitive cells (Bhunia et al. 1991; Chikindas et al. 1993). Pediocin-PA-1/AcH also dissipates the $\Delta\Psi$ and the ΔpH from *Pediococcus pentosaceus* and *L. monocytogenes* (Chikindas et al. 1993; Christensen and Hutkins 1992) and induces rapid depletion of intracellular ATP in these bacteria. When pediocin PA-1/AcH was added to carboxyfluorescein-loaded vesicles derived from membranes from sensitive and insensitive cells, only vesicles made from membranes of pediocin-sensitive cells became leaky (Chikindas et al. 1993). The authors proposed that a docking molecule was needed for pediocin PA-1/AcH to be fully efficient. Subsequently, Chen et al. reported that pediocin PA-1/AcH induced leakage of carboxyfluorescein from artificial liposomes, but possibly at higher concentrations than those needed to induce leakage from membranes of sensitive cells (Chen et al. 1997a; Chen et al. 1997b). Furthermore, a chimeric protein synthesized after joining the N-terminal domain of pediocin PA-1 with the C-terminal domain of the maltose-binding protein exhibited antimicrobial activity, suggesting that the N-terminal β-sheet-like domain should remain outside of the cell, while the C-terminal hairpin domain must, consequently, be that which penetrates into the core of target membranes (Miller et al. 1998b). To develop an improved version of pediocin PA-1, reciprocal chimeras between pediocin PA-1 and enterocin A, another class IIa bacteriocin, were constructed. Chimera EP, which consisted of the C-terminal half of pediocin PA-1 fused to the N-terminal half of enterocin A, showed increased activity against a strain of *Leuconostoc lactis* isolated from a sour-spoiled dairy product. To develop an even more effective version of this chimera, a DNA-shuffling library was constructed, wherein four specific regions within the N-terminal half of pediocin PA-1 were shuffled with the corresponding sequences from 10 other class IIa bacteriocins. Activity screening indicated that 63 out of 280 shuffled mutants had antimicrobial activity (Tominaga and Hatakeyama 2007).

Studies conducted on other models of class II a bacteriocins such as bavaricin MN (Kaiser and Montville 1996), enterocin P (Herranz et al. 2001a, b), mesentericin Y105 (Maftah et al. 1993), and mundticin (Bennik et al. 1998) also pointed out a membrane-permeabilizing mode of action similar to that of pediocin PA-1/AcH, but this is not the case of mesentericin Y105, which is able to uncouple mitochondrial respiration (Maftah et al. 1993). Tryptophan fluorescence spectroscopy studies on pediocin PA-1/AcH and mesentericin Y105 in the presence of liposomes have shown that at least the C-terminal tryptophan-containing parts of class IIa bacteriocins penetrate into liposomes/membranes, but the extent of penetration depends on the charge of the liposomes used and the location of the tryptophan residue in the peptide (Chen et al. 1997a; Chikindas et al. 1993; Morisset et al. 2004).

However, these results do not report (or mention only briefly) the structure of the suggested bacteriocin pores or whether the peptides act as monomers or oligomers, nor do they reveal the proposed docking molecule. The requirement for a docking molecule became more evident after the discovery that a 15-mer peptide fragment derived from the C-terminal half of pediocin PA-1/AcH inhibited pediocin PA-1/AcH activity and, to a lesser extent, the activities of other class IIa bacteriocins (Fimland et al. 1998). The authors hypothesized that this 15-mer peptide fragment might interact with a docking molecule and thus compete with pediocin PA-1/AcH for recognition/binding to the target molecule. Yan et al. (2000) showed that an all-D leucocin A enantiomer was inactive and thereby convincingly demonstrated that permeabilization of target cells by leucocin A involves chiral interactions, possibly with a receptor at the surface of the target cell (Yan et al. 2000). This receptor has been proposed to be the mannose permease (Héchard et al 2001), specifically an extracellular loop of the membrane-located protein MptC that is responsible for specific target recognition by the class IIa bacteriocins (Kjos et al. 2010).

The orientation of class IIa bacteriocins into target cell membranes has been subjected to practical and simulations analysis. A study conducted on curvacin A and 55 variants generated by site-directed mutagenesis suggested a specific orientation and that the hydrophilic short central helix (residues 19–24), along with the N-terminal β-sheet-like structure (residues 1–16), inserts into the interface region of the target cell membrane, with Ala22 close to the hydrophobic core of the membrane (Haugen et al. 2008). Then, the hinge region, with Gly28 as an important residue, may form a turn wherein Gly28 becomes positioned near the border between the interface and the hydrophobic regions, thus permitting the longer and more-hydrophobic C-terminal helix (residues 29–41) to insert into the hydrophobic core of the membrane (Haugen et al. 2008). Accordingly, this helix contains three glycine residues at positions 33, 37, and 40, leading to a putative helix–helix-interacting GxxxGxxG motif. The replacement of any of these glycine residue with a larger residue was very detrimental, suggesting thereof their possible implication in helix–helix interactions with a membrane-embedded receptor protein (Haugen et al. 2008).

Studies on structure–activity relationships were performed on the recombinant form of DvnV41 peptide carrying four additional amino-acid residues (AMDP) located in the N-terminal domain. The recombinant DvnRV41 retained the initial antibacterial activity (Richard et al. 2004). The relationship between the structure of DvnRV41 and its particularly elevated antibacterial activity was investigated by creating eight structural variants in which a specific amino acid was replaced in the sequence by another one of a same group, in agreement with the amino acid Venn relation (Taylor 1986). Eight amino acids were targeted for change in the DvnRV41 variants: Trp19, Gln21, Ala22, Ser23, Val30, Leu35, Ala38, and Pro40. These amino acids were selected because they are (1) common to class IIa bacteriocins (Trp19, Ala22) or (2) essentially specific for the group of class IIa bacteriocins closely related to DvnV41, which possess two disulfide bonds.

The antibacterial activity of all of the variants was diminished in comparison to that of unmodified DvnRV41, and this drop in antibacterial activity was not ascribed to any global conformational change (Rihakova et al. 2009a). The effects of DvnRV41 in combination with K-channel activators argue for the formation of ATP-dependent K^+-channels based on the strong synergistic/additive effect observed between DvnRV41 and cromakalim (Rihakova et al. 2010); cromakalim is a K^+ channel opener that causes smooth muscle relaxation by activating ATP-sensitive K^+ (K_{ATP}) channels and producing membrane hyperpolarization (Smoaki 1999). Divergicin M35 exhibited activity against *L. monocytogenes* with sensitive (DivS) and resistant (DivM) phenotypes, as well as on synthetic phospholipid liposomes, resulting in the release of 1,6-diphenyl-1,3,5-hexatriene (DPH) from zwitterionic (DMPC) and anionic (DMPC/DMPG, 4:1) liposomes upon treatment with 5 μg/ml of this bacteriocin (Naghmouchi et al. 2007).

Short 15-mer synthetic peptides derived from the N-terminal half of enterocin CRL35, mesentericin Y105, pediocin PA-1/AcH and piscicolin 126 exhibited antilisterial activity in synthetic and minimal media and this anti-*Listeria* activity was exerted through dissipation of the membrane potential (Salvucci et al. 2007). Another study by Minahk et al. (2004) revealed that the combination of enterocin CRL35 with various antibiotics, such as erythromycin, chloramphenicol, and tetracycline, induced a significant membrane gradient dissipation without appreciable cell death. A plausible explanation is that membrane depolarization is necessary, but not sufficient to provoke cell death, and that another concentration-dependent step might be required.

Bacteriocin J46 is a 27-residue peptide produced by *Lactococcus lactis* subsp. *cremoris* J46 (*L. lactis* subsp. *cremoris* J46) in fermented milk. The natural form of J46 (nJ46) exhibited a broad antimicrobial spectrum, while the synthetic form of J46 (sJ46) exhibited biochemical and physicochemical properties identical to those of its natural counterpart. It showed a potent antimicrobial activity against both lactic acid bacteria and other Gram-positive bacteria (Lasta et al. 2008).

Immunity Against Class IIa Bacteriocins

It is known that bacteria producing pediocin-like bacteriocins also produce cognate immunity proteins that protect them from being killed by their own bacteriocins. About twenty immunity proteins (IP) of class IIa bacteriocins have been deduced from genome sequences (Fimland et al. 2005). Overall, they (1) exhibit a degree of similarity ranking from 5 to 85 %, (2) contain a number of amino acids ranging between 88 and 115, and (3) show a high level of specificity, since they confer resistance solely to their cognate class IIa bacteriocin and to bacteriocins that are closely related to the cognate bacteriocin (Fimland et al. 2002; Johnsen et al. 2005). To strengthen the idea of high sequence similarity between immunity proteins, it was recently shown by Feng et al. (2009) that IP of mundticin L has 98 amino acids, the same length as those of mundticin KS and enterocin CRL35, but it differs from them by the amino acids located at positions 30 (Ile), 31 (Glu), 46 (Lys), and 89 (Ile). The probable function of protein UbaI from *Streptococcus uberis* E, which is constituted of 99 amino acids and displays 52% identity (71% similarity) to MunC, the protein which confers immunity to the class IIa peptide mundticin ERL35, is very likely to protect the ubericin A producer strain from the lethal effects of its own bacteriocin (Heng et al. 2007). In bacteria closely related to lactic acid bacteria (LAB), such as *Streptococcus pyogenes*, which is a common human pathogen, the crystal structure of the gene product of locus "Spy_2152" was found to comprise an antiparallel four-helix bundle that is structurally similar to other bacteriocin immunity proteins (Chang et al. 2009).

Moreover, the nuclear magnetic resonance and crystal structures studies of immunity proteins of carnobacteriocin B2, enterocin A, and pediocin PP1 revealed an antiparallel four-helix bundle with a C-terminal region that appears to be structurally somewhat flexible (Sprules et al. 2004; Kim et al. 2007). The immunity proteins are a means of protection and are anticipated to act through association with the internal side of the cell membrane, since it has been demonstrated using hybrid pediocin-like bacteriocins and hybrid immunity proteins that the membrane-penetrating C-terminal bacteriocin domain specifically interacts with the C-terminal half of the immunity protein, possibly via the structurally somewhat flexible C-terminal tail of the immunity proteins (Johnsen et al. 2005). Nevertheless, studies aiming to unravel direct interactions between immunity proteins and cognate bacteriocins were unsuccessful (Sprules et al. 2004). The strong binding of immunity proteins to the mannose phosphotransferase permease occurs if the cognate bacteriocin binds to the permease system as well. The process of interaction is complex and consists of binding of an immunity protein to the permease–bacteriocin complex and blockage of bacteriocin-induced membrane leakage. However, this mechanism of action requires that the strain-dependent variation in the receptor (mannose phosphotransferase permease) is large enough to induce sufficient variations in the receptor binding of (1) the various class IIa bacteriocins (to account for the different target cell specificities of the class IIa bacteriocins) and (2) the various immunity proteins (to account for specificity of immunity proteins for their cognate bacteriocins) (Diep et al. 2007).

Immunity proteins are categorized into three groups named A, B, and C, based on sequence homologies. The structures of immunity proteins from group C, such as ImB2, and two from group A (EntA-im and PedB) have previously been reported. Recently, the solution structure of a group B immunity protein named PisI, which confers immunity to piscicolin 126 (PisA), was investigated by nuclear magnetic resonance. PisI folds into a globular protein in aqueous solution and contains an antiparallel four-helix bundle (Martin-Visscher et al. 2008). This type of folding is considered as a conserved motif within class IIa immunity proteins. Furthermore, PisA exhibits a substantially longer and more flexible N-terminal domain, but a shorter C-terminal domain, when compared to ImB2 and EntA-im. No direct interaction between the bacteriocin (piscicocin 126) and the immunity protein (PisI) was observed by NMR in either aqueous or membrane-mimicking environments, arguing that the mechanism that mediates immunity is not due to a direct bacteriocin-immunity protein interaction, but rather would be receptor mediated.

The data from molecular dynamics indicate that while carnobacteriocinB2 remained embedded in the bilayer, it tends to move toward the interface and in the presence of carnobacteriocinB2 and in the DPPC bilayer attracts immunity protein B2 toward the bilayer. Overall, in one of the orientations in DPPC bilayer system (simulation 1), immunity protein B2 penetrates the bilayer and interacts with carnobacteriocinB2 by ion-pair interaction. Simulation in POPG bilayer displayed strong interaction between the positively charged immunity protein B2 and the negatively charged polar head groups of the POPG molecules at the lipid–water interface. In this case, immunity protein B2 was not able to penetrate the bilayer, thereby preventing any interaction between immunity protein B2 and carnobacteriocinB2 (Soliman et al. 2007).

Genetic Organization of DNA Coding for Class IIa Bacteriocins

The bacterial molecular machinery displayed in the frame of production of class IIa bacteriocins usually requires four genes: a bacteriocin gene (which encodes the bacteriocin precursor), an immunity gene (which protects the producer from its bacteriocin), and the ABC transporter and transport accessory genes (Nes et al. 1996; Skaugen et al. 2003). Most of the genes coding for bacteriocins are plasmid encoded, except for enterocin A (Aymerich et al. 1996), divercin V41 (Métivier et al. 1998), sakacin P (Hühne et al. 1996), and carnobacteriocins B2 and BM1 (Quadri et al. 1997), for which genes are located on chromosomes. The class IIa bacteriocin genes are most often arranged in one or a few operons and their common organization has been compared and reviewed by Ennahar et al. (2000). However, an exception to the common gene organization is observed for the divercin V41 bacteriocin genes, which present a particular organization attributed to DNA rearrangements (Ennahar et al. 2000). Genetically characterized class IIa bacteriocin gene clusters are composed of one or three gene modules. In the cases of pediocin PA-1, plantaricin 423, and coagulin, all four genes needed for bacteriocin production and secretion are located in one operon. In other cases, the genes are distributed throughout several operons, where one operon carries the structural and immunity genes, a second operon carries genes for bacteriocin secretion, and a third operon carries genes involved in regulation of bacteriocin production (Ennahar et al. 2000). Genes implicated in biosynthesis and production of class IIa bacteriocins are not necessarily located on the same locus or even the same DNA determinant.

As recently reported, the structural and immunity genes on one hand and the genes for bacteriocin secretion on the other hand were found to reside together on individual plasmids in *Pediococcus parvulus* and *Lactobacillus plantarum* WHE92 (Miller et al. 2005).

As described above, the organization of genes involved in the production of plantaricin 423, pediocin PA-1/AcH, and coagulin are almost identical (Van Reenen et al. 2003). More specifically, the coagulin-codifying DNA (*coaABCD* operon) is identically organized and displayed high sequence similarity to that of the *pap* operon encoding the pediocin PA-1/AcH genes (Miller et al. 2005), despite the fact that coagulin is produced by the relatively unrelated non-LAB *Bacillus coagulans* I4 (Le Marrec et al. 2000). This finding could be attributed to a plasmid-borne gene transfer between *Pediococcus acidilactici* and *Bacillus coagulans*, since horizontal gene/operon transfer between bacteria seems to be more common than was originally thought (De la Cruz and Davies 2000). Finally, the genetic organization of the mundticin KS gene cluster (*mun* locus) has been established (Kawamoto et al. 2002). It consists of three genes encoding (1) the 58-amino-acid mundticin precursor (*munA*), (2) a 674-amino acid polypeptide with similarity to the ABC transporter proteins EntT (Nes et al. 1996) and CbnT (O'Keeffe et al. 1999), and (3) the peptide modulating mundticin KS immunity (*munC*).

Plantaricin 423 is a class IIa bacteriocin produced by *Lb. plantarum* that is encoded by a plasmid designated pPLA4. The plantaricin 423 operon shares high sequence similarity with the operons of coagulin and pediocin PA-1/AcH, with small differences in the DNA sequence encoding the mature bacteriocin peptide and the immunity protein. Apart from the bacteriocin operon, no significant sequence similarity can be detected between the DNA or translated sequence of pPLA4 and the available DNA or translated sequences of the plasmids encoding pediocin AcH, pediocin PA-1, and coagulin, possibly indicating a different origin. In addition to the bacteriocin operon, sequence analysis of pPLA4 revealed the presence of two open reading frames (ORFs). ORF1 encodes a putative mobilization (Mob) protein that is homologous to the pMV158 superfamily of mobilization proteins. Highest sequence similarity occurs between this protein and the Mob protein of *Lb. plantarum* NCDO 1088. ORF2 encodes a putative replication protein that has low sequence similarity to replication proteins of plasmids pLME300 from *Lb. plantarum* and pYIT356 from *Lb. casei*. The immunity protein of plantaricin 423 contains 109 amino acids. Although plantaricin 423 shares high sequence similarity with the pediocin PA-1 operon, no cross-reactivity was detected between the immunity proteins and plantaricin PA1. The information related to genetic organization of class IIa bacteriocins was extensively debated in previous reviews (Ennahar et al. 2000; Drider et al. 2006). The novel information presented here is related to the genetic organization of ubericin A and avicin A. The genetic locus *uba* responsible for ubericin A production is composed of two genes referred to as *ubaA* and *ubaI*, which overlap by one base pair (i.e., the last nucleotide of the *ubaA* stop codon is the first nucleotide of the *ubaI* start codon) (Heng et al. 2007). The *ubaA* gene encodes the ubericin A propeptide consisting of the mature ubericin A C-terminal region (49 amino acids) preceded by a 21-amino acid leader peptide secretion signal peptide containing a "double-glycine" (GG) motif, a distinctive characteristic of many bacteriocins produced by Gram-positive bacteria.

DNA coding for avicin A production is contained within eight putative ORFs, seven of which are related to bacteriocin production. The seven ORFs include the genes encoding a structural bacteriocin (*avcA*) and its immunity protein (*avcI*), a divergicin A-like bacteriocin (*avcB*), a dedicated ABC-type transporter (*avcT*), a peptide induced two-component regulatory system (histidine kinase [*avcK*], a response regulator [*avcR*], a peptide pheromone [*avcP*]), and a transport accessory protein (*avcD*). All these ORFs seem to be unidirectionally oriented, and the bacteriocin gene cluster is organized into four predicted operon structures. The first operon (*avcAI*) includes the structural gene for avicin A (*avcA*) and the immunity gene (*avcI*). The second operon was predicted to be a monocistronic transcript unit consisting of a single ORF, whose product is significantly similar to divergicin A. The third operon (*avcTKR*) contains genes coding for an ABC transporter (*avcT*), a histidine protein kinase (*avcK*), and a response regulator (*avcR*). Finally, the fourth operon comprises genes encoding a putative peptide pheromone (*avcP*) and a transport accessory protein (*avcD*) (Birri et al. 2010).

Resistance Mechanism to Class IIa Bacteriocins

The mode of action of class IIa bacteriocins has been investigated at the molecular level. An early genetic study using transposon-induced mutants of *L. monocytogenes* showed that the inactivation of the *rpoN* gene resulted in resistance to mesentericin Y105 (Robichon et al. 1997). According to the same authors, the complementation of this mutant with the wild-type *rpoN* gene restored the sensitivity. This result indicated that *rpoN* was involved in the sensitivity of *L. monocytogenes*. The *rpoN* gene encodes the σ^{54} subunit of the bacterial RNA polymerase. This subunit is an alternative sigma factor responsible for the transcription of a specific set of genes. The σ^{54}-dependent transcription displays three specific features. First, σ^{54} recognizes specific promoters with a particular consensus sequence, which is different from those recognized by the housekeeping σ^{70} factor. Second, the promoter sequence is located at position −12/−24. Third, σ^{54} requires an interaction with an activator protein to initiate the transcription. It has been shown that the interruption of *rpoN* in another class IIa-sensitive species, *Enterococcus faecalis*, also leads to resistance (Dalet et al. 2000). Therefore, *rpoN* might be involved directly or indirectly in a general mechanism of sensitivity to class IIa bacteriocins. It has been hypothesized that *rpoN* is involved in the expression of a target molecule for class IIa bacteriocins, whose loss of expression leads to resistance.

Furthermore, directed mutagenesis of the *mpt* operons of both *L. monocytogenes* and *E. faecalis* led to resistance of these bacteria to class IIa bacteriocins (Dalet et al. 2001; Héchard et al. 2001). Factor σ^{54} directs the expression of the *mpt* operon, and high-level resistance to mesentericin Y105 as well as other class IIa bacteriocins results from the loss of *mpt* expression, either in defined mutants or in spontaneous resistant strains (Gravesen et al. 2002). The *mpt* operon encodes a mannose permease, named EII^{Man}, which belongs to the phosphotransferase system (PTS). The PTS is responsible for the transport and concomitant phosphorylation of sugars inside both Gram-negative and Gram-positive bacteria (Postma et al. 1993). The PTS permeases of the mannose family are composed of four domains IIA, IIB, IIC, and IID arranged in two to four subunits. The IIA and IIB cytoplasmic domains are involved in phosphorylation, whereas the IIC and IID membrane domains are involved in transport. The EII^{Man} permease of *L. monocytogenes* is a complex of three subunits, as IIA and IIB are fused. In addition, glucose and mannose induce the expression of *mpt*, while other sugars, such as cellobiose or fructose, do not. The level of *mpt* expression is correlated to the level of sensitivity (Dalet et al. 2001, Héchard et al. 2001) and these results suggest that the EII^{Man} permease might be a target molecule for class IIa bacteriocins. The EII^{Man} phosphotransferase system permease encoded by the *mpt* operon is the principal glucose transporter in *L. monocytogenes*. EII^{Man} participates in glucose-mediated carbon catabolite repression (CCR) and downregulation of virulence gene expression (Vu-khac and Miller 2009). As the IIC and IID subunits are likely present in the membrane, they are ideal targets for class IIa bacteriocins. Finally, the *mptACD* operon of *Listeria monocytogenes* was heterologously expressed in an insensitive species, *L. lactis* (Ramnath et al. 2004). Upon induction of the *mpt* operon, the recombinant *L. lactis* became sensitive to various class IIa bacteriocins. Furthermore, each gene of the *mptACD* operon has been expressed independently in *L. lactis*. These results showed that the expression of *mptC* alone is sufficient to confer sensitivity to *L. lactis*. Accordingly, it has been proposed that the IIC subunit is the target molecule of the class IIa bacteriocins (Ramnath et al. 2004). Membrane-located proteins (IIC and IID) of the mannose-phosphotransferase system (man-PTS) are described as target receptors for several bacteriocins including class IIa bacteriocins and could exist in the genome of many bacteria (Kjos et al. 2009). A multiple sequence alignment of IIC and IID proteins revealed three sequence regions (two in IIC and one in IID) that distinguish members of the bacteriocin-susceptible group from those of the other groups, suggesting that these amino acid regions confer the specific bacteriocin receptor function (Kjos et al. 2009).

Other investigators have reported a correlation between bacteriocin resistance and cell surface modifications, such as the lipid composition of the membrane, alanine content, and surface charge (Vadyvaloo et al. 2002, 2004). The authors concluded that membrane adaptation might be part of a resistance mechanism but that several resistance mechanisms may contribute to a resistance phenotype. The levels of resistance may vary according to the type of mechanism involved. It has been suggested that resistance to class IIa bacteriocins occurs at either a low or a high level. In listerial strains, low-level resistance (2-4-fold) to class IIa bacteriocins is attributed to alterations in membrane lipid composition. In *L. monocytogenes* and *E. faecalis*, high-level resistance (1000-fold) correlates with inactivation of the *mptACD* operon, which encodes the EII^{Man} mannose permease of the phosphotransferase system. Studies performed on *L. monocytogenes* further revealed the implication of σ^{54} factor and the ManR activator in high-level resistance. Recently, three genes associated with the resistance of *E. faecalis* JH2-2 to divercin V41, a class IIa bacteriocin from

Carnobacterium divergens V41, were clearly identified by screening an insertional mutant library of *E. faecalis* JH2-2. These three genes counted the well-known *rpoN* gene, which encodes σ^{54} factor, *glpQ* encoding a glycerophosphoryl diester phosphodiesterase (GlpQ), and *pde* encoding a protein with a putative phosphodiesterase function (PDE). Enterococci defective in the aforementioned genes exhibited gradual resistance, since the *rpoN* mutant was more resistant than the *glpQ* mutant, which was in turn more resistant than the *pde* mutant (Calvez et al. 2007). The impact of lysozyme on the growth of these strains was assessed. This inhibitory substance did not show any cross-resistance but has an additive effect with recombinant divercin V41 (DvnRV41). Fourier transform infrared spectroscopy (FT-IR) pointed out some differences at the structural level between the aforementioned strains. Specifically, mutants deficient in synthesis of σ^{54} factor and glycerophosphoryl diester phosphodiesterase differ from both the mutant deficient in putative phosphoesterase and the wild-type strain in the fatty acid region and polysaccharide composition (Calvez et al. 2010). *L. monocytogenes* resistance to sakacin P has revealed a great diversity and substantial variations in the stability of the acquired resistance to sakacin P, growth fitness, food-related stress tolerance, and biofilm-forming ability. Fourier transform infrared spectroscopy revealed differences between wild-type and resistant strains in polysaccharide, fatty acid, and protein regions (Tessema et al. 2009).

The relative expression of genes (*mptABCD* operon, *glpQ*, *pde*, *rpoN*, and *mptR*) implicated in resistance of *E. faecalis* as studied by reverse transcription (RT) real-time polymerase chain reaction (PCR) showed that in the presence of DvnRV41, the *rpoN* and *glpQ* genes were downregulated, *mptR*, *mptC*, and *pde* genes were upregulated, while expression of *mptB* and *mptD* genes remained unmodified (Calvez et al. 2008a). On the basis of the implication of the *pde* and *glpQ* genes in intermediate resistance of *E. faecalis* to DvnV41, DNA sequence alignments were performed in order to identify putative genes orthologs of *glpQ* and *pde* in the genome of *L. monocytogenes* EGDe (Glaser et al. 2001) This *in silico* analysis led to identification of *lmo0052* and *lmo1292* open reading frames as possible orthologs of enterococcal *glpQ* and *pde*. To confirm whether *lmo0052* and *lmo1292* were involved in the mechanism of resistance of *L. monocytogenes* EGDe to class IIa bacteriocins, they were inactivated by homologous recombination (Calvez et al. 2008b), yielding listerial mutant strains inactivated in either the putative *glpQ* gene or the *pde* gene. While the listerial mutant inactivated in the putative *glpQ* gene was slightly resistant to DvnV41, the mutant strain inactivated in the putative *pde* gene remained, as the wild-type strain, sensitive to DvnV41, but was affected in its growth parameters (Calvez et al. 2008b).

Medical and Food Applications of Class IIa Bacteriocins and Future Prospects

Are bacteriocins the new generation of antimicrobials? This question comes from a previous review by Joerger (2003) who indicated that "purified or partially purified bacteriocins could be used for the reduction or elimination of certain pathogens." Certainly, alternative and novel treatment regimens are of major interest to address

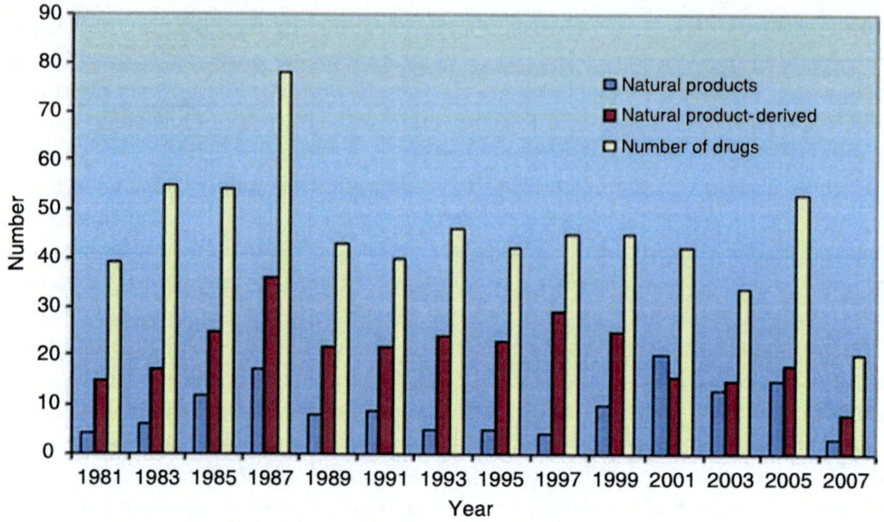

Fig. 10.3 Number of drugs approved in USA from 2001 to 2007

the growing threat of antimicrobial-resistant pathogens. Clearly, there is a real need for novel antibacterials in the medical arena to face the growing phenomenon of bacterial multidrug resistance. The emergence of antibiotic-resistant bacteria has promoted research, which is aimed at the development of novel and potent antimicrobial agents at a time when there is a decreasing number of novel antibiotics marketed over the last decade (Li and Vederas 2009 and Fig. 10.3). Bacteriocins differ from antibiotics in several aspects (Cleveland et al. 2001). They have the potential to employ antimicrobial therapies with bactericidal outcomes differing dramatically from those of traditional antibiotic regimens. Further, bacteriocins are active (bactericidal) at nanomolar concentrations, while classical antibiotics are needed in much higher concentrations. Additionally, there is no information suggesting toxicity of bacteriocins for humans and animals and their accumulation in the treated subject (Belguesmia et al. 2010; Rea et al. 2010). Considering all of these properties, bacteriocins appear to be strongly advantageous compared to conventional antibiotics. Bacteriocins are promising antimicrobial agents to inhibit pathogenic growth despite many disappointing efforts in the past to implement them into practical use. Class IIa bacteriocins, as well as other natural or modified bacteriocins, have potential in medical, veterinary, and food applications.

The use of class IIa bacteriocins as therapeutic molecules was reported for pediocin PA-1, piscicolin 126, bacteriocin OR7, enterocin E50-52, and divercin RV41. Pediocin PA-1 offered protection against *Listeria monocytogenes* infection when administered into the gastrointestinal tract of ICR mice (Dabour et al. 2009), while piscicolin 126 and recombinant divercin RV41 were able to relieve *Listeria* infection in various tissues when injected intravenously into the tail vein of BALB/c mice (Ingham et al. 2003; Rihakova et al. 2010). Remarkably, structural variants gathered from recombinant divercin V41 were able to retain antilisterial activity but remained less efficient than recombinant divercin RV41 (Rihakova et al. 2010). The strategy to combine antibiotics and bacteriocins might be of interest, as observed for pediocin PA-1, whose spectrum of activity was broadened when it was associated with antibiotics. In light of this information, the combination of pediocin PA-1 with polymyxin E was shown to act in a synergistic fashion against Gram-positive *L. monocytogenes* and Gram-negative *E. coli* with resistant phenotypes (Naghmouchi et al. 2010a, b). Finally, the effects of bacteriocin OR7 and enterocin E50-52 were successfully tested *in vitro* against resistant clinical isolates composed of *Citrobacter freundii*, *Klebsiella pneumoniae*, *Proteus* spp., *Acetobacter baumanii, Pseudomonas aeruginosa, E. coli*, and *Staphylococcus aureus* (Svetoch et al. 2008b). All these examples emphasize that class IIa bacteriocins may play an important role as part of a nonantibiotic cyclic rotation with antibiotics and merit further study.

Notwithstanding the fact that antiviral activities against HSV-1, an important human pathogen, were also reported for enterocin CRL 35 (Wachsman et al. 1999, 2003) and enterocin ST5H, with a selectivity index of 173 (Todorov et al. 2010), the mode of action of bacteriocins against viruses remains unknown. Possible explanations could be aggregation of virus particles, blockage of receptor sites on the host cell, or inhibition of key reactions in the multiplication cycle (Wachsman et al. 2003).

In terms of food applications, bacteriocins are historically linked to many types of food productions, and many foodborne LAB are known to be bacteriocin producers. As production of bacteriocin is an important criterion to design bacteria for food-grade probiotics, we anticipate more future beneficial applications of bacteriocins. Class IIa bacteriocins can be used as purified ingredients or through selected starter cultures. Pediocin PA-1 is undoubtedly the model of class II bacteriocins likely to be utilized. Currently, pediocin PA-1 exists in a commercial form in Alta™ products. Pediocin PA-1 offers advantages in dairy foods due to its ability to reduce the levels of *L. monocytogenes* in dressed cottage cheese, half-and-half cream, and cheese sauce (Rodríguez et al. 2002). Another matrix to be studied is raw meats. Indeed its treatment with pediocin PA-1/AcH has delayed growth of spoilage by Gram-positive bacteria such as *Brochotrix thermosphacta* and reduced cells counts of *L. monocytogenes* and *Clostridium perfringens* (Rodríguez et al. 2002; Nieto-Lozano et al. 2006). Carnobacteriocins, or bacteriocins produced by LAB belonging to the *Carnobacterium* genus, first gained interest for protection of seafood products after they were shown to reduce *L. monocytogenes* populations in smoked salmon through in situ production of divercin V41 (Richard et al. 2003). When provided in the bird's drinking water, purified bacteriocin OR-7 markedly reduced the cell load of *Campylobacter jejuni* among inoculated chicks. This intervention measure is aimed at controlling the pathogen in poultry production and therefore reducing the risk of campylobacteriosis (Svetoch and Stern 2010).

Research on class IIa bacteriocins has completed an important step over the last decade and it is anticipated to continue by providing novel academic information and applied demonstration. In terms of academic research, one must expect to acquire more knowledge on the resistance mechanisms of pathogens to these peptides and a better understanding of the immunity characteristics. Applications of class IIa bacteriocins as food additives or alternatives to antibiotics are the next important stages of industrial applications of these beneficial proteins.

Acknowledgments The authors would like to thank Prof. Dzung Diep, Dr. Bruce Seal, Prof. Mike Chikindas, and Dr. Sutyak Noll for critical reading of the manuscript and stimulating discussions. YB was recipient of PhD scholarship awarded by la Région des Pays de la Loire http://www.paysdelaloire.fr/

References

Atrih A, Rekhif N, Moir AJ, Lebrihi A, Lefebvre G (2001) Mode of action, purification and amino acid sequence of plantaricin C19, an anti-*Listeria* bacteriocin produced by *Lactobacillus plantarum* C19. Int J Food Microbiol 68:93–104

Aymerich T, Holo H, Håvarstein LS, Hugas M, Gariga M, Nes IF (1996) Biochemical and genetic characterization of enterocin A from Enterococcus faecium, a new antilisterial bacteriocin in the pediocin family of bacteriocins. Appl Environ Microbiol 62:1676–1682

Belguesmia Y, Madi A, Sperandio D, Merieau A, Feuilloley M, Prévost H, Drider D, Connil N (2010) Growing insights into safety of bacteriocins: case of enterocin S37. Res Microbiol In press

Bennik MHJ, Vanloo B, Brasseur R, Gorris LGM, Smid EJ (1998) A novel bacteriocin with a YGNGV motif from vegetable-associated Enterococcus mundtii: full characterization and interaction with target organisms. Biochim Biophys Acta 1373:47–58

Bhunia AK, Johnson MC, Ray B, Kalchayanand N (1991) Mode of action of pediocin AcH from *Pediococcus acidilactici* H on sensitive bacterial strains. J Appl Bacteriol 70:25–33

Birri DJ, Brede DA, Forberg T, Holo H, Nes IF (2010) Molecular and genetic characterization of a novel bacteriocin locus in *Enterococcus avium* isolates from infants. Appl Environ Microbiol 76:483–492

Calvez S, Rincé A, Auffray Y, Prévost H, Drider D (2007) Identification of new genes associated with intermediate resistance of Enterococcus faecalis to divercin V41, a pediocin-like bacteriocin. Microbiology 153:1609–1618

Calvez S, Prévost H, Drider D (2008a) Relative expression of genes involved in the resistance/sensitivity of *Enterococcus faecalis* JH2-2 to recombinant divercin RV41. Biotechnol Lett 30:1795–1800

Calvez S, Prévost H, Drider D (2008b) Identification of a new molecular target of class IIa bacteriocins in *Listeria monocytogenes* EGDe. Folia Microbiol (Praha) 53:417–422

Calvez S, Kohler A, Prévost H, Møretrø T, Drider D (2010) Physiological and structural differences between *Enterococcus faecalis* JH2-2 and mutant strains resistant to (P)-divercin RV41. Probiotics Antimicro Prot. doi:DOI 10.1007/s12602-010-9048-1 (Published online)

Chang C, Cogill P, Bateman A, Finn RD, Cymorowski M, Otwinowski Z, Minor W, Volkart L, Joachim A (2009) The structure of pyogenecin immunity proytein, a novel bacteriocn-like immunity protein from *Streptococcus pyogenes*. BMC Struct Biol 17:9–75

Chen Y, Ludescher RD, Montville TJ (1997a) Electrostatic interactions, but not the YGNGV consensus motif, govern the binding of pediocin PA-1 and its fragments to phospholipid vesicles. Appl Environ Microbiol 63:4770–4777

Chen Y, Shapira R, Eisenstein M, Montville TJ (1997b) Functional characterization of pediocin PA-1 binding to liposomes in the absence of a protein receptor and its relationship to a predicted tertiary structure. Appl Environ Microbiol 63:524–531

Chikindas ML, Garcia-Garcera MJ, Driessen AJ, Ledeboer AM, Nissen-Meyer J, Nes IF, Abee T, Konings WN, Venema G (1993) Pediocin PA-1, a bacteriocin from *Pediococcus acidilactici* PAC1.0, forms hydrophilic pores in the cytoplasmic membrane of target cells. Appl Environ Microbiol 59:3577–3584

Cho HS, Pelton JG, Yan D, Kustu S, Wemmer DE (2001) Phosphoaspartates in bacterial signal transduction. Curr Opin Struct Biol 11:679–684

Christensen DP, Hutkins RW (1992) Collapse of the proton motive force in *Listeria monocytogenes* caused by a bacteriocin produced by *Pediococcus acidilactici*. Appl Environ Microbiol 58:3312–3315

Cintas LM, Casaus P, Håvarstein LS, Hernández PE, Nes IF (1997) Biochemical and genetical characterization of enterocin P, a novel *sec*-dependent bacteriocin from *Enterococcus faecium* P13 with a broad antimicrobial spectrum. Appl Environ Microbiol 63:4321–4330

Cleveland J, Montville TJ, Nes IF, Chikindas ML (2001) Bacteriocins: safe, natural antimicrobials for food preservation. Int J Food Microbiol 71:1–20

Dabour N, Zihler A, Kheadr E, Lacroix C, Fliss I (2009) In vivo study on the effectiveness of pediocin PA-1 and *Pediococcus acidilactici* UL5 at inhibiting *Listeria monocytogenes*. Int J Food Microbiol 133:225–233

Dalet K, Briand C, Cenatiempo Y, Héchard Y (2000) The *rpoN* gene of *Enterococcus faecalis* directs sensitivity to subclass IIa bacteriocins. Curr Microbiol 41:441–443

Dalet K, Cenatiempo Y, Cossart P, Hechard Y (2001) European *Listeria* Genome Consortium. A sigma (54)-dependent PTS permease of the mannose family is responsible for sensitivity of *Listeria monocytogenes* to mesentericin Y105. Microbiology 147:3263–3269

De Kwaadsteniet M, Fraser T, Van Reenen CA, Dicks LMT (2006) Bacteriocin T8, a novel class Iia sec-dependent bacteriocin produced by *Enterococcus faecium* T8, isolated from vaginal secretions of children infected with human immunodeficiency virus. Appl Environ Microbiol 72:4761–4766

De la Cruz I, Davies I (2000) Horizontal gene transfer and the origin of species: lessons from bacteria. Trends Microbiol 8:128–133

Del Campo R, Tenorio C, Jimenez-Diaz R, Rubio C, Gomez-Lus R, Baquero F, Torres C (2001) Bacteriocin production in vancomycin-resistant and vancomycin-susceptible *Enterococcus* isolates of different origins. Antimicrob Agents Chemother 45:905–912

Diep DB, Godager L, Brede D, Nes IF (2006) Data mining and characterization of a novel pediocin-like bacteriocin system from the genome of *Pediococcus pentosaceus* ATCC 25745. Microbiology 152:1649–1659

Diep DB, Skaugen M, Salehian Z, Holo H, Nes IF (2007) Common mechanisms of target cell recognition and immunity for class II bacteriocins. Proc Natl Acad Sci USA 104:2384–2389

Drider D, Fimland G, Héchard Y, McMullen LM, Prévost H (2006) The continuing story of class IIa bacteriocins. Microbiol Mol Biol Rev 70:564–582

Eguchi T, Kaminaka K, Shima J, Kawamoto S, Mori K, Choi SH, Doi K, Ohmomo S, Ogata S (2001) Isolation and characterization of enterocin SE-K4 produced by thermophilic enterococci, *Enterococcus faecalis* K-4. Biosci Biotechnol Biochem 65:247–253

Ennahar S, Sashihara T, Sonomoto K, Ishizaki A (2000) Class IIa bacteriocins: biosynthesis, structure and activity. FEMS Microbiol Rev 24:85–106

Feng G, Guron GK, Churey JJ, Worobo RW (2009) Characterization of mundticin L, a class IIa anti-*Listeria* bacteriocin from *Enterococcus mundtii* CUGF08. Appl Environ Microbiol 75: 5708–5713

Ferchichi M, Frère J, Mabrouk K, Manai M (2001) Lactococcin MMFII, a novel class IIa bacteriocin produced by *Lactococcus lactis* MMFII, isolated from a Tunisian dairy product. Lett Appl Microbiol 205:49–55

Fimland G, Jack R, Jung G, Nes IF, Nissen-Meyer J (1998) The bactericidal activity of pediocin PA-1 is specifically inhibited by a 15-mer fragment that spans the bacteriocin from the center toward the C terminus. Appl Environ Microbiol 64:5057–5060

Fimland G, Eijsink VG, Nissen-Meyer J (2002) Comparative studies of immunity proteins of pediocin-like bacteriocins. Microbiology 148:3661–3670

Fimland G, Johnsen L, Dalhus B, Nissen-Meyer J (2005) Pediocin-like antimicrobial peptides (class IIa bacteriocins) and their immunity proteins: biosynthesis, structure and mode of action. J Pept Sci 11:688–696

Fimland G, Pirneskoski J, Kaewsrichan J, Jutila A, Kristiansen PE, Kinnunen PKJ, Nissen-Meyer J (2006) Mutational analysis and membrane-interactions of the beta-sheet-like N-terminal domain of the pediocin-like antimicrobial peptide sakacin P. Biochim Biophys Acta 1764:1132–1140

Fleury Y, Dayem MA, Montagne JJ, Chaboisseau E, Le Caer JP, Nicolas P, Delfour A (1996) Covalent structure, synthesis, and structure-function studies of mesentericin Y 105(37), a defensive peptide from gram-positive bacteria *Leuconostoc mesenteroides*. J Biol Chem 271:114421–114429

Frégeau Gallagher NL, Sailer M, Niemczura WP, Nakashima TT, Stiles ME, Vederas JC (1997) Three-dimensional structure of leucocin A in trifluoroethanol and dodecylphosphocholine micelles: spatial location of residues critical for biological activity in type IIa bacteriocins from lactic acid bacteria. Biochemistry 36:15062–15072

Glaser P, Frangeul L, Buchrieser C, Rusniok C, Amend A, Baquero F, Berche P, Bloecker H, Brandt P, Chakraborty T, Charbit A, Chetouani F, Couvé E, de Daruvar A, Dehoux P, Domann F, Domínguez-Bernal G, Duchaud E, Durant L, Dussurget O, Entian KD, Fsihi H, García-del Portillo F, Garrido P, Gautier L, Goebel W, Gómez-López N, Hain T, Hauf J, Jackson D, Jones LM, Kaerst U, Kreft J, Kuhn M, Kunst F, Kurapkat G, Madueno F, Maitournam A, Vicente JM, Ng E, Nedjari H, Nordsiek G, Novella S, de Pablos B, Pérez-Diaz JC, Purcell R, Remmel B, Rose M, Schlueter T, Simoes N, Tierrez A, Vázquez-Boland JA, Voss H, Wehland J, Cossart P (2001) Comparative genomics of *Listeria* species. Science 294:849–852

Gravesen A, Ramnath M, Rechinger KB, Andersen N, Jansch L, Héchard Y, Hastings JW, Knochel S (2002) High-level resistance to class IIa bacteriocins is associated with one general mechanism in *Listeria monocytogenes*. Microbiology 148:2361–2369

Hauge HH, Mantzilas D, Moll GN, Konings WN, Driessen AJM, Eijsink VGH, Nissen-Meyer J (1998) Plantaricin A is an amphiphilic α-helical bacteriocin-like pheromone which exerts antimicrobial and pheromone activities through different mechanisms. Biochemistry 37:16026–16032

Haugen HS, Fimland G, Nissen-Meyer J, Kristiansen PE (2005) Three-dimensional structure in lipid micelles of the pediocin-like antimicrobial peptide curvacin A. Biochemistry 44: 16149–16157

Haugen HS, Kristiansen PE, Fimland G, Nissen-Meyer J (2008) Mutational analysis of the class IIa bacteriocin curvacin A and its orientation in target cell membranes. Appl Environ Microbiol 74:6766–6773

Héchard Y, Pelletier C, Cenatiempo Y, Frère J (2001) Analysis of sigma(54)-dependent genes in *Enterococcus faecalis*: a mannose PTS permease (EII(Man)) is involved in sensitivity to a bacteriocin, mesentericin Y105. Microbiology 147:1575–1580

Henderson JT, Chopko AL, van Wassenaar PD (1992) Purification and primary structure of pediocin PA-1 produced by *Pediococcus acidilactici* PAC-1.0. Arch Biochem Biophys 295:5–12

Heng NCK, Burtenshaw GA, Jack RW, Tagg JR (2007) Ubericin A, a Class IIa bacteriocin produced by *Streptococcus uberis*. Appl Environ Microbiol 73:7763–7766

Herranz C, Chen Y, Chung HJ, Cintas LM, Hernández PE, Montville TJ, Chikindas ML (2001a) Enterocin P selectively dissipates the membrane potential of *Enterococcus faecium* T136. Appl Environ Microbiol 67:1689–1692

Herranz C, Cintas LM, Hernández PE, Moll GN, Driessen AJ (2001b) Enterocin P causes potassium ion efflux from *Enterococcus faecium* T136 cells. Antimicrob Agents Chemother 45:901–904

Hühne K, Axelsson L, Holck A, Kröckel L (1996) Analysis of the sakacin P gene cluster from *Lactobacillus sake* Lb674 and its expression in sakacin-negative *Lb. sake* strains. Microbiology 142:1437–1448

Ingham A, Ford M, Moore RJ, Tizard M (2003) The bacteriocin piscicolin 126 retains antilisterial activity in vivo. J Antimicrob Chemother 51:1365–1371

Jack RW, Wan J, Gordon J, Harmark K, Davidson BE, Hillier AJ (1996) Characterization of the chemical and antimicrobial properties of piscicolin 126, a bacteriocin produced by *Carnobacterium piscicola* JG126. Appl Environ Microbiol 62:2897–2903

Joerger RD (2003) Alternatives to antibiotics: bacteriocins, antimicrobial peptides and bacteriophages. Poult Sci 82:640–647

Johnsen L, Dalhus B, Leiros I, Nissen-Meyer J (2005) 1.56 Å crystal structure of entA-im: a bacterial immunity protein conferring immunity to the antimicrobial activity of the pediocin-like bacteriocin enterocin A. J Biol Chem 280:19045–19050

Kaiser AL, Montville TJ (1996) Purification of the bacteriocin bavaricin MN and characterization of its mode of action against *Listeria monocytogenes* Scott A cells and lipid vesicles. Appl Environ Microbiol 62:4529–4535

Kalmokoff ML, Banerjee SK, Cyr TD, Hefford MA, Gleeson T (2001) Identification of a new plasmid-encoded *sec*-dependent bacteriocin produced by *Listeria innocua* 743. Appl Environ Microbiol 67:4041–4047

Kaur K, Andrew LC, Wishart DS, Vederas JC (2004) Dynamic relationships among type IIa bacteriocins: temperature effects on antimicrobial activity and on structure of the C-terminal amphipathic_helix as a receptor binding region. Biochemistry 43:9009–9020

Kawamoto S, Shima J, Sato R, Eguchi T, Ohmomo S, Shibato J, Horikoshi N, Takeshita K, Sameshima T (2002) Biochemical and genetic characterization of mundticin KS, an antilisterial peptide produced by *Enterococcus mundtii* NFRI 7393. Appl Environ Microbiol 68:3830–3840

Kazazic M, Nissen-Meyer J, Fimland G (2002) Mutational analysis of the role of charged residues in target-cell binding, potency and specificity of the pediocin-like bacteriocin sakacin P. Microbiology 148:2019–2027

Kim IK, Kim MK, Kim JH, Yim HS, Cha SS, Kang SO (2007) High resolution crystal structure of PedB: a structural basis for the classification of pediocin-like immunity proteins. BMC Struct Biol 30(7):35

Kjos M, Nes IF, Diep DB (2009) Class II one-peptide bacteriocins target a phylogenetically defined subgroup of mannose phosphotransferase systems on sensitive cells. Microbiology 155:2949–2961

Kjos M, Salehian Z, Nes IF, Diep DB (2010) An extracellular loop of the mannose phosphotransferase 1 system component IIC is responsible for specific targeting by class IIa bacteriocins. J Bacteriol

Kristiansen PE, Fimland G, Mantzilas D, Nissen-Meyer J (2005) Structure and mode of action of the membrane permeabilizing antimicrobial peptide pheromone plantaricin A. J Biol Chem 280:22945–22950

Lasta S, Fajloum Z, Darbon H, Mansuelle P, Andreotti N, Sabatier JM, Boudabous A, Sampieri F (2008) Chemical synthesis and characterization of J46 peptide, an atypical class IIa bacteriocin from *Lactococcus* lactis subsp. *cremoris* J46 strain. J Antibiot (Tokyo) 61:89–93

Le Marrec C, Hyronimus B, Bressollier P, Verneuil B, Urdaci MC (2000) Biochemical and genetic characterization of coagulin, a new antilisterial bacteriocin in the pediocin family of bacteriocins, produced by *Bacillus coagulans*. Appl Environ Microbiol 66:5213–5220

Li JW, Vederas JC (2009) Drug discovery and natural products: end of an era or an endless frontier? Science 325:161–165

Line JE, Svetoch EA, Eruslanov BV, Perelygin VV, Motsevich EV, Mitsevich IP, Levchuk VP, Svetoch OE, Seal BE, Siragusa GR, Stern NJ (2008) Isolation and purification of enterocin E-760 with broad antimicrobial activity against gram-positive and gram-negative bacteria. Antimicrob Agents Chemother 52:1094–1100

Maftah A, Renault D, Vignoles C, Héchard Y, Bressollier P, Ratinaud MH, Cenatiempo Y, Julien R (1993) Membrane permeabilization of *Listeria monocytogenes* and mitochondria by the bacteriocin mesentericin Y105. J Bacteriol 175:3232–3235

Martin-Visscher LA, Sprules T, Gursky LJ, Vederas J (2008) Nuclear magnetic resonance solution structure of PisL, a group B immunity protein that provides protection against the type IIa bacteriocin piscicolin 126, PisA. Biochemistry 47:6427–6436

Métivier A, Pilet MF, Dousset X, Sorokine O, Anglade P, Zagorec M, Picard JC, Marion D, Cenatiemp Y, Frémaux C (1998) Divercin V41, a new bacteriocin with two disulphide bonds produced by *Carnobacterium divergens* V41: primary structure and genomic organization. Microbiology 144:2837–2844

Miller KW, Schamber R, Osmanagaoglu O, Ray B (1998a) Isolation and characterization of pediocin AcH chimeric protein mutants with altered bactericidal activity. Appl Environ Microbiol 64:1997–2005

Miller KW, Schamber R, Chen Y, Ray B (1998b) Production of active chimeric pediocin AcH in *Escherichia coli* in the absence of processing and secretion genes from the *Pediococcus pap* operon. Appl Environ Microbiol 64:14–20

Miller KW, Ray P, Steinmetz T, Hanekamp T, Ray B (2005) Gene organization and sequences of pediocin AcH/PA-1, production operons in *Pediococcus* and *Lactococcus* plasmids. Lett Appl Microbiol 40:52–62

Minahk CJ, Dupuy F, Morero RD (2004) Enhancement of antibiotic activity by sub-lethal concentrations of enterocin CRL35. J Antimicrob Chemother 54:240–246

Morisset D, Berjeaud JM, Marion D, Lacombe C, Frère J (2004) Mutational analysis of mesentericin Y105 an anti-*Listeria* bacteriocin, for determination of impact on bactericidal activity, in vitro secondary structure, and membrane interaction. Appl Environ Microbiol 70:4672–4680

Naghmouchi K, Drider D, Fliss I (2007) Action of divergicin M35, a class IIa bacteriocin, on liposomes and *Listeria*. J Appl Microbiol 102:1508–1517

Naghmouchi K, Belguesmia Y, Baah J, Ron T, Drider D (2010a) Antibacterial activity of class I and IIa bacteriocins combined with polymyxin E against resistant variants of *Listeria monocytogenes* and *Escherichia coli*. Res Microbiol In press

Naghmouchi K, Drider D, Baah J, Teather R (2010b) Nisin A and polymyxin B as synergistic inhibitors of Gram-positive and Gram-negative bacteria. Probiotics Antimicrob Proteins 1–2:98–103

Nes IF, Eijsink VGH (1999) Regulation of group II peptide bacteriocin synthesis by quorum-sensing mechanisms. In: Dunny GM, Winans SC (eds) Cell–cell signalling in bacteria. American Society for Microbiology, Washington, DC, pp 175–192

Nes IF, Diep DB, Havarstein LS, Brurberg MB, Eijsink V, Holo H (1996) Biosynthesis of bacteriocins in lactic acid bacteria. Antoine Van Leeuwenhoek 70:113–128

Nieto-Lozano JC, Reguera-Useros JI, Peláez-Martínez MC, Hardisson de la Torre A (2006) Effect of a bacteriocin produced by *Pediococcus acidilactici* against *Listeria monocytogenes* and *Clostridium perfringens* on Spanish raw meat. Meat Sci 72:57–61

Nissen-Meyer J, Rogne P, Oppegard C, Haugen HS, Kristiansen PE (2009) Structure-function relationships of the non-lanthionine-containing peptide (class II) bacteriocins produced by gram-positive bacteria. Curr Pharm Biotechnol 10:19–37

O'Keeffe T, Hill C, Ross R (1999) Characterization and heterologous expression of the genes encoding enterocin A production, immunity, and regulation in *Enterococcus faecium* DPC1146. Appl Environ Microbiol 65:1506–1515

O'Mahony AT, O'Sullivan FO, Walsh Y, Vaughan A, Maher M, Fitzerald GF, Van Sindersen D (2000) Characterization and heterologous of antimicrobial producing lactic acid bacteria from malted barley. J Inst Brew 106:403–410

Papathanasopoulos MA, Dykes GA, Revol-Junelles AM, Delfour A, von Holy A, Hastings JW (1998) Sequence and structural relationships of leucocins A-, B-, and C-TA33a from *Leuconostoc mesenteroides* TA33a. Microbiology 144:1343–1348

Postma PW, Langeler JW, Jacobson GR (1993) Phosphoenolpyruvate: carbohydrate phosphotransferase systems of bacteria. Microbiol Rev 57:543–594

Quadri LEN, Sailer M, Roy KL, Vederas JC, Stiles ME (1994) Chemical and genetic characterization of bacteriocins produced by *Carnobacterium piscicola* LV17B. J Biol Chem 269:12204–12211

Quadri LEN, Kleerebezem M, Kuipers OP, de Vos WM, Roy KL, Vederas JC, Stiles ME (1997) Characterization of a locus from *Carnobacterium piscicola* LV17B involved in bacteriocin production and immunity: evidence for global inducer-mediated transcriptional regulation. J Bacteriol 179:6163–6171

Ramnath M, Arous S, Gravesen A, Hastings JW, Héchard Y (2004) Expression of *mptC* of *Listeria monocytogenes* induces sensitivity to class IIa bacteriocins in *Lactococcus lactis*. Microbiology 150:2663–2668
Rea MC, Sit CS, Clayton E, O'Connor PM, Whittal RM, Zheng J, Vederas JC, Ross RP, Hill C (2010) Thuricin CD, a posttranslationally modified bacteriocin with a narrow spectrum of activity against *Clostridium difficile*. Proc Natl Acad Sci U S A 107:9352–9357
Richard C, Brillet A, Pilet MF, Prévost H, Drider D (2003) Evidence on inhibition of *Listeria monocytogenes* by divercin V41 action. Lett Appl Microbiol 36:288–292
Richard C, Drider D, Elmorjani K, Marion D, Prévost H (2004) Heterologous expression and purification of active divercin V41, a class IIa bacteriocin encoded by a synthetic gene in *Escherichia coli*. J Bacteriol 186:4276–4284
Richard C, Cañon R, Naghmouchi K, Bertrand D, Prévost H, Drider D (2006) Evidence on correlation between number of disulfide bridge and toxicity of class IIa bacteriocins. Food Microbiol 23:175–183
Rihakova J, Petit VW, Demnerova K, Prévost H, Rebuffat S, Drider D (2009a) Insights into structure-activity relationships in the C-terminal region of divercin V41, a class IIa bacteriocin with high-level antilisterial activity. Appl Environ Microbiol 75:1811–1819
Rihakova J, Belguesmia Y, Petit VW, Pilet MF, Prévost H, Dousset X, Drider D (2009b) Divercin V41 from gene characterization to food applications: 1998–2008, a decade of solved and unsolved questions. Lett Appl Microbiol 48:1–7
Rihakova J, Cappelier JM, Hue I, Demnerova K, Fédérighi M, Prévost H, Drider D (2010). Antimicrob Agents Chemother. 54: 563–564
Robichon D, Gouin E, De'barbouille M, Cossart P, Cenatiempo Y, Héchard Y (1997) The rpoN (σ54) gene from Listeria monocytogenes is involved in resistance to mesentericin Y105, an antibacterial peptide from Leuconostoc mesenteroides. J Bacteriol 179:7591–7594
Rodríguez JM, Martínez MI, Kok J (2002) Pediocin PA-1, a wide-spectrum bacteriocin from lactic acid bacteria. Crit Rev Food Sci Nutr 42:91–121
Salvucci E, Saavedra L, Sesma F (2007) Short peptides derived from the NH2-terminus of subclass IIa bacteriocin enterocin CRL35 show antimicrobial activity. J Antimicrob Chemother 59:1102–1108
Simon L, Frémaux C, Cenatiempo Y, Berjeaud JM (2002) Sakacin G, a new type of antilisterial bacteriocin. Appl Environ Microbiol 12:6416–6420
Skaugen M, Cintas LM, Nes IF (2003) Genetics of bacteriocin production in lactic acid bacteria. In: Wood BJB, Warner PJ (eds) Genetics of lactic acid bacteria. Kluwer Academic/Plenum Publishers, New York, NY, pp 225–260
Smoaki W (1999) Cromakalim: Embryonic effects and reduction of tolbutamide-induced dysmorphogenesis in vitro. Teratology 60:260–264
Soliman W, Bhattacharjee S, Kaur K (2007) Molecular dynamics simulation study of interaction between a class IIa bacteriocin and its immunity protein. Biochim Biophys Acta 1774:1002–10013
Sprules T, Kawulka KE, Gibbs AC, Wishart DS, Vederas JC (2004) NMR solution structure of the precursor for carnobacteriocin B2, an antimicrobial peptide from *Carnobacterium piscicola*. Eur J Biochem 271:1748–1756
Stern NJ, Svetoch EA, Eruslanov BV, Perelygin VV, Mitsevich EV, Mitsevich IP, Pokhilenko VD, Levchuk VP, Svetoch OE, Seal BS (2006) Isolation of a Lactobacillus salivarius strain and purification of its bacteriocin, which is inhibitory to campylobacter jejuni in the chicken gastrointestinal system. Antimicrob Agents Chemother 50:3111–3116
Straume D, Kjos M, Nes IF, Diep DB (2007) Quorum-sensing based bacteriocin production is down-regulated by N-terminally truncated species of gene activators. Mol Genet Genomics 278:283–293
Svetoch EA, Stern NJ (2010) Bacteriocins to control *Campylobacter spp.* in poultry. Poult Sci 89:1763–1768
Svetoch EA, Eruslanov BV, Perelygin VV, Mitsevich EV, Mitsevich IP, Borzenkov VN, Levchuk VP, Svetoch OE, Kovalev YN, Stepanshin YG, Siragusa GR, Seal BS, Stern NJ (2008a) Diverse

antimicrobial killing by *Enterococcus faecium* E 50-52 bacteriocin. J Agric Food Chem 56:1942–1948
Svetoch EA, Levchuk VP, Pokhilenko VD, Eruslanov BV, Mitsevich EV, Mitsevich IP, Perelygin VV, Stepanshin YG, Stern NJ (2008b) Inactivating methicillin-resistant *Staphylococcus aureus* and other pathogens by use of bacteriocins OR-7 and E 50-52. J Clin Microbiol 46:3863–3865
Tahiri I, Desbiens M, Benech R, Kheadr E, Lacroix C, Thibault S, Ouellet D, Fliss I (2004) Purification, characterization and amino acid sequencing of divergicin M35: a novel class IIa bacteriocin produced by *Carnobacterium divergens* M35. Int J Food Microbiol 97:123–136
Taylor WR (1986) The classification of amino acid conservation. J Theor Biol 119:205–218
Tessema GT, Moretro T, Kohler A, Axelsson L, Naterstad K (2009) Complex phenotypic and genotypic responses of *Listeria monocytogenes* strains exposed to the class IIa bacteriocin sakacin P. Appl Environ Microbiol 75:6973–6980
Tichaczek PS, Nissen-Meyer J, Nes IF, Vogel RF, Hammes WP (1992) Characterization of the bacteriocins curvacin A from *Lactobacillus curvatus* LTH 1174 and Sakacin P from *L. sake* LTH 673. Syst Appl Microbiol 15:460–468
Tichaczek PS, Vogel RF, Hammes WP (1994) Cloning and sequencing of sakP encoding sakacin P, the bacteriocin produced by *Lactobacillus sake* LTH 673. Microbiology 140:361–367
Todorov SD, Wachsman M, Tomé E, Dousset X, Destro MT, Dicks LM, Franco BD, Vaz-Velho M, Drider D (2010) Characterisation of an antiviral pediocin-like bacteriocin produced by *Enterococcus faecium*. Food Microbiol 27:869–879
Tominaga T, Hatakeyama Y (2007) Development of innovative pediocin PA-1 by DNA shuffling among class IIa bacteriocins. Appl Environ Microbiol 73:5292–5299
Tomita H, Fujimoto S, Tanimoto K, Ike Y (1996) Cloning and genetic organization of the bacteriocin 31 determinant encoded on the *Enterococcus faecalis* pheromone-responsive conjugative plasmid pYI17. J Bacteriol 178:3585–3593
Uteng M, Hauge HH, Markwick PR, Fimland G, Mantzilas D, Nissen-Meyer J, Muhle-Goll C (2003) Three-dimensional structure in lipid micelles of the pediocin-like antimicrobial peptide sakacin P and a sakacin P variant that is structurally stabilized by an inserted C-terminal disulfide bridge. Biochemistry 42:11417–11426
Vadyvaloo V, Hastings JW, Van der Merwe MJ, Rautenbach M (2002) Membranes of class IIa bacteriocin-resistant *Listeria monocytogenes* cells contain increased levels of desaturated and short-acyl-chain phosphatidylglycerols. Appl Environ Microbiol 68:5223–5230
Vadyvaloo V, Arous S, Gravesen A, Héchard Y, Chauhan-Haubrock R, Hastings JW, Rautenbach M (2004) Cell-surface alterations in class IIa bacteriocin-resistant *Listeria monocytogenes* strains. Microbiology 150:3025–3033
Van Reenen CA, Dicks LMT, Chikindas ML (1998) Isolation, purification and partial characterization of plantaricin 423, a bacteriocin produced by *Lactobacillus plantarum*. J Appl Microbiol 84(84):1131–1137
Van Reenen CA, Chikindas ML, Van Zyl WH, Dicks LMT (2003) Characterization and heterologous expression of class IIa bacteriocin, plantaricin 423 from *Lactobacillus plantarum* in *Saccharomyces cerevisiae*. Int J Food Microbiol 81:29–40
Vaughan A, Eijsink VGH, O'Sullivan TF, O'Hanlon K, van Sinderen D (2001) An analysis of bacteriocins produced by lactic acid bacteria isolated from malted barley. J Appl Microbiol 91:131–138
Vu-Khac H, Miller KW (2009) Regulation of mannose phosphotransferase system permease and virulence gene expression in *Listeria monocytogenes* by the EII(t)Man transporter. Appl Environ Microbiol 75:6671–6678
Wachsman MB, Farías ME, Takeda E, Sesma F, De Ruiz Holdago A, De Torres RA, Coto CE (1999) Antiviral activity of enterocin CRL35 against herpes viruses. Int J Antimicrob Agents 12:293e299
Wachsman MB, Castilla V, Holgado APD, De Torres RA, Sesma F, Coto CE (2003) Enterocin CRL35 inhibits late stages of HSV-1 and HSV-2 replication in vitro. Antiviral Res 58:17–24

Watson RM, Woody RW, Lewis RV, Bohle DS, Andreotti AH, Ray B, Miller KW (2001) Conformational changes in pediocin AcH upon vesicle binding and approximation of the membrane-bound structure in detergent micelles. Biochemistry 40:14037–14046

Yamazaki K, Suzuki M, Kawai Y, Inoue N, Montville TJ (2005) Purification and characterization of a novel class IIa bacteriocin, piscicocin CS526, from surimi-associated *Carnobacterium piscicola* CS526. Appl Environ Microbiol 71:554–557

Yan LZ, Gibbs AC, Stiles ME, Wishart DS, Vederas JC (2000) Analogues of bacteriocins: antimicrobial specificity and interactions of leucocin A with its enantiomer, carnobacteriocin B2, and truncated derivatives. J Med Chem 43:4579–8451

Zhang J, Liu G, Shang N, Cheng W, Chen S, Li P (2009) Purification and partial amino acid sequence of pentocin 31-1, an anti-*Listeria* bacteriocin produced by *Lactobacillus pentosus* 31-1. J Food Prot 72:2524–529

Chapter 11
The Two-Peptide (Class-IIb) Bacteriocins: Genetics, Biosynthesis, Structure, and Mode of Action

Jon Nissen-Meyer, Camilla Oppegård, Per Rogne, Helen Sophie Haugen, and Per Eugen Kristiansen

Abstract The two-peptide (class-IIb) bacteriocins consist of two different peptides, both of which are required to obtain high antimicrobial activity. These bacteriocins kill target-cells by inducing membrane-leakage and they seem to display some specificity with respect to the molecules they transfer across membranes. The genes encoding the two peptides of two-peptide bacteriocins are next to each other on the same operon. In the same or a nearby operon are genes encoding (i) the immunity protein that protects the bacteriocin-producer from its own bacteriocin, (ii) a dedicated ABC-transporter that exports the bacteriocin from cells and cleaves off the N-terminal bacteriocin leader sequence, and (iii) an accessory protein whose exact function has not been fully clarified. Some two-peptide bacteriocins appear to be produced constitutively, whereas the production of other two-peptide bacteriocins is regulated through a three-component regulatory system that consists of a peptide pheromone, a membrane-associated histidine protein kinase, and response regulators. It has recently been proposed that the two peptides of (some) two-peptide bacteriocins may form a membrane-penetrating helix-helix structure involving helix-helix interacting GxxxG-motifs present in all currently characterized two-peptide bacteriocins. It has also been suggested that the helix-helix structure interacts with an integrated membrane (transport) protein, thus inducing a conformational change in the protein, which in turn causes membrane-leakage. This proposed mode-of-action is similar to that of the pediocin-like (class-IIa) bacteriocins and lactococcin A, which bind to a part of the mannose phosphotransferase permease that is embedded in the cell membrane, thereby altering the conformation of the permease in a manner that causes membrane-leakage and cell death.

This chapter focuses on the two-peptide (class-IIb) bacteriocins. As the name suggests, the two-peptide bacteriocins are novel in that they consist of two very different peptides and optimal activity requires the presence of both peptides in about equal amounts (Garneau, Martin and Vederas 2002; Oppegård et al. 2007b; Nissen-Meyer et al. 2009, 2010). Since the first isolation of a two-peptide bacteriocin, lactococcin G, in 1992 (Nissen-Meyer et al. 1992), at least 15 two-peptide bacteriocins produced by lactic acid

J. Nissen-Meyer (✉)
Department of Molecular Biosciences, University of Oslo,
Pb 1041 Blindern, 0316 Oslo, Norway
e-mail: jon.nissen-meyer@imbv.uio.no

D. Drider and S. Rebuffat (eds.), *Prokaryotic Antimicrobial Peptides: From Genes to Applications*, DOI 10.1007/978-1-4419-7692-5_11,

Lactococcin G	LcnG-α LcnG-β	GTWDDIGQGIGRVAYWVGKAMGNMSDVNQASRINRKKKH KKWGWLAWVDPAYEFIKGFGKGAIKEGNKDKWKNI
Lactococcin Q	LcnQα LcnQβ	SIWGDIGQGVGKAAYWVGKAMGNMSDVNQASRINRKKKH KKWGWLAWVEPAGEFLKGFGKGAIKEGNKDKWKNI
Enterocin 1071	Entα Entβ	ESVFSKIGNAVGPAAYWILKGLGNMSDVNQADRINRKKH GPGKWLPWLQPAYDFVTGLAKGIGKEGNKNKWKNV
Plantaricin E/F	PlnE PlnF	FNRGGYNFGKSVRHVVDAIGSVAGIRGILKSIR VFHAYSARGVRNNYKSAVGPADWVISAVRGFIHG
Plantaricin J/K	PlnJ PlnK	GAWKNFWSSLRKGFYDGEAGRAIRR RRSRKNGIGYAIGYAFGAVERAVLGGSRDYNK
Plantaricin S	Plsα Plsβ	RNKLAYNMGHYAGKATIFGLAAWALLA KKKKQSWYAAAGDAIVSFGEGFLNAW
Plantaricin NC8	PLNC8α PLNC8β	DLTTKLWSSWGYYLGKKARWNLKHPYVQF SVPTSVYTLGIKILWSAYKHRKTIEKSFNKGFYH
Lactacin F	LafA LafX	RNNWQTNVGGAVGSAMIGATVGGTICGPACAVAGAHYLPILWTGVTAATGGFGKIRK NRWGDTVLSAASGAGTGIKACKSFGPWGMAICGVGGAAIGGYFGYTHN
Brochocin-C	BrcA BrcB	YSSKDCLKDIGKGIGAGTVAGAAGGGLAAGLGAIPGAFVGAHFGVIGGSAACIGGLLGN KINWGNVGGSCVGGAVIGGALGGLGGAGGGCITGAIGSIWDQW
Thermophilin 13	ThmA ThmB	YSGKDCLKDMGGYALAGAGSGALWGAPAGGVGALPGAFVGAHVGAIAGGFACMGGMIGNKFN QINWGSVVGHCIGGAIIGGAFSGGAAAGVGCLVGSGKAIINGL
ABP-118	Abp118α Abp118β	KRGPNCVGNFLGGLFAGAAAGVPLGPAGIVGGANLGMVGGALTCL KNGYGGSGNRWVHCGAGIVGGALIGAIGGPWSAVAGGISGGFTSCR
Salivaricin P	Sln1 Sln2	KRGPNCVGNFLGGLFAGAAAGVPLGPAGIVGGANLGMVGGALTCL KNGYGGSGNRWVHCGAGIVGGALIGAIGGPWSAVAGGISGGFASCH
Mutacin IV	NlmA NlmB	KVSGGEAVAAIGICATASAAIGGLAGATLVTPYCVGTWGLIRSH DKQAADTFLSAVGGAASGFTYCASNGVWHPYILAGCAGVGAVGSVVFPH
Lactocin 705	705α 705β	GMSGYIQGIPDFLKGYLHGISAANKHKKGRLGY GFWGGLGYIAGRVGAAYGHAQASANNHHSPING

Fig. 11.1 Amino acid sequences of two-peptide (class-IIb) bacteriocins. The GxxxG-motifs are marked with *black* background. Plantaricin Sβ and plantaricin NC8β have the GxxxG-like motifs AxxxA-and SxxxS-motif, respectively, instead of the GxxxG-motif. The two-peptide bacteriocins lactococcin MN (van Belkum et al. 1991), lactococcin MMT24 (Ghrairi et al. 2005), and leucocin H (Blom et al. 1999) are not included in the figure. Lactococcin MN has putative GxxxG-motifs, but is not included since only the sequences of the pre-forms are known and the cleavage site is unknown. Lactococcin MMT24 has not been sequenced and leucocin H has not been completely sequenced, but the partial sequence reveals a putative GxxxG-motif. The recently characterized two-peptide bacteriocin enterocin C (Maldonado-Barragan et al. 2009) is also not included. Its sequence is identical to enterocin 1071, except that enterocin C has an alanine residue instead of a threonine residue at position 17 in its β peptide (Maldonado-Barragan et al. 2009). References for the sequences are as follows: lactococcin G (Nissen-Meyer et al. 1992), lactococcin Q (Zendo et al. 2006), enterocin 1071 (Balla et al. 2000; Franz et al. 2002; Balla and Dicks 2005), plantaricin E/F and J/K (Diep, Håvarstein and Nes 1996; Anderssen et al. 1998), plantaricin S (Jiménez-Diaz et al. 1995; Stephens et al. 1998), plantaricin NC8 (Maldonado, Ruiz-Barba and Jiménez-Diaz 2003), lactacin F (Fremaux, Ahn and Klaenhammer 1993; Allison, Fremaux and Klaenhammer 1994), brochocin C (McCormick et al. 1998), thermophilin 13 (Marciset et al. 1997), ABP-118 (Flynn et al. 2002), salivaricin P (differs from ABP-118 in only two residues) (Barrett et al. 2007), mutacin IV (Qi, Chen and Caufield 2001), and lactocin 705 (Cuozzo et al. 2000). The figure is a modified version of figure 1 in Oppegård et al. (2008) and in Nissen-Meyer et al. (2010)

bacteria have been identified and characterized (Fig. 11.1). It should be noted that also two-peptide lantibiotics (i.e. class I bacteriocins) have been identified (Cotter, Hill and Ross 2005a, b), but they are not discussed in this chapter.

The Two-Peptide Bacteriocins Render Target-Cell Membranes Permeable to Small Molecules

All two-peptide bacteriocins whose mode-of-action has been analyzed, this includes lactacin F (Abee, Klaenhammer and Letellier et al. 1994), lactococcin G (Moll et al. 1996, 1998), thermophilin 13 (Marciset et al. 1997), plantaricin E/F (Moll et al. 1999), plantaricin J/K (Moll et al. 1999), and lactocin 705 (Castellano, Raya and Vignolo et al. 2003; Cuozzo et al. 2003), induce membrane-leakage in sensitive bacteria. Interestingly, the bacteriocins seem to display specificity with respect to the molecules they transfer across membranes, and they appear to some extent to vary in their specificities. For instance, both plantaricin E/F and plantaricin J/K render target-cell membranes permeable for monovalent ions, including H^+, but not for divalent ions such as Mg^{2+} and phosphate (Moll et al. 1999). Moreover, it seems that plantaricin E/F conducts cations with greater efficiency than plantaricin J/K and vice versa for anions (Moll et al. 1999). Also lactococcin G renders target-cell membranes permeable for different monovalent cations, such as Na^+, K^+, Li^+, Cs^+, Rb^+, and choline, but not to H^+ (in contrast to plantaricin E/F and plantaricin J/K), nor to divalent cations such as Mg^{2+} and anions such as phosphate (Moll et al. 1996, 1998). This capacity to discriminate between molecules they transfer through membranes suggests that two-peptide bacteriocins do not cause membrane leakage through simply a detergent-like disruption of membranes, but form (or trigger the formation of) rather sophisticated pores that show specificity with respect to transport of molecules. Furthermore, the high potency of two-peptide bacteriocins indicates that bacteriocin-induced membrane leakage depends on a low number of peptides, in contrast to what is expected if leakage is due to a detergent-like disruption of membranes.

The Two Peptides of Two-Peptide Bacteriocins Interact and Function as One Entity

The individual peptides of two-peptide bacteriocins often have features in common with one-peptide bacteriocins; they are usually (1) 30–50 residues long, (2) cationic, (3) membrane-active, (4) amphipathic and/or hydrophobic, and (5) synthesized as prepeptides with a 15- to 30-mer N-terminal leader sequence that is cut off by a dedicated ABC transporter upon export of the peptides from cells (Garneau et al. 2002; Oppegård et al. 2007b; Nissen-Meyer et al. 2009, 2010). Furthermore, one or both peptides of a few two-peptide bacteriocins [for instance, plantaricin E/F, plantaricin J/K (Anderssen et al. 1998), and lactacin F (Allison et al. 1994)] may separately have some – although very low – antimicrobial activity. Nevertheless, two-peptide bacteriocins should not be considered as simply being two one-peptide bacteriocins that function in a synergistic manner. In fact, the individual peptides of many two-peptide bacteriocins do not exhibit any significant activity. For example, the two lactococcin G peptides exhibit no activity when assayed separately at concentrations up to 50 μM but are in combination active at pico- to nanomolar concentrations (Nissen-Meyer et al. 1992; Moll et al. 1996). Moreover, a peptide from a two-peptide bacteriocin shows high antimicrobial activity only when combined with the complementary peptide from the same two-peptide bacteriocin, or in some instances when combined with a peptide from a similar two-peptide bacteriocin. For example, the two lactococcin G peptides exhibit no activity when combined with either the E- or F-peptide of plantaricin E/F or the J- or K-peptide of plantaricin J/K (Anderssen et al. 1998). High activity is, however, attained when one of the lactococcin G peptides is combined with the complementary peptide from enterocin 1071 or lactococcin Q (Zendo et al. 2006; Oppegård et al. 2007a, b), apparently because these three two-peptide bacteriocins have similar sequences: almost 60% sequence identity between lactococcin Q and enterocin 1071, and between lactococcin G and enterocin 1071, and almost 90% identity between lactococcin G and lactococcin Q, (Balla et al. 2000; Zendo et al. 2006; Oppegård et al. 2007a, b) (Fig. 11.1). The facts that (1) both complementary peptides are required to attain high antimicrobial activity, (2) peptides function together only if they belong to the same or a similar two-peptide bacteriocin, (3) the genes encoding the two peptides of two-peptide bacteriocins are next to each other on the same operon (for references, see legend to Fig. 11.1), and (4) there is only one immunity gene for each two-peptide bacteriocin (for references, see legend to Fig. 11.1) clearly demonstrate that the two peptides of two-peptide bacteriocins should in fact be thought of as one functional antimicrobial entity. Furthermore, circular dichroism (CD) structural studies done on three two-peptide bacteriocins (plantaricin E/F, plantaricin J/K, and lactococcin G) revealed that the two peptides of these bacteriocins interact and structure each other upon exposure to target membranes (Hauge et al. 1998b, 1999). This is probably also the case for the peptides that constitute lactococcin Q, enterocin 1071, and enterocin C, since these three bacteriocins have more than 55% sequence identity with lactococcin G (see Fig. 11.1). The synergistic action of the two peptides of many, if not all, two-peptide bacteriocins is thus due to the fact that the two peptides interact with each other and form one antibacterial entity, rather than that they act separately at two different sites on target cells. It is not clear at what exact stage the two peptides interact with each other. It does not, however, occur before the peptides come in contact with the target cells, since both peptides of two-peptide bacteriocins are unstructured (and thus do not interact and structure each other) in the absence of membrane-like entities (Hauge et al. 1998b, 1999; Fimland et al. 2008; Rogne et al. 2008, 2009). Moreover, antimicrobial activity of lactococcin G is attained when sensitive cells are first treated with one of the lactococcin G peptides, washed, and subsequently treated with the other lactococcin G peptide, indicating that these peptides need not interact with each other before they interact with target cells. No antimicrobial activity is, however, obtained when target cells that have been treated with one of the lactococcin G peptides are mixed with target cells that have been treated with the other lactococcin G peptide (Moll et al. 1998), indicating that the peptides are not able to diffuse to another cell once bound to the cell surface.

Although two-peptide bacteriocins are obviously not two synergistically acting one-peptide bacteriocins, one might speculate that some two-peptide bacteriocins have evolved from two one-peptide bacteriocins (Oppegård et al. 2007b). If two one-peptide bacteriocins that function in a synergistic manner were produced by the same bacteria, there could be a selection pressure for improvement of the synergistic effect – possibly at the expense of the activity of the individual peptides. This might in turn create selection pressure for genetically linking the two peptides, with the formation of a two-peptide bacteriocin.

Genes and Proteins Required for Production of Two-Peptide Bacteriocins

At least five genes found in either one or two operons are needed for the production of two-peptide bacteriocins. These genes are (1) the two structural genes encoding the preforms of the two peptides that constitute the bacteriocin, (2) the immunity gene encoding the immunity protein that protects the bacteriocin producer from its own bacteriocin, (3) the gene encoding the dedicated ABC transporter that exports the bacteriocin from cells, and (4) a gene encoding an accessory protein whose function has not been fully clarified, which might be involved in immunity and or export of the bacteriocin. For all two-peptide bacteriocins that have been characterized genetically, the two structural genes are found to be next to each other on the same operon and the two peptides that constitute these bacteriocins are thus thought to be produced in about equal amounts. This operon also contains the immunity gene, whereas the genes that encode the dedicated ABC transporter and the accessory protein are either also in this operon [this is the case for lactococcin G (Nes et al. 1995)] or on a separate – but nearby – operon [this is the case for plantaricin E/F, plantaricin J/K, and enterocin 1071 (Diep et al. 1996, 2009; Balla et al. 2000; Franz et al. 2002; Balla and Dicks 2005)].

Most, if not all, two-peptide bacteriocins (Fig. 11.1) are initially synthesized with a 15- to 30-mer N-terminal leader sequence of the double-glycine type. This leader sequence is apparently cut off at the C-terminal side of two glycine residues by the dedicated ABC transporter when the bacteriocin is exported out of cells. A special feature of the dedicated ABC transporter is an N-terminal region of approximately 150 residues that is absent in other ABC transporters. Studies using the N-terminal region of the ABC transporter that exports lactococcin G revealed that this region specifically cuts off the lactococcin G leader sequence at the C-terminal side of the double-glycine motif (Håvarstein et al. 1995). The bacteriocin leader sequence consequently appears to facilitate interactions with the ABC transporter, as has been shown to be the case for some one-peptide bacteriocins (van Belkum et al. 1997; Horn et al. 1998). The leader sequence might possibly also function to keep the bacteriocin inactive until it has been exported.

It is not entirely clear how immunity proteins protect cells from two-peptide bacteriocins. Recent results have indicated, however, that the immunity proteins for lactococcin G and enterocin 1071 function through a cellular component, since the ability of these immunity proteins to recognize their cognate bacteriocin is highly dependent on the cells in which the immunity proteins are expressed (Oppegård et al. 2010). Also, the functionality of the immunity proteins for the one-peptide pediocin-like bacteriocins depends on a cellular component (Fimland et al. 2002a; Johnsen et al. 2004), and this component appears to be the bacteriocin receptor, the mannose phosphotransferase permease (Dalet et al. 2001; Gravesen et al. 2002; Ramnath et al. 2004; Diep et al. 2007). It has thus been hypothesized that the cellular component through which the lactococcin G and enterocin 1071 immunity proteins function might be the receptors for these two bacteriocins. This receptor has, however, not yet been identified. The immunity proteins for two-peptide bacteriocins might possibly interact with the cellular component via the cell membrane, since structure predictions indicate that some immunity proteins for two-peptide bacteriocins have transmembrane helices. The putative immunity proteins of the two-peptide bacteriocins brochocin-C (McCormick et al. 1998; Nes et al. 2002), plantaricin S (Stephens et al. 1998; Nes et al. 2002), and thermophilin 13 (Marciset et al. 1997; Nes et al. 2002) are predicted to contain two transmembrane helices, while the immunity proteins of lactococcin G (Nes et al. 1995, 2002), lactococcin MN (van Belkum et al. 1991; Nes et al. 2002), plantaricin E/F (Diep et al. 1996; Nes et al. 2002) and plantaricin J/K (Diep et al. 1996; Nes et al. 2002) may contain 4–5 transmembrane helices. The number of transmembrane helices thus seems to differ, but a common mechanism for bacteriocin-immunity-involving interactions with membrane proteins – as is the case for the immunity proteins for the pediocin-like (class-IIa) bacteriocins (Diep et al. 2007) – may nevertheless exist.

Production of Two-Peptide Bacteriocins in Some Bacteria Is Regulated Through Three-Component Regulatory Systems

Some two-peptide bacteriocins appear to be produced constitutively, while the production of other two-peptide bacteriocins [such as NC8 (Maldonado et al. 2004), ABP-118 (Flynn et al. 2002), plantaricin E/F, plantaricin J/K (Diep et al. 1995, 1996, 2003)] is regulated transcriptionally through a three-component regulatory system that consists of a peptide pheromone, a membrane-associated histidine protein kinase, and response regulators (Kleerebezem and Quadri 2001). Upon export from bacteria, the peptide pheromone interacts with the membrane-associated histidine kinase and thus triggers the kinase to phosphorylate the intracellular response regulator, thereby enabling the response regulator to activate the operons that are required for bacteriocin synthesis and secretion (Kleerebezem and Quadri 2001).

The regulation of bacteriocin production by a three-component regulatory system has been thoroughly studied in *Lactobacillus plantarum* C11 (Diep et al.

1995, 1996, 2003). This strain produces the two two-peptide bacteriocins plantaricin E/F and plantaricin J/K, as well as the peptide pheromone plantaricin A that triggers the production of the two bacteriocins (Diep et al. 1995, 1996, 2003; Anderssen et al. 1998; Hauge et al. 1998a). The gene encoding plantaricin A is positioned in a regulatory operon that also contains the genes that encode the histidine protein kinase and two response regulators. The production of plantaricin E/F and plantaricin J/K is apparently triggered when the concentration of plantaricin A attains a threshold value due to high cell density. In addition to inducing bacteriocin production, plantaricin A stimulates its own production. An autoinduction loop is thus formed, resulting in rapid enhancement of the transcription of all genes involved in the production of plantaricin E/F and plantaricin J/K.

Signal transduction is evidently initiated by the binding of plantaricin A to the membrane-associated histidine protein kinase, and this induces the kinase to phosphorylate the two response regulators. The phosphorylated response regulators then activate the genes that encode plantaricin E/F and plantaricin J/K, as well as the gene that encodes plantaricin A (Diep et al. 2003). Structure and mode-of-action studies have revealed a novel membrane-interacting mechanism by which plantaricin A (and possibly other membrane-active peptide pheromones) functions (Kristiansen et al. 2005). Plantaricin A apparently initially interacts in a nonchiral manner with membrane lipids. This interaction induces α-helical structuring in a region of the peptide (Kristiansen et al. 2005), thus enabling the peptide to interact in a chiral manner with the histidine protein kinase receptor in or near the membrane–water interface.

The Structure of Two-Peptide Bacteriocins

The structures of three two-peptide bacteriocins (lactococcin G, plantaricin E/F, and plantaricin J/K) have been studied by use of circular dichroism (CD) and nuclear magnetic resonance (NMR) spectroscopy (Hauge et al. 1998b, 1999; Fimland et al. 2008; Rogne et al. 2008, 2009). The CD studies revealed that the peptides of these three bacteriocins are unstructured in aqueous solution (Hauge et al. 1998b, 1999). Moreover, there are no apparent structural interactions between the peptides. Helical structuring is, however, attained upon exposing the peptides individually to membrane-like environments, such as solutions containing micelles or liposomes (Hauge et al. 1998b, 1999; Fimland et al. 2008; Rogne et al. 2008, 2009). Furthermore, additional structuring is attained upon jointly exposing complementary peptides to membrane-like entities such as liposomes, suggesting that the peptides of two-peptide bacteriocins interact in a structure-inducing manner upon contact with target-cell membranes (Hauge et al. 1998b, 1999).

Interestingly, all presently characterized two-peptide bacteriocins contain GxxxG motifs (Fig. 11.1). This motif is often involved in helix–helix interactions between transmembrane helices in membrane proteins (Senes et al. 2000, 2001, 2004). When a GxxxG motif is present in an α-helix, the two glycine residues in the motif will be on the same side of the helix and thus form a flat surface that permits close interhelical contact. This enables optimal interhelical van der Waals interactions and formation of stabilizing interhelical backbone Cα–H···O hydrogen bonds (Senes et al. 2001). The high helical content and presence of GxxxG motifs in the peptides of two-peptide bacteriocins thus suggest that a membrane-penetrating helix–helix structure involving GxxxG motifs may be a dominant structural element in many, if not all, two-peptide bacteriocins.[1]

Structure and Membrane-Orientation of Lactococcin G

Lactococcin G is produced by several strains of *Lactococcus lactis* and is presently the two-peptide bacteriocin that is best characterized (Nissen-Meyer et al. 1992; Moll et al. 1996, 1998; Hauge et al. 1998b; Oppegård et al. 2007a, 2008; Rogne et al. 2008). The bacteriocin consists of the 39-mer α-peptide (often termed LcnG-α) and the 35-mer β-peptide (often termed LcnG-β) (Fig. 11.1). CD- and NMR-structural studies revealed extensive α-helical structuring in both the α- and β-peptides upon exposure to either dodecylphosphocholine (DPC) micelles or the structure-inducing compound trifluoroethanol (TFE) (Hauge et al. 1998b; Rogne et al. 2008). Based on the NMR structures of the two peptides (Rogne et al. 2008) as well as on recent mutagenesis studies (Oppegård et al. 2008), a structural model of lactococcin G has been proposed (Oppegård et al. 2008; Rogne et al. 2008; Nissen-Meyer et al. 2009, 2010). The structural model is expected to be valid also for lactococcin Q, enterocin C, and enterocin 1071, because of their sequence similarities to lactococcin G. The structural model entails that the α- and β-peptides form a membrane-spanning helix–helix structure that is stabilized by helix–helix interacting GxxxG motifs and that the two peptides lie in a parallel and staggered fashion relative to each other (Oppegård et al. 2008; Rogne et al. 2008; Nissen-Meyer et al. 2009, 2010) (Fig. 11.2). The helix–helix region includes the N-terminal part of the α-peptide (covering approximately residues 3–22) and the C-terminal part of the β-peptide (covering approximately residues 13–32) (Fig. 11.2). Furthermore, it is proposed that the positively charged and unstructured C-terminal end (Arg-Lys-Lys-Lys-His; residues 35–39) of the α-peptide is pulled across the target-cell membrane by the transmembrane potential, while the relatively unstructured tryptophan-rich N-terminal part (covering approximately residues 1–10) of the β-peptide remains in the outer membrane interface (Fig. 11.2). The α- and β-peptides thereby form a helix–helix structure that spans the target-cell membrane with the N-termini of the two peptides outside and the C-termini inside the cell (Fig. 11.2). This structural model can account for recent results that imply that the lactococcin G immunity protein recognizes both a

[1] Although the two-peptide bacteriocin brochocin-C contains several GxxxG motifs (Fig. 11.1), it has been reported that brochocin-C – and possibly also thermophilin 13, since its sequence is similar to that of brochocin-C (Fig. 11.1) – might contain β-sheet structure (Garneau et al. 2003).

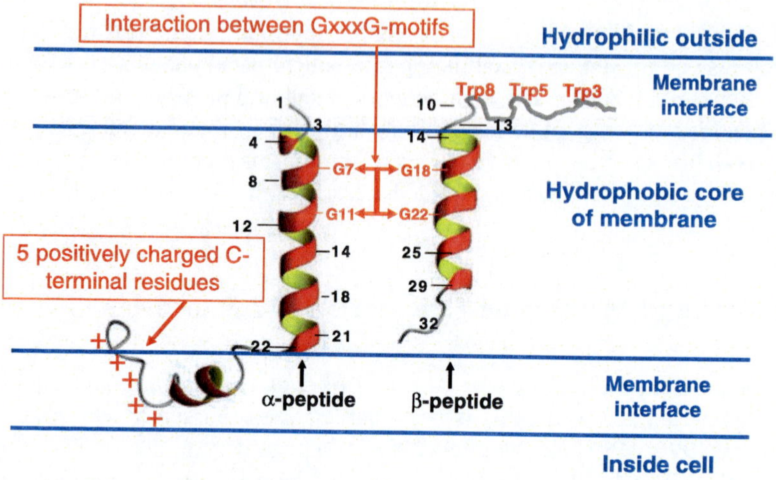

Fig. 11.2 Cartoon presentation of the proposed structural model of lactococcin G and its orientation in cell membranes. The two peptides interact through the G_7xxxG_{11} motif in the α-peptide and the $G_{18}xxxG_{22}$ motif in the β-peptide and form a transmembrane helix–helix structure. The tryptophan residues in the structurally flexible N-terminal region of the β-peptide are in or near the outer membrane interface, whereas the positively charged and structurally flexible C-terminal end of the α-peptide is forced through the membrane by the transmembrane potential (negative inside). The figure is from Nissen-Meyer et al. (2010)

sequence in the N-terminal part (residues 1–13) of the α-peptide and a sequence on the C-terminal side of residue 13 in the β-peptide (Oppegård et al. 2010). In the proposed structure, these regions in the α- and β-peptides are adjacent to each other, and this would thus enable the two regions to simultaneously interact with the immunity protein.

Structure of Plantaricin J/K

Also, the three-dimensional structures of the 25-mer J-peptide and 32-mer K-peptide of plantaricin J/K have been studied by CD and NMR spectroscopy (Rogne et al. 2009). Upon exposure to either DPC micelles or structure-inducing TFE, the J-peptide has an N-terminal amphipathic helical region from residue 3 to 15, whereas the K-peptide has a central amphipathic helical region from residue 9 to 24. There is one GxxxG motif ($G_{13}xxxG_{17}$) in the J-peptide and two (G_9xxxG_{13} and $G_{13}xxxG_{17}$) in the K-peptide (Fig. 11.1). Replacements of glycine residues in these peptides with large residues were very detrimental when the glycine residues were part of the $G_{13}xxxG_{17}$ motifs in the J and K peptides, whereas the replacements were fairly well tolerated when they were outside these two $G_{13}xxxG_{17}$ motifs (Rogne et al. 2009). The mutagenesis data and the NMR structures are consistent with a structural model in which also the J and K peptides of plantaricin J/K interact and form a transmembrane helix–helix structure involving their $G_{13}xxxG_{17}$ motifs (Rogne et al. 2009). The results do not, however, reveal which relative orientation of the two peptides, the parallel or the antiparallel, is the most probable orientation (Rogne et al. 2009).

Structure of Plantaricin E/F

CD and NMR analysis of the 33-mer E-peptide and 34-mer F-peptide of plantaracin E/F (in the presence of micelles) revealed that the E-peptide has two helical regions separated by a flexible GxxxG motif ($G_{20}xxxG_{24}$), while the F-peptide has one long helical region (Fimland et al. 2008). The E-peptide has altogether two GxxxG motifs (G_5xxxG_9 and $G_{20}xxxG_{24}$) that might be involved interpeptide helix–helix interactions, whereas the F-peptide has one such motif ($G_{30}xxxG_{34}$). As is the case for lactococcin G, it has been proposed that the E- and F-peptides of plantaricin E/F interact in a staggered and parallel fashion relative to each other and form a transmembrane helix–helix structure that involves their GxxxG motifs (Fimland et al. 2008). There are, however, alternatives to how these GxxxG motifs might combine. It remains to be determined which of the possible alternatives are most likely.

Bacteriocins as Peptide Ligands that Interact with Membrane-Associated Receptors

Structure–function studies on the one-peptide pediocin-like (class-IIa) bacteriocins have shown that these bacteriocins have in their C-terminal half a helix-containing segment that penetrates into target-cell membranes (Fregeau Gallagher et al. 1997; Wang et al. 1999; Uteng et al. 2003; Haugen et al. 2005; Fimland et al. 2005; Drider et al. 2006; Nissen-Meyer et al. 2009). The membrane-penetrating helix-containing region is involved in determining the target-cell specificity of these bacteriocins and is also the region that is recognized by their immunity proteins (Fimland et al. 1996, 2002b, 2005; Johnsen et al. 2005; Drider et al. 2006; Nissen-Meyer et al. 2009). It has also been shown that the pediocin-like bacteriocins kill target cells by interacting with the membrane-associated mannose phosphotransferase permease (Diep et al. 2007). A part, probably the membrane-penetrating helix, of these bacteriocins apparently binds to the permease subunits (the MptC and/or MptD subunits) that are embedded in the membrane, thereby altering the conformation of the permease in a manner that leads to membrane leakage and cell death. Immunity proteins that protect cells from pediocin-like bacteriocins sense the altered conformation and bind to the bacteriocin–permease complex, thereby preventing membrane leakage (Diep et al. 2007). The fact that lactococcin A, a one-peptide class IId bacteriocin that is entirely different from the pediocin-like bacteriocins, functions in a similar manner (i.e., by binding to the membrane-embedded mannose permease subunits) (Diep et al. 2007), suggests that several very different membrane-active peptide bacteriocins may induce membrane leakage in basically the same way; the common theme is that membrane leakage is caused by structural alterations in an integrated target-cell membrane (transport) protein and that these structural alterations are triggered by interactions between the target-cell membrane protein and a membrane-penetrating helical segment of the bacteriocin. For two-peptide bacteriocins, these interactions might involve the binding of the transmembrane helix–helix structure to an integrated membrane protein/receptor. A membrane-associated receptor for a two-peptide bacteriocin has, however, yet to be identified. It has nevertheless been proposed that the lactococcin G immunity protein interacts with lactococcin G via a lactococcin G receptor in an analogous manner as the immunity proteins for the pediocin-like bacteriocins, since the functionality of the lactococcin G immunity protein depends on a cellular component (Oppegård et al. 2010) and this component might well be the lactococcin G receptor. It should be noted that some studies suggest that some two-peptide bacteriocins, in particular thermophilin 13 and brochocin-C, might not need membrane receptors to exert their activity (Marciset et al. 1997; Gao et al. 1999). It is important that more extensive structure–function studies on peptide bacteriocins are carried out, since that will give us even better insight into how peptide bacteriocins function and reveal structural features that are especially important for their potency. Such insight is invaluable for rational design and construction of novel antibacterial peptides that may especially be useful for medical and biotechnological applications.

References

Abee T, Klaenhammer TR, Letellier L (1994) Kinetic studies of the action of lactacin F, a bacteriocin produced by *Lactobacillus johnsonii* that forms poration complexes in the cytoplasmic membrane. Appl Environ Microbiol 60:1006–1013

Allison GE, Fremaux C, Klaenhammer TR (1994) Expansion of bacteriocin activity and host range upon complementation of two peptides encoded within the lactacin F operon. J Bacteriol 176:2235–2241

Anderssen EL, Diep DB, Nes IF, Eijsink VGH, Nissen-Meyer J (1998) Antagonistic activity of *Lactobacillus plantarum* C11: two new two-peptide bacteriocins, plantaricin EF and JK, and the induction factor, plantaricin A. Appl Environ Microbiol 64:2269–2272

Balla E, Dicks LMT (2005) Molecular analysis of the gene cluster involved in the production and secretion of enterocins 1071A and 1071B and of the genes responsible for the replication and transfer of plasmid pEF1071. Int J Food Microbiol 99:33–45

Balla E, Dicks LMT, Du Toit M, van der Merwe MJ, Holzapfel WH (2000) Characterization and cloning of the genes encoding enterocin 1071A and enterocin 1071B, two antimicrobial peptides produced by *Enterococcus faecalis* BFE 1071. Appl Environ Microbiol 66:1298–1304

Barrett E, Hayes M, O'Connor P, Gardiner G, Fitzgerald GF, Stanton C, Ross RP, Hill C (2007) Salivaricin P, one of a family of two-component antilisterial bacteriocins produced by intestinal isolates of *Lactobacillus salivarius*. Appl Environ Microbiol 73:3719–3723

Blom H, Katla T, Holck A, Sletten K, Axelsson L, Holo H (1999) Characterization, production, and purification of leucocin H, a two-peptide bacteriocin from *Leuconostoc* MF215B. Curr Microbiol 39:43–48

Castellano P, Raya R, Vignolo G (2003) Mode of action of lactocin 705, a two-component bacteriocin from *Lactobacillus casei* CRL705. Int J Food Microbiol 85:35–43

Cotter PD, Hill C, Ross RP (2005a) Bacterial lantibiotics: strategies to improve therapeutic potential. Curr Protein Pept Sci 6:61–75

Cotter PD, Hill C, Ross RP (2005b) Bacteriocins: developing innate immunity for food. Nat Rev Microbiol 3:777–788

Cuozzo SA, Castellano P, Sesma FJM, Vignolo GM, Raya RR (2003) Differential roles of the two-component peptides of lactocin 705 in antimicrobial activity. Curr Microbiol 46:180–183

Cuozzo SA, Sesma F, Palacios JM, de Ruiz Holgado AP, Raya RR (2000) Identification and nucleotide sequence of genes involved in the synthesis of lactocin 705, a two-peptide bacteriocin from *Lactobacillus casei* CRL 705. FEMS Microbiol Lett 185:157–161

Dalet K, Cenatiempo Y, Cossart P, Héchard Y, The European Listeria Genome Consortium (2001) A σ^{54}-dependent PTS permease of the mannose family is responsible for sensitivity of *Listeria monocytogenes* to mesentericin Y105. Microbiology 147:3263–3269

Diep DB, Håvarstein LS, Nes IF (1995) A bacteriocin-like peptide induces bacteriocin synthesis in *Lactobacillus plantarum* C11. Mol Microbiol 18:631–639

Diep DB, Håvarstein LS, Nes IF (1996) Characterization of the locus responsible for the bacteriocin production in *Lactobacillus plantarum* C11. J Bacteriol 178:4472–4483

Diep DB, Myhre R, Johnsborg O, Aakra Å, Nes IF (2003) Inducible bacteriocin production in *Lactobacillus* is regulated by differential expression of *pln* operons and by two antagonizing response regulators, the activity of which is enhanced upon phosphorylation. Mol Microbiol 47:483–494

Diep DB, Skaugen M, Salehian Z, Holo H, Nes IF (2007) Common mechanisms of target cell recognition and immunity for class II bacteriocins. Proc Natl Acad Sci U S A 104:2384–2389

Diep DP, Straume D, Kjos M, Torres C, Nes IF (2009) An overview of the mosaic bacteriocin *pln* loci from *Lactobacillus plantarum*. Peptides 30:1562–1574

Drider D, Fimland G, Héchard Y, McMullen LM, Prévost H (2006) The continuing story of class IIa bacteriocins. Microbiol Mol Biol Rev 70:564–582

Fimland G, Blingsmo OR, Sletten K, Jung G, Nes IF, Nissen-Meyer J (1996) New biologically active hybrid bacteriocins constructed by combining regions from various pediocin-like bacteriocins: the C-terminal region is important for determining specificity. Appl Environ Microbiol 62:3313–3318

Fimland G, Eijsink VGH, Nissen-Meyer J (2002a) Comparative studies of immunity proteins of pediocin-like bacteriocins. Microbiology 148:3661–3670

Fimland G, Eijsink VGH, Nissen-Meyer J (2002b) Mutational analysis of the role of tryptophan residues in an antimicrobial peptide. Biochemistry 41:9508–9515

Fimland G, Johnsen L, Dalhus B, Nissen-Meyer J (2005) Pediocin-like antimicrobial peptides (class IIa bacteriocins) and their immunity proteins: biosynthesis, structure, and mode of action. J Pept Sci 11:688–696

Fimland N, Rogne P, Fimland G, Nissen-Meyer J, Kristiansen PE (2008) Three-dimensional structure of the two peptides that constitute the two-peptide bacteriocin plantaricin EF. Biochim Biophys Acta 1784:1711–1719

Flynn S, van Sinderen D, Thornton GM, Holo H, Nes IF, Collins JK (2002) Characterization of the genetic locus responsible for the production of ABP-118, a novel bacteriocin produced by the probiotic bacterium *Lactobacillus salivarius* subsp. salivarius UCC118. Microbiology 148:973–984

Franz CMAP, Grube A, Herrmann A, Abriouel H, Stärke J, Lombardi A, Tauscher B, Holzapfel WH (2002) Biochemical and genetic characterization of the two-peptide bacteriocin enterocin 1071 produced by *Enterococcus faecalis* FAIR-E 309. Appl Environ Microbiol 68:2550–2554

Fregeau Gallagher NL, Sailer M, Niemczura WP, Nakashima TT, Stiles ME, Vederas JC (1997) Three-dimensional structure of leucocin A in trifluoroethanol and dodecylphosphocholine micelles: spatial location of residues critical for biological activity in type IIa bacteriocins from lactic acid bacteria. Biochemistry 36:15062–15072

Fremaux C, Ahn C, Klaenhammer TR (1993) Molecular analysis of the lactacin F operon. Appl Environ Microbiol 59:3906–3915

Gao Y, van Belkum MJ, Stiles ME (1999) The outer membrane of gram-negative bacteria inhibits antibacterial activity of brochocin-C. Appl Environ Microbiol 65:4329–4333

Garneau S, Ference CA, van Belkum MJ, Stiles ME, Vederas JC (2003) Purification and characterization of brochocin A and brochocin B(10-43), a functional fragment generated by heterologous expression in *Carnobacterium piscicola*. Appl Environ Microbiol 69:1352–1358

Garneau S, Martin NI, Vederas JC (2002) Two-peptide bacteriocins produced by lactic acid bacteria. Biochimie 84:577–592

Ghrairi T, Frére J, Berjeaud JM, Manai M (2005) Lactococcin MMT24, a novel two-peptide bacteriocin produced by *Lactococcus lactis* isolated from rigouta cheese. Int J Food Microbiol 105:389–398

Gravesen A, Ramnath M, Rechinger KB, Andersen N, Jänsch L, Héchard Y, Hastings JW, Knøchel S (2002) High-level resistance to classIIa bacteriocins is associated with one general mechanism in *Listeria monocytogenes*. Microbiology 148:2361–2369

Hauge HH, Mantzilas D, Eijsink VGH, Nissen-Meyer J (1999) Membrane-mimicking entities induce structuring of the two-peptide bacteriocins plantaricin E/F and plantaricin J/K. J Bacteriol 181:740–747

Hauge HH, Mantzilas D, Moll GN, Konings WN, Driessen AJ, Eijsink VGH, Nissen-Meyer J (1998a) Plantaricin A is an amphiphilic α-helical bacteriocin-like pheromone which exerts antimicrobial and pheromone activities through different mechanisms. Biochemistry 37:16026–16032

Hauge HH, Nissen-Meyer J, Nes IF, Eijsink VGH (1998b) Amphiphilic α-helices are important structural motifs in the α and β peptides that constitute the bacteriocin lactococcin G: enhancement of helix formation upon α-β interaction. Eur J Biochem 251:565–572

Haugen HS, Fimland G, Nissen-Meyer J, Kristiansen PE (2005) Three-dimensional structure in lipid micelles of the pediocin-like antimicrobial peptide curvacin A. Biochemistry 44:16149–16157

Horn N, Martínez M, Martinez JM, Hernández PE, Gasson MJ, Rodríguez JM, Dodd H (1998) Production of pediocin PA-1 by *Lactococcus lactis* using the lactococcin A secretory apparatus. Appl Environ Microbiol 64:818–823

Håvarstein LS, Diep DB, Nes IF (1995) A family of bacteriocin ABC transporters carry out proteolytic processing of their substrates concomitant with export. Mol Microbiol 16:229–240

Jiménez-Diaz R, Ruiz-Barba JL, Cathcart DP, Holo H, Nes IF, Sletten KH, Warner PJ (1995) Purification and partial amino acid sequence of plantaricin S, a bacteriocin produced by *Lactobacillus plantarum* LPCO10, the activity of which depends on the complementary action of two peptides. Appl Environ Microbiol 61:4459–4463

Johnsen L, Fimland G, Mantzilas D, Nissen-Meyer J (2004) Structure-function analysis of immunity proteins of pediocin-like bacteriocins: C-terminal parts of immunity proteins are involved in specific recognition of cognate bacteriocins. Appl Environ Microbiol 70:2647–2652

Johnsen L, Fimland G, Nissen-Meyer J (2005) The C-terminal domain of pediocin-like antimicrobial peptides (class IIa bacteriocins) is involved in specific recognition of the C-terminal part of cognate immunity proteins and in determining the antimicrobial spectrum. J Biol Chem 280:9243–9250

Kleerebezem M, Quadri LE (2001) Peptide pheromone-dependent regulation of antimicrobial peptide production in gram-positive bacteria: a case of multicellular behavior. Peptides 22:1579–1596

Kristiansen PE, Fimland G, Mantzilas D, Nissen-Meyer J (2005) Structure and mode of action of the membrane-permeabilizing antimicrobial peptide pheromone plantaricin A. J Biol Chem 280:22945–22950

Maldonado-Barragan A, Caballero-Guerrero B, Jiménez E, Jimenez-Diaz R, Ruiz-Barba JL, Rodriguez JM (2009) Enterocin C, a class IIb bacteriocin produced by *E. faecalis* C901, a strain isolated from human colostrum. Int J Food Microbiol 133:105–112

Maldonado A, Jiménez-Diaz R, Ruiz-Barba JL (2004) Induction of plantaricin production in *Lactobacillus plantarum* NC8 after coculture with specific gram-positive bacteria is mediated by an autoinduction mechanism. J Bacteriol 186:1556–1564

Maldonado A, Ruiz-Barba JL, Jiménez-Diaz R (2003) Purification and genetic characterization of plantaricin NC8, a novel coculture-inducible two-peptide bacteriocin from *Lactobacillus plantarum* NC8. Appl Environ Microbiol 69:383–389

Marciset O, Jewronimus-Stratingh MC, Mollet B, Poolman B (1997) Thermophilin 13, a nontypical antilisterial poration complex bacteriocin, that functions without a receptor. J Biol Chem 272:14277–14284

McCormick JK, Poon A, Sailer M, Gao Y, Roy KL, McMullen LM, Vederas JC, Stiles ME, van Belkum MJ (1998) Genetic characterization and heterologous expression of brochocin-C, an antibotulinal, two-peptide bacteriocin produced by *Brochothrix campestris* ATCC 43754. Appl Environ Microbiol 64:4757–4766

Moll G, Hauge HH, Nissen-Meyer J, Nes IF, Konings WN, Driessen AJM (1998) Mechanistic properties of the two-component bacteriocin lactococcin G. J Bacteriol 180:96–99

Moll G, Ubbink-Kok T, Hauge HH, Nissen-Meyer J, Nes IF, Konings WN, Driessen AJM (1996) Lactococcin G is a potassium ion-conducting, two component bacteriocin. J Bacteriol 178:600–605

Moll G, van der Akker HE, Hauge HH, Nissen-Meyer J, Nes IF, Konings WN, Driessen AJM (1999) Complementary and overlapping selectivity of the two-peptide bacteriocins EF and JK. J Bacteriol 181:4848–4852

Nes IF, Holo H, Fimland G, Hauge HH, Nissen-Meyer J (2002) Unmodified peptide-bacteriocins (class II) produced by lactic acid bacteria. In: Dutton CJ, Haxell MA, McArthur HAI, Wax RG (eds) Peptide antibiotics, discovery, modes of action, and application. Dekker, New York

Nes IF, Håvarstein LS, Holo H (1995) Genetics of non-lantibiotic bacteriocins. In: Ferretti JJ, Gilmore MS, Klaenhammer TR, Brown F (eds) Genetics of Streptococci, Enterococci, and Lactococci. Developments in biological standards, vol 85. Karger, Basel, pp 645–651

Nissen-Meyer J, Holo H, Håvarstein LS, Sletten K, Nes IF (1992) A novel lactococcal bacteriocin whose activity depends on the complementary action of two peptides. J Bacteriol 174:5686–5692

Nissen-Meyer J, Oppegård C, Rogne P, Haugen HS, Kristiansen PE (2010) Structure and mode-of-action of two-peptide (class IIb) bacteriocins. Probiotics Antimicrob Proteins 2:52–60

Nissen-Meyer J, Rogne P, Oppegård C, Haugen HS, Kristiansen PE (2009) Structure-function relationships of non-lanthionine-containing peptide (class II) bacteriocins produced by gram-positive bacteria. Curr Pharm Biotechnol 10:19–37

Oppegård C, Emanuelsen L, Thorbek L, Fimland G, Nissen-Meyer J (2010) The lactococcin G immunity protein recognizes specific regions in both peptides constituting the two-peptide bacteriocin lactococcin G. Appl Environ Microbiol 76:1267–1273

Oppegård C, Fimland G, Thorbæk L, Nissen-Meyer J (2007a) Analysis of the two-peptide bacteriocins lactococcin G and enterocin 1071 by site-directed mutagenesis. Appl Environ Microbiol 73:2931–2938

Oppegård C, Rogne P, Emanuelsen L, Kristiansen PE, Fimland G, Nissen-Meyer J (2007b) The two-peptide class II bacteriocins: structure, production, and mode of action. J Mol Microbiol Biotechnol 13:210–219

Oppegård C, Schmidt J, Kristiansen PE, Nissen-Meyer J (2008) Mutational analysis of putative helix-helix interacting GxxxG-motifs and tryptophan residues in the two-peptide bacteriocin lactococcin G. Biochemistry 47:5242–5249

Qi F, Chen P, Caufield PW (2001) The group I strain of *Streptococcus mutans*, UA140, produces both the lantibiotic mutacin I and a nonlantibiotic bacteriocin, mutacin IV. Appl Environ Microbiol 67:15–21

Ramnath M, Arous S, Gravesen A, Hastings JW, Héchard Y (2004) Expression of *mptC* of *Listeria monocytogenes* induces sensitivity to class IIa bacteriocins in *Lactococcus lactis*. Microbiology 150:2663–2668

Rogne P, Fimland G, Nissen-Meyer J, Kristiansen PE (2008) Three-dimensional structure of the two peptides that constitute the two-peptide bacteriocin lactococcin G. Biochim Biophys Acta 1784:543–554

Rogne P, Haugen C, Fimland G, Nissen-Meyer J, Kristiansen PE (2009) Three-dimensional structure of the two-peptide bacteriocin plantaricin JK. Peptides 30:1613–1621

Senes A, Engel DE, DeGrado WF (2004) Folding of helical membrane proteins: the role of polar, GxxxG-like and proline motifs. Curr Opin Struct Biol 14:465–479

Senes A, Gerstein M, Engelman DM (2000) Statistical analysis of amino acid patterns in transmembrane helices: the GxxxG motif occurs frequently and in association with beta-branched residues at neighboring positions. J Mol Biol 296:921–936

Senes A, Ubarretxena-Belandia I, Engelman DM (2001) The Cα–H···O hydrogen bond: a determinant of stability and specificity in transmembrane helix interactions. Proc Natl Acad Sci U S A 98:9056–9061

Stephens SK, Floriano B, Cathcart DP, Bayley SA, Witt VF, Jiménez-Díaz R, Warner PJ, Ruiz-Barba JL (1998) Molecular analysis of the locus responsible for production of plantaricin S, a two-peptide bacteriocin produced by *Lactobacillus plantarum* LPCO10. Appl Environ Microbiol 64:1871–1877

Uteng M, Hauge HH, Markwick PRL, Fimland G, Mantzilas D, Nissen-Meyer J, Muhle-Goll C (2003) Three-dimensional structure in lipid micelles of the pediocin-like antimicrobial peptide sakacin P and a sakacin P variant that is structurally stabilized by an inserted C-terminal disulfide bridge. Biochemistry 42:11417–11426

van Belkum MJ, Hayema BJ, Jeeninga RE, Kok J, Venema G (1991) Organization and nucleotide sequence of two lactococcal bacteriocin operons. Appl Environ Microbiol 57:492–498

van Belkum MJ, Worobo RW, Stiles ME (1997) Doble-glycine-type leader peptides direct secretion of bacteriocins by ABC transporters: colicin V secretion in *Lactococcus lactis*. Mol Microbiol 23:1293–1301

Wang Y, Henz ME, Fregeau Gallagher NL, Chai S, Gibbs AC, Yan LZY, Stiles ME, Wishart DS, Vederas JC (1999) Solution structure of carnobacteriocin B2 and implications for structure-activity relationships among type IIa bacteriocins from lactic acid bacteria. Biochemistry 38:15438–15447

Zendo T, Koga S, Shigeri Y, Nakayama J, Sonomoto K (2006) Lactococcin Q, a novel two-peptide bacteriocin produced by *Lactococcus lactis* QU 4. Appl Environ Microbiol 72:3383–3389

Chapter 12
Class IIc or Circular Bacteriocins

Leah A. Martin-Visscher, Marco J. van Belkum, and John C. Vederas

Abstract The circular bacteriocins produced by Gram-positive bacteria represent a diverse class of antimicrobial peptides. These bacteriocins display enhanced stability compared to linear bacteriocins, which arises from their characteristic circular backbone. Currently, eight unique circular bacteriocins have been identified, and analysis of their gene clusters indicates that they likely utilize complex mechanisms for maturation and secretion, as well as for immunity. These bacteriocins target the cytoplasmic membrane of sensitive cells, leading to pore formation that results in loss of ions, dissipation of membrane potential, and ultimately, cell death. Structural studies suggest that despite variation in their sequences, most of these bacteriocins likely adopt a common three-dimensional architecture, consisting of four or five tightly packed helices encompassing a hydrophobic core. There are many mysteries surrounding the biosynthesis of these peptides, particularly in regard to the mechanism by which they are cyclized. Elucidation of such a mechanism may provide exciting new approaches to the bioengineering of new, stable, and antimicrobially active circular peptides.

Introduction

The bacteriocins produced by Gram-positive bacteria are a diverse group of peptides that typically display potent antimicrobial activity toward closely related bacteria. Within this group of peptides, a new class of bacteriocins has recently emerged: the circular bacteriocins. The circular bacteriocins are characterized by the head-to-tail cyclization of their backbone (Craik et al. 2003; Maqueda et al. 2004, 2008). Although the lantibiotics and other bacteriocins may contain cyclic

J.C. Vederas (✉)
Department of Chemistry, University of Alberta, Edmonton, AB, Canada T6G 2G2
e-mail: john.vederas@ualberta.ca

D. Drider and S. Rebuffat (eds.), *Prokaryotic Antimicrobial Peptides: From Genes to Applications*, DOI 10.1007/978-1-4419-7692-5_12,

elements (due to lanthionine residues or the presence of disulfide bridges), they are not considered to be circular bacteriocins since their backbones are not N- to C-cyclized. In addition, many examples of cyclic antimicrobial peptides of microbial origin have been reported (such as the polymyxins, cyclosporin A, and gramicidin S) that are not classified as bacteriocins, since they are not ribosomally synthesized. Rather, they are constructed by complex, multidomain nonribosomal peptide synthetases (NRPS) (Koglin and Walsh 2009; Marahiel 2009; Walsh 2004). Ribosomally synthesized circular peptides have also been discovered in plants and animals and are known to exhibit a diverse range of bioactivities. Typically, the circular peptides from these organisms are shorter in length and contain at least one disulfide bond, further enhancing their structural integrity (Craik 2006, 2009). In particular, the cyclotides comprise a large family of well-studied and characterized circular peptides produced by numerous plants in the *Violaceae* and *Rubiaceae* families (Craik 2006, 2009; Daly et al. 2009).

Although this chapter focuses on the circular bacteriocins produced by Gram-positive bacteria, it should also be noted that microcin J25 (MccJ25), produced by *Escherichia coli* AY25 (Salomón and Farías 1992), is also a cyclic peptide. Initially, it was thought that this 21-amino acid bacteriocin was head-to-tail cyclized, and thus, was a true circular bacteriocin (Blond et al. 1999, 2001). However, in 2003, three different groups reported that instead of a backbone cyclization, MccJ25 had a lariat-type structure in which a backbone-to-side chain ring lassoed the C-terminus of the peptide (Bayro et al. 2003; Rosengren et al. 2003; Wilson et al. 2003). The ring structure is formed by a peptide bond between the N-terminus (Gly-1) and the carboxylate side chain of Glu-8. The C-terminus loops through this ring and is held in place by noncovalent interactions and steric trapping due to the side chains of Phe-19 and Tyr-20, which flank either side of the ring (Fig. 12.1). Since MccJ25 does not contain a head-to-tail cyclization, it does not adhere to the canonical definition of the circular bacteriocins.

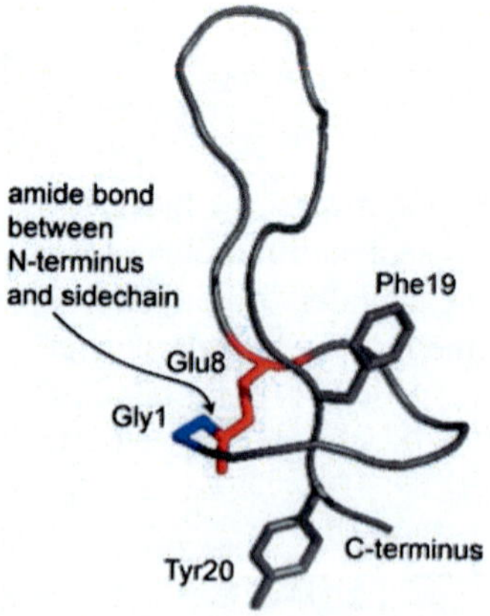

Fig. 12.1 3D NMR solution structure of microcin J25 (pdb code 1Q71). Gly-1 and Glu-8 are colored *blue* and *red*, respectively. The amide bond between the amino group of Gly-1 and the carboxylate of the Glu-8 side chain is indicated. The side chains of Phe-19 and Tyr-20, which trap the C-terminus within the ring, are indicated

The classification of the circular bacteriocins has been debated in recent years. Traditionally, they were considered to be members of the class II bacteriocins, in accordance with Klaenhammer's classification scheme. Within this grouping, they have been referred to as type IIc (Cotter et al. 2005), IId (Nes et al. 2007a), and IIf (van Belkum and Stiles 2000) bacteriocins. However, since the type II bacteriocins are not posttranslationally modified, several groups have suggested moving the circular bacteriocins to a unique class (Franz et al. 2007; Heng and Tagg 2006; Kemperman et al. 2003b; Maqueda et al. 2004, 2008). Regardless of the classification, it is clear that these peptides are a unique class of bacteriocins that must undergo interesting posttranslational modifications.

The Circular Bacteriocins Produced by Gram-Positive Bacteria

The circular bacteriocins from Gram-positive bacteria are produced by both lactic acid bacteria (LAB) and non-LAB. To date, eight different circular bacteriocins have been reported. These include enterocin AS-48 (Samyn et al. 1994), butyrivibriocin AR10 (Kalmokoff et al. 2003), gassericin A (Kawai et al. 1998a), circularin A (Kemperman et al. 2003b), subtilosin A (Babasaki et al. 1985; Zheng et al. 1999), uberolysin (Wirawan et al. 2007), carnocyclin A (Martin-Visscher et al. 2008), and, most recently, lactocyclicin Q (Sawa et al. 2009). Acidocin B is considered a putative circular bacteriocin, as it shows a high degree of sequence identity (98%) to gassericin A (Leer et al. 1995). However, its circular nature has not yet been confirmed. Previously, reutericin 6 was also classified as a circular bacteriocin. Reutericin 6 and gassericin A have identical primary structures, but they were reported to have different spectra of activity, different secondary structure profiles and to contain different amounts of D–Ala (Itoh et al. 1995; Kawai et al. 2004; Toba et al. 1991). However, it has recently been reported that the gene cluster encoding reutericin 6 is 100% identical to that of gassericin A (Ito et al. 2009). Furthermore, reevaluation of the stereochemical analysis of these two peptides has revealed that neither contains D-Ala, and as such, they are identical bacteriocins (Arakawa et al. 2010). The amino-acid sequences of the circular bacteriocins and their leader peptides are listed in Fig. 12.2.

Aside from having a high degree of hydrophobicity, little sequence homology exists among the circular bacteriocins. Nonetheless, it has been suggested the peptides from LAB be subdivided into two groups, according to sequence similarity and biochemical characteristics (Cotter et al. 2005; Kawai et al. 2009). Group i consists of the "AS-48 like" bacteriocins (AS-48, UblA, CirA, LycQ, CclA), whereas group ii consists of the "gassericin A-like" bacteriocins (GaaA, BviA). These two groups also show differences in their physicochemical properties (Martin-Visscher et al. 2008). The group i peptides are cationic and predicted to have high isoelectric points (pI>9) at physiological pH. In contrast, the group ii peptides are highly hydrophobic and are either neutral or anionic, with lower pI values (pI<7). Table 12.1 lists the physical properties and producer organisms of the various

Leader peptide:

```
              -30       -20       -10        -1↓
AS-48     MVKENKFSKIFILMALSFLGLALFSASLQFLPIAH-
UblA                                   MDILLE-
CirA                                      MFL-
LycQ                                       MK-
CclA                                     MLYE-
GaaA        MVTKYGRNLGLNKVELFAIWAVLVVALLLTTAN-
BviA                    MSKKQIMSNCISIALLIALIPN-
SubA                                 MKKAVIVE-
```

Bacteriocin sequence:

```
         1        10        20        30        40        50        60
AS-48    MAKEFGIPAAVAGTVLNVVEAGGWVTTIVSILTAVGSGGLSLLAAAGRESIKAYLKKEIKKKGKRAVIAW
UblA     LAGYTGIASGTAKKVVDAIDKGAAAFVIISIISTVISAGALGAVSASADFIILTVKNYISRNLKAQAVIW
CirA     VAGALGVQTAAATTIVNVILNAGTLVTVLGIIASIASGGAGTLMTIGWATFKATVQKLAKQSMARAIAY
LycQ     LIDHLGAPRWAVDTILGAIAVGNLASWVLALVPGPGWAVKAGLATAAAIVKHQGKAAAAAW
CclA     LVAYGIAQGTAEKVVSLINAGLTVGSIISILGGVTVGLSGVFTAVKAAIAKQGIKKAIQL
GaaA     IYWIADQFGIHLATGTARKLLDAMASGASLGTAFAAILGVTLPAWALAAAGALGATAA
BviA     IYFIADKMGIQLAPAWYQDIVNWVSAGGTLTTGFAIIVGVTVPAWIAEAAAAFGIASA
SubA     NKGCATCSIGAACLVDGPIPDFEIAGATGLFGLWG
```

Fig. 12.2 Amino-acid sequences of the circular bacteriocins and their leader peptides (*bolded*). The *red arrow* indicates the point of cleavage between the leader peptide and the linear bacteriocin. Amino-acid positions are numbered. The abbreviations AS-48, UblA, CirA, LycQ, CclA, GaaA, BviA, and SubA refer to enterocin AS-48 (Martínez-Bueno et al. 1994), uberolysin (Wirawan et al. 2007), circularin A (Kemperman et al. 2003b), lactocyclicin Q (Sawa et al. 2009), carnocyclin A (Martin-Visscher et al. 2008), gassericin A (Kawai et al. 1998b), butyrivibriocin AR10 (Kalmokoff et al. 2003), and subtilosin A (Zheng et al. 1999), respectively

circular bacteriocins. Like many other bacteriocins, the circular bacteriocins are highly thermostable. However, they also display enhanced stability to pH variation and are generally resistant to degradation by numerous proteases. It is believed that the circular structure of these peptides helps provide structural integrity and may enhance the bioactivity of the peptide (Craik et al. 2003; Kawai et al. 2003; Maqueda et al. 2008; Montalbán-López et al. 2008).

Subtilosin A is atypical when compared to other circular bacteriocins, as it is significantly shorter (35 aa) and also contains three unique thioether cross-links. These unusual bridges are formed between the sulfur atom of cysteine residues and the α-carbon of phenylalanine and threonine residues (Kawulka et al. 2003, 2004). These modifications are distinct from those found in the lantibiotics, and as such, SubA should not be considered a lantibiotic. The lantibiotics are characterized by the presence of lanthionine and methyllanthionine residues, which are formed by the dehydration of gene-encoded serine or threonine residues followed by the addition of a cysteine thiol onto the β-carbon of these dehydrated residues. Both steps are enzyme-mediated by either two separate enzymes or one bifunctional enzyme (Chatterjee et al. 2005; Chen and Hoover 2003; Cotter et al. 2005; Nes et al. 2007b). Since the thioether linkages in SubA represent a unique type of modification, it has been suggested that SubA actually belongs to a new class of bacteriocins (Kawulka et al. 2003, 2004).

Table 12.1 Physical properties of the circular bacteriocins

Bacteriocin (abbr.)	Leader (aa)	Bacteriocin (aa)	MW[a] (Da)	pI[b]	Net charge[b]	Producer organism
Enterocin AS-48 (AS-48)	35	70	7,150	10.1	+5.9	*Enterococcus faecalis* S-48
Uberolysin (UblA)	6	70	7,048	9.6	+2.9	*Streptococcus uberis* 42
Circularin A (CirA)	3	69	6,771	10.5	+3.9	*Clostridium beijerinckii* ATCC 25752
Lactocyclicin Q (LycQ)	2	61	6,060	9.7	+2.4	*Lactococcus* sp. strain QU12
Carnocyclin A (CclA)	4	60	5,862	10.0	+3.9	*Carnobacterium maltaromaticum* UAL307
Gassericin A (GaaA)	33	58	5,654	6.8	0.2	*Lactobacillus gasseri* LA39 *L. gasseri* LA6[c]
Butyrivibriocin AR10 (BviA)	22	58	5,982	4.0	−2.1	*Butyrivibrio fibrisolvens* AR10
[d]Acidocin B (AcdB)	33	58	–	6.8	0.2	*Lactobacillus acidophilus* M46
Subtilosin A (SubA)	8	35	3,402[e]	4.0	−2.2	*Bacillus subtilis* strain 168

[a]Calculated molecular weight of the cyclized product (Gasteiger et al. 2005)
[b]Predicted for the linear bacteriocin at physiological pH (Gasteiger et al. 2005)
[c]Originally reported to produce reutericin 6
[d]Putative circular bacteriocin
[e]In addition to backbone cyclization (−18 Da), the peptide contains three thioether bridges (−6 Da). See text for references

Genetics

The gene clusters for several of the circular bacteriocins have been reported. For a detailed discussion and comparison of the genetic features of these peptides, the reader is directed to a recent review covering this topic (Maqueda et al. 2008). The genetic determinants for these bacteriocins are located either on the chromosome or on plasmids. Aside from the structural gene, many of the genes encode for basic and hydrophobic proteins that are predicted to contain multiple transmembrane domains. In general, the function of these proteins is inferred, based on sequence homology. Figure 12.3 compares the known gene clusters of several of the circular bacteriocins, and Table 12.2 lists the physical properties and known or putative functions of the gene products. For carnocyclin A and lactocyclicin Q, only the sequences for the structural genes have been published thus far.

The most extensively studied circular bacteriocin, from both a genetic and biochemical perspective, is enterocin AS-48. Complete (wild-type) expression and immunity of enterocin AS-48 requires the coordinated expression of 10 genes: *as-48ABCC$_1$DD$_1$EFGH* (Diaz et al. 2003; Martínez-Bueno et al. 1998). The genetic locus for enterocin AS-48 is plasmid-associated (Martínez-Bueno et al. 1990). *as-48A* encodes for the bacteriocin precursor, while *as-48D$_1$* encodes for a dedicated immunity protein. It is believed that maturation and secretion depend upon *as-48BCC$_1$D*, as mutations in these genes result in decreased AS-48 production (Martínez-Bueno et al. 1998). According to sequence homology, the gene products of *as-48C$_1$D* likely encode an ABC-transporter and are implicated in secretion of the peptide. Their expression also enhances immunity to AS-48. As-48C is predicted to be an integral membrane protein and may function as an accessory protein in the maturation and secretion of the bacteriocin. It has been proposed that *as-48B* is involved in the cyclization of the peptide since inactivation of this gene was sufficient in preventing production of AS-48 (Martínez-Bueno et al. 1998). Full immunity against AS-48 also requires the expression of *as-48EFGH*, which likely encode for another ABC-transporter (Diaz et al. 2003).

The butyrivibriocin AR10 gene cluster consists of six genes, *bviBCDAE* and *orf6*, where *bviA* encodes for the bacteriocin precursor (Kalmokoff et al. 2003). Based on sequence homology, it has been proposed that BviB is the ATP-binding domain of an ABC-transporter and that BviE is the immunity protein for butyrivibriocin AR10, although neither of these functions has been confirmed. The functions of the other gene products are unknown. Immediately upstream of the bacteriocin operon are two orfs, which likely encode for a response regulator and protein kinase and may be involved in regulation of bacteriocin expression.

For circularin A, the chromosomally located gene cluster consists of *cirABCDEGHI*, although minimal production and immunity requires the expression of just five essential genes (*cirABCDE*) as confirmed by heterologous expression (Kemperman et al. 2003a). Several of the genes within the cluster are overlapping, suggesting that translational coupling ensures proper ratios of gene expression. *cirA* encodes for the bacteriocin precursor, while *cirBD*, according to sequence homology, appears to encode an ABC-transporter and is likely required for secretion of the

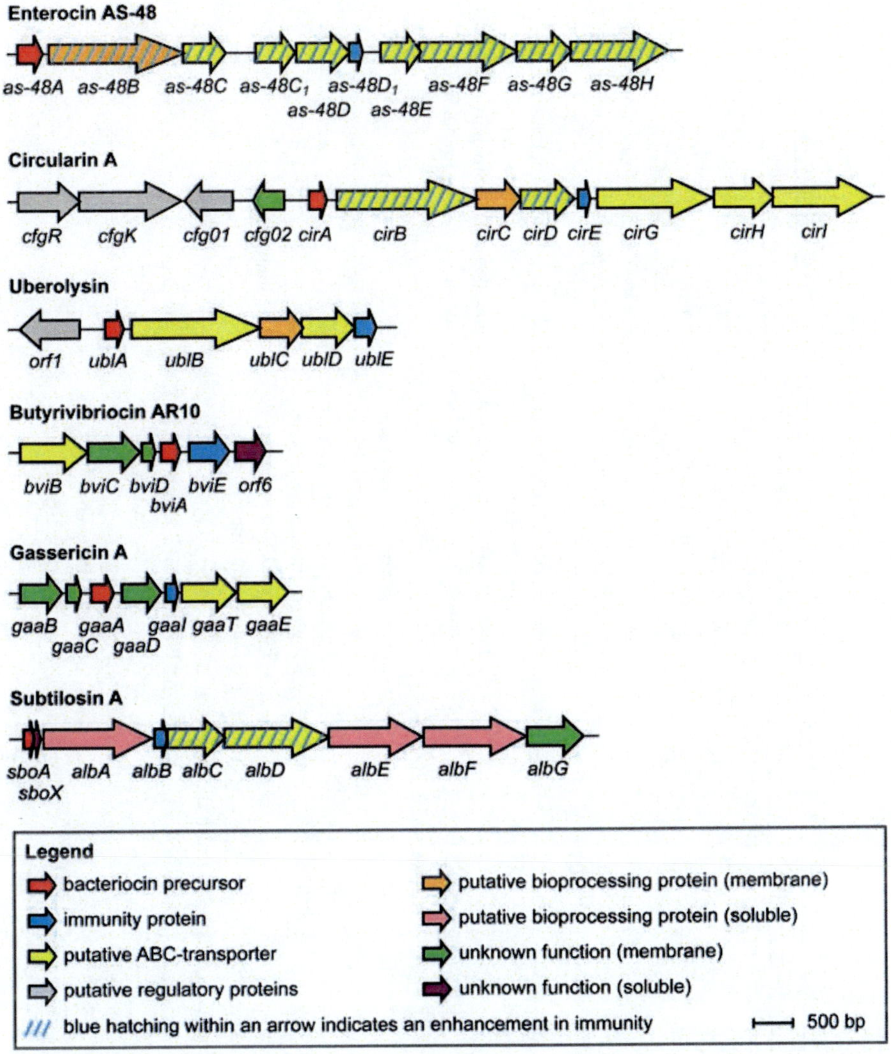

Fig. 12.3 Gene clusters of the circular bacteriocins. Genes are colored according to their known or putative functions (see the legend), as described in the literature (adapted from Maqueda et al. 2008)

peptide. It is believed that CirC is required for maturation of CirA and may be involved in cyclization of the peptide. CirE is required for immunity and is likely a dedicated immunity protein, although full immunity also depends upon CirBD, suggesting that circularin A utilizes multiple mechanisms of immunity, as does enterocin AS-48. Based on homology, *cirGHI* likely represents another ABC-transporter. Whether or not this additional transporter enhances immunity to CirA has not yet been established. In addition to the genes responsible for maturation, secretion, and immunity, there appears to be several genes related to regulation of

Table 12.2 Characteristics and known or putative functions of the proteins encoded by the gene clusters of the various circular bacteriocins

ORF	Size (aa)	MW [a] (kDa)	pI [b]	TM [c]	Known or putative functions
Enterocin AS-48 (GenBank accession numbers Y12234 and AJ438950) (Diaz et al. 2003; Martínez-Bueno et al. 1998)					
as-48A	105	11.1	10.1	2	Bacteriocin precursor
as-48B	563	66.0	9.7	11	Putative maturation (cyclization) protein
as-48C	178	20.3	9.2	4	Auxiliary immunity protein
as-48C$_1$	164	18.8	10.0	5	Putative ABC transporter (transmembrane domain) – secretion/immunity
as-48D	219	25.7	8.6	0	Putative ABC transporter (ATP-binding domain) – secretion/immunity
as-48D$_1$	56	6.3	10.0	2	Immunity protein
as-48E	169	19.1	9.4	4	Putative ABC transporter (transmembrane domain) – immunity
as-48F	407	45.1	5.5	1	Accessory protein
as-48G	227	25.7	5.7	0	Putative ABC transporter (ATP-binding domain) – immunity
as-48H	399	44.1	9.2	4	Putative ABC transporter (transmembrane domain) – immunity
Uberolysin (GenBank accession number DQ650653) (Wirawan et al. 2007)					
orf1	251	29.6	9.4	0	Putative response regulator
ublA	76	7.8	8.5	1	Bacteriocin precursor
ublB	535	63.5	9.5	12	Putative protein involved in secretion
ublC	181	21.3	8.7	4	Putative maturation protein
ublD	218	25.5	4.8	0	Putative ABC transporter (ATP-binding domain)
ublE	88	10.2	9.2	3	Putative immunity protein
Butyrivibriocin AR10 (GenBank accession number AF076529) (Kalmokoff et al. 2003)					
bviB	280	31.9	9.0	0	Putative ABC transporter (ATP-binding domain)
bviC	216	23.7	9.0	6	?
bviD	53	6.1	9.4	2	?
bviA	80	8.4	5.9	2	Bacteriocin precursor
bviE	166	18.8	9.1	4	Putative immunity protein
orf6	126	14.8	5.0	0	?

Gassericin A (GenBank accession number AB007043) (Ito et al. 2009; Kawai et al. 1998b, 2009)					
gaaB	174	20.3	9.6	5	?
gaaC	60	7.3	9.6	2	?
gaaA	91	9.3	9.4	2	Bacteriocin precursor
gaaD	162	18.4	6.7	4	?
gaaI	53	6.1	11.1	2	Immunity protein
gaaT	226	25.1	4.7	0	Putative ABC transporter (ATP-binding domain)
gaaE	212	23.7	9.5	6	Putative ABC transporter (transmembrane domain)
Circularin A (GenBank accession number AJ566621) (Kemperman et al. 2003a)					
cfgR	258	30.4	6	0	Putative response regulator
cfgK	425	49.8	6.7	7	Putative histidine protein kinase
cfg01	199	23.2	9.6	5	AgrB regulatory protein
cfg02	130	14.4	9.8	4	?
cirA	72	7.2	10.5	2	Bacteriocin precursor
cirB	581	68.8	9.2	11	Putative protein involved in secretion/immunity
cirC	185	20.9	9.8	4	Putative maturation (cyclization) protein
cirD	221	25.7	6.3	0	Putative ABC transporter (ATP-binding domain)
cirE	49	5.7	10.2	2	Putative immunity protein based on homology to AS-48D_1
cirG	475	51.7	4.8	1	Accessory protein
cirH	248	27.6	6.1	0	Putative ABC transporter (ATP-binding domain)
cirI	422	45.7	9.4	4	Putative ABC transporter permease
Subtilosin A (GenBank accession number NC_000964) (Zheng et al. 1999, 2000)					
sboA	43	4.3	4.8	0	Bacteriocin precursor
sboX	50	5.8	9.5	0	Putative type II bacteriocin precursor
albA	448	51.5	6.20	0	Cofactor synthesis
albB	53	6.1	10.5	2	Dedicated immunity protein
albC	239	27.3	5.4	0	Putative ABC transporter (ATP-binding domain) – secretion/immunity

(continued)

Table 12.2 (continued)

ORF	Size (aa)	MW [a] (kDa)	pI [b]	TM [c]	Known or putative functions
albD	436	49.5	9.9	10	Putative protein involved in secretion/immunity
albE	394	44.1	6.1	0	Putative processing protease
albF	426	49.0	5.5	0	Zinc endoproteinase
albG	233	26.3	9.3	6	?

[a] Calculated molecular weight based on the translated gene product
[b] Predicted by ProtParam (Gasteiger et al. 2005)
[c] TM; number of putative transmembrane sequences, predicted by TMHMM (Krogh et al. 2001) (available on the ExPASy Proteomics Server)

bacteriocin production. These genes (*cfgR, cfgK*) encode for proteins that show homology to response regulators and histidine protein kinases, which have been shown to play a role in bacteriocin production via a quorum-sensing pathway.

The uberolysin locus, located on the chromosome, contains six genes, *orf1* and *ublABCDE* (Wirawan et al. 2007). Based on homology, *orf1* appears to encode for a response regulator protein and may be involved in regulation of bacteriocin expression, although an accompanying protein kinase has not yet been identified. *ublA* encodes for the bacteriocin precursor, and it is believed that *ublBD* and *ublC* encode for an ABC-transporter and accessory protein, respectively, which are all involved in bacteriocin secretion. It has been suggested that *ublC* may be involved in maturation of the bacteriocin. UblE is likely a dedicated immunity protein, although this has not yet been confirmed.

The complete gene cluster required for production of gassericin A has recently been described (Ito et al. 2009; Kawai et al. 2009). Initially, it was believed that the genetic determinants were chromosomally located (Kawai et al. 1998b), but it is now known that the seven genes required for bacteriocin production, *gaaBCADITE*, reside on a large, 33-kb conjugative plasmid (Ito et al. 2009). *gaaA* encodes for the bacteriocin precursor, whereas *gaaI* is the dedicated immunity gene (Kawai et al. 2009). According to sequence homology, *gaaTE* likely encode an ABC-transporter, presumably for bacteriocin transport. GaaB, GaaC, and GaaD are predicted to be membrane-associated proteins, but their functions are unknown.

The production of subtilosin A requires the expression of genes from the *sbo-alb* locus, located on the chromosome of *Bacillus subtilis*. *sboA* is the gene for the subtilosin A precursor, whereas *albABCDEFG* are required for the full production of and immunity to SubA (Zheng et al. 1999). Mutational analysis of the *sbo-alb* locus (Zheng et al. 2000) revealed that *albA* and *albF* are crucial for the production of SubA. AlbA shows homology to members of the MoaA/NifB/PqqE family of proteins (which are involved in cofactor biosynthesis), whereas the N-terminal half of AlbF shows homology to zinc metalloproteases. As such, it is believed that AlbA and AlbF are required for constructing the unique posttranslational modifications of SubA. It has also been shown that full immunity depends upon the expression of *albBCD*, where *albB* likely encodes for a dedicated immunity protein and AlbCD comprise an ABC-transporter that enhances immunity. An additional open reading frame, *sboX*, was also found to overlap with the SubA structural gene (*sboA*). *sboX* likely encodes for a bacteriocin-like peptide, as the gene product appears to have a double-glycine-type leader sequence. An in-frame deletion of *sboX* did not impair production of SubA. At present, the role of the *sboX* gene is not known.

Biosynthesis

For both the lantibiotics (class I) and type IIa bacteriocins (class II), much is known regarding the regulation of expression, posttranslational modifications, maturation, secretion, and mechanisms of immunity for these peptides. In contrast, relatively little is known about the biosynthesis of the circular bacteriocins. The circular bacteriocins are synthesized as linear precursor peptides, with N-terminal extensions. As depicted in Fig. 12.2, there is little sequence similarity between the leader peptides, some of which are very short, while others are much longer, making it difficult to predict what the role of the leader peptide is. For bacteriocins, in general, it is believed that leader peptides render the bacteriocin inactive while in the cytoplasm and also help target the bacteriocin to the enzymes responsible for posttranslational modifications (in the case of lantibiotics) and export it out of the cell (Drider et al. 2006; Håvarstein et al. 1995; Oman and van der Donk 2010; van Belkum et al. 1997). In the case of the circular bacteriocins, since there is such variation in both length and sequence among the leader peptides, it is unknown what role these leaders play.

Likewise, since the bacteriocins display such varied sequences, it is unclear if their cyclization is controlled by a conserved mechanism. In most cases, it appears that the head-to-tail ligation occurs between two hydrophobic residues. For the group i circular bacteriocins, a bulky aromatic or hydrophobic residue (Tyr, Trp, or Leu) from the C-terminus condenses with a hydrophobic residue on the N-terminus (Met, Leu, or Val). For the group ii circular bacteriocins, a smaller hydrophobic residue (Ala) is observed as the C-terminal residue. Subtilosin A is atypical in this regard, as polar residues (Gly and Asn) are involved in the cyclization.

Perhaps the greatest mystery surrounding the biosynthesis of the circular bacteriocins is how the cyclization occurs. From a chemical perspective, it is unclear how a "free", or unactivated C-terminus is subject to attack by the N-terminal amino group. In contrast to the circular bacteriocins, many other ribosomally synthesized circular peptides, such as the cyclotides, are synthesized as precursor peptides with both N- and C-terminal extensions (Daly et al. 2009; Gillon et al. 2008). For the cyclotides, it is believed that an asparaginyl endoproteinase is responsible for cleaving this C-terminal extension and facilitating cyclization (Saska et al. 2007), according to the following mechanism. After loss of the N-terminal leader peptide, it is believed that the protease recognizes a conserved Asn or Asp residue, immediately preceding the C-terminal extension of the cyclotide. A cysteine residue of the protease attacks this amino acid, cleaving the C-terminal extension. The cyclotide remains covalently bound to the enzyme active site, as an acyl intermediate. The amino group of the N-terminus of the peptide then enters the active site and attacks the thioester, thus regenerating the free enzyme and releasing the cyclized peptide. It has been found that mutations to the conserved Asn or Asp residues fail to generate circular cyclotides (Ireland et al. 2006). In addition, a high degree of sequence similarity is observed between the N-terminus of the cyclotide and the initial residues of their C-terminal extensions, suggesting that the N-terminus of the cyclotide is a likely substrate for the enzyme active site (Saska et al. 2007). Another method by which amino acids may be activated for attack is adenylation. Such an approach is utilized in the nonribosomal assemblage of peptides by NRPS, as well as the initial activation of amino acids by aminoacyl tRNA synthetases. The biosynthesis of cyclodipeptides has been shown to utilize amino acids that have been activated as aminoacyl-tRNAs (Gondry et al. 2009). Whether or not similar adenylation and activation type mechanisms are employed in the construction of the circular bacteriocins is unknown.

At present, it is unknown how the circular bacteriocins are secreted from producer cells. Numerous ABC-transporters are encoded within the gene clusters of the circular bacteriocins and it is tempting to speculate that they utilize such systems for translocation, as is the case for many other bacteriocins produced by Gram-positive bacteria (Cotter et al. 2005; van Belkum and Stiles 2000). It is also unclear as to whether or not secretion of the peptide precedes, follows, or is coupled to removal of the leader peptide and cyclization of the bacteriocin.

In order to protect themselves, bacteriocin producers also produce immunity proteins. In some cases, a dedicated immunity protein is sufficient in providing full immunity (exemplified by the type IIa bacteriocins) (Drider et al. 2006), whereas in other cases (such as with nisin and several other lantibiotics), a two-fold mechanism is employed, wherein a dedicated immunity and an ABC-transporter are required for full immunity (Draper et al. 2008). At present, it is not known if the circular bacteriocins share a common mechanism of immunity. However, analysis of the gene clusters for enterocin AS-48 (Diaz et al. 2003; Martínez-Bueno et al. 1998), subtilosin A (Zheng et al. 2000), and circularin A (Kemperman et al. 2003a) suggest that these bacteriocins use a two-fold mechanism of immunity. In these cases, a dedicated immunity protein has been identified, but full resistance also depends on the expression of additional ABC-transporters. No structural studies of the immunity proteins for these bacteriocins have yet been reported.

Structure

As mentioned earlier, it is believed that the remarkable stability of the circular bacteriocins is due, in part, to their cyclic structure. To investigate the role of the circular backbone on both stability and bioactivity, a linear variant of enterocin AS-48 was produced by digestion of AS-48 with thermolysin (cleavage between Ala-10 and Val-11) (Montalbán-López et al. 2008). Although the linear peptide was largely helical, as indicated by circular dichroism, its thermal denaturation occurred at lower temperatures than wild-type AS-48, and its antimicrobial activity was abolished upon freezing and thawing. In addition, the activity of the linear peptide was found to be 300 times less than that of the circular bacteriocin. Likewise, although linear gassericin A (produced by heterologous expression in *E. coli*, as a fusion protein) appears to assume a compact, globular structure (evidenced by migration on SDS-PAGE comparable to that of the wild-type peptide), its bioactivity was found to be almost 200 times less than that of the circular bacteriocin (Kawai et al. 2003). Thus, cyclization of the peptides is a key factor in both their stability and activity.

Currently, the 3D structures for three circular bacteriocins are known: subtilosin A, enterocin AS-48, and carnocyclin A. Initial sequencing and mass spectral analysis of SubA indicated that the peptide was likely backbone-cyclized and contained unusual intramolecular cross-links. In 2001, Marx et al. (2001) proposed an initial structure for SubA and determined that the cross-links involved thioether bridges between cysteine and phenylalanine or threonine residues. However, they were

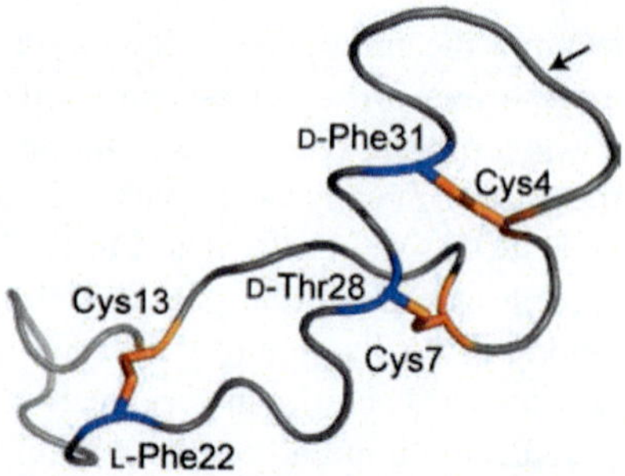

Fig. 12.4 3D NMR solution structure of subtilosin A (pdb code 1PQX). The point of backbone cyclization is indicated with *an arrow*. The amino acids involved in the thioether bridges are labeled. Cysteine main-chain and side-chain atoms are colored *orange*, whereas Phe or Thr main-chain atoms are colored *blue*

unable to determine the connectivity of these bridges. Shortly thereafter, Kawulka et al. (2003, 2004) elucidated the identity of these modifications, revealing that they arose from the coupling of the cysteine thiols of Cys-13, Cys-7, and Cys-4 onto the α-carbons of Phe-22, Thr-28, and Phe-31, respectively. In addition to solving the complete primary structure of SubA, they reported the 3D NMR solution structure of the bacteriocin. SubA adopts a rigid, bowl-like structure, in which most of the side chains are pointed outward, as depicted in Fig. 12.4. The stereochemistry of the α-carbons was determined by performing structure calculations on all eight possible stereoisomers. The isomer that fit the NOE data the best and gave the lowest energy ensemble of structures with the best rmsd was found to be the stereoisomer with (L)-Phe-22, (D)-Thr-28, and (D)-Phe-31 (the LDD isomer). Despite being anionic, the structure of the peptide revealed that SubA is amphipathic, as its three acidic residues (Asp-16, Asp-21, and Glu-23) are located toward one end of the molecule, whereas its basic residue (Lys-2) and several hydrophobic residues, including Trp-34, are located at the other end (Fig. 12.8).

The 3D structure of AS-48 has been solved by both NMR (González et al. 2000; Langdon et al. 1998) and X-ray crystallography (Sánchez-Barrena et al. 2003), under different pH conditions. At acidic pH, AS-48 is monomeric, but at physiological pH, it exists as a dimer (Abriouel et al. 2001). The NMR solution structure of the AS-48 monomer showed that it consists of a globular arrangement of five helices, encompassing a compact hydrophobic core and the covalent bond linking the N- and C-termini resides within the fifth helix (Fig. 12.5a) (González et al. 2000). The 3D structure of AS-48 showed significant homology to NK-lysin, a saposin-like peptide that exhibits cytotoxic and antimicrobial activity and is known to interact with the cytoplasmic membrane. Analysis of the electrostatic surface potential of AS-48 shows significant charge separation across the molecule, as a cluster of seven lysines along helices 4 and 5 impart a high degree of positive charge to one surface of the peptide (Fig. 12.5a). Initially, the authors surmised that this charge separation was crucial for pore formation through a mechanism known as molecular electroporation, which was also used to account for the

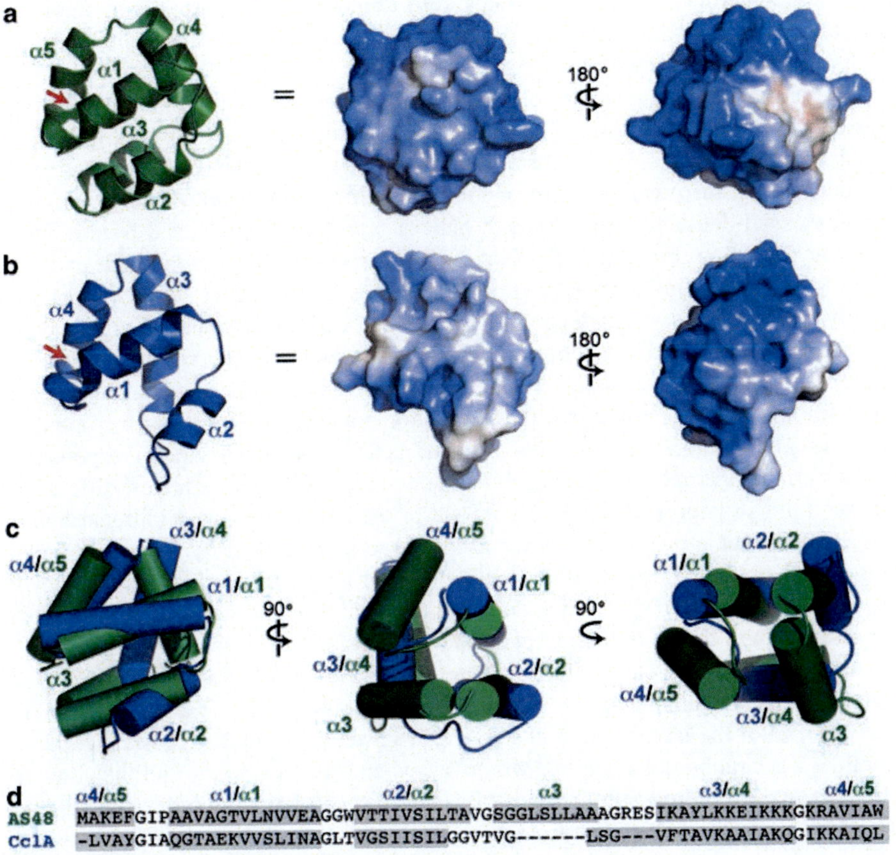

Fig. 12.5 Comparing the structures of AS-48 (pdb code 1E68) and CclA (pdb code 2KJF). Ribbon diagrams and electrostatic surfaces of AS-48 (**a**) and CclA (**b**). Helices are labeled appropriately. The *red arrow* indicates the point of cyclization. In the electrostatic surface diagrams, *red* and *blue* refer to negatively and positively charged regions, respectively. (**c**) Structural overlay of AS-48 (*green*) and CclA (*blue*). (**d**) Alignment of the primary structures, with helical sections highlighted in *gray* (adapted from Martin-Visscher et al. 2009)

pore-forming abilities of NK-lysin (González et al. 2000; Miteva et al. 1999). In 2003, the crystal structure of dimeric AS-48 was reported, confirming the basic architecture of the individual AS-48 units (Sánchez-Barrena et al. 2003). However, the crystal structure also revealed two different forms of the dimer. In aqueous solvent, AS-48 exists as a water-soluble dimer (DF-I), in which the hydrophobic faces of the individual monomers (particularly along helices 1 and 2) are in contact, and polar interactions with the solvent are maximized. Upon interaction with a membrane, the dimer inverts, exposing its hydrophobic faces (DF-II) (Fig. 12.7). The authors proposed that it is a combination of molecular electroporation and dimeric reorganization that facilitates insertion of AS-48 into the membrane (Jiménez et al. 2005; Maqueda et al. 2004, 2008; Sánchez-Barrena et al. 2003), as is discussed in the following section.

The NMR solution structure of carnocyclin A has also been reported (Martin-Visscher et al. 2009). Although this bacteriocin is monomeric at physiological pH and is shorter than AS-48, its 3D structure is remarkably similar. CclA is comprised of four helices, surrounding a hydrophobic core. Most of the helices are connected by short, well-defined loops, although helices 2 and 3 are connected by a longer, more flexible loop (Fig. 12.5b). Like AS-48, the point of cyclization resides within one of the helices. In addition, a cluster of lysines along helices 3 and 4 impart a highly localized positive charge on one end of CclA and may also help facilitate pore formation via molecular electroporation. Several hydrophobic residues are also solvent-exposed and are likely crucial for interaction with the cytoplasmic membrane. It has been suggested that the hydrophobic patches surrounding the point of head-to-tail cyclization may be crucial for interaction with the enzyme(s) responsible for performing the cyclization (Martin-Visscher et al. 2009). The greatest difference between the structures of CclA and AS-48 is that AS-48 has an additional helix. Inspection of the structural overlay and sequence alignment (Fig. 12.5c, d) shows this extra helix coincides with the flexible loop of CclA. Like AS-48, CclA shows structural similarity to NK-lysin and several other saposin and saposin-like peptides, all of which are known to interact with the cytoplasmic membrane of target cells.

It is likely that the other circular bacteriocins within both group i and group ii adopt similar 3D structures. Circular dichroism of gassericin A (Kawai et al. 2004) and analysis of the lactocyclicin Q sequence (Sawa et al. 2009) suggests that these peptides are largely alpha helical. Moreover, when the sequences of all the circular bacteriocins (excluding SubA) were subjected to secondary structure prediction, it was found that they all displayed a high degree of helicity, in approximately the same positions along their sequences (Martin-Visscher et al. 2009). Figure 12.6 shows the predicted helical elements for the circular bacteriocins. For the longer peptides within group i, they appear to contain an extra helix, whereas for the shorter peptides of group i, as well as for group ii, it is likely that they contain four helices. Overall, it is likely that this family of bacteriocins shares a common structural motif, indicative of the saposin fold.

Group i:

```
AS-48  MAKEFGIPAAVAGTVLNVVEAGGWVTTIVSILTAVGSGGLSLLAAAGRESIKAYLKKEIKKKGKRAVIAW
CirA   VAGALGVQTAAATTIVNVILNAGTLVTVLGIIA-SIASGGAGTLMTIGWATFKATVQKLAKQSMARAIAY
UblA   LAGYTGIASGTAKKVVDAIDKGAAAFVIISIISTVISAGALGAVSASADFIILTVKNYISRNLKAQAVIW
CclA   LVAYGIAQGTAEKVVSLINAGLTVGSIISILGGVTVGLSGVFTAVKAAIAKQGIKKAIQL
LycQ   LIDHLGAPRWAVDTILGAIAVGNLASWVLALVPGPGWAVKAGLATAAAIVKHQGKAAAAAW
```

Group ii:

```
GaaA   IYWIADQFGIHLATGTARKLLDAMASGASLGTAFAAILGVTLPAWALAAAGALGATAA
BviA   IYFIADKMGIQLAPAWYQDIVNWVSAGGTLTTGFAIIVGVTVPAWIAEAAAAFGIASA
```

Fig. 12.6 Predicted secondary structure of the circular bacteriocins (excluding subtilosin A) as described by Martin-Visscher et al. (2009). Residues highlighted in *gray* are known to be involved in helical segments, whereas those colored *pink* are predicted to be helical and those colored *green* are predicted to be either extended strand or helical (adapted from Martin-Visscher et al. 2009)

Mode of Action

The circular bacteriocins display broad spectra of activity toward various Gram-positive bacteria, including many serious food-spoilage pathogens. In addition, enterocin AS-48 (Gálvez et al. 1989), lactocyclicin Q (Sawa et al. 2009), and subtilosin A (Shelburne et al. 2007) are active against Gram-negative bacteria, although much higher concentrations of bacteriocin are required to elicit activity. In the case of enterocin AS-48 (Abriouel et al. 1998; Ananou et al. 2005) and subtilosin A (Shelburne et al. 2007), activity toward Gram-negative bacteria is enhanced when the integrity of the outer membrane of is compromised (i.e., treatment with EDTA or other chelating agents, pH or temperature variation). It has been reported that gassericin A can also kill Gram-negative bacteria when used in combination with glycine (Arakawa et al. 2009). At present, little is known about the effect of these bacteriocins against eukaryotic membranes, although it has been reported that AS-48 displays no activity against a variety of eukaryotic cell lines or erythrocytes, even at high (100 μg/mL) concentrations (Maqueda et al. 2004). It has been shown that SubA exhibits spermicidal activity toward bovine, horse, and rat sperm (Silkin et al. 2008). Recently, a natural variant of SubA with hemolytic activity has been discovered (Huang et al. 2009).

The circular bacteriocins are bactericidal toward their target cells. It has also been reported that enterocin AS-48 and uberolysin exhibit bacteriolytic properties. In the case of uberolysin, it was found that certain sensitive strains were susceptible to lysis when in early or mid-exponential growth phase (Wirawan et al. 2007). However, for enterocin AS-48, bacteriolytic activity was not growth dependent, as cell lysis was observed at all stages of growth. It was also reported that for AS-48, not all sensitive strains were subject to bacteriolytic effects. As such, it is likely that the lytic activity of AS-48 is a secondary effect, arising from the weakening of the cell membrane upon exposure to the bacteriocin.

Like many other bacteriocins produced by Gram-positive bacteria, it appears that the mode of action of the circular bacteriocins is permeation of the cell membrane, resulting in leakage of ions, dissipation of membrane potential, and eventually, cell death. The modes of action of gassericin A (Kawai et al. 2004), enterocin AS-48 (Gálvez et al. 1991; González et al. 2000; Sánchez-Barrena et al. 2003), CclA (Gong et al. 2009), and subtilosin A (Thennarasu et al. 2005) have all been investigated. To study the effect of gassericin A on bacterial cells and liposomes, ATP and potassium efflux were measured (Kawai et al. 2004). When treated with gassericin A, sensitive cells showed no efflux of ATP. However, release of potassium was observed for both sensitive cells and liposomes treated with the bacteriocin, suggesting that gassericin A targets the membrane of sensitive cells, causing cell death by release of potassium ions (Kawai et al. 2004).

By examining the effect of enterocin AS-48 on bacterial cells, membrane vesicles, and lipid bilayers, it has been shown that AS-48 induces permeation of bacterial cells and is able to form nonselective ion channels in the cytoplasmic membrane. This results in the free diffusion of ions and low-molecular-weight solutes across the membrane, ultimately leading to collapse of membrane potential and cell death

(Gálvez et al. 1991). As mentioned previously, it was initially proposed that AS-48 induced permeation of the membrane via molecular electroporation (González et al. 2000; Miteva et al. 1999). However, after the two dimeric forms of AS-48 (DF-I and DF-II) were elucidated by X-ray crystallography (Sánchez-Barrena et al. 2003), a more complex mechanism of action was described. In addition, it was found that a 21-residue fragment of AS-48, which encompassed the positive charge and assumed a conformation similar to the native peptide, was devoid of antimicrobial activity, suggesting that mere molecular electroporation could not account for cell permeation (Jiménez et al. 2005). It is now proposed that after approaching the membrane surface due to electrostatic interactions, AS-48 interconverts from its water-soluble form (DF-I) to its membrane-bound form (DF-II), thus exposing the hydrophobic faces of H1 and H2 and facilitating insertion into the membrane via molecular electroporation and hydrophobic interactions (Fig. 12.7) (Jiménez et al. 2005; Maqueda et al. 2004, 2008; Sánchez-Barrena et al. 2003).

Lipid bilayer and single-channel recording techniques were used to probe the interaction of carnocyclin A with the membrane (Gong et al. 2009). These studies revealed that CclA was able to interact directly with the membrane, indicating that like gassericin A and enterocin AS-48, a cell surface receptor may not be required for bacteriocin activity. CclA formed anion-selective channels in the lipid bilayers,

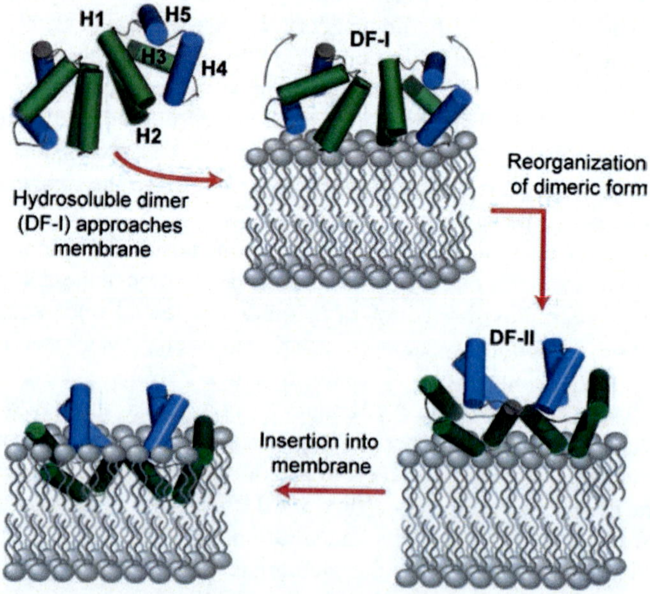

Fig. 12.7 Proposed interaction of AS-48 with the cytoplasmic membrane. The hydrosoluble dimeric form DF-I (pdb code 1083) approaches the surface of the membrane, upon which the dimer reorganizes into its hydrophobic and membrane-soluble form, DF-II (pdb code 1084). Insertion into the membrane entails. *Green* and *blue* represent hydrophobic and polar faces of the helices, respectively (adapted from Sánchez-Barrena et al. 2003)

and channel formation was voltage-dependent, as a negative membrane potential was required for pore formation, although channel gating was voltage-independent. It was also found that acidic conditions enhanced the stability of the CclA pores, as greater conductance across the membrane was observed. According to these studies, it is likely that CclA approaches the surface of a bacterial cell (due to electrostatic interactions), wherein it is exposed to the negative membrane potential of the bacterial cell. At this potential difference, CclA creates anion selective pores across the membrane, and channel activation occurs. As anions flow through the channel, depolarization occurs, leading to the inactivation of the CclA channels. In contrast, the ion channels formed by AS-48 were found to be voltage-independent and nonselective.

To examine the mode of action of subtilosin A with phospholipid bilayers, a variety of experimental techniques were employed, including fluorescence spectroscopy, solid-state NMR, and differential scanning calorimetry (Thennarasu et al. 2005). These studies revealed that SubA is able to interact with a membrane, upon which it assumes an orientation in which the side chain of Trp-2 partially inserts into the membrane, and the anionic end of the molecule is directed away from the surface of the membrane (Fig. 12.8). This results in a conformational change in the phosphate head groups and a slight disordering of the hydrophobic core of the membrane. However, these studies also revealed that in order to induce leakage of small unilamellar vesicles (SUVs), high concentrations of SubA were required – much higher than the observed MICs for SubA (Thennarasu et al. 2005). Thus, although SubA is able to interact directly with model membranes and induce leakage at high concentrations, it is likely that a cell surface receptor is involved in the interaction between SubA and sensitive cells (Thennarasu et al. 2005).

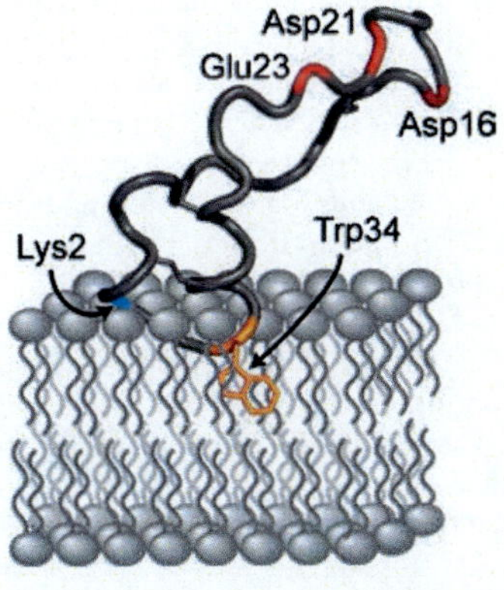

Fig. 12.8 Proposed interaction of SubA with the cytoplasmic membrane. The basic and acidic residues are labeled and colored *blue* and *red*, respectively. The side chain of Trp-34, which inserts into the membrane, is labeled and is colored *orange* (adapted from Thennarasu et al. 2005)

Conclusions

The circular bacteriocins produced by Gram-positive bacteria are an exciting new class of bacteriocins. Owing to their enhanced structural stability and broad spectra of activity, there is tremendous potential for the application of these peptides in food safety. Currently, the producer organism of CclA is already approved for use by the US Food and Drug Administration (Martin-Visscher et al. 2008). In addition, a vast amount of work has examined the use of AS-48 as a food preservative. Furthermore, upon solving the mysteries governing the cyclization of these peptides, such knowledge may allow for the engineering of an arsenal of new circular bacteriocins that act as highly stable antimicrobial agents for use as human therapeutics or as stable scaffolds for drug delivery.

References

Abriouel H, Valdivia E, Gálvez A, Maqueda M (1998) Response of *Salmonella choleraesuis* LT2 spheroplasts and permeabilized cells to the bacteriocin AS-48. Appl Environ Microbiol 64:4623–4626

Abriouel H, Valdivia E, Gálvez A, Maqueda M (2001) Influence of physico-chemical factors on the oligomerization and biological activity of bacteriocin AS-48. Curr Microbiol 42:89–95

Ananou S, Gálvez A, Martínez-Bueno M, Maqueda M, Valdivia E (2005) Synergistic effect of enterocin AS-48 in combination with outer membrane permeabilizing treatments against *Escherichia coli* O157: H7. J Appl Microbiol 99:1364–1372

Arakawa K, Kawai Y, Ito Y, Nakamura K, Chujo T, Nishimura J, Kitazawa H, Saito T (2010) HPLC purification and re-evaluation of chemical identity of two circular bacteriocins, gassericin A and reutericin 6. Lett Appl Microbiol 50:406–411

Arakawa K, Kawai Y, Iioka H, Tanioka M, Nishimura J, Kitazawa H, Tsurumi K, Saito T (2009) Effects of gassericins A and T, bacteriocins produced by *Lactobacillus gasseri*, with glycine on custard cream preservation. J Dairy Sci 92:2365–2372

Babasaki K, Takao T, Shimonishi Y, Kurahashi K (1985) Subtilosin A, a new antibiotic peptide produced by *Bacillus subtilis* 168: isolation, structural analysis, and biogenesis. J Biochem 98:585–603

Bayro MJ, Mukhopadhyay J, Swapna GVT, Huang JY, Ma LC, Sineva E, Dawson PE, Montelione GT, Ebright RH (2003) Structure of antibacterial peptide microcin J25: A 21-residue lariat protoknot. J Am Chem Soc 125:12382–12383

Blond A, Peduzzi J, Goulard C, Chiuchiolo MJ, Barthélémy M, Prigent Y, Salomón RA, Farías RN, Moreno F, Rebuffat S (1999) The cyclic structure of microcin J25, a 21-residue peptide antibiotic from *Escherichia coli*. Eur J Biochem 259:747–755

Blond A, Cheminant M, Ségalas-Milazzo I, Peduzzi J, Barthélémy M, Goulard C, Salomón R, Moreno F, Farías R, Rebuffat S (2001) Solution structure of microcin J25, the single macrocyclic antimicrobial peptide from *Escherichia coli*. Eur J Biochem 268:2124–2133

Chatterjee C, Paul M, Xie LL, van der Donk WA (2005) Biosynthesis and mode of action of lantibiotics. Chem Rev 105:633–683

Chen H, Hoover DG (2003) Bacteriocins and their food applications. Compr Rev Food Sci Food Saf 2:82–100

Cotter PD, Hill C, Ross RP (2005) Bacteriocins: developing innate immunity for food. Nat Rev Microbiol 3:777–788

Craik DJ (2006) Chemistry – seamless proteins tie up their loose ends. Science 311:1563–1564

Craik DJ (2009) Circling the enemy: cyclic proteins in plant defence. Trends Plant Sci 14:328–335

Craik DJ, Daly NL, Saska I, Trabi M, Rosengren KJ (2003) Structures of naturally occurring circular proteins from bacteria. J Bacteriol 185:4011–4021

Daly NL, Rosengren KJ, Craik DJ (2009) Discovery, structure and biological activities of cyclotides. Adv Drug Deliv Rev 61:918–930

Diaz M, Valdivia E, Martínez-Bueno M, Fernández M, Soler-González AS, Ramírez-Rodrigo H, Maqueda M (2003) Characterization of a new operon, *as-48EFGH*, from the *as-48* gene cluster involved in immunity to enterocin AS-48. Appl Environ Microbiol 69:1229–1236

Draper LA, Ross RP, Hill C, Cotter PD (2008) Lantibiotic immunity. Curr Protein Pept Sci 9:39–49

Drider D, Fimland G, Héchard Y, McMullen LM, Prévost H (2006) The continuing story of class IIa bacteriocins. Microbiol Mol Biol Rev 70:564–582

Franz CMAP, van Belkum MJ, Holzapfel WH, Abriouel H, Gálvez A (2007) Diversity of enterococcal bacteriocins and their grouping in a new classification scheme. FEMS Microbiol Rev 31:293–310

Gálvez A, Maqueda M, Martínez-Bueno M, Valdivia E (1989) Bactericidal and bacteriolytic action of peptide antibiotic AS-48 against gram-positive and gram-negative bacteria and other organisms. Res Microbiol 140:57–68

Gálvez A, Maqueda M, Martínez-Bueno M, Valdivia E (1991) Permeation of bacterical cells, permeation of cytoplasmic and artificial membrane vesicles, and channel formation on lipid bilayers by peptide antibiotic AS-48 J. Bacteriology 173:886–892

Gasteiger E, Hoogland C, Gattiker A, Duvaud S, Wilkins MR, Appel RD, Bairoch A (2005) Protein identification and analysis tools on the ExPASy server, p. 571-607. In: Walker JM (ed) The proteomics protocols handbook. Humana, NJ

Gillon AD, Saska I, Jennings CV, Guarino RF, Craik DJ, Anderson MA (2008) Biosynthesis of circular proteins in plants. Plant J 53:505–515

Gondry M, Sauguet L, Belin P, Thai R, Amouroux R, Tellier C, Tuphile K, Jacquet M, Braud S, Courcon M, Masson C, Dubois S, Lautru S, Lecoq A, Hashimoto S, Genet R, Pernodet JL (2009) Cyclodipeptide synthases are a family of tRNA-dependent peptide bond-forming enzymes. Nat Chem Biol 5:414–420

Gong XD, Martin-Visscher LA, Nahirney D, Vederas JC, Duszyk M (2009) The circular bacteriocin, carnocyclin A, forms anion-selective channels in lipid bilayers. Biochim Biophys Acta 1788:1797–1803

González C, Langdon GM, Bruix M, Gálvez A, Valdivia E, Maqueda M, Rico M (2000) Bacteriocin AS-48, a microbial cyclic polypeptide structurally and functionally related to mammalian NK-lysin. Proc Natl Acad Sci USA 97:11221–11226

Håvarstein LS, Diep DB, Nes IF (1995) A family of bacteriocin ABC transporters carry out proteolytic processing of their substrates concomitant with export. Mol Microbiol 16:229–240

Heng NCK, Tagg JR (2006) What's in a name? Class distinction for bacteriocins. Nat Rev Microbiol 4:160

Huang T, Geng H, Miyyapuram VR, Sit CS, Vederas JC, Nakano MM (2009) Isolation of a variant of subtilosin A with hemolytic activity. J Bacteriol 191:5690–5696

Ireland DC, Colgravel ML, Nguyencong P, Daly NL, Craik DJ (2006) Discovery and characterization of a linear cyclotide from *Viola odorata*: Implications for the processing of circular proteins. J Mol Biol 357:1522–1535

Ito Y, Kawai Y, Arakawa K, Honme Y, Sasaki T, Saito T (2009) Conjugative plasmid from *Lactobacillus gasseri* LA39 that carries genes for production of and immunity to the circular bacteriocin gassericin A. Appl Environ Microbiol 75:6340–6351

Itoh T, Fujimoto Y, Kawai Y, Toba T, Saito T (1995) Inhibition of food-borne pathogenic bacteria by bacteriocins from *Lactobacillus gasseri*. Lett Appl Microbiol 21:137–141

Jiménez MA, Barrachi-Saccilotto AC, Valdivia E, Maqueda M, Rico M (2005) Design, NMR characterization and activity of a 21-residue peptide fragment of bacteriocin AS-48 containing its putative membrane interacting region. J Pept Sci 11:29–36

Kalmokoff ML, Cyr TD, Hefford MA, Whitford MF, Teather RM (2003) Butyrivibriocin AR10, a new cyclic bacteriocin produced by the ruminal anaerobe *Butyrivibrio fibrisolvens* AR10: characterization of the gene and peptide. Can J Microbiol 49:763–773

Kawai Y, Saito T, Kitazawa H, Itoh T (1998a) Gassericin A; an uncommon cyclic bacteriocin produced by *Lactobacillus gasseri* LA39 linked at N- and C-terminal ends. Biosci Biotechnol Biochem 62:2438–2440

Kawai Y, Saito T, Suzuki M, Itoh T (1998b) Sequence analysis by cloning of the structural gene of gassericin A, a hydrophobic bacteriocin produced by *Lactobacillus gasseri* LA39. Biosci Biotechnol Biochem 62:887–892

Kawai Y, Arakawa K, Itoh A, Saitoh B, Ishii Y, Nishimura J, Kitazawa H, Itoh T, Saito T (2003) Heterologous expression of gassericin A, a bacteriocin produced by *Lactobacillus gasseri* LA39. Anim Sci J 74:45–51

Kawai Y, Ishii Y, Arakawa K, Uemura K, Saitoh B, Nishimura J, Kitazawa H, Yamazaki Y, Tateno Y, Itoh T, Saitoh T (2004) Structural and functional differences in two cyclic bacteriocins with the same sequences produced by lactobacilli. Appl Environ Microbiol 70:2906–2911

Kawai Y, Kusnadi J, Kemperman R, Kok J, Ito Y, Endo M, Arakawa K, Uchida H, Nishimura J, Kitazawa H, Saito T (2009) DNA sequencing and homologous expression of a small peptide conferring immunity to gassericin A, a circular bacteriocin produced by *Lactobacillus gasseri* LA39. Appl Environ Microbiol 75:1324–1330

Kawulka K, Sprules T, McKay RT, Mercier P, Diaper CM, Zuber P, Vederas JC (2003) Structure of subtilosin A, an antimicrobial peptide from *Bacillus subtilis* with unusual posttranslational modifications linking cysteine sulfurs to α-carbons of phenylalanine and threonine. J Am Chem Soc 125:4726–4727

Kawulka KE, Sprules T, Diaper CM, Whittal RM, McKay RT, Mercier P, Zuber P, Vederas JC (2004) Structure of subtilosin A, a cyclic antimicrobial peptide from *Bacillus subtilis* with unusual sulfur to α-carbon cross-links: Formation and reduction of α-thio-α-amino acid derivatives. Biochemistry 43:3385–3395

Kemperman R, Jonker M, Nauta A, Kuipers OP, Kok J (2003a) Functional analysis of the gene cluster involved in production of the bacteriocin circularin A by *Clostridium beijerinckii* ATCC 25752. Appl Environ Microbiol 69:5839–5848

Kemperman R, Kuipers A, Karsens H, Nauta A, Kuipers O, Kok J (2003b) Identification and characterization of two novel clostridial bacteriocins, circularin A and closticin 574. Appl Environ Microbiol 69:1589–1597

Koglin A, Walsh CT (2009) Structural insights into nonribosomal peptide enzymatic assembly lines. Nat Prod Rep 26:987–1000

Krogh A, Larsson B, von Heijne G, Sonnhammer ELL (2001) Predicting transmembrane protein topology with a hidden Markov model: application to complete genomes. J Mol Biol 305:567–580

Langdon GM, Bruix M, Gálvez A, Valdivia E, Maqueda M, Rico M (1998) Sequence-specific ^{1}H assignment and secondary structure of the bacteriocin AS-48 cyclic peptide. J Biomol NMR 12:173–175

Leer RJ, van der Vossen JMBM, van Giezen M, van Noort JM, Pouwels PH (1995) Genetic analysis of acidocin B, a novel bacteriocin produced by *Lactobacillus acidophilus*. Microbiology 141:1629–1635

Maqueda M, Gálvez A, Martínez-Bueno M, Sanchez-Barrena MJ, González C, Albert A, Rico M, Valdivia E (2004) Peptide AS-48: prototype of a new class of cyclic bacteriocins. Curr Protein Pept Sci 5:399–416

Maqueda M, Sánchez-Hidalgo M, Fernández M, Montalbán-López M, Valdivia E, Martínez-Bueno M (2008) Genetic features of circular bacteriocins produced by Gram-positive bacteria. FEMS Microbiol Rev 32:2–22

Marahiel MA (2009) Working outside the protein-synthesis rules: insights into non-ribosomal peptide synthesis. J Pept Sci 15:799–807

Martínez-Bueno M, Gálvez A, Valdivia E, Maqueda M (1990) A transferable plasmid associated with AS-48 production in *Enterococcus faecalis*. J Bacteriol 172:2817–2818

Martínez-Bueno M, Maqueda M, Gálvez A, Samyn B, Van Beeumen J, Coyette J, Valdivia E (1994) Determination of the gene sequence and the molecular structure of the enterococcal peptide antibiotic AS-48. J Bacteriol 176:6334–6339

Martínez-Bueno M, Valdivia E, Gálvez A, Coyette J, Maqueda M (1998) Analysis of the gene cluster involved in production and immunity of the peptide antibiotic AS-48 in *Enterococcus faecalis*. Mol Microbiol 27:347–358

Martin-Visscher LA, van Belkum MJ, Garneau-Tsodikova S, Whittal RM, Zheng J, McMullen LM, Vederas JC (2008) Isolation and characterization of carnocyclin A, a novel circular bacteriocin produced by *Carnobacterium maltaromaticum* UAL307. Appl Environ Microbiol 74:4756–4763

Martin-Visscher LA, Gong XD, Duszyk M, Vederas JC (2009) The three-dimensional structure of carnocyclin a reveals that many circular bacteriocins share a common structural motif. J Biol Chem 284:28674–28681

Marx R, Stein T, Entian KD, Glaser SJ (2001) Structure of the *Bacillus subtilis* peptide antibiotic subtilosin A determined by ^{1}H-NMR and matrix assisted laser desorption/ionization time-of-fight mass spectrometry. J Protein Chem 20:501–506

Miteva M, Andersson M, Karshikoff A, Otting G (1999) Molecular electroporation: a unifying concept for the description of membrane pore formation by antibacterial peptides, exemplified with NK-lysin. FEBS Lett 462:155–158

Montalbán-López M, Spolaore B, Pinato O, Martínez-Bueno M, Valdivia E, Maqueda M, Fontana A (2008) Characterization of linear forms of the circular enterocin AS-48 obtained by limited proteolysis. FEBS Lett 582:3237–3242

Nes IF, Diep DB, Holo H (2007a) Bacteriocin diversity in *Streptococcus* and *Enterococcus*. J Bacteriol 189:1189–1198

Nes IF, Yoon SS, Diep WB (2007b) Ribosomally synthesiszed antimicrobial peptides (Bacteriocins) in lactic acid bacteria: A review. Food Sci Biotechnol 16:675–690

Oman TJ, van der Donk WA (2010) Follow the leader: the use of leader peptides to guide natural product biosynthesis. Nat Chem Biol 6:9–18

Rosengren KJ, Clark RJ, Daly NL, Göransson U, Jones A, Craik DJ (2003) Microcin J25 has a threaded sidechain-to-backbone ring structure and not a head-to-tail cyclized backbone. J Am Chem Soc 125:12464–12474

Salomón RA, Farías RN (1992) Microcin-25, a novel antimicrobial peptide produced by *Escherichia coli*. J Bacteriol 174:7428–7435

Samyn B, Martínez-Bueno M, Devreese B, Maqueda M, Gálvez A, Valdivia E, Coyette J, Van Beeumen J (1994) The cyclic structure of the enterococcal peptide antibiotic AS-48. FEBS Lett 352:87–90

Sánchez-Barrena MJ, Martínez-Ripoll M, Gálvez A, Valdivia E, Maqueda M, Cruz V, Albert A (2003) Structure of bacteriocin AS-48: from soluble state to membrane bound state. J Mol Biol 334:541–549

Saska I, Gillon AD, Hatsugai N, Dietzgen RG, Hara-Nishimura I, Anderson MA, Craik DJ (2007) An asparaginyl endopeptidase mediates *in vivo* protein backbone cyclization. J Biol Chem 282:29721–29728

Sawa N, Zendo T, Kiyofuji J, Fujita K, Himeno K, Nakayama J, Sonomoto K (2009) Identification and characterization of lactocyclicin Q, a novel cyclic bacteriocin produced by *Lactococcus* sp. strain QU 12. Appl Environ Microbiol 75:1552–1558

Shelburne CE, An FY, Dholpe V, Ramamoorthy A, Lopatin DE, Lantz MS (2007) The spectrum of antimicrobial activity of the bacteriocin subtilosin A. J Antimicrob Chemother 59:297–300

Silkin L, Hamza S, Kaufman S, Cobb SL, Vederas JC (2008) Spermicidal bacteriocins: Lacticin 3147 and subtilosin A. Bioorg Med Chem Lett 18:3103–3106

Thennarasu S, Lee DK, Poon A, Kawulka KE, Vederas JC, Ramamoorthy A (2005) Membrane permeabilization, orientation, and antimicrobial mechanism of subtilosin A. Chem Phys Lipids 137:38–51

Toba T, Samant SK, Yoshioka E, Itoh T (1991) Reutericin 6, a new bacteriocin produced by *Lactobacillus reuteri* LA6. Lett Appl Microbiol 13:281–286

van Belkum MJ, Stiles ME (2000) Nonlantibiotic antibacterial peptides from lactic acid bacteria. Nat Prod Rep 17:323–335

van Belkum MJ, Worobo RW, Stiles ME (1997) Double-glycine-type leader peptides direct secretion of bacteriocins by ABC transporters: colicin V secretion in *Lactococcus lactis*. Mol Microbiol 23:1293–1301

Walsh CT (2004) Polyketide and nonribosomal peptide antibiotics: modularity and versatility. Science 303:1805–1810

Wilson KA, Kalkum M, Ottesen J, Yuzenkova J, Chait BT, Landick R, Muir T, Severinov K, Darst SA (2003) Structure of microcin J25, a peptide inhibitor of bacterial RNA polymerase, is a lassoed tail. J Am Chem Soc 125:12475–12483

Wirawan RE, Swanson KM, Kleffmann T, Jack RW, Tagg JR (2007) Uberolysin: a novel cyclic bacteriocin produced by *Streptococcus uberis*. Microbiology 153:1619–1630

Zheng G, Yan LZ, Vederas JC, Zuber P (1999) Genes of the *sbo-alb* locus of *Bacillus subtilis* are required for production of the antilisterial bacteriocin subtilosin. J Bacteriol 181:7346–7355

Zheng GL, Hehn R, Zuber P (2000) Mutational analysis of the *sbo-alb* locus of *Bacillus subtilis*: identification of genes required for subtilosin production and immunity. J Bacteriol 182:3266–3273

Chapter 13
Class IId or Linear and Non-Pediocin-Like Bacteriocins

Shun Iwatani, Takeshi Zendo, and Kenji Sonomoto

Abstract Class IId bacteriocins are one of the subclasses of class II bacteriocins produced by lactic acid bacteria. This class of bacteriocins, however, show a great diversity in their primary structures and modes of action. This chapter focuses on two aspects: (1) the description of those heterogeneous bacteriocins with the concept of three potential subgroups and (2) the modes of action of lactococcin A, lactococcin 972, and lacticin Q, each of which belongs to a different subgroup and is well characterized in its unique mode of action.

Introduction

Class IId bacteriocins are categorized as bacteriocins that have no significant sequence similarity to the other class II bacteriocins; in other words, they are "non-pediocin-like one-peptide linear" bacteriocins according to the classification proposed by Cotter et al. (2005). However, the recognition of this class is rather broad and variable compared to that of the other subclasses, and thus, has been changing along with the information available on bacteriocins. In 1996, Nes et al. proposed a subgroup of bacteriocins that contain a typical signal peptide and are secreted by the general secretory (*sec*-) pathway of the producer cell (Nes et al. 1996). This concept of *sec*-dependent bacteriocins was followed by Diep et al. who also listed an additional group of bacteriocins that contain no leader sequence (Diep and Nes 2002). This subgroup of leaderless bacteriocins has also been recognized in two more recent reviews (Franz et al. 2007; Nes et al. 2007), although both of these reviews focused mainly on the bacteriocins from enterococci or streptococci. In this context, based on the above understanding, we consider class IId

K. Sonomoto (✉)
Laboratory of Microbial Technology, Division of Applied Molecular Microbiology and Biomass Chemistry, Department of Bioscience and Biotechnology, Faculty of Agriculture Graduate School, Kyushu University, 6-10-1 Hakozaki, Higashi-ku, 812-8581, Fukuoka, Japan
e-mail: sonomoto@agr.kyushu-u.ac.jp

D. Drider and S. Rebuffat (eds.), *Prokaryotic Antimicrobial Peptides: From Genes to Applications*, DOI 10.1007/978-1-4419-7692-5_13,

bacteriocins as a broad category that is potentially subdivided into (1) *sec*-dependent bacteriocins, (2) leaderless bacteriocins, and (3) nonsubgrouped bacteriocins (Table 13.1).

Table 13.1 Class IId bacteriocins

Bacteriocin		Producer strain	Mass (amino acids)[a], Da	Reference
Sec-dependent bacteriocins				
Divergicin A		*C. divergens* LV13	4,223 (46)	(Worobo et al. 1995)
Propionicin T1		Strains of *P. thoenii*	7,130 (65)	(Faye et al. 2000)
Lactococcin 972		*L. lactis* IPLA972	7,381 (66)	(Martínez et al. 1999)
The ones with the pediocin-box sequence				
Enterocin P		*E. faecium* P13	4,493 (44)	(Cintas et al. 1997)
Bacteriocin 31		*E. faecalis* 31	5,008 (43)	(Tomita et al. 1996)
Listeriocin 743A		*L. innocua* 743	4,484 (43)	(Kalmokoff et al. 2001)
Bacteriocin T8		*E. faecium* T8	5,090 (44)	(De Kwaadsteniet et al. 2006)
Hiracin JM79		*E. hirae* DCH5	5,093 (44)	(Sánchez et al. 2008)
Enterocin SE-K4		*E. faecalis* K-4	5,356 (43)	(Doi et al. 2002)
Leaderless bacteriocins				
Enterocin L50A		*E. faecium* L50	5,190 (44)	(Cintas et al. 1998)
Enterocin L50B		*E. faecium* L50	5,178 (43)	(Cintas et al. 1998)
Enterocin MR10A		*E. faecalis* MRR10-3	5,202 (44)	(Martín-Platero et al. 2006)
Enterocin MR10B		*E. faecalis* MRR10-3	5,208 (43)	(Martín-Platero et al. 2006)
Enterocin Q		*E. faecium* L50	3,980 (34)	(Cintas et al. 2000)
Enterocin EJ97		*E. faecalis* EJ97	5,322 (44)	(Sánchez-Hidalgo et al. 2003)
Aureocin A70	AurA	*S. aureus* A70	2,924 (31)	(Netz et al. 2001)
	AurB		2,797 (30)	
	AurC		2,955 (31)	
	AurD		3,087 (31)	
Aureocin A53		*S. aureus* A53	6,012 (51)	(Netz et al. 2002)
BHT-B		*S. rattus* strain BHT	5,195 (45)	(Hyink et al. 2005)
Lacticin Q		*L. lactis* QU 5	5,926 (53)	(Fujita et al. 2007)
Lacticin Z		*L. lactis* QU 14	5,971 (53)	(Iwatani et al. 2007)
LsbB		A natural isolate of *L. lactis*	3,407 (30)	(Gajic et al. 2003)
Nonsubgrouped bacteriocins				
Lactococcin A		Strains of *L. lactis*	5,778 (54)	(Holo et al. 1991)
Lactococcin M		Strains of *L. lactis*	4,325 (48)	(van Belkum et al. 1991a)
Lactococcin B		Strains of *L. lactis*	5,327 (47)	(van Belkum et al. 1992)

(continued)

Table 13.1 (continued)

Bacteriocin	Producer strain	Mass (amino acids)[a], Da	Reference
Carnobacteriocin A	*C. piscicola* LV17A	5,052 (53)	(Worobo et al. 1994)
Piscicolin 61	*C. piscicola* LV61	5,052 (53)	(Holck et al. 1994)
Enterocin B	*E. faecium* T136	5,479 (53)	(Casaus et al. 1997)
Durancin TW-49M	*E. durans* QU 49	5,228 (54)	(Hu et al. 2008)
Divergicin 750	*C. divergens* 750	3,447 (34)	(Holck et al. 1996)
LsbA	A natural isolate of *L. lactis*	5,226 (44)	(Gajic et al. 2003)
Bovicin 255	*S. gallolyticus* LRC0255	5,967 (56)	(Whitford et al. 2001)
Thermophilin A	*S. thermophilus* ST134	5,776 (62)	(Ward and Somkuti 1995)

[a]Theoretical molecular masses calculated with the determined primary structures. The numbers in the parentheses show the number of the amino-acid residues of mature peptides

Sec-*Dependent Bacteriocins*

The N-terminal extensions of most class II bacteriocins (nonlantibiotics) are what is termed double-glycine-type leader sequences, which contain a conserved processing site of Gly-Gly residues at positions –1 and –2, and are cleaved off concomitantly with export across the cytoplasmic membrane by dedicated ATP-binding cassette (ABC)-transporters (Nes et al. 1996). However, some class II bacteriocins contain an N-terminal extension of the so-called *sec*-type signal peptide, which ensures the proper targeting to the *sec*-pathway and translocation across the cytoplasmic membrane without any dedicated transporters.

Divergicin A, produced by *Carnobacterium divergens* LV13, is the first bacteriocin from lactic acid bacteria that is reported not to have a dedicated transporter (Worobo et al. 1995). It is synthesized as a prepeptide of 75 amino acids consisting of a 29-amino-acid N-terminal sequence and a mature peptide of 46 amino acids. The N-terminal sequence of divergicin A has a cleavage site of Ala-Ser-Ala residues at positions –3 to –1 and acts as a signal peptide that accesses the general *sec*-pathway (Worobo et al. 1995). Propionicin T1 is produced by two strains of *Propionibacterium thoenii* as a prepeptide of 96 amino acids with a *sec*-type leader sequence, which is cleaved to give a mature peptide of 65 amino acids (Faye et al. 2000). Although the two strains of *P. thoenii* 419 and *P. thoenii* LMG 2792 produce an identical bacteriocin, propionicin T1, they show great differences in their optimal conditions for the production of bacteriocin (Faye et al. 2000). Lactococcin 972, produced by *Lactococcus lactis* IPLA972, is encoded as a 91-amino-acid prepeptide that contains a 25-amino-acid *sec*-dependent signal peptide to produce the mature peptide of 66 amino acids (Martínez et al. 1999). The active form of this bacteriocin seems to be homodimeric, and the primary target of the active peptide is not the cytoplasmic membrane (Martínez et al. 1996). This seemingly unique mode of action is discussed later.

There are some bacteriocins that have a *sec*-dependent N-terminal leader as well as a "pediocin-box" motif containing a consensus sequence of YGNGVXC in the N-terminus of their mature forms. These bacteriocins include enterocin P from *Enterococcus faecium* P13 (Cintas et al. 1997), bacteriocin 31 from *Enterococcus faecalis* 31 (Tomita et al. 1996), listeriocin 743A from *Listeria innocua* 743 (Kalmokoff et al. 2001), bacteriocin T8 from *E. faecium* T8 (De Kwaadsteniet et al. 2006), hiracin JM79 from *Enterococcus hirae* DCH5 (Sánchez et al. 2008), and enterocin SE-K4 from *E. faecalis* K-4 (Doi et al. 2002). In some instances, the above-mentioned bacteriocins have been considered to be included in class IIa, namely, the pediocin-like bacteriocins.

Acidocin B, which is produced by *Lactobacillus acidophilus* M46, is also an early-identified *sec*-dependent bacteriocin (Leer et al. 1995). The structural gene of acidocin B suggests that this bacteriocin is synthesized as a prepeptide, which contains a *sec*-type N-terminal leader and a mature peptide deduced from 59 amino acids (Leer et al. 1995). However, it has recently been reported that the primary structure and gene locus of gassericin A, a circular bacteriocin produced by *Lactobacillus gasseri* LA39, are almost identical to those of acidocin B (Kawai et al. 2009); therefore, this bacteriocin is appropriately classified in the family of cyclic bacteriocins.

Leaderless Bacteriocins

Some class IId bacteriocins are synthesized without an N-terminal leader sequence or signal peptide, thus are called "leaderless bacteriocins." These kinds of bacteriocins have been found in relatively recent years, and most of them still remain to be studied in more detail to clarify their biosynthesis mechanism.

Enterocin L50, produced by *E. faecium* L50, is an early-identified and well-characterized leaderless bacteriocin that consists of two highly similar peptides, enterocin L50A (EntL50A) and enterocin L50B (EntL50B), which share 72% sequence similarity (Cintas et al. 1998). The structural genes encoding 44-amino-acid EntL50A and 43-amino-acid EntL50B are tandemly located on a 50-kb plasmid, pCIZ1, and thus seem to be cotranscribed in the same unit (Cintas et al. 1998). In addition to the fact that each peptide of EntL50A and EntL50B exhibits considerable antimicrobial activity, this activity was synergistically increased against certain microbial strains when these two peptides were used in combination (Cintas et al. 1998). In this regard, enterocin L50 may be considered to belong to the class IIb bacteriocins that consist of two complementary peptides with synergistic activity (Cotter et al. 2005); however, the characteristics of EntL50A and EntL50B, such as the significant activity of each peptide alone and being synthesized without a leader sequence, distinguish them from general class IIb bacteriocins. The genes encoding EntL50A and EntL50B are also identified on a 21-kb plasmid, pEF1, harbored in *E. faecium* 6T1a, although the peptides in this strain were termed as enterocin I and enterocin J, respectively

(Floriano et al. 1998; Ruiz-Barba et al. 2006). Moreover, enterocin L50 homologues were found in *E. faecalis* MRR 10-3, which produces two antimicrobial peptides, namely, enterocin MR10A and MR10B, which are almost identical to EntL50A and EntL50B, respectively (Martín-Platero et al. 2006). There is a single conservative change (Glu^{38} to Asp) in enterocin MR10A and two changes (Thr^{9} to Ala, Leu^{15} to Phe) in enterocin MR10B. These two peptides also exhibit both individual and synergistic activity, as in the case of enterocin L50. In contrast with plasmid-encoded enterocin L50, the structural genes of enterocin MR10 seem to be located on the chromosome (Martín-Platero et al. 2006), and these structural genes are also found in another isolate of *E. faecalis* SL-5 (Kang et al. 2009). Enterocin RJ-11, produced by *E. faecalis* RJ-11, is also highly homologous to EntL50A, with slight differences in the N-terminal and C-terminal regions, in addition to the replacements of five internal amino-acid residues (Yamamoto et al. 2003). However, because of a lack of genetic information, it is unclear whether this peptide contains a leader sequence or if there is another peptide with which enterocin RJ-11 synergistically acts. Enterocin EJ97, composed of 44 amino-acid residues, is encoded on the 60-kb conjugative plasmid pEJ97 of the strain *E. faecalis* EJ97 (Sánchez-Hidalgo et al. 2003). Although this peptide is not identical to any of the EntL50-homologues, it shares some conserved residues (Lys^{6}, Lys^{10}, Lys^{21}, Ile^{27}, and Trp^{39}) with each peptide of EntL50A/B and enterocin MR10A/B (Franz et al. 2007). In addition to the production of enterocin L50, *E. faecium* L50 also produces another leaderless bacteriocin, enterocin Q, whose structural gene is located not on pCIZ1, but on the 7.4-kb plasmid pCIZ2 (Cintas et al. 2000; Criado et al. 2006).

Apart from enterococcal bacteriocins, *Staphylococcus aureus* A70 produces a multiple peptide bacteriocin, aureocin A70, all of which are composed of 30 or 31 residues without an N-terminal leader sequence (Netz et al. 2001). The structural genes of the four peptides (AurA, AurB, AurC, and AurD) are located on the 8.0-kb plasmid pRJ6, which has been shown to be mobilized by staphylococcal conjugative plasmids (Oliveia et al. 1998). Three out of the four peptides, AurA, AurB, and AurC, exhibit individual antimicrobial activity. The fourth peptide, AurD, could not be purified separately, and thus, the antimicrobial activity of this peptide has yet to be evaluated. However, the high sequence similarity between AurD and other peptides suggests the potential antimicrobial activity of AurD (Oliveia et al. 1998). Synergistic activity among these peptides is still not recognized. Another *S. aureus* strain A53 produces aureocin A53, a single leaderless peptide, whose structural gene is located on the 10.4-kb plasmid pRJ9 (Netz et al. 2002). Aureocin A53 is a highly cationic and Trp-rich peptide of 51 amino-acid residues, and the majority of the five Trp residues have been suggested to be exposed at the surface of the folded aureocin A53 (Netz et al. 2002). Since Trp residues of antimicrobial peptides play a key role in the interaction between the peptide and biological membranes (Fimland et al. 2002), the structural properties of aureocin A53 possibly enable it to target cell membranes. It actually shows antimicrobial activity against a broad range of Gram-positive bacteria with nanomolar minimal inhibitory concentration (MIC) values (Netz et al. 2002).

Lacticin Q (Fujita et al. 2007) and lacticin Z (Iwatani et al. 2007) are 53-amino-acid homologous bacteriocins, which are produced by *L. lactis* QU 5 and *L. lactis* QU 14, respectively. These two peptides differ in three amino-acid residues at positions 10, 33, and 44 in their primary structures (lacticin Q; Leu^{10}, Leu^{33}, Ser^{44}/lacticin Z; Ile^{10}, Ile^{33}, Glu^{44}). The N-termini of the two peptides are estimated to be a formylated-methionine because of the gap between the observed molecular mass and the theoretical mass of the deduced amino-acid sequence (Fujita et al. 2007; Iwatani et al. 2007). For a similar reason, it is expected that N-terminal formylation is retained in the primary structure of EntL50A/B (Izquierdo et al. 2008) and aureocin A53 (Netz et al. 2002). There is so far no report focusing on the significance of N-terminal formylation in the antimicrobial activity or in the biosynthetic process of these bacteriocins. However, the antimicrobial activities of lacticins Q/Z remained even after the removal of N-formyl-methionine by the treatment of cyanogen bromide. Moreover, in vitro-synthesized lacticin Q, which is translated within an insect cell extract and thus is theoretically not formylated at N-terminus, exhibited the antimicrobial activity (unpublished data). Together, an N-terminal formylation is not crucial for the activity of lacticins Q/Z. These two bacteriocins exhibit nanomolar MIC values against a variety of Gram-positive bacteria including *Bacillus*, *Lactobacillus*, *Enterococcus*, and *Lactococcus*, whereas their activities are relatively low against *Streptococcus* (Fujita et al. 2007; Iwatani et al. 2007). The primary structures of lacticins Q/Z show considerable similarity to aureocin A53 (Fimland et al. 2002) and BHT-B, produced by *Streptococcus rattus* strain BHT (Hyink et al. 2005), with a highly conserved region (Ala^{12} to Leu^{30}) and relatively variable C-terminal (Iwatani et al. 2007, Fig. 13.1).

A natural isolate of *L. lactis* also produces a 30-amino-acid leaderless bacteriocin LsbB, which is coproduced with LsbA, another class IId bacteriocin synthesized with a leader peptide as described below (Gajic et al. 2003). There is no synergistic activity between these two peptides and both of their activities show narrow antimicrobial spectra, as they only inhibit the growth of closely related *L. lactis* strains (Gajic et al. 2003). There is an ABC-type multidrug resistance transporter, LmrB, encoded nearby the structural genes of LsbB and LsbA. This transporter is identified to be involved in both secretion and self-immunity of these two peptides (Gajic et al. 2003).

Nonsubgrouped Bacteriocins

Class IId bacteriocins that do not fit into either of the two above-mentioned subgroups are described here. These bacteriocins are non-pediocin-like linear peptides that are synthesized with a *sec*-independent double-glycine-type leader sequence and are transported by dedicated ABC-transporters. Among these bacteriocins, lactococcin A, produced by some strains of *L. lactis*, is an early-identified and supposedly the best-characterized bacteriocin. Lactococcin A is initially

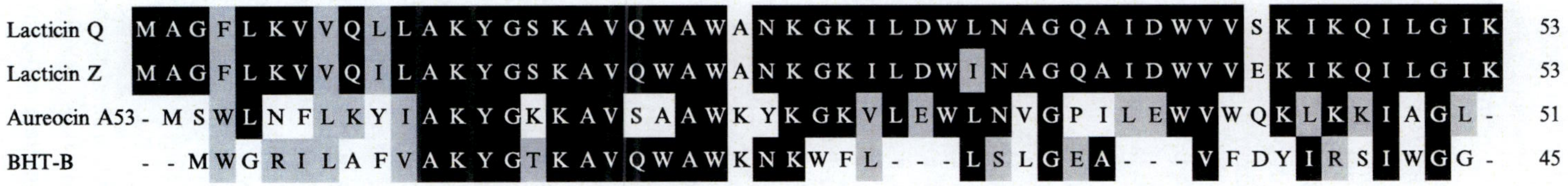

Fig. 13.1 Amino-acid sequence alignment of lacticin Q (Fujita et al. 2007), lacticin Z (Iwatani et al. 2007), aureocin A53 (Netz et al. 2002), and BHT-B (Hyink et al. 2005). The alignments were performed using the BLAST program. Identical residues are shown in a *black* background, and conservative substitutions are shown in a *gray* background

synthesized as a 75-amino-acid prepeptide that consists of a 21-amino-acid double-glycine-type leader peptide and a 54-amino-acid mature peptide (Holo et al. 1991). The genetic determinant for this bacteriocin is located on a 60-kb conjugative plasmid p9B4-6 from *L. lactis* subsp. *cremoris* 9B4 (van Belkum et al. 1991a), which is also responsible for the production of two other class IId bacteriocins, lactococcin B (van Belkum et al. 1992) and lactococcin M (van Belkum et al. 1991a). Despite a lack of significant similarities in their mature forms, the N-terminal extensions of these three lactococcins are almost identical, which indicates a common mechanism for the processing and export of these bacteriocins. Two other bacteriocins that share homology with lactococcin A are bovicin 255 from *Streptococcus gallolyticus* LRC0255 (29% identity; Whitford et al. 2001) and thermophilin A from *Streptococcus thermophilus* ST134 (22% identity; Ward and Somkuti 1995), both of which share 35% identity along the length of each prepeptide.

Enterocin B, which is coproduced with pediocin-like bacteriocin enterocin A from *E. faecium* T136, is synthesized as a 71-amino-acid prepeptide containing an N-terminal leader peptide of double-glycine type, which is then processed to give a mature peptide of 53 amino acids (Casaus et al. 1997). The 18-amino-acid N-terminal leader sequence contains the consensus elements of a double-glycine-type leader peptide that are common among other class II bacteriocins (Håvarstein et al. 1994). Enterocin B shows sequence similarity to carnobacteriocin A and piscicolin 61, which are two identical bacteriocins produced by *Carnobacterium piscicola* LV17A (Worobo et al. 1994) and *C. piscicola* LV61 (Holck et al. 1994), respectively. Moreover, it has recently been reported that *Enterococcus durans* QU 49 produces another enterocin B-homologue, durancin TW49-M (Hu et al. 2008). The N-terminal region of mature durancin TW49-M is almost identical to that of enterocin B, in contrast to the variability found in their C-terminal parts (Hu et al. 2008). An overview of these homologues suggests that they share a conserved region "LSKGGAKC" in the middle portion of mature peptides (Holck et al. 1994, Fig. 13.2). Divergicin 750, produced by *C. divergens* 750, is a relatively low-molecular-weight bacteriocin, the mature peptide of which is a 34-amino-acid residue after removal of 29-amino-acid double-glycine-type leader (Holck et al. 1996).

As an exception, LsbA, which is coproduced with LsbB from a strain of *L. lactis*, has a 23-amino-acid leader peptide that is not applicable to either the double-glycine-type leader or the consensus signal peptides of the *sec*-pathway (Gajic et al. 2003). Alternatively, the prepeptide of LsbA has a possible cleavage site for HtrA (Gajic et al. 2003), which is a surface housekeeping protease in *L. lactis* and is responsible for natural protein processing (Poquet et al. 2000).

Modes of Action

The modes of action of bacteriocins are linked to the primary structures of the peptides. Since bacteriocin classification is established on the basis of the structure, bacteriocins belonging to the same class have similar modes of action. This

Fig. 13.2 Amino-acid sequence alignment of enterocin B (Casaus et al. 1997), carnobacteriocin A (Worobo et al. 1994), piscicolin 61 (Holck et al. 1994), and durancin TW49-M (Hu et al. 2008). The alignments were performed using the BLAST program. Identical residues are shown in a *black* background, and conservative substitutions are shown in a *gray* background. Carnobacteriocin A and piscicolin 61 are identical bacteriocins and thus are shown in the third row collectively

similarity is demonstrated clearly by the fact that bacteriocins of the same class share a common receptor in the target cell membrane, such as lipid II molecules for lantibiotics (Hasper et al. 2006) and the mannose phosphotransferase permease (man-PTS) components for pediocin-like bacteriocins (Gravesen et al. 2002; Ramnath et al. 2004). This fact is, however, not applicable to class IId bacteriocins because of the fundamental diversity in their primary structures. Thus, in this section, we limit our discussion to the cases of lactococcin A, lactococcin 972, and lacticin Q, each of which is well characterized in its unique mode of action.

Lactococcin A (LcnA), a member of the nonsubgrouped class IId bacteriocins, has a relatively narrow antimicrobial spectrum, which mainly includes other lactococci. In the early 1990s, LcnA was shown to increase the permeability of the cytoplasmic membrane and dissipate the membrane potential of the target cells in a voltage-independent, protein-mediated manner (van Belkum et al. 1991b). An immunity factor of LcnA, termed LciA, was identified and characterized as being partly associated with the cell membranes (Nissen-Meyer et al. 1993), and it was suggested that LciA renders the cells immune by blocking the interaction between LcnA and its putative receptor (Venema et al. 1994). It was eventually proved by Diep et al. that LcnA binds to the components of man-PTS as a receptor in susceptible cells, just as the case of pediocin-like bacteriocins and that the immunity protein LciA strongly binds to the LcnA-receptor complex, thereby preventing cell membranes from LcnA-induced permeabilization (Diep et al. 2007). Interestingly, the complex between LciA and the receptor proteins forms in a bacteriocin-dependent manner, as the complex is recognized only in the presence of LcnA. Moreover, among the components of man-PTS (IIAB, IIC, and IID), membrane-located IIC and IID are the components sufficient for both the sensitivity to LcnA and the complex formation with LciA (Diep et al. 2007). It is noteworthy that not only LcnA but also lactococcin B (LcnB), another class IId bacteriocin, is suggested to use the man-PTS components IIC and IID as the receptor, although LcnA and LcnB share no sequence similarity to each other and neither shares any similarity to class IIa pediocin-like bacteriocins (Diep et al. 2007).

In contrast to other class II bacteriocins, lactococcin 972 (Lcn972), a member of the *sec*-dependent group of class IId bacteriocins, does not target the plasma membrane; instead, it inhibits the growth of susceptible cells by interfering with septum formation (Martínez et al. 2000). It was also shown that Lcn972 activates the two-component system CesSR that triggers the response of susceptible cells to cell envelope stress (Martínez et al. 2007), inducing the release of a prophage (Madera et al. 2009). The inhibition of cell wall biosynthesis is also recognized in the actions of type B lantibiotic mersacidin, which forms a complex with lipid II and inhibits peptidoglycan biosynthesis at the level of transglycosylation (Brötz et al. 1998). On the other hand, it has also been revealed that Lcn972, as the first nonlantibiotic bacteriocin, specifically interacts with the cell wall precursor lipid II (Martínez et al. 2008). The activity of Lcn972 against the target strain was antagonized in the presence of external lipid II; in contrast, antagonization was not observed when treated with other cell-wall precursors, such as lipid I, C_{55}-P, or C_{55}-PP (Martínez et al. 2008). Moreover, Lcn972 is coprecipitated only with lipid II micelles, and it inhibits the activity of two enzymes that use lipid II as a substrate (Martínez et al. 2008). Since Lcn972 only consists of unmodified amino-acid residues, it apparently lacks the structural motif, a lanthionine ring, which is found in nisin as a binding cage for lipid II (Hsu et al. 2004). In addition, the antimicrobial activity of Lcn972 is restricted to lactococci, which is in contrast to the wide antimicrobial activity of nisin. These observations strongly suggest that Lcn972 has a novel motif for recognizing lipid II and/or other factors that facilitate binding.

Another lactococcal bacteriocin, lacticin Q (LnqQ), a member of the leaderless group of class IId bacteriocins, is a cationic 53-amino-acid peptide that shows bactericidal activity against a wide range of Gram-positive bacteria (Fujita et al. 2007). LnqQ was predicted to contain two amphiphilic α-helixes at positions 4–13 and 38–49 (see Fig. 13.1) and, by recording the CD spectrum of the peptide, was confirmed to form an α-helical structure in the presence of negatively charged liposomes (Yoneyama et al. 2009a). LnqQ equally disrupted the membrane potentials of intact and vancomycin-treated cells, in contrast to the finding that the membrane-permeabilizing activity of nisin A was interfered by vancomycin treatment. Together, it is considered that LnqQ does not use lipid II as a target receptor (Yoneyama et al. 2009a). Furthermore, a drastic leakage of entrapped indicator from negatively charged liposome, large unilamellar vesicles (LUVs) was observed when the LUVs were treated with very low concentrations of LnqQ at which neither of nisin A nor pediocin PA-1 had previously been reported to cause a considerable leakage (Breukink et al. 1997; Chen et al. 1997); thus, it appears that LnqQ, in contrast to other bacteriocins, does not require a receptor for its membrane-permeabilizing activity (Yoneyama et al. 2009a). In more detailed studies, peptide translocation of LnqQ from the outer to the inner membrane leaflets was detected by fluorescent resonance energy transfer from the tryptophan residues of LnqQ to dansyl-labeled LUVs (Yoneyama et al. 2009b). In addition, lipid flip-flop of LUVs was observed when the liposomes were treated with LnqQ. Finally, both the peptide translocation and the lipid flip-flop were estimated to occur synchronously with pore formation of the LUVs (Yoneyama et al. 2009b). These phenomena are observed in the "toroidal-pore model" of the membrane-permeation mechanism (Brogden 2005). Some antimicrobial peptides from multicellular eukaryotes, such as magainin 2, induce a toroidal-pore (Yang et al. 1998), but none of the Gram-positive bacteria are reported to induce the same pore formation. Thus, LnqQ is the first bacteriocin from Gram-positive bacteria that has been shown to act via a toroidal-pore mechanism. Moreover, the size of the pore formed by LnqQ treatment was estimated to be 4.6–6.6 nm in diameter, which leads to protein leakage from bacterial cells (Yoneyama et al. 2009b). Considering all the above-mentioned findings, Yoneyama et al. proposed the mode of action of LnqQ as the "Huge Toroidal Pore (HTP)" model (Yoneyama et al. 2009b, Fig. 13.3).

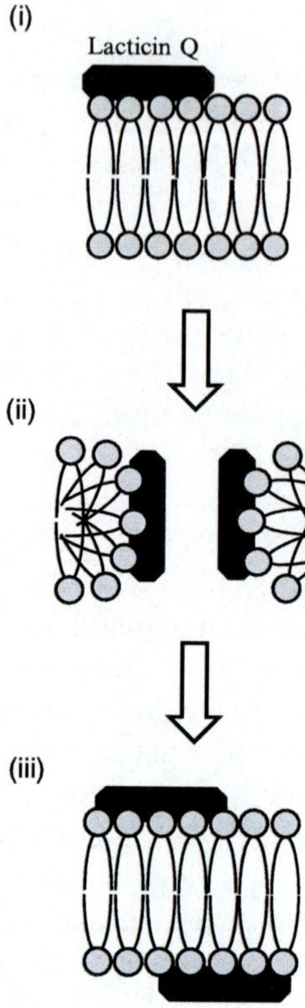

Fig. 13.3 The mode of action of lacticin Q (HTP model). Lacticin Q binds to the negatively charged phospholipid bilayer (i) and forms a huge toroidal pore (4.6–6.6 nm in diameter) accompanied by lipid flip-flop (ii), which leads to a leakage of intracellular molecules from target cells. Peptide translocation of lacticin Q occurs synchronously with the pore formation (iii) (This figure is a modified version of Fig. 7 of Yoneyama et al. 2009b)

Concluding Remarks

A great number of bacteriocins from Gram-positive bacteria have been identified since nisin was firstly discovered in 1928. A series of research on bacteriocins has been developed to such an extent that they are now classified into some groups according to their primary structures and biological features. There are, however, a considerable number of bacteriocins found to be unusual and not to fall into the majority. These heterogeneous bacteriocins, namely, the class IId bacteriocins, offer a wide variety of structures, which are potentially accompanied by the same variety of antimicrobial functions. In this regard, three lactococcal bacteriocins, such as lactococcin A, lactococcin 972, and lacticin Q were exemplified in this chapter to show the uniqueness of its structure and antimicrobial mechanism. This is, of course, only a small part of the whole thing. Considering the diverse structures of class IId bacteriocins, they can be an amazing source of novel antimicrobial peptides. Additionally, since all of them are "unmodified linear peptides," they can also be a model for artificially synthesized antimicrobial agents. There are, however, a lot of steps remaining toward the practical use of these bacteriocins or producers. The basal understandings such as the mechanisms of biosynthesis and gene regulation, modes of action, or structure–function relationships are still unexplained in most of these bacteriocins. To deepen the understandings and find out the missing piece of information is strongly required not only for the utilization of these bacteriocins but also for the development of the possibility of bacteriocins.

References

Breukink E, van Kraaij C, Demel RA, Siezen RJ, Kuipers OP, de Kruijff B (1997) The C-terminal region of nisin is responsible for the initial interaction of nisin with the target membrane. Biochemistry 36:6968–6976

Brogden KA (2005) Antimicrobial peptides: pore formers or metabolic inhibitors in bacteria? Nat Rev Microbiol 3:777–788

Brötz H, Josten M, Wiedemann I, Schneider U, Götz F, Bierbaum G, Sahl H-G (1998) Role of lipid-bound peptidoglycan precursors in the formation of pores by nisin, epidermin and other lantibiotics. Mol Microbiol 30:317–327

Casaus P, Nilsen T, Cintas LM, Nes IF, Hernández PE, Holo H (1997) Enterocin B, a new bacteriocin from *Enterococcus faecium* T136 which can act synergistically with enterocin A. Microbiology 143:2287–2294

Chen Y, Shapira R, Eisenstein M, Montville TJ (1997) Functional characterization of pediocin PA-1 binding to liposomes in the absence of a protein receptor and its relationship to a predicted tertiary structure. Appl Environ Microbiol 63:524–531

Cintas LM, Casaus P, Håvarstein LS, Hernández PE, Nes IF (1997) Biochemical and genetic characterization of enterocin P, a novel *sec*-dependent bacteriocin from *Enterococcus faecium* P13 with a broad antimicrobial spectrum. Appl Environ Microbiol 63:4321–4330

Cintas LM, Casaus P, Holo H, Hernández PE, Nes IF, Håvarstein LS (1998) Enterocins L50A and L50B, two novel bacteriocins from *Enterococcus faecium* L50, are related to staphylococcal hemolysins. J Bacteriol 180:1988–1994

Cintas LM, Casaus P, Herranz C, Håvarstein LS, Holo H, Hernández PE, Nes IF (2000) Biochemical and genetic evidence that *Enterococcus faecium* L50 produces enterocins L50A and L50B, the *sec*-dependent enterocin P, and a novel bacteriocin secreted without an N-Terminal extension termed enterocin Q. J Bacteriol 182:6806–6814

Cotter PD, Hill C, Ross P (2005) Bacteriocins: developing innate immunity for food. Nat Rev Microbiol 3:777–788

Criado R, Diep DB, Aakra Å, Gutiérrez J, Nes IF, Hernández PE, Cintas LM (2006) Complete sequence of the enterocin Q-encoding plasmid pCIZ2 from the multiple bacteriocin producer *Enterococcus faecium* L50 and genetic characterization of enterocin Q production and immunity. Appl Environ Microbiol 72:6653–6666

De Kwaadsteniet M, Fraser T, Van Reenen CA, Dicks LMT (2006) Bacteriocin T8, a novel class IIa *sec*-dependent bacteriocin produced by *Enterococcus faecium* T8, isolated from vaginal secretions of children infected with human immunodeficiency virus. Appl Environ Microbiol 72:4761–4766

Diep DB, Nes IF (2002) Ribosomally synthesized antibacterial peptides in gram positive bacteria. Curr Drug Targets 3:107–122

Diep DB, Skaugen M, Salehian Z, Holo H, Nes IF (2007) Common mechanism of target cell recognition and immunity for class II bacteriocins. Proc Natl Acad Sci U S A 104:2384–2389

Doi K, Eguchi T, Choi S-H, Iwatake A, Ohmomo S, Ogata S (2002) Isolation of enterocin SE-K4-encoding plasmid and a high enterocin SE-K4 producing strain of *Enterococcus faecalis* K-4. J Biosci Bioeng 93:434–436

Faye T, Langsrud T, Nes IF, Holo H (2000) Biochemical and genetic characterization of propionicin T1, a new bacteriocin from *Propionibacterium thoenii*. Appl Environ Microbiol 66: 4230–4236

Fimland G, Eijsink VGH, Nissen-Meyer J (2002) Mutational analysis of the role of tryptophan residues in an antimicrobial peptide. Biochemistry 41:9508–9515

Floriano B, Ruiz-Barba JL, Jiménez-Díaz R (1998) Purification and genetic characterization of enterocin I from *Enterococcus faecium* 6T1a, a novel antilisterial plasmid-encoded bacteriocin which does not belong to the pediocin family of bacteriocins. Appl Environ Microbiol 64:4883–4890

Franz CMA, Belkum MJ, Holzapfel WH, Abriouel H, Gálvez A (2007) Diversity of enterococcal bacteriocins and their grouping in a new classification scheme. FEMS Microbiol Rev 31:293–310

Fujita K, Ichimasa S, Zendo T, Koga S, Yoneyama F, Nakayama J, Sonomoto K (2007) Structural analysis and characterization of lacticin Q, a novel bacteriocin belonging to a new family of unmodified bacteriocins of gram-positive bacteria. Appl Environ Microbiol 73:2871–2877

Gajic O, Buist G, Kojic M, Topisirovic L, Kuipers OP, Kok J (2003) Novel mechanism of bacteriocin secretion and immunity carried out by lactococcal multidrug resistance proteins. J Biol Chem 278:34291–34298

Gravesen A, Ramnath M, Rechinger KB, Andersen N, Jänsch L, Héchard Y, Hastings JW, Knøchel S (2002) High-level resistance to class IIa bacteriocins is associated with one general mechanism in *Listeria monocytogenes*. Microbiology 148:2361–2369

Hasper HE, Kramer NE, Smith JL, Hillman JD, Zachariah C, Kuipers OP, de Kruijff B, Breukink E (2006) An alternative bactericidal mechanism of action for lantibiotic peptides that target lipid II. Science 313:1636–1637

Håvarstein LS, Holo H, Nes IF (1994) The leader peptide of colicin V shares consensus sequences with leader peptides that are common amongst peptide bacteriocins produced by gram-positive bacteria. Microbiology 140:2383–2389

Holck AL, Axelsson L, Schillinger U (1994) Purification and cloning of piscicolin 61, a bacteriocin from *Carnobacterium piscicola* LV61. Curr Microbiol 29:63–68

Holck A, Axelsson L, Schillinger U (1996) Divergicin 750, a novel bacteriocin produced by *Carnobacterium divergens* 750. FEMS Microbiol Lett 136:163–168

Holo H, Nilssen Ø, Nes IF (1991) Lactococcin A, a new bacteriocin from *Lactococcus lactis* subsp. *cremoris*: isolation and characterization of the protein and its gene. J Bacteriol 173:3879–3887

Hsu ST, Breukink E, Tischenko E, Lutters MA, de Kruijff B, Kaptein R, Bonvin AM, van Nuland NA (2004) The nisin-lipid II complex reveals a pyrophosphate cage that provides a blueprint for novel antibiotics. Nat Struct Mol Biol 11:963–967

Hu C-B, Zendo T, Nakayama J, Sonomoto K (2008) Description of durancin TW-49M, a novel enterocin B-homologous bacteriocin in carrot-isolated *Enterococcus durans* QU 49. J Appl Microbiol 105:681–690

Hyink O, Balakrishnan M, Tagg JR (2005) *Streptococcus rattus* strain BHT produces both a class I two-component lantibiotic and a class II bacteriocin. FEMS Microbiol Lett 252:235–241

Iwatani S, Zendo T, Yoneyama F, Nakayama J, Sonomoto K (2007) Characterization and structure analysis of a novel bacteriocin, lacticin Z, produced by *Lactococcus lactis* QU 14. Biosci Biotechnol Biochem 71:1984–1992

Izquierdo E, Bednarczyk A, Schaeffer C, Cai Y, Marchioni E, Van Dorsselaer A, Ennahar S (2008) Production of enterocin L50A, L50B, and IT, a new enterocin, by *Enterococcus faecium* IT62, a strain isolated from italian ryegrass in Japan. Antimicrob Agents Chemother 52:1917–1923

Kalmokoff ML, Banerjee SK, Cyr T, Hefford MA, Gleeson T (2001) Identification of a new plasmid-encoded *sec*-dependent bacteriocin produced by *Listeria innocua* 743. Appl Environ Microbiol 67:4041–4047

Kang BS, Seo JG, Lee GS, Kim JH, Kim SY, Han YW, Kang H, Kim HO, Rhee JH, Chung MJ, Park YM (2009) Antimicrobial activity of enterocins from *Enterococcus faecalis* SL-5 against Propionibacterium acnes, the causative agent in acne vulgaris, and its therapeutic effect. J Microbiol 47:101–109

Kawai Y, Kusnadi J, Kemperman R, Kok J, Ito Y, Endo M, Arakawa K, Uchida H, Nishimura J, Kitazawa H, Saito T (2009) DNA sequence and homologous expression of a small peptide conferring immunity to gassericin A, a circular bacteriocin produced by *Lactobacillus gasseri* LA39. Appl Environ Microbiol 75:1324–1330

Leer RJ, van der Vossen JMBM, van Giezen M, van Noort JM, Pouwels PH (1995) Genetic analysis of acidocin B, a novel bacteriocin produced by *Lactobacillus acidophilus*. Microbiology 141:1629–1635

Madera C, Carcía P, Rodríguez A, Suárez JE, Martínez B (2009) Prophage induction in *Lactococcus lactis* by the bacteriocin Lactococcin 972. Int J Food Microbiol 129:99–102

Martínez B, Suárez JE, Rodríguez A (1996) Lactococcin 972: a homodimeric lactococcal bacteriocin whose primary target is not the plasma membrane. Microbiology 142:2393–2398

Martínez B, Fernández M, Suárez JE, Rodríguez A (1999) Synthesis of lactococcin 972, a bacteriocin produced by *Lactococcus lactis* IPLA 972, depends on the expression of a plasmid-encoded bicistronic operon. Microbiology 145:3155–3161

Martínez B, Rodríguez A, Suárez JE (2000) Lactococcin 972, a bacteriocin that inhibits septum formation in lactococci. Microbiology 146:949–955

Martínez B, Zomer AL, Rodríguez A, Kok J, Kuipers OP (2007) Cell envelope stress induced by the bacteriocin Lcn972 is sensed by the lactococcal two-component system CesSR. Mol Microbiol 64:473–486

Martínez B, Böttiger T, Schneider T, Rodríguez A, Sahl H-G, Wiedemann I (2008) Specific interaction of the unmodified bacteriocin lactococcin 972 with the cell wall precursor lipid II. Appl Environ Microbiol 74:4666–4670

Martín-Platero AM, Valdivia E, Ruíz-Rodríguez M, Soler JJ, Martín-Vivaldi M, Maqueda M, Martínez-Bueno M (2006) Characterization of antimicrobial substances produced by *Enterococcus faecalis* MRR 10-3, isolated from the uropygial gland of the hoopoe (*Uupa epop*). Appl Environ Microbiol 72:4245–4249

Nes IF, Diep DB, Håvarstein LS, Brurberg MB, Eijsing V, Holo H (1996) Biosynthesis of bacteriocins in lactic acid bacteria. Antonie Leeuwenhoek 70:113–128

Nes IF, Diep DB, Holo H (2007) Bacteriocin diversity in *Streptococcus* and *Enterococcus*. J Bacteriol 189:1189–1198

Netz DJA, Sahl H-G, Marcolino R, Nascimento JS, de Oliveira SS, Soares MB, Bastos MCF (2001) Molecular characterisation of aureocin A70, a multi-peptide bacteriocin isolated from *Staphylococcus aureus*. J Mol Biol 311:939–949

Netz DJA, Pohl R, Beck-Sickinger AG, Selmer T, Pierik AJ, Bastos MCF, Sahl H-G (2002) Biochemical characterisation and genetic analysis of aureocin A53, a new, atypical bacteriocin from *Staphylococcus aureus*. J Mol Biol 319:745–756

Nissen-Meyer J, Håvarstein LS, Holo H, Sletten K, Nes IF (1993) Association of the lactococcin A immunity factor with the cell membrane: purification and characterization of the immunity factor. J Gen Microbiol 139:1503–1509

Oliveia SS, Nascimento JS, Póvoa DC, de Araújo SA, Gamon MR, Bastos MCF (1998) Genetic analysis of the bacteriocin-encoding plasmids pRJ6 and pRJ9 of *Staphylococcus aureus* by transposon mutagenesis and cloning of genes involved in bacteriocin production. J Appl Microbiol 85:972–984

Poquet I, Saint V, Seznec E, Simoes N, Bolotin A, Gruss A (2000) HtrA is the unique surface housekeeping protease in *Lactococcus lactis* and is required for natural protein processing. Mol Microbiol 35:1042–1051

Ramnath M, Arous S, Gravesen A, Hastings JW, Héchard Y (2004) Expression of mptC of *Listeria monocytogenes* induces sensitivity to class IIa bacteriocins in *Lactococcus lactis*. Microbiology 150:2663–2668

Ruiz-Barba JL, Floriano B, Maldonado-Barragán A, Jiménez-Díaz R (2006) Molecular analysis of the 21-kb bacteriocin-encoding plasmid pEF1 from *Enterococcus faecium* 6T1a. Plasmid 57:175–181

Sánchez J, Borrero J, Gómez-Sala B, Basanta A, Herranz C, Cintas LM, Hernández PE (2008) Cloning and heterologous production of hiracin JM79, a sec-dependent bacteriocin produced by *Enterococcus hirae* DCH5, in lactic acid bacteria and *Pichia pastoris*. Appl Environ Microbiol 74:2471–2479

Sánchez-Hidalgo M, Maqueda M, Gálvez A, Abriouel H, Valdivia E, Martínez-Bueno M (2003) The genes coding for enterocin EJ97 production by *Enterococcus faecalis* EJ97 are located on a conjugative plasmid. Appl Environ Microbiol 69:1633–1641

Tomita H, Fujimoto S, Tanimoto K, Ike Y (1996) Cloning and genetic organization of the bacteriocin 31 determinant encoded on the *Enterococcus faecalis* pheromone-responsive conjugative plasmid pYI17. J Bacteriol 178:3585–3593

van Belkum MJ, Hayema BJ, Jeeninga RE, Kok J, Venema G (1991a) Organization and nucleotide sequences of two lactococcal bacteriocin operons. Appl Environ Microbiol 57:492–498

van Belkum MJ, Kok J, Venema G, Holo H, Nes IF, Konings WN, Abee T (1991b) The bacteriocin lactococcin A specifically increases permeability of lactococcal cytoplasmic membranes in a voltage-independent, protein-mediated manner. J Bacteriol 173:7934–7941

van Belkum MJ, Kok J, Venema G (1992) Cloning, sequencing, and expression in *Escherichia coli* of *lcnB*, a third bacteriocin determinant from the lactococcal bacteriocin plasmid p9B4-6. Appl Environ Microbiol 58:572–577

Venema K, Haverkort RE, Abee T, Haandrikman AJ, Leenhouts KJ, de Leij L, Venema G, Kok J (1994) Mode of action of LciA, the lactococcin A immunity protein. Mol Microbiol 14:521–532

Ward DJ, Somkuti GA (1995) Characterization of a bacteriocin produced by *Streptococcus thermophilus* ST134. Appl Microbiol Biotechnol 43:330–335

Whitford MF, Mcpherson MA, Forster RJ, Teather RM (2001) Identification of bacteriocin-like inhibitors from rumen *Streptococcus* spp. and isolation and characterization of bovicin 255. Appl Environ Microbiol 67:569–574

Worobo RW, Henkel T, Sailer M, Roy KL, Vederas JC, Stiles ME (1994) Characteristics and genetic determinant of a hydrophobic peptide bacteriocin, carnobacteriocin A, produced by *Carnobacterium piscicola* LV17A. Microbiology 140:517–526

Worobo RW, van Belkum MJ, Sailer M, Roy KL, Vederas JC, Stiles ME (1995) A signal peptide secretion-dependent bacteriocin from *Carnobacterium divergens*. J Bacteriol 177:3143–3149

Yamamoto Y, Togawa Y, Shimosaka M, Okazaki M (2003) Purification and characterization of a novel bacteriocin produced by *Enterococcus faecalis* strain RJ-11. Appl Environ Microbiol 69:5746–5753

Yang L, Harroun TA, Heller WT, Weiss TM, Huang HW (1998) Neutron off-plane scattering of aligned membranes. I. Method of measurement. Biophys J 75:641–645

Yoneyama F, Imura Y, Ichimasa S, Fujita K, Zendo T, Nakayama J, Matsuzaki K, Sonomoto K (2009a) Lacticin Q, a lactococcal bacteriocin, causes high-level membrane permeability in the absence of specific receptors. Appl Environ Microbiol 75:538–541

Yoneyama F, Imura Y, Ohono K, Zendo T, Nakayama J, Matsuzaki K, Sonomoto K (2009b) Peptide-lipid huge toroidal pore, a new antimicrobial mechanism mediated by a lactococcal bacteriocin, lacticin Q. Antimicrob Agents Chemother 53:3211–3217

Part V
Genetics, Biosynthesis, Structure, and Mode of Action of AMP from Gram-Negative Bacteria

Chapter 14
Colicin Killing: Foiled Cell Defense and Hijacked Cell Functions

Miklos de Zamaroczy and Mathieu Chauleau

Abstract The study of bacteriocins, notably those produced by *E. coli* (and named colicins), was initiated in 1925 by Gratia, who first discovered "*un remarquable exemple d'antagonisme entre deux souches de colibacilles*". Since this innovating observation, the production of toxic exoproteins has been widely reported in all major lineages of Eubacteria and in Archaebacteria. Bacteriocins belong to the most abundant and most diverse group of these bacterial defense systems. Paradoxically, these antimicrobial cytotoxins are actually powerful weapons in the intense battle for bacterial survival. They are also biotechnologically useful since several bacteriocins are used as preservatives in the food industry or as antibiotics or as potential antitumor agents in human health care. Most colicins kill bacteria in one of two ways. The first type is those that form pores in the phospholipid bilayer of the inner membrane. They are active immediately after their translocation across the outer membrane. The translocation pathway requires generally either the BtuB receptor and the Tol (OmpF/TolABQR) complex, or the FepA, FhuA, or Cir receptor and the Ton (TonB/ExbBD) system. The second type of colicins encodes specific endonuclease activities that target DNA, rRNA, or tRNAs in the cytoplasm. To be active, these colicins require translocation across both the outer and inner membranes. The molecular mechanisms implicated in the complex cascade of interactions, required for the transfers of colicin molecules from the extracellular medium through the different "cellular compartments" (outer membrane, periplasm, inner membrane, and cytoplasm), are still incompletely understood. It is clear, however, that the colicins "hijack" specific cellular functions to facilitate access to their target. In this chapter, following a general presentation of colicin biology, we describe, compare, and update several of the concepts related to colicin toxicity and discuss recent, often unexpected findings, which help to advance our understanding of the molecular events governing colicin import. In particular, our review includes the following: (1) Structural data on the tripartite interaction of a colicin with the

M. de Zamaroczy (✉)
Institut de Biologie Physico-Chimique, CNRS, UPR 9073,
13, rue Pierre et Marie Curie, Paris 75005, France
e-mail: zamaroczy@ibpc.fr

D. Drider and S. Rebuffat (eds.), *Prokaryotic Antimicrobial Peptides: From Genes to Applications*, DOI 10.1007/978-1-4419-7692-5_14,

outer membrane receptor and the translocation machinery, (2) Comparison of the normal cellular functions of the Tol and Ton systems of the inner membrane with their "hijacked" roles during colicin import, (3) An analysis of the interaction of a nuclease-type colicin with its cognate immunity protein in the context of the immunity of producer cells, and of the dissociation of this complex in the context of the attack of the colicin on target cells, (4) Information on the endoproteolytic cleavage, which presumably accompanies the penetration of nuclease-type colicins into the cytoplasm. The new data presented here provides further insight into cellular functions "hijacked" or "borrowed" by colicins to permit their entry into target cells.

Introduction

Bacteriocins are toxic exoproteins synthesized by bacteria. Bacteriocins that are produced by *Escherichia coli* strains are named colicins (Gratia 1946). Colicins are proteins (30–75 kDa) that are released into the culture medium and are toxic against the same or closely related species, but not against colicinogenic (i.e., colicin producer) cells, which stay immune. Colicins that have a molecular weight below 10 kDa were designated as microcins (or antimicrobial peptides) (Baquero et al. 1978; Braun et al. 2002). The term "bacteriocin" was preferentially assigned to antibiotic peptides produced by Gram-positive bacteria, which are toxic against a wide spectrum of bacteria (Cotter et al. 2005).

There are two major classes of colicins based on the type of activity involved in cell killing, either a nuclease or a pore-forming activity. Nuclease colicins target a particular type of nucleic acid (e.g., chromosomal DNA, 16S rRNA, or tRNAs) (Bowman et al. 1971; Senior and Holland 1971; Ogawa et al. 1999; James et al. 2002; Masaki and Ogawa 2002), while ionophoric colicins make pores (i.e., voltage-dependent ion channels) in the phospholipid bilayer of the inner membrane, and thus, provoke an irreversible depolarization of the membrane followed by a depletion of cytoplasmic ATP (Guihard et al. 1993; Lakey and Slatin 2001).

A specific plasmid is required for the expression of each colicin, and it carries at least two and sometimes three colicin-related genes. The first gene (*cxa*, where "x" corresponds to the specific letter of each colicin) encodes the colicin, while the second one (*cxi*) encodes the immunity protein (ImmX). In the case of nuclease colicins, these two genes are organized in an operon followed by a third structural gene (*cxl*) encoding a BRP *(bacteriocin release protein*, so-called "lysis protein") (Fig. 14.1a1). When the BRP synthesis is induced, it allows the release of colicin into the medium and causes the death of the producer cell (Jakes and Zinder 1984; van der Wal et al. 1995). In the case of pore-forming colicins, the structural gene encoding the cognate Imm protein is located downstream of *cxa* gene, but in an inverted orientation, and thus, it is only transcribed from its own constitutive promoter (Hardy and Meynell 1972; Little and Mount 1982; Lloubes et al. 1988) (Fig. 14.1a2). Moreover, there is no BRP encoding gene in some pore-forming-colicin-encoding operons, (e.g., colicins Ia, Ib, and B) (Braun et al. 2002).

Colicin production is inducible by environmental stress. The three structural genes belonging to the operon of nuclease-type colicins are cotranscribed from a SOS promoter so that their expression is enhanced simultaneously during the stress response of the cell, (Masaki et al. 1985; Roos et al. 1989; James et al. 1996) (Fig. 14.1a1). In the absence of any SOS stimuli (for example, DNA damage triggered by UV light or the antibiotic mitomycin C), the transcription of colicin operon is repressed by LexA. Two dimers of the LexA repressor bind to specific sites in the promoter region. The stress response activates the RecA protein, which stimulates the self-cleavage of LexA and the subsequent release of LexA from the SOS promoter region. This in turn induces the transcription of the colicin operon (Little 1982; Lloubes et al. 1988).

The nuclease colicins (E2–E9 and D) form a tight (1:1 M) heterodimeric complex with their cognate immunity proteins (Kleanthous et al. 1998; Graille et al. 2004). The structural gene encoding the Imm protein is expressed constitutively from its own promoter as well as part of the three gene operon, thus protecting the colicin producer cells against exogenous colicin molecules (Fig. 14.1a1). The Imm protein of pore-forming colicins is only transcribed from its own constitutive promoter (Hardy and Meynell 1972; Little and Mount 1982; Lloubes et al. 1988) (Fig. 14.1a2). It interacts directly, in the cytoplasmic membrane of the infected cell, with the specific, pore-forming domain of the exogenous colicin molecule (Goldman et al. 1985; Geli and Lazdunski 1992). Pore-forming colicin producer cells do not need to be protected against their own colicin molecules, since the polarity of the cytoplasmic membrane is opposite to that necessary for opening the pore.

The particularity of colicin biology, from the point of view of bacterial cell physiology, is that the process of colicin import, which precedes the killing of the target cell, makes use of a series of the target cells' own functions, which the colicins "hijack" for their own use. To some extent, this exploitation of the cells' own machinery is also true for colicin release into the extracellular medium, even though the release process frequently requires a colicin-specific BRP protein (Fig. 14.1a).

During the import, colicins parasitize different multiprotein systems of the cell envelope (e.g., ligand-gated receptors for uptake of vitamin B12 (cobalamin), iron siderophores or nucleoside, and outer membrane porins) and hijack the Tol or Ton membrane system coupled to the proton-motive force (PMF) of the cytoplasmic membrane. Normally, the Tol system plays a role in cell envelope stability, while the Ton system transduces energy for the transport of some essential nutrients. Other than the Tol or Ton system, several LPS biogenesis proteins (Sharma et al. 2009), a periplasmic chaperone FkpA (Hullmann et al. 2008) and two essential proteins of the cytoplasmic membrane, LepB and FtsH (de Zamaroczy et al. 2001; Walker et al. 2007), are occasionally recruited during colicin import.

A historical classification of colicins derives from the cross-sensitivity of the target cells, which is dependent on the translocation pathway required for the colicin to cross the outer membrane and transit the periplasm. Accordingly, the colicins defined as belonging to the group A require generally the BtuB receptor and the Tol (OmpF/TolABQR) system, whereas colicins of group B are mainly dependent on the FepA, FhuA, or Cir receptor and the Ton (TonB /ExbBD) system.

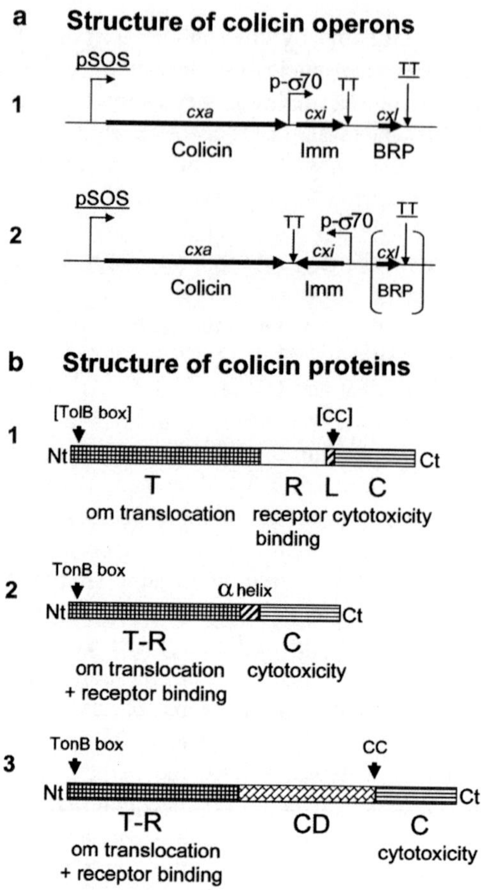

Fig. 14.1 (**a**) Structural organization of colicin operons and (**b**) modular domain-organization of the colicin encoding structural genes. (**a1**) Typical gene organization of nuclease colicins of groups A and B, e.g., colicins E2–E9 and colicin D, respectively. Bent arrows show the position of transcription promoters: pSOS and p-σ70 indicate inducible (SOS response dependent) and constitutive promoters, respectively. "TT" indicates transcription terminator. The colicin is expressed from the *cxa* structural gene, Imm (the immunity protein) is encoded by the *cxi* gene and BRP (*bacteriocin release protein*) is expressed from the *cxl*. (**a2**) Typical gene organization of pore-forming colicins of groups A (e.g., colicins A and E1). In all cases, the Imm encoding structural gene has an inverted orientation, and it is only expressed from its own constitutive promoter. [BRP] indicates the absence of gene encoding a BRP in several operons for pore-forming colicins of group B, (e.g., colicins Ia, Ib, and B). (**b1**) The canonical domain organization of group-A colicins (e.g., colicins E and colicin A) and group-B colicins Ia and Ib. The R- (central receptor-binding) domain is used to recognize the specific TonB gated receptor, hijacked for colicin import. The N-terminal T- (translocation) domain is usually required for the interaction with the secondary om (outer membrane) translocator. [TolB box] indicates the specific site of interaction with TolB, in the case of several colicins of group A. (The TonB box of colicins Ia and Ib is not shown.) The C- (cytotoxicity domain) carries the pore-forming or nuclease activity used for cell killing. [CC] indicates the consensus 30-amino-acid sequence, only present in nuclease colicins, located at the junction between the R- and C-domains [including the short N-terminal L- (linker) region just upstream of the C-domain]. (**b2**) The domain organization of group-B

Colicins of group A (A, E1 to E9, K, N, U, and cloacin DF13) and some colicins of group B (e.g., colicins Ia, Ib, and probably colicin M) generally contain structurally identifiable N-terminal, central, and C-terminal domains, which are implicated in translocation (T-domain), receptor binding (R-domain), and cytotoxicity (C-domain), respectively, and thus correspond to sequential steps of colicin import and cell killing (Ohno-Iwashita and Imahori 1980; Ohno-Iwashita and Imahori 1982; Benedetti et al. 1991a; Braun et al. 2002) (Fig. 14.1b1). Since their targets are Gram-negative bacteria, pore-forming colicins (e.g., colicins E1, A, N, Ia, B) need to be translocated across the outer membrane barrier and the periplasmic space (transit) to reach the cytoplasmic membrane to be active. In addition, the nuclease colicins (E2 to E9, D), cloacin DF13, and klebicins C and D (Chavan et al. 2005) also require translocation across the inner membrane to reach their targets in the cytoplasm.

One of the oldest areas of prokaryotic research is the biology of colicins. It started more than 70 years ago and is still subject to intense investigation. A lot of data concerning the export of colicins, the import and mechanisms of toxicity in target cells, and the self-immunity of colicin producers has accumulated during the last few decades and has been reported in previous reviews (Pugsley 1984; James et al. 1996; Lazdunski et al. 1998; Braun et al. 2002; de Zamaroczy and Buckingham 2002; James et al. 2002; Masaki and Ogawa 2002; Riley and Wertz 2002; Postle and Kadner 2003; Cascales et al. 2007). Despite the recent advances in our knowledge of colicin biology, and in particular concerning the molecular mechanisms for the successive transfer of colicin molecules between different "cellular compartments", the complete import mechanisms remain poorly understood, notably in the case of the colicins of group B. We describe and update some intriguing aspects of colicin acquisition and highlight recently reported data and concepts concerning the cascade of interactions between colicins and the cell envelope. The import processes are presented for each of the A- and B-group colicins and are described following the chronological steps of colicin entry into the periplasm of the target cell. Where necessary, a description of the normal functions of hijacked membrane components is given before its role in colicin uptake is described. For some steps, specific "strategies" developed by some colicins are also discussed. The last section concerns the translocation mechanisms across the inner membrane, which are specific to nuclease colicins, belonging to both colicin groups A and B.

Fig. 14.1 (continued) colicin B. The combined multifunctional (T-R) domain is separated by a 74 Å long α helix from the C-terminal pore-forming domain. "TonB box" (called the "TonB box"-like sequence in the text) indicates the specific interaction site with the TonB protein. (**b3**) The domain organization of the group-B colicin D. The 313-amino-acid long N-terminal (T-R) domain is 93% identical to that of colicin B, but in contrast to colicin B (**B2**), a large CD- (central) domain precedes the C-terminal tRNase domain. The CD plays a role in the recognition of TonB, together with the N-terminal part of the T-R domain, and in the formation of the complex with ImmD, together with the C-terminal tRNase domain. All references are given in the text

Translocation Machinery of Colicins of Group A: Physiological Function and Hijacked Role of the BtuB/Tol System

Substrate Transport or Colicin Reception by BtuB

The biological function of BtuB is to transport vitamin B12 (cyanocobalamin) into the periplasm. Its activity is strictly dependent on the TonB energy transduction system, which is driven by the proton-motive force (PMF) across the inner membrane (Postle and Kadner 2003). BtuB, like the other structurally homologous TonB-dependent outer membrane receptors (FhuA, FepA, FecA used by colicins of group B, see below), is a porin-like protein involved in the uptake of essential nutriments. It is formed by a 22-stranded C-terminal β-barrel domain and a globular N-terminal "plug" or "cork" domain, formed by the first 132 amino acids, which is inserted inside the barrel, (Ferguson et al. 1998; Locher et al. 1998; Buchanan et al. 1999; Chimento et al. 2003a). The channel in this state is said to be occluded and is not able to translocate either cobalamin or colicins. A structurally disordered peptide motif of BtuB is the major recognition element for TonB (the so-called "TonB box"). It is located close to the N terminus of the cork and is oriented toward the periplasm. The interaction between the TonB box of BtuB and TonB is dependent upon a transmembrane signalling, which is induced by the substrate binding to its cognate receptor (Chimento et al. 2003b; Cherezov et al. 2006). Thus, in the presence of its natural substrate, the interaction of BtuB with TonB induces the partial unfolding or other localized conformational rearrangements in the cork domain. In fact, the crystal structure of TonB in complex with FhuA or BtuB showed that this interaction allows TonB to apply a "mechanical pulling force" to the central β sheet of the cork domain, which provokes its disruption (Pawelek et al. 2006; Shultis et al. 2006). The consequence of this is that either a small transient channel opens within the barrel, which still contains the reorganized cork domain, or the cork domain is pulled partially out of the barrel sothat in either case the substrate can be transported into the periplasm.

The initial step of colicin import is the recognition of target cells, mediated by colicin binding to a specific outer membrane receptor. In the case of colicin A and colicins E, the primary cognate receptor, the BtuB protein, is recognized by the apex of the central receptor-binding R-domain, which has an extended coiled-coil structure (Fig. 14.2a–c). The dissociation constant for the receptor domain of colicins E2, E7, and E3 to the receptor BtuB, determined by biosensor assay and qualitatively confirmed by microbiological spot tests, is $K_d = 10^{-9}$ M, whereas the binding affinities of colicins A and E1 were significantly lower (Kurisu et al. 2003; Sharma et al. 2009). Importantly and in contrast with substrate uptake, the import of A-group colicins does not necessitate the Ton system. In agreement with this, the addition of the RNase colicin E3 to BtuB, incorporated into planar bilayers, did not reveal any ion channel activity in the receptor (Kurisu et al. 2003). Thus, during A-group colicin import, there is no displacement of the plug from inside the receptor, which indicates that the colicin is not translocated through the pore formed by the β-barrel of the receptor. Thus, the primary colicin receptor BtuB is only involved in high-affinity colicin recognition and not in translocation.

Colicin Translocation Across the Outer Membrane: the Passage Through OmpF Porin

The crystal structures of BtuB in complex with the R-domain of the RNase colicin E3 or DNase colicin E2 revealed a probably conserved mechanism for the translocation of colicins of group A across the outer membrane (Kurisu et al. 2003; Sharma et al. 2007). The central R-domain of colicin E3 and of pore-forming colicin Ia (which belongs to colicins of group B and is presented in detail in "The Outer Membrane Translocation Machinery of B-Group Colicins: The Function") consist essentially of a long coiled-coil structure of 100 and 160 Å, respectively (Wiener et al. 1997; Soelaiman et al. 2001). Both are very tightly anchored to long extracellular loops on the cell surface of their high affinity receptors (BtuB and Cir, respectively) (Figs. 14.2a and 14.3a), but in symmetrically opposite directions. The orientation of these R-domains is at an angle of about 45° above the lipid bilayer in both complexes. Moreover, both colicins enter the binding pocket of their receptor to a similar depth, but only in the case of the colicin Ia bound to Cir is there a conformational change in two extracellular loops of the receptor, as a consequence of the colicin's entry into the vestibule of the pore (Buchanan et al. 2007). The highly flexible N-terminal translocation domain of both these colicins is slanting and directed toward a "secondary receptor". The latter acts as the real colicin translocator, and in the case of colicins E1 to E9, it is the trimeric outer membrane OmpF porin (Benedetti et al. 1989; Zakharov et al. 2006). The BtuB receptor with the bound colicin and the recruited OmpF porin together constitute the outer membrane translocon for the import of most A-group colicins (Fig. 14.2a) (Kurisu et al. 2003; Zakharov et al. 2004; Housden et al. 2005). (For the B-group colicin Ia the secondary receptor/translocator is presented in Colicin of Group B: Reception by FepA, FhuA, or Cir.)

The attachment of the R-domain to BtuB was predicted to trigger colicin unfolding, which is thought to be an early step of colicin import and which is necessary for the passage of the colicin across the outer membrane (Kurisu et al. 2003). Paradoxically, calorimetric analysis (ITC) of the binding of the DNase colicin E9 to BtuB showed that this interaction does not induce any conformational changes in the colicin domains, and so colicin unfolding must take place after the assembly of the tripartite colicin/BtuB/OmpF complex (Housden et al. 2005). However, the interaction of colicin E3 with either BtuB or OmpF reduced the efficiency of FRET between T- and C-domains of the colicin, linked with a flexible region. The present view is that unfolding of the colicin seems to be induced by both the recognition of BtuB by the R-domain and subsequent insertion of the T-domain into OmpF (Zakharov et al. 2008). In the case of colicins E3 and E9, about 80 residues of the N terminus of the T-domain, which is disordered in the native colicins, occlude one of the pores of the trimeric OmpF, whereas the distal part of the T-region remains

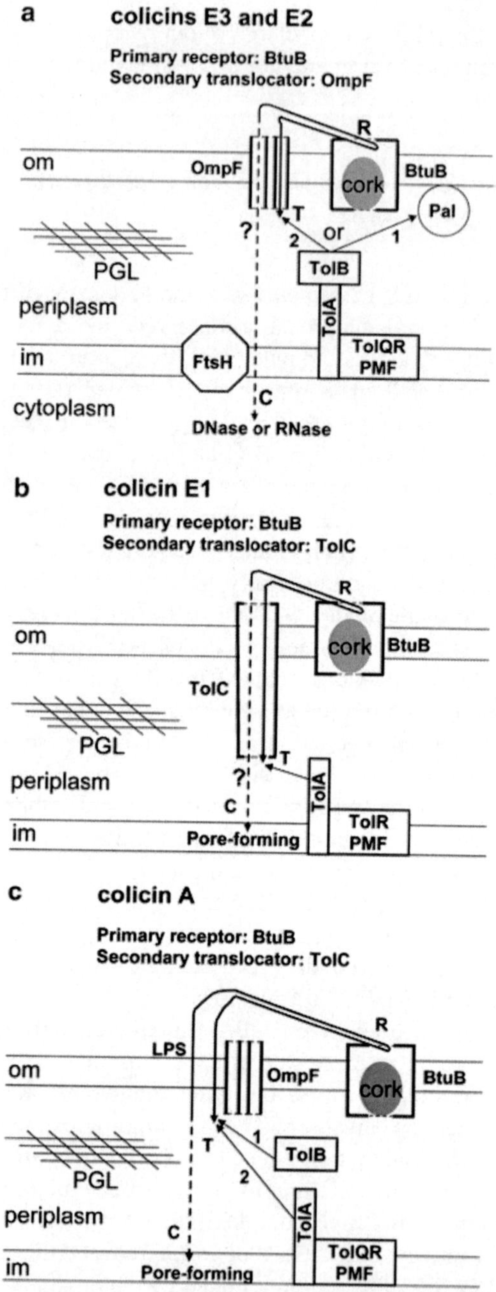

Fig. 14.2 Tentative models of the import mechanisms for different colicins of group A. Each diagram shows the proteins known to be involved in the passage of the colicin across the outer membrane (om), and the periplasm, to the cytoplasmic membrane (im) [pore forming colicins, (**b–c**)], and across the im into the cytoplasm [nuclease colicins (**a**)]. The three domains of the colicins are indicated as "T" (N-terminal: translocation), "R" (central: receptor-binding), and "C"

surface-exposed, until the unfolding of the receptor-binding domain bound to BtuB. The length of the N-terminal subdomain inserted into OmpF is critical, since it allows the TolB recognition sequence, located close to the colicin N terminus and named TolB box, to reach the periplasm and then interact with the periplasmic TolB protein (Fig. 14.2a) (Sharma and Cramer 2007; Zhang et al. 2008). This key interaction of colicins of group A with the Tol system is described in detail below.

Some colicins of group A use variations of these processes to cross the outer membrane. For instance, the pore-forming colicin N uses OmpF as the sole receptor-translocator (Bourdineaud et al. 1990; Jeanteur et al. 1994). While colicin E1, another pore-forming colicin, once bound to BtuB, requires neither OmpF nor TolB but interacts specifically with a cognate secondary receptor-translocator, identified as TolC (Fig. 14.2b) (Benedetti et al. 1991b; Zakharov et al. 2004). TolC is an outer membrane "channel-tunnel" over 140 Å long that spans both the outer membrane and periplasmic space and which is required for the export of xenobiotics and hemolysin (Koronakis et al. 1997; Koronakis et al. 2000). Unexpectedly, TolC was also shown to be essential for the toxicity of colicin A (Sharma et al. 2009). But in contrast with its function as a channel for colicin E1 translocation, the effect of TolC on colicin A toxicity is presumably indirect and was attributed to a TolC-mediated control of OmpF synthesis (Misra and Reeves 1987). OmpF is the primary high-affinity receptor for colicin N, which remains toxic against Δ*tolC* target cells. In contrast, colicin A, which uses OmpF as secondary receptor for its

Fig. 14.2 (continued) (C-terminal: cytotoxicity). Cork refers to the N-terminal "cork" domains of the TonB-gated receptor, BtuB, which carries the TonB box, the site of interaction of the receptors with TonB. The TolQ, TolR proteins associated with TolA are essential to the energy transduction of the im proton-motive force (PMF) to the om. The three channels of the trimeric porin OmpF are indicated, as are Pal (*peptidoglycan associated lipoprotein*) and PGL (*peptidoglycan layer*). The *arrows* indicate the sites of interaction between the colicins and different proteins of the cell envelope, or the location of the targets of colicin activity domains. A "?" on arrows with dashed lines indicates a putative route for the translocation of the C-domains of the colicins across the cell envelope. See text for full details and references. (**a**) Translocon of colicins E3 and E2 via BtuB, OmpF, and TolB. The R-domain of the colicin makes contact with BtuB, the primary receptor, while the T-domain (and possibly the C-domain) enters via the secondary translocator OmpF. The TolA(QR)/TolB network is required either for membrane integrity control, via the interaction with Pal (*arrow* "1") "or" for the "energization" of colicin import, via the interaction with the T-domain (*arrow* "2"). For the sake of clarity, the release of the immunity protein from the nuclease-colicin bound to BtuB is not shown. FtsH is an essential ATPase/protease required for the translocation of all nuclease colicins across the inner membrane. (**b**) Translocon of colicin E1 via BtuB, TolC, and TolA(R). BtuB acts as the primary receptor by interacting with the R-domain. TolC, as the secondary translocator is energized via TolA and allows the passage of the T and possibly the C domains of colicin. (**c**) Translocon of colicin A via BtuB, OmpF, TolB, and TolA(QR). After a primary contact between the R-domain and the BtuB receptor, the T-domain is proposed to slide down the OmpF–lipopolysaccharide (LPS) interface toward the periplasm. This process requires OmpF and successive interactions of the colicin with TolB and TolA (*arrows* "1" and "2", respectively). The essential, but presumably indirect, role of TolC in colicin A toxicity is not indicated but is described in the text

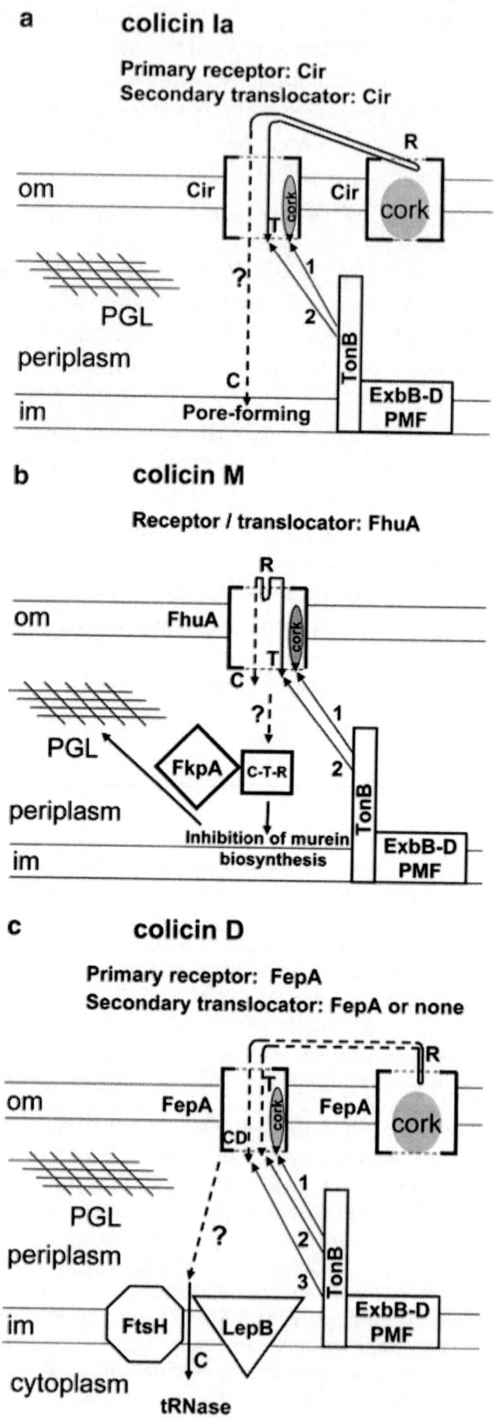
a
colicin Ia
Primary receptor: Cir
Secondary translocator: Cir
R
om
Cir
cork
Cir
cork
T
?
1
PGL
2
periplasm
TonB
C
im
Pore-forming
ExbB-D
PMF
b
colicin M
Receptor / translocator: FhuA
R
om
FhuA
cork
T
C
?
1
PGL
FkpA
C-T-R
2
periplasm
TonB
Inhibition of murein
biosynthesis
im
ExbB-D
PMF
c
colicin D
Primary receptor: FepA
Secondary translocator: FepA or none
R
T
om
FepA
cork
FepA
cork
CD
1
PGL
?
periplasm
2
3
TonB
im
FtsH
LepB
ExbB-D
PMF
C
cytoplasm
tRNase

translocation, has a much lower affinity for OmpF than colicin N (Fig. 14.2c). Thus, colicin A is more sensitive to a decreased level of OmpF, and therefore, the deletion of *tolC* renders target cells completely resistant to colicin A (Sharma et al. 2009).

The Tol System of the Cell Envelope in Gram-Negative Bacteria

The Tol system is a multiprotein complex of the cell envelope involved in maintaining outer membrane stability (Guihard et al. 1994; Lazzaroni et al. 1995). It is composed of TolQ, an integral inner membrane protein, and TolR with TolA, which are anchored in the inner membrane and have large periplasmic domains. The transmembrane domains of these three proteins interact with each other, and in addition, TolA contacts outer membrane porins through an elongated coiled-coil structure. The primary role of TolQR is to couple the energy of the PMF across the inner membrane to TolA. Thus, TolQR were proposed to form an ion-conducting channel involved in the transduction of conformational changes to TolA (Germon et al. 2001). The soluble periplasmic protein TolB completes the Tol system. TolB, as well as being associated with the outer membrane and interacting with TolA, also interacts with both the peptidoglycan layer and outer membrane protein Pal (peptidoglycan-associated lipoprotein) (Lazdunski et al. 2000; Sturgis 2001). The association of Pal with the peptidoglycan layer and its binding to TolB are mutually exclusive (Bouveret et al. 1999). It is noteworthy that the C-terminal part of TolB forms a six-bladed β-propeller, each blade of which consists of twisted β-sheets. This conserved structure binds substrates or ligands at or above the central chamber, which is formed at the junction of the "propeller blades" (Abergel et al. 1999; Ponting and Pallen 1999). In the case of TolB, the association of its

Fig. 14.3 Tentative models of the import mechanisms for different colicins of group B. Each diagram shows the proteins known to be involved in the passage of each colicin across the outer membrane (om), and the periplasm, to the cytoplasmic membrane (im) [pore-forming colicin Ia, **(a)**, and murein synthesis inhibiting colicin M, **(b)**] and across the im into the cytoplasm [tRNase colicin D, **(c)**]. **(a)** Translocon of colicin Ia via two copies of Cir and TonB. Cir acts as both the primary receptor and secondary translocator. *Arrows* "1" and "2" indicate two sequential interactions with TonB, first with the cork domain of the receptor and then with the T-domain of the colicin. **(b)** Import pathway of colicin M via FhuA, which acts presumably as both primary receptor and secondary translocator, and TonB. *Arrows* "1" and "2" indicate two interactions with TonB, first with the cork domain of the receptor, and then with the T-domain of the colicin. FkpA is a periplasmic chaperone required for refolding the whole colicin M molecule, in the periplasm, into an active toxin. **(c)** Colicin D uptake via FepA and TonB. FepA is the primary receptor and possibly a second copy of FepA might act as the secondary translocator. There are three interactions with TonB involving the cork domain of the receptor (*arrow* "1"), the TonB-box like sequence in the T-domain (*arrow* "2"), and a sequence in colicin D-specific central domain (CD) (*arrow* "3"). The entrance of colicin D tRNase into the cytoplasm requires both the FtsH protease and specifically the signal peptidase (LepB) of the inner membrane. "R-, T-, C-" domains of colicins, "cork" of the receptors Cir, FhuA, or FepA, the *arrows*, and "?" are indicated in the legend of Fig. 14.2. The ExbB, ExbD proteins, associated with TonB, are essential for the energy transduction of the im proton-motive force (PMF) to the TonB-gated receptors. See text for full details and references β-propeller domain with Pal (Ray et al. 2000) is a key interaction ensuring outer membrane stability. The interaction of TolB with the energized TolA (Walburger et al. 2002), as well as a probable direct interaction between the "energized" TolA and Pal, ensures the maintenance of the outer membrane integrity (Cascales et al. 2000; Lazzaroni et al. 2002). The Tol network has also been reported to have another physiological role: it may contribute to the adaptation of the cell envelope structure prior to the formation of the septal wall during cell division (Gerding et al. 2007).

Colicin Uptake Necessitates Interactions with the Tol System

Colicins of group A appear to parasitize the Tol system to energize their translocation. The translocation involves sequential interactions in the periplasm between the N-terminal translocation domains of the colicins, as they pass through OmpF, and the TolA-B proteins (Fig. 14.2a) (Cheng et al. 2006). Most colicins of group A use the TolABQR proteins of the Tol network, but not TolC. However, the pore-forming colicin E1 requires TolAQR and specifically TolC as described above (Fig. 14.2b), while the pore-forming colicin N, as well as the RNase cloacin DF13, needs only the TolAQR proteins. In addition, some of colicin/Tol system interactions are very subtle and/or specific. For instance, a recent genome-wide screening using the "Keio collection" (a systematic single-gene knockout library of all nonessential *E. coli* genes) revealed that (i) colicin E1 does not require absolutely TolQ for cytotoxicity, in contrast with a previous observation, and confirmed that (ii) despite the strong association of Pal with TolB and its key role in membrane stability (Clavel et al. 1998), Pal is not essential for the cytotoxicity of A-group colicins (Sharma et al. 2009).

Cascade Interactions with the Tol System for Colicin Transport Across the Outer Membrane and Transit Through the Periplasm

There is a large amount of data concerning the passage of colicins of group A through the outer membrane and periplasmic space, but it is only recently that a composite picture can be drawn from the sometimes disparate results. After binding of the central R-domain to the receptor, the T-domain of colicin E3 (Fig. 14.2a) or E9 binds to the β propeller of TolB, in the same way as Pal. Thus, colicin E9 was found to compete efficiently with Pal (Carr et al. 2000). But the comparison of the dissociation constants of the interaction of colicin E9 and Pal with TolB is clearly not in favor of such a competition. However, the presence of Ca^{2+} ions increases the affinity of the colicin E9 T-domain for TolB >10-fold, and it becomes able to dislodge Pal from its complex with TolB. Ca^{2+} ions provoke a reversal of the surface potential of the TolB β-propeller channel producing a global positive charge, and this improves its interaction with the negatively charged T-domain of colicin. The same charge switch does not critically influence the affinity of Pal to TolB (Loftus et al. 2006). In summary, the competitive recruitment of TolB by the TolB box of colicins E, translocated through the OmpF lumen, leads to the disruption of TolB/Pal complex. A key intermediary role of TolB, in connecting the inner membrane TolA to the outer membrane Pal has recently been demonstrated by a variety of in vitro techniques, combining crystallography, NMR, ITC, and cross-linking (Bonsor et al. 2009). An allosteric signalling through the TolB β-propeller was found to affect the conformational equilibrium between the Pal–TolB interaction and the TolB–TolA complex. The precise molecular nature of the signal is not yet known, but it is supposed to be linked to outer membrane stability (Bonsor et al. 2009). Pal induces the allosteric signal as long-range conformational changes in TolB that stabilize the sequestration of its binding site (called the TolA box) for TolA. In contrast, when the TolB box in the T-domain of colicin E9 is translocated into the periplasm, it not only recruits competitively TolB from its complex with Pal (see above) but also stabilizes the interaction of TolB with TolA, by increasing the affinity of this complex. Even if colicin E9 mimicks Pal binding to TolB, its complex with TolB overrides the expected displacement of TolB from TolA, when Pal binds to TolB. Consequently, colicin bound to TolB has direct access to the PMF because the TolB–TolA–TolQR network is maintained (Bonsor et al. 2009) (Fig. 14.2a). This cascade should provide energy from the PMF for the release of the immunity protein (see next section) and the translocation of the C-terminal killing domain after partial or even full unfolding of colicin. The interaction cascade described above is not compatible with a direct interaction of TolA with Pal, as previously reported by in vivo data (Cascales et al. 2000; Cascales and Lloubes 2004). But the structural network described here is in agreement with the absence of any direct interaction between TolA and nuclease colicins of group A during the import. In contrast, the interaction of the N terminus of colicin E1 or A with the C-terminal domain of TolA was demonstrated by several in vitro and in vivo techniques (Derouiche et al. 1997). The possibility of forming a tripartite complex in vivo, including TolA-colicin A (T-domain)–TolB (Fig. 14.2c), was invoked by an antibody overlay experiment (Bouveret et al. 1998). In addition, a weak and transient interaction was reported between the T-domain of colicins A and E3, but not of colicin E1, and the central, periplasmic domain of TolR (Journet et al. 2001), whereas no interaction between TolQ-requiring colicins and the TolQ protein has been demonstrated.

The hijacking of the Tol system during colicin entry into the cell results in a miscoupled PMF that may be responsible for a partial loss in the outer membrane integrity, comparable to that observed in *tol* gene deletion strains, which exhibit sensitivity to vancomycin, detergents (SDS), and bile salts (Bernstein et al. 1972; Lazzaroni and Portalier 1981; Prouty et al. 2002) and which are also characterized by a defect in the LPS (lipopolysaccharide) on the cell surface (Gaspar et al. 2000). An analysis of these *tol*-defect-like phenotypes and a site-directed mutagenesis of the extended TolB box region of colicin E9 and colicin A revealed differences in the recruitment of the TolB, necessary for the import of A-group colicins (Zhang et al. 2010). It has been proposed that the T-domain of colicin A does not enter by a pore of OmpF, unlike colicin E9, but in contrast slides down the outside of OmpF (Fig. 14.2c) to reach the periplasm. The primary receptor BtuB as well as the secondary receptor OmpF form therefore an essential scaffold for the import of colicin A (Lazzaroni et al. 2002), dependent on its successive interactions with TolB and TolA (Zhang et al. 2010). In agreement with this hypothesis, a decrease in colicin A toxicity was found in strains deleted for *gmhB* or *rffT*, involved in the biosynthesis of LPS and the enterobacterial common antigen, respectively (Fig. 14.2c) (Sharma et al. 2009). A similar translocation model had been previously proposed for colicin N to cross the outer membrane at the OmpF–LPS interface (Baboolal et al. 2008). In agreement with this model, mutations affecting the biosynthesis of LPS also conferred resistance to colicin N (Sharma et al. 2009).

The major unanswered question concerning translocation is how the C-terminal killing domains of the colicins enter the periplasm after the proceeding cascade of interactions between the T-domain of the colicin and the Tol system has taken place. The OmpF trimer consists of three ion-permeable channels (Cowan et al. 1992; Alcaraz et al. 2004) of which one is occluded by the entry of the N-terminal T-domain. An exciting idea is that the unfolded C-terminal, cytotoxic domain of the nuclease colicin E could penetrate into the periplasmic space via one of the other unoccupied OmpF pores (Fig. 14.2a). This may occur following the T-domain-mediated colicin association with the Tol system through OmpF while central receptor domain remains bound to BtuB as a scaffold on the surface of the cell (Zakharov et al. 2006; Yamashita et al. 2008). Alternatively, the rupture of the Pal–TolB interaction provoked by colicin binding could destabilize the outer membrane and allow the C-terminal domain to cross the outer membrane close to but outside of OmpF. Consistent with this hypothesis, colicin A was found to span the cell envelope and to be still bound to its BtuB receptor, when its killing activity became detectable (Fig. 14.2c) (Benedetti et al. 1992). The C-terminal pore-forming domains of colicins A and N presumably cross the outer membrane outside the OmpF lumen, like their T-domain. On the other hand, colicin E1, including its C-terminal pore-forming domain, may pass to the periplasm through the TolC channel in a mostly unfolded form, and its translocation presumably requires a specific interaction of colicin E1 with TolA (Fig. 14.2b) (Schendel et al. 1997).

Release of the Imm Protein of Nuclease Colicins Occurs at a Late Stage of Translocation

The nuclease colicins E form high-affinity complexes ($K_d \sim 10^{-14}$ M) with their cognate immunity protein (Imm) in producer cells, but Imm must be released during colicin import so that the colicin can kill target cells. The binding per se of the colicin–Imm complex to BtuB is not sufficient to release the Imm protein into the medium. The release of the immunity protein from the colicins E9– and E2–Imm complexes requires the unfolding of the colicin as it penetrates into OmpF and makes contact with the Tol system (Duche et al. 2006; Zhang et al. 2008). In addition, the loss of ImmE9 from its association with the nuclease domain was shown to be energy dependent. Recently, a sensitive fluorescence assay has, for the first time, implicated TolA and thus the PMF of the cytoplasmic membrane, in the "energization" of the TolB-bound colicin E9, allowing the release of ImmE9 from the DNase domain during cell entry (Vankemmelbeke et al. 2009).

Colicin Transit in the Periplasm

The precise molecular mechanism of colicin transit through the periplasmic space is not known, but it does not seem to be an energy-dependent process. One model invokes Brownian Ratcheting, guided by a complex and sequential network of protein–protein interactions in the periplasm and involving colicin domains and membrane components (Lazdunski et al. 1998; Journet et al. 2001). For the colicin to progress through the periplasm, to either form a pore in the inner membrane or be translocated across the inner membrane, there should be a constant increase in the affinity of the subsequent interactions (Hands et al. 2005), but no data are available.

The Outer Membrane Translocation Machinery of B-Group Colicins: The Function of the FepA, FhuA, or FecA Receptors and the Ton System

For all colicins the initial import step is the binding of the colicin to its cognate outer membrane receptor. Among colicins of group B, the pore-forming colicins Ia and Ib use Cir, the tRNase colicin D and pore-forming colicin B use FepA, and colicin M, a unique colicin which acts as an inhibitor of murein biosynthesis, requires the outer membrane receptor FhuA, to bind to their target cell (Fig. 14.3a–c). These receptors, like the previously described BtuB (used by most colicins of group A), are energy-coupled transporters dependent on the TonB system for the translocation of the different essential nutrients, which are their natural substrates (Braun et al. 2002). Although the entry of the colicins of group B into the cell is conceptually similar to that of the group A (hijacking of outer membrane receptors leading ultimately to the translocation of the C-terminal cytotoxic domain), it is functionally different due to the strict necessity of the active transport function of the colicin-bound receptor. For colicins of group A, the function of the Tol system appears to be limited to providing energy to release the immunity protein from the nuclease domain and to unfold the colicin protein prior to the translocation. The important difference for the translocation of the B-group colicins across the outer membrane compared to A-group is that in addition to a direct contact of TonB with the colicin molecule itself, the PMF coupled to the receptor by the Ton system was shown to be essential for the B-group colicins. This is a strong argument in favor of B-group colicins directly using their receptors to penetrate the target cell, in a way similar to that of the normal substrates.

Colicin of Group B: Reception by FepA, FhuA, or Cir

The binding of colicins Ia and M to Cir and FhuA, respectively, was shown not to be dependent on the interaction of their receptor with TonB, unlike the subsequent translocation step (Buchanan et al. 2007; Zeth et al. 2008). Thus, in the absence of energy provided by the PMF, the binding of the R-domain of colicin Ia to Cir (Fig. 14.3a) neither displaces the plug from the β-barrel nor does it open the β-barrel by disrupting the plug, for the passage of the colicin from the exterior into the periplasm (Buchanan et al. 2007). But the observation that the binding of colicin Ia to Cir stabilized large conformational changes in the extracellular loops of the receptor, which exposed the interior of the β barrel to the incoming colicin, is in favor of the hypothesis that the receptor Cir has also a "transporter" function in colicin translocation (Buchanan et al. 2007). In contrast, as reported above, the binding of A-group colicin E3 to BtuB did not induce any conformational changes in the cell surface receptor, even though the crystal structures of the two colicin/receptor complexes are similar. The role of BtuB in A-group colicin import is limited to the colicin recognition followed by the recruitment of a secondary receptor/transporter (Kurisu et al. 2003). The hypothesis that the TonB-gated receptor plays a dual role (as receptor and transporter) is strengthened by the fact that no secondary receptor-translocator has ever been identified as necessary for the import of B-group colicins. However, as in the case of A-group colicins, there are some variations to the general pattern. The pore-forming colicins 5 and 10 are exceptions in the sense that these colicins of group B require TonB for their uptake, even though they bind to the TonB-independent outer membrane receptor, Tsx (Braun et al. 2002; Sharma et al. 2009). Unexpectedly, the translocation of colicins 5 and 10 requires also a secondary translocator TolC, like that of A-group colicin E1, but no other protein of the Tol system (Braun et al. 2002).

A recent study of chimeric Ia/E3 colicins has provided evidence that two distinct Cir molecules act as receptor and translocator for the import of colicin Ia. When the R-domain of colicin Ia was replaced with that of colicin E3 (Fig. 14.1b1), the hybrid colicin required both the primary receptor BtuB of colicin E3 and the Cir/TonB system, necessary for the translocation of wild-type colicin Ia. This result is consistent with the involvement of two distinct Cir receptors for the translocation of wild-type colicin Ia across the outer membrane (Fig. 14.3a). The first Cir acts as the high-affinity primary receptor, which is replaced by BtuB when the Ia R-domain is replaced with the E3 R-domain. The Cir-bound colicin Ia forms a stable scaffolding, from which the T-domain of colicin Ia seeks out and then appropriates a second colicin-free Cir receptor with a lower affinity. This second copy presumably functions as the translocator for the entry of the T-domain and possibly also for the pore-forming C-terminal domain of colicin Ia, into the periplasm through its β-barrel (Jakes and Finkelstein 2010). This translocation model of colicin Ia strongly resembles that of A-group colicin E3 via BtuB and OmpF (Figs. 14.2a and 14.3a). The two main differences concern the fact that both the primary receptor and the secondary translocator required for the uptake of colicin Ia correspond to two distinct copies of the same outer membrane receptor (Cir), and the translocation step across the membrane is dependent on the energy provided by the Ton system, presumably to the secondary translocator (Cir). It was also shown that the N-terminal T-domain of colicin Ia may interact with Cir independently of the primary interaction of the central R-domain with Cir. This could explain the residual killing activity of colicin Ia in the absence of any R-domain. Thus, the killing capacity of colicin Ia is greatly enhanced by its local concentration at the cell surface, mediated by its primary binding to the Cir receptor. High concentrations of colicin Ia were shown to prevent cell killing. Cir receptors are indeed saturated by their extensive interaction with R-domains of colicin Ia, and this apparently leaves no Cir proteins available to be recruited as the secondary translocator through an interaction with the T-domains (Jakes and Finkelstein 2010).

Several observations argue against a conserved mechanism used by all the different colicins of group B for their translocation across the outer membrane. Only two colicins of group B, the colicins Ia and likely M, have a modular T-R-C type domain structure, resembling that found with colicins of group A (Fig. 14.1b1). Moreover, the crystal structures of only one of them, colicin Ia, either alone (Wiener et al. 1997) or in complex (R-domain) with the Cir receptor (PDB accession number: 2HDI) (Buchanan et al. 2007), are reminiscent of the previous structures of A-group colicins (e.g., colicin E3 R-domain/BtuB; PDB accession number: 1UJW) (Kurisu et al. 2003). Colicin M architecture shows a unique fold in which the functional domains may form structural entities, but seem to be much less well separated from each other than in colicin Ia or E3 structures (Zeth et al. 2008; Barreteau et al. 2010). It is noticeable that the C-terminal domain of colicin M was reported to form a novel kind of phosphatase, since it exhibits no peptide sequence similarity to known phosphatases (Zeth et al. 2008), nor to proteins with canonical phosphatase-type catalytic mechanisms (Barreteau et al. 2010). At the moment, there is no rational way that colicin M (even if it is the smallest colicin known) can pass across the outer membrane while preserving the compact structural form, found by the crystallographic study. Moreover, a weak association between the presumed individual domains might facilitate their unfolding during translocation through the outer membrane (Zeth et al. 2008).

In the case of colicins B and D, their first 313 N-terminal amino acids are 96% identical (Fig. 14.1b2, 3). The remaining C-terminal part of colicin B corresponds essentially to its pore-forming domain, i.e., colicin B has no distinct central R-domain. In fact, the N-terminal domains of colicins B and D are necessary and sufficient to bind the FepA receptor and to allow the translocation of colicin B (and presumably of colicin D) across the outer membrane (Roos et al. 1989; de Zamaroczy and Buckingham 2002). The concept of a combined and multifunctional N-terminal domain used for both receptor binding and translocation is supported by the dumbbell-shaped crystal structure of colicin B, in which a single,

74-Å-long α-helix separates the N-terminal domain and C-terminal killing domain (Hilsenbeck et al. 2004) (Fig. 14.1b2). This structure does not appear to be sufficiently flexible to scan for a secondary translocator (Buchanan et al. 2007). One model of colicin B translocation through FepA initially suggested that the internal plug domain of FepA leaves the barrel entirely to allow colicin B translocation in its compact folded form (Hilsenbeck et al. 2004). During colicin B stimulation, the plug domain of the FepA receptor was shown to become accessible to a labelling reagent, biotin maleimide, in the periplasm. Even though residues (replaced with cysteines) near the N-terminal of the FepA reacted strongly, other residues belonging to the distal part of the plug reacted poorly. These results implied that the plug only partially comes out from the barrel when colicin B crosses the FepA channel to enter the periplasm. This, in turn, implies that if the β-barrel remains partially occluded by its plug, then large conformational rearrangements of colicin B are necessary for its translocation (Devanathan and Postle 2007). Nevertheless, recent data obtained with the same technique, although with different labelling reagents and experimental conditions, have found no significant reactivity for any of 35 genetically engineered cysteine derivatives covering the whole FepA plug domain (Smallwood et al. 2009). This observation does not support the previous interpretation of the passage of colicin B through the FepA β-barrel (Hilsenbeck et al. 2004). It is not yet possible to reconcile these opposite conclusions. Indeed, the latter results seem only to be compatible with the putative entry of colicin B into the periplasm along the lipid–barrel interface, where a few colicin B molecules might "breach" the impermeability of the outer membrane (Smallwood et al. 2009). However, in this model, the role of the sequential interactions with TonB in colicin B import is not really understood.

TonB-Dependent Colicin Transit into the Periplasm

The N terminus of TonB protein is anchored to the inner membrane and is separated by a proline-rich linker from its C terminus, which extends 100 Å toward the outer membrane (Postle and Good 1983; Pawelek et al. 2006). TonB has multiple contacts with the integral, inner-membrane proteins ExbB and ExbD, while its C-terminal domain can interact with the outer membrane receptors. ExbB-D couple the PMF to TonB, in the form of conformational changes, and hence to the outer membrane receptors (Braun et al. 1996; Larsen et al. 1999). ExbB, as a chaperone-like protein, is also involved in a transient interaction with TonB in the cytoplasm during its synthesis and assembly (Karlsson et al. 1993). During the uptake of iron siderophore complexes by FepA, FhuA, or Cir, TonB is converted to its energized conformation and either "spans" or "crosses" the periplasm, according to the currently proposed "propeller" or "shuttling" models, respectively (Letain and Postle 1997; Chang et al. 2001; Wiener 2005). Subsequently, TonB contacts the outer membrane receptor (transporter) close to the receptor's N terminus, at a short conserved peptide motif of the plug domain, termed the "TonB box". This interaction allows the energy, derived from the PMF of the cytoplasmic membrane, to be harnessed for the active transport of the ferric siderophore (Postle and Kadner 2003). In vivo cross-linking experiments show a direct interaction between TonB and FepA (Skare et al. 1993), and genetic studies confirm the importance of this interaction in the transport of ferric enterobactin. Other investigations including site-directed disulfide or formaldehyde cross-links led to similar conclusions in the case of the FhuA, FecA, and BtuB receptors and showed that their interaction with TonB is enhanced when they are loaded with substrate (Cadieux and Kadner 1999; Moeck and Letellier 2001; Endriss et al. 2003; Ogierman and Braun 2003).

Cross-complementation studies between the Tol and Ton systems revealed that *tolQ-R* mutants transformed with plasmids carrying *exbB-D* became sensitive to colicins of group A, and reciprocally, *exbB-D* mutants transformed with plasmid-encoded *tolQ-R* became sensitive to colicins of group B. In contrast, no cross-complementation has been observed between the very distinct TonB and TolA proteins, which mediate exclusive interactions with B-group or A-group colicins, respectively (Braun and Herrmann 1993).

"TonB box"-like sequences have been described in the natively disordered N-termini of all TonB-dependent colicins (Figs. 14.1b2–3 and 14.3a–c) (Braun et al. 2002). The five amino-acid long consensus for TonB box sequences, (D/E) Tx(V/T)V, deduced from those of receptors and colicins of group B, is not strictly conserved, except for the Val residue in the fifth position (Mora et al. 2005). Genetic studies show that TonB boxes contact the central region of TonB. Mutations affecting the Q160 residue of TonB can restore activity to receptors with mutated TonB boxes and restore some sensitivity to inactive derivatives of colicins B and M, carrying mutations in their TonB box-like sequences (Mende and Braun 1990; Pilsl et al. 1993). Spontaneous *tonB* mutations, affecting R158 or P161, completely abolished colicin D toxicity but did not affect either the sensitivity to other colicins (in particular to colicin B) or the FepA-dependent siderophore uptake capacity (Mora et al. 2005). Moreover, the sensitivity of these *tonB* mutants to colicin D was fully restored by compensatory suppressor mutations introduced in the TonB box-like sequence of colicin D. This result demonstrated that the interaction of colicins with TonB depends not only on the local conformation of the energy coupling TonB box motif, as previously concluded (Bell et al. 1990; Postle and Kadner 2003), but importantly also on the identity of the side chains of the individual residues in the motif (Mora et al. 2005). Thus, the high-affinity binding step of B-group colicins to their cognate receptor is followed by two distinct interactions with TonB, involving successively the colicin-bound outer membrane receptor ("cork" domain) and the N terminus of the colicin itself (Fig. 14.3a–c), as originally suggested by Braun et al 2002. The resistance to colicins B and D of a mutant strain in which the "cork" domain of FepA was deleted is consistent with this hypothesis (Vakharia and Postle 2002).

The recognition of the T-domain of colicin Ia with the second copy of Cir could initiate the productive interaction of the Cir TonB box with TonB (Fig. 14.3a) (Jakes and Finkelstein 2010), by mimicking the effect of the binding of the natural substrate to the receptor (Pawelek et al. 2006; Shultis et al. 2006).

Similarly, it was suggested that the binding of colicin B to FepA induced the receptor's TonB box "to become exposed in the periplasm" and to interact transiently with TonB (Cascales et al. 2007). The structure of the BtuB/TonB complex may provide a structural rationalization for a substrate- or colicin-induced order-to-disorder transition in the receptor TonB box (Shultis et al. 2006). Such a disordered TonB box is able to bind to TonB, which in turn promotes the entry of the colicin into the periplasm after the disruption of the "cork" domain (Fig. 14.3a–c). Such a disordered (i.e., high-entropy) structure should permit only transient interactions of the TonB boxes with TonB, especially if the affinities vary with the ligand-bound state of the receptor. In turn, a cascade of transient interactions should lead to serial contacts between TonB and the various receptor TonB boxes and so allow a continuous recycling of TonB for the uptake of substrates (or colicin).

The rearrangement of the plug domain inside the β-barrel of the second copy of Cir, as a consequence of the receptor interaction with TonB, may allow colicin Ia to penetrate further inside the β-barrel until the TonB-box like motif of the colicin meets a second copy of TonB at the periplasmic side (Fig. 14.3a). This second interaction between a disordered TonB-box of colicin T-domain and TonB may provide the energy required to unwind the rest of the colicin Ia molecule, which is presumably necessary to drive the translocation of the C-terminal pore-forming domain. Similarly, in the case of colicin M, following its binding to FhuA, the energy-transducing TonB would be required first to open the β-barrel of FhuA, and second, to progressively unfold the colicin so that the whole 30-kDa colicin M molecule can pass inside FhuA and reach the periplasm (Fig. 14.3b) (Zeth et al. 2008). On the contrary, during siderophore transport through FhuA, it was suggested that the rearranged cork domain remains within the β-barrel rather than being displaced (Eisenhauer et al. 2005).

Compared to colicin B, the 75-kDa colicin D, the largest colicin, possesses an additional long-central domain of 280 residues (Fig. 14.1b3), which was shown to be involved in another interaction with TonB and which is particularly required for colicin D import (Fig. 14.3c). This domain does not have a second canonic "TonB-box like" sequence (Mora et al. 2008) (Fig. 14.1b3). The crystal structure of BtuB/TonB is compatible with the possible interaction of BtuB with two TonB molecules (Shultis et al. 2006), and a dimerization of TonB has been observed in vivo (Sauter et al. 2003; Ghosh and Postle 2005). The large molecular size of colicin D and the possibility that it might interact with a TonB dimer support the previous hypothesis that a single FepA receptor alone would not be sufficient to allow the tRNase domain of colicin D to cross the outer membrane (de Zamaroczy and Buckingham 2002). In the same way as two Cir receptors are required to transport colicin Ia (Jakes and Finkelstein 2010) (see above), the translocation of colicin D might involve two copies of its FepA receptor (Fig. 14.3c). As well as being essential for uptake and thus cell killing, the central domain of colicin D also interacts with the immunity protein and so participates in the production of the immunity complex, which is released into the medium by cells producing nucleolytic colicins (Graille et al. 2004; Mora et al. 2008).

Passage of Colicins of Group B in the Periplasm

A 160 Å long coiled-coil structure, formed by two long linker helices, separates the T- and C-domains of colicin Ia from its R-domain. This length is sufficient to span the periplasm (average width of 150 Å) so that the pore-forming amino acids of the C-domain could be inserted in the inner membrane, while the R-domain remains attached to the Cir receptor (Fig. 14.3a) (Wiener et al. 1997; Buchanan et al. 2007). Consistent with this hypothesis, another pore-forming A-group colicin, the BtuB/OmpF-dependent colicin A is still bound to the outer membrane when pore formation occurs (Fig. 14.2c) (Benedetti et al. 1992). Similarly, colicin E2 remains bound to its BtuB receptor and to the translocation machinery after the colicin has initiated cell killing (via random degradation of chromosomal DNA) in the cytoplasm (Fig. 14.2a) (Duche 2007; Sharma et al. 2007). But in contrast with the colicin Ia structure, the colicin B linker structure (Hilsenbeck et al. 2004) is not long enough to span the cell envelope (Buchanan et al. 2007) (Fig. 14.1b2). Thus, this is consistent with the possibility of a complete release of colicin B into the periplasm (Devanathan and Postle 2007).

Once the entire colicin molecule or at least its C-terminal cytotoxic domain has entered the periplasm, the partially unfolded colicin needs to be refolded to become toxic. In the case of pore-forming colicins, their active conformation may occur simultaneously with their insertion into the inner membrane. There is no current data about the spontaneous (or cytoplasmic-chaperone assisted) folding of nuclease colicins. In contrast, it was shown that the periplasmic chaperone FkpA is essential for the specific toxicity of colicin M (Hullmann et al. 2008), which involves inhibition of murein biosynthesis (Schaller et al. 1982). Colicin M probably has a phosphodiesterase activity, and thus, catalyzes the hydrolysis of peptidoglycan–lipid intermediates I and II, by cleaving the pyrophosphate linkage between the carrier lipid and the murein precursor. This activity, which probably occurs on the periplasmic side of the inner membrane, has been demonstrated in vivo and in vitro. The depletion of these intermediates prevents peptidoglycan polymerization and leads to cell lysis (El Ghachi et al. 2006). FkpA, which accelerates the refolding of the colicin M molecule in the periplasm to an active form (Fig. 14.3b), represents yet another example of a cellular function hijacked by a colicin (Hullmann et al. 2008; Barreteau et al. 2010). Colicin M producer cells are protected by a coexpressed, cognate immunity protein, which is anchored to the periplasmic side of inner membrane and inactivates exogenous and endogenous colicin M molecules (Olschlager et al. 1991).

Final Translocation Step of Nuclease Colicins: Crossing the Cytoplasmic Membrane

Nuclease colicins, once they have reached the periplasm, need to parasitize more of the cell's functions than the pore-forming colicins, in order to transfer their cytotoxic domain across the inner membrane into the cytoplasm (de Zamaroczy and Buckingham 2002). However, to cross the cytoplasmic membrane colicins, do not necessitate any retrograde transport via the Sec or Tat pathways of the target cell (Walker et al. 2007). The first question was whether there is a common mechanism for all nuclease colicins or if it could be specific for those belonging to the A- or B-group, or alternatively for the DNase colicins (E2, E7, E8, E9) or RNase colicins (E3, E5, E6, DF13, and D). The second major question was whether an endoproteolytic processing step, occurring either before or during colicin translocation through the inner membrane so that only the C-terminal catalytic domain is liberated in the cytoplasm. It should be noted that the DNase or RNase catalytic domains alone are sufficient for an efficient catalytic activity in vitro.

Self-Propulsion Across the Inner Membrane Mediated by Channel-Forming Activity of DNase Colicins

Enzymatic A–B toxins (e.g., diphtheria, cholera, pertussis, and anthrax) transport their lethal A-domain into the cytosol with the aid of their B-subunit. The B-domain of diphtheria forms a conducting channel (in planar lipid bilayers), which is used by the toxic ADP-ribosyl transferase to enter the cytoplasm and then kill the mammalian target cell (Oh et al. 1999; Falnes and Sandvig 2000). Thus, the possibility that nuclease colicins have a pore-forming activity, in addition to their nuclease activity, which allows the translocation of the toxic domain across the inner membrane, was investigated. The catalytic domain of the DNase colicin E9 exhibited a nonvoltage-gated, random, channel-forming activity in planar lipid bilayers, which involved considerable conformational changes. Under the same experimental conditions, a similar channel-forming activity was also found with the closely related DNase colicins E2, E7, and E8 (Mosbahi et al. 2002; Mosbahi et al. 2004). These channels, unlike those formed by pore-forming colicins, are not directly responsible for cell killing, but may allow "self-propulsion" of the toxic domains into the cytoplasm, possibly monitored by the electrostatic association of the DNase colicin with the inner membrane (Walker et al. 2007). Indeed, the intrinsic channel-forming activity of the DNase domain was shown to be essential for colicin cytotoxicity. Although an association between the RNase colicin E3 and anionic phospholipid surfaces was also reported (Mosbahi et al. 2006), no r- or t-RNase colicins have been shown to exhibit any channel-forming activity. This suggests that RNase colicins should utilize alternative mechanisms for their final translocation step across the inner membrane.

Proteolytic Cleavage of Nuclease Colicins May Accompany Their Entry into the Cytoplasm

As described above, several colicins remain in tight contact with their receptor and the import machinery, during pore formation in the inner membrane (colicin Ia and A) (Figs. 14.2c and 14.3a) (Benedetti et al. 1992; Duche et al. 1995; Wiener et al.

1997) or when the nuclease domain (colicins E2 and E3) enters the cytoplasm (Fig. 14.2a) (Kurisu et al. 2003; Duche 2007; Sharma et al. 2007). This is a strong argument in favor of an endoproteolytic cleavage step prior to or during the release of the C-terminal toxic domain of nuclease colicins into the cytoplasm. Such a cleavage allows the DNase domain, compared to the entire colicin molecule, displays a net global positive charge, essential for the attachment to the anionic phospholipid of the inner membrane (Walker et al. 2007). Moreover, the short lifetime of the translocation channel formed by the DNase colicins in the cytoplasmic membrane, compared to those made by pore-forming colicins (milliseconds rather than seconds) (Mosbahi et al. 2002), is consistent with the cleavage of the catalytic domain from the rest of colicin molecule.

There is good genetic evidence that tRNase colicin D and the DNase colicin E7 undergo proteolytic processing during penetration into the cell such that only the C-terminal catalytic domain of the molecule reaches the cytoplasm. Specific point mutations were found, which altered the cleavage of these colicins in vitro. These mutations did not affect the nucleolytic activities in vitro, but did abolish cytotoxicity in vivo (de Zamaroczy et al. 2001; Shi et al. 2005). Similarly, a point mutation introduced at the putative cleavage site located upstream of the nuclease domain of colicins E2 caused a partial loss of cytotoxicity (Sharma et al. 2007) and seems to support the processing model.

The inner membrane signal peptidase LepB (Dalbey and Wickner 1985) was shown to be specifically required for cell killing by colicin D (Fig. 14.3c). Moreover, in the presence of a cell extract, a LepB-dependent cleavage occurred in vitro quickly (30–60 min) on the colicin D molecule freed of the immunity protein. This indicates that ImmD, in addition to its canonical role in the inhibition of the toxic tRNase activity, is also involved in protecting the colicin molecule against proteolytic cleavage during its secretion. However, the inability of purified LepB to cleave colicin D indicated the need for another membrane or periplasmic protease, which is presumably responsible for the catalysis of endoproteolytic cleavage (de Zamaroczy et al. 2001). The role of LepB in cell killing by colicin D may be structural so that it is necessary to maintain this toxin in a form recognized by the true endoproteolytic enzyme. Such a chaperone-like function was already reported for another membrane protease, FtsH (Schumann 1999). No enzyme has yet been identified for the in vitro proteolytic cleavage of the other nuclease colicins, such as colicins E2, E3, and E7, which do not require LepB (de Zamaroczy and Buckingham 2002 and unpublished data). In the presence of concentrated periplasmic proteins, a set of C-terminal cleavage products derived from colicin E7 (in complex with its Imm protein) were observed in vitro but only after a very long incubation time (8–12 h) (Liao et al. 2001; Shi et al. 2005). The intracellular presence of some cleaved colicin E7 forms was also reported by these authors. Nevertheless, such long-time incubation and the observation that colicin D and more recently colicin E2 (i.e., a DNase colicin closely related to colicin E7) are *de facto* efficiently protected from endoproteolytic degradation when they are complexed to their Imm proteins (de Zamaroczy et al. 2001; Duche et al. 2009) renders it difficult to understand the biological significance of the cleavage results obtained with colicin E7 in the presence of its Imm protein.

Among the currently known nuclease colicins, the vast majority (including A-group DNase and RNase colicins E, the B-group tRNase colicin D, the rRNase cloacin DF13, and the tRNase klebicin D, but not the rRNase klebicin C) have an approximately 30-amino-acid long conserved sequence overlapping the end of the central domain and the linker region at the start of the C-terminal catalytic domain (Fig. 14.1b1, b3). The sequence identity of this consensus is very high among the DNase colicins E but is less pronounced in the case of rRNase colicins E, tRNase colicin D, and klebicin D. Significantly, this consensus sequence is absent from all other colicins, notably the pore-forming species. The consensus is rich in lysine and arginine and located near the presumed endoproteolytic cleavage site of colicins D, E7, and E2. Thus, we, and others, have proposed that proteolytic processing may be a common step in cell killing by nuclease-type colicins, prior to or concomitant with the translocation of the catalytic domain across the inner membrane (de Zamaroczy et al. 2001; de Zamaroczy and Buckingham 2002; Shi et al. 2005; Sharma et al. 2007).

Following earlier observations (Krone et al. 1986), the outer membrane endopeptidase OmpT was shown to cleave receptor-bound colicins, and it is thought to improve bacterial survival in the presence of colicins or antimicrobial peptides (Cavard and Lazdunski 1990; Vandeputte-Rutten et al. 2001). When incubated with *E. coli*, OmpT cleaves the receptor-bound colicin E1, in its N-terminal translocation domain, and colicin E2, in the C-terminal catalytic domain, as detected by the presence of cleaved colicin forms in the supernatant (Masi et al. 2007; Duche et al. 2009). Nuclease colicins E3, E7, and D are cleaved in vitro in the presence of purified OmpT, upstream of their catalytic domain (de Zamaroczy et al. unpublished data). However, these cleavages appear to have no relationship to colicin import, since the inactivation of the *ompT* structural gene has no effect on colicin toxicity. This implies that the proteolytic activity responsible for colicin processing should be dependent on another periplasmic or inner membrane protease. This protease may be essential, since no candidate was identified during the screening of Keio library of nonessential genes (Sharma et al. 2009).

Translocation of All Nuclease Colicins Through the Inner Membrane Needs the ATPase/protease FtsH

In addition to the various effectors described above, an important new enzyme has recently joined the list of those affecting toxicity. The inner membrane-bound, ATP-dependent, Zn-metalloendoprotease, FtsH, belonging to the AAA^+ family (ATPases associated with diverse cellular activities) was found to be required for toxicity of all nuclease colicins (Walker et al. 2007). In contrast with earlier results (Matsuzawa et al. 1984; Qu et al. 1996), FtsH was found not to affect the sensitivity of either pore-forming colicins or colicin M (Walker et al. 2007). FtsH is universally conserved in eubacteria, chloroplasts, and mitochondria and is essential for maintaining the LPS and phospholipid ratio in *E. coli*. In its absence, lipid A (and LPS) are overproduced due to the loss of the FtsH-mediated cleavage of LpxC, a key enzyme of LPS regulation, which leads to cell death (Ogura et al. 1999). FtsH has a limited spectrum of known substrates. In the presence of Zn^{2+} and ATP, FtsH degrades soluble short-lived regulatory proteins (e.g., the alternative sigma factor σ32 in *E. coli*), and more generally FtsH orthologs, dislocate misfolded, unassembled, or damaged membrane proteins by "pulling" them out of the lipid bilayer into the cytoplasm. Thus, it has an essential function in quality and regulatory control of protein turnover and processing (Striebel et al. 2009). The proteolytic reaction itself is not energy-driven. ATP hydrolysis may be required for conformational rearrangements that provide the mechanical force to allow the substrate access to the FtsH active site and then translocate it through the central pore formed by the homohexameric FtsH complex (Bieniossek et al. 2009). The cleavage specificity of FtsH is weak, and its endoproteolytic reaction is processive. Proteins are completely degraded from the starting point of the hydrolysis (Ito and Akiyama 2005).

The function of FtsH in colicin import is still unclear. It might be direct, acting as the protease required to liberate the nuclease domain, since both the ATPase and protease activities of FtsH are required for the toxicity of nuclease colicins (Figs. 14.2a and 14.3c) irrespective of whether they are dependent on the Tol or Ton system for their translocation across the outer membrane (Walker et al. 2007). Although the unfolding activity of FtsH has been shown to be essential for the proteolytic activity (Bieniossek et al. 2009), previous data obtained both in vivo and in vitro led to the conclusion that the unfoldase activity of FtsH is low. This opens the possibility that FtsH may especially recognize (partially) unfolded proteins because they exhibit a reduced intrinsic thermodynamic stability (Herman et al. 2003). The interaction between the nuclease domains of group-A colicins and anionic phospholipid vesicles (which mimic the cytoplasmic membrane) has been investigated by ITFE (intrinsic tryptophan fluorescence emission) and CDS (circular dichroism spectroscopy) (Mosbahi et al. 2004; Mosbahi et al. 2006). These experiments suggested that colicin molecules in contact with the cytoplasmic membrane may be partially unfolded and might therefore be recognized by FtsH as a possible "substrate". Assuming that the nuclease colicins are direct targets of FtsH, the subsequent cleavage and translocation of their C-terminal nuclease domain across the inner membrane could also be mediated by FtsH. In the cytoplasm it might require a final refolding step, which is thought to take place spontaneously and rapidly, as has been observed in vitro with whole colicins D and E3 or their RNase domains alone (Ogawa et al. 1999; de Zamaroczy et al. 2001 and unpublished data).

Conclusion

The details of mechanisms for the import of colicins have considerably diversified within each of the two main Tol and Ton pathways. This could mean that each colicin has individually optimized its import pathway, but this hypothesis is not consistent with either the high level of conservation in the domain organization of colicins or the high level of similarity characterizing the crystal structures of A-group colicins. However, because the colicins are highly efficient killing machines, such an optimization, which logically should have been reached during the evolution from a common conserved mechanism, may not be necessary. The fact that colicins are highly toxic indeed may compensate for their weaker capacity to enter the cell. In fact, both the pore-forming and nuclease-killing activity may be lethal, when only one or at most a few colicin molecules reach their cellular target (Pugsley 1984; Cascales et al. 2007). This supports the hypothesis that the yield of the translocated colicins which reach their target is rather low, probably due to the large molecular size of these toxic "suicide-proteins". However to override the natural impermeability of the cell envelope, all colicins start by adhering to an outer membrane receptor, which is present in greater numbers (e.g., the number of FepA receptors can range from about 500 to 15,000 under iron-excess or iron-limited conditions, respectively) (Higgs et al. 2002). Another factor affecting the fratricide killing efficiency is the time required for the colicin to reach its target. It is possible to detect killing activity by pore-forming colicins within a minute (colicin A) (Duche et al. 1995), while for the nuclease colicins, it can take as long as 20 min (colicin E2; Duche 2007). It is tempting to speculate that the difference is due to the necessity of the nuclease colicins to cross the inner membrane, but the fact that the release of ImmE2 protein from the DNase domain at the cell surface necessitates a similar long period (Duche et al. 2006) suggests that other factors may be responsible for the longer time delay before cell killings occur by nuclease colicins.

Acknowledgements We gratefully thank J. Plumbridge for many helpful comments and in-depth discussions on the manuscript. We also thank J.-C. Lazzaroni and L. Mora for critical reading of the review. This work was supported by the Centre National de la Recherche Scientifique (UPR 9073).

References

Abergel C, Bouveret E, Claverie JM et al (1999) Structure of the *Escherichia coli* TolB protein determined by MAD methods at 1.95 A resolution. Structure 7:1291–1300

Alcaraz A, Nestorovich EM, Aguilella-Arzo M et al (2004) Salting out the ionic selectivity of a wide channel: the asymmetry of OmpF. Biophys J 87:943–957

Baboolal TG, Conroy MJ, Gill K et al (2008) Colicin N binds to the periphery of its receptor and translocator, outer membrane protein F. Structure 16:371–379

Baquero F, Bouanchaud D, Martinez-Perez MC et al (1978) Microcin plasmids: a group of extrachromosomal elements coding for low-molecular-weight antibiotics in *Escherichia coli*. J Bacteriol 135:342–347

Barreteau H, Bouhss A, Gerard F et al (2010) Deciphering the catalytic domain of colicin M, a peptidoglycan lipid II degrading enzyme. J Biol Chem 285:12378–12389

Bell PE, Nau CD, Brown JT et al (1990) Genetic suppression demonstrates interaction of TonB protein with outer membrane transport proteins in *Escherichia coli*. J Bacteriol 172:3826–3829

Benedetti H, Frenette M, Baty D et al (1989) Comparison of the uptake systems for the entry of various BtuB group colicins into *Escherichia coli*. J Gen Microbiol 135:3413–3420

Benedetti H, Frenette M, Baty D et al (1991a) Individual domains of colicins confer specificity in colicin uptake, in pore-properties and in immunity requirement. J Mol Biol 217:429–439
Benedetti H, Lazdunski C, Lloubes R (1991b) Protein import into *Escherichia coli*: colicins A and E1 interact with a component of their translocation system. EMBO J 10:1989–1995
Benedetti H, Lloubes R, Lazdunski C et al (1992) Colicin A unfolds during its translocation in *Escherichia coli* cells and spans the whole cell envelope when its pore has formed. EMBO J 11:441–447
Bernstein A, Rolfe B, Onodera K (1972) Pleiotropic properties and genetic organization of the tolA, B locus of *Escherichia coli* K-12. J Bacteriol 112:74–83
Bieniossek C, Niederhauser B, Baumann UM (2009) The crystal structure of apo-FtsH reveals domain movements necessary for substrate unfolding and translocation. Proc Natl Acad Sci USA 106:21579–21584
Bonsor DA, Hecht O, Vankemmelbeke M et al (2009) Allosteric beta-propeller signalling in TolB and its manipulation by translocating colicins. EMBO J 28:2846–2857
Bourdineaud JP, Fierobe HP, Lazdunski C et al (1990) Involvement of OmpF during reception and translocation steps of colicin N entry. Mol Microbiol 4:1737–1743
Bouveret E, Rigal A, Lazdunski C et al (1998) Distinct regions of the colicin A translocation domain are involved in the interaction with TolA and TolB proteins upon import into *Escherichia coli*. Mol Microbiol 27:143–157
Bouveret E, Benedetti H, Rigal A et al (1999) *In vitro* characterization of peptidoglycan-associated lipoprotein (PAL)-peptidoglycan and PAL–TolB interactions. J Bacteriol 181:6306–6311
Bowman CM, Sidikaro J, Nomura M (1971) Specific inactivation of ribosomes by colicin E3 *in vitro* and mechanism of immunity in colicinogenic cells. Nat New Biol 234:133–137
Braun V, Herrmann C (1993) Evolutionary relationship of uptake systems for biopolymers in *Escherichia coli*: cross-complementation between the TonB–ExbB–ExbD and the TolA–TolQ–TolR proteins. Mol Microbiol 8:261–268
Braun V, Gaisser S, Herrmann C et al (1996) Energy-coupled transport across the outer membrane of *Escherichia coli*: ExbB binds ExbD and TonB *in vitro*, and leucine 132 in the periplasmic region and aspartate 25 in the transmembrane region are important for ExbD activity. J Bacteriol 178:2836–2845
Braun V, Patzer SI, Hantke K (2002) Ton-dependent colicins and microcins: modular design and evolution. Biochimie 84:365–380
Buchanan SK, Smith BS, Venkatramani L et al (1999) Crystal structure of the outer membrane active transporter FepA from *Escherichia coli*. Nat Struct Biol 6:56–63
Buchanan SK, Lukacik P, Grizot S et al (2007) Structure of colicin I receptor bound to the R-domain of colicin Ia: implications for protein import. EMBO J 26:2594–2604
Cadieux N, Kadner RJ (1999) Site-directed disulfide bonding reveals an interaction site between energy-coupling protein TonB and BtuB, the outer membrane cobalamin transporter. Proc Natl Acad Sci USA 96:10673–10678
Carr S, Penfold CN, Bamford V et al (2000) The structure of TolB, an essential component of the tol-dependent translocation system, and its protein-protein interaction with the translocation domain of colicin E9. Structure Fold Des 8:57–66
Cascales E, Gavioli M, Sturgis JN et al (2000) Proton motive force drives the interaction of the inner membrane TolA and outer membrane pal proteins in *Escherichia coli*. Mol Microbiol 38:904–915
Cascales E, Lloubes R (2004) Deletion analyses of the peptidoglycan-associated lipoprotein Pal reveals three independent binding sequences including a TolA box. Mol Microbiol 51:873–885
Cascales E, Buchanan SK, Duche D et al (2007) Colicin biology. Microbiol Mol Biol Rev 71:158–229
Cavard D, Lazdunski C (1990) Colicin cleavage by OmpT protease during both entry into and release from *Escherichia coli* cells. J Bacteriol 172:648–652
Chang C, Mooser A, Pluckthun A et al (2001) Crystal structure of the dimeric C-terminal domain of TonB reveals a novel fold. J Biol Chem 276:27535–27540

Chavan M, Rafi H, Wertz J et al (2005) Phage associated bacteriocins reveal a novel mechanism for bacteriocin diversification in *Klebsiella*. J Mol Evol 60:546–556

Cheng YS, Shi Z, Doudeva LG et al (2006) High-resolution crystal structure of a truncated ColE7 translocation domain: implications for colicin transport across membranes. J Mol Biol 356:22–31

Cherezov V, Yamashita E, Liu W et al (2006) *In meso* structure of the cobalamin transporter, BtuB, at 1.95 A resolution. J Mol Biol 364:716–734

Chimento DP, Kadner RJ, Wiener MC (2003a) The *Escherichia coli* outer membrane cobalamin transporter BtuB: structural analysis of calcium and substrate binding, and identification of orthologous transporters by sequence/structure conservation. J Mol Biol 332:999–1014

Chimento DP, Mohanty AK, Kadner RJ et al (2003b) Substrate-induced transmembrane signaling in the cobalamin transporter BtuB. Nat Struct Biol 10:394–401

Clavel T, Germon P, Vianney A et al (1998) TolB protein of *Escherichia coli* K-12 interacts with the outer membrane peptidoglycan-associated proteins Pal, Lpp and OmpA. Mol Microbiol 29:359–367

Cotter PD, Hill C, Ross RP (2005) Bacteriocins: developing innate immunity for food. Nat Rev Microbiol 3:777–788

Cowan SW, Schirmer T, Rummel G et al (1992) Crystal structures explain functional properties of two *E. coli* porins. Nature 358:727–733

Dalbey RE, Wickner W (1985) Leader peptidase catalyzes the release of exported proteins from the outer surface of the *Escherichia coli* plasma membrane. J Biol Chem 260:15925–15931

de Zamaroczy M, Mora L, Lecuyer A et al (2001) Cleavage of colicin D is necessary for cell killing and requires the inner membrane peptidase LepB. Mol Cell 8:159–168

de Zamaroczy M, Buckingham RH (2002) Importation of nuclease colicins into *E. coli* cells: endoproteolytic cleavage and its prevention by the Immunity protein. Biochimie 84:423–432

Derouiche R, Zeder-Lutz G, Bénédetti H et al (1997) Binding of colicins A and E1 to purified TolA domains. Microbiology 143:3185–3192

Devanathan S, Postle K (2007) Studies on colicin B translocation: FepA is gated by TonB. Mol Microbiol 65:441–453

Duche D, Letellier L, Geli V et al (1995) Quantification of group A colicin import sites. J Bacteriol 177:4935–4939

Duche D, Frenkian A, Prima V et al (2006) Release of immunity protein requires functional endonuclease colicin import machinery. J Bacteriol 188:8593–8600

Duche D (2007) Colicin E2 is still in contact with its receptor and import machinery when its nuclease domain enters the cytoplasm. J Bacteriol 189:4217–4222

Duche D, Issouf M, Lloubes R (2009) Immunity protein protects colicin E2 from OmpT protease. J Biochem 145:95–101

Eisenhauer HA, Shames S, Pawelek PD et al (2005) Siderophore transport through *Escherichia coli* outer membrane receptor FhuA with disulfide-tethered cork and barrel domains. J Biol Chem 280:30574–30580

El Ghachi M, Bouhss A, Barreteau H et al (2006) Colicin M exerts its bacteriolytic effect via enzymatic degradation of undecaprenyl phosphate-linked peptidoglycan precursors. J Biol Chem 281:22761–22772

Endriss F, Braun M, Killmann H et al (2003) Mutant analysis of the *Escherichia coli* FhuA protein reveals sites of FhuA activity. J Bacteriol 185:4683–4692

Falnes PO, Sandvig K (2000) Penetration of protein toxins into cells. Curr Opin Cell Biol 12:407–413

Ferguson AD, Hofmann E, Coulton JW et al (1998) Siderophore-mediated iron transport: crystal structure of FhuA with bound lipopolysaccharide. Science 282:2215–2220

Gaspar JA, Thomas JA, Marolda CL et al (2000) Surface expression of O-specific lipopolysaccharide in *Escherichia coli* requires the function of the TolA protein. Mol Microbiol 38:262–275

Geli V, Lazdunski C (1992) An alpha-helical hydrophobic hairpin as a specific determinant in protein-protein interaction occurring in *Escherichia coli* colicin A and B immunity systems. J Bacteriol 174:6432–6437

Gerding MA, Ogata Y, Pecora ND et al (2007) The trans-envelope Tol-Pal complex is part of the cell division machinery and required for proper outer-membrane invagination during cell constriction in *E. coli.* Mol Microbiol 63:1008–1025

Germon P, Ray MC, Vianney A et al (2001) Energy-dependent conformational change in the TolA protein of *Escherichia coli* involves its N-terminal domain, TolQ, and TolR. J Bacteriol 183:4110–4114

Ghosh J, Postle K (2005) Disulphide trapping of an *in vivo* energy-dependent conformation of *Escherichia coli* TonB protein. Mol Microbiol 55:276–288

Goldman K, Suit JL, Kayalar C (1985) Identification of the plasmid-encoded immunity protein for colicin E1 in the inner membrane of *Escherichia coli.* FEBS Lett 190:319–323

Graille M, Mora L, Buckingham RH et al (2004) Structural inhibition of the colicin D tRNase by the tRNA-mimicking immunity protein. EMBO J 23:1474–1482

Gratia A, Fredericq P (1946) Diversités des souches antibiotiques de *Bacterium coli* et étendue variable de leur champ d'action. C R Soc Biol (Paris) 140:1032–1033

Guihard G, Benedetti H, Besnard M et al (1993) Phosphate efflux through the channels formed by colicins and phage T5 in *Escherichia coli* cells is responsible for the fall in cytoplasmic ATP. J Biol Chem 268:17775–17780

Guihard G, Boulanger P, Benedetti H et al (1994) Colicin A and the Tol proteins involved in its translocation are preferentially located in the contact sites between the inner and outer membranes of *Escherichia coli* cells. J Biol Chem 269:5874–5880

Hands SL, Holland LE, Vankemmelbeke M et al (2005) Interactions of TolB with the translocation domain of colicin E9 require an extended TolB box. J Bacteriol 187:6733–6741

Hardy KG, Meynell GG (1972) "Induction" of colicin factor E2–P9 by mitomycin C. J Bacteriol 112:1007–1009

Herman C, Prakash S, Lu CZ et al (2003) Lack of a robust unfoldase activity confers a unique level of substrate specificity to the universal AAA protease FtsH. Mol Cell 11:659–669

Higgs PI, Larsen RA, Postle K (2002) Quantification of known components of the *Escherichia coli* TonB energy transduction system: TonB, ExbB, ExbD and FepA. Mol Microbiol 44:271–281

Hilsenbeck JL, Park H, Chen G et al (2004) Crystal structure of the cytotoxic bacterial protein colicin B at 2.5 Å resolution. Mol Microbiol 51:711–720

Housden NG, Loftus SR, Moore GR et al (2005) Cell entry mechanism of enzymatic bacterial colicins: porin recruitment and the thermodynamics of receptor binding. Proc Natl Acad Sci USA 102:13849–13854

Hullmann J, Patzer SI, Romer C et al (2008) Periplasmic chaperone FkpA is essential for imported colicin M toxicity. Mol Microbiol 69:926–937

Ito K, Akiyama Y (2005) Cellular functions, mechanism of action, and regulation of FtsH protease. Annu Rev Microbiol 59:211–231

Jakes KS, Zinder ND (1984) Plasmid ColE3 specifies a lysis protein. J Bacteriol 157:582–590

Jakes KS, Finkelstein A (2010) The colicin Ia receptor, Cir, is also the translocator for colicin Ia. Mol Microbiol 75:567–578

James R, Kleanthous C, Moore GR (1996) The biology of E colicins: paradigms and paradoxes. Microbiology 142:1569–1580

James R, Penfold CN, Moore GR et al (2002) Killing of *E. coli* cells by E group nuclease colicins. Biochimie 84:381–389

Jeanteur D, Schirmer T, Fourel D et al (1994) Structural and functional alterations of a colicin-resistant mutant of OmpF porin from *Escherichia coli.* Proc Natl Acad Sci USA 91:10675–10679

Journet L, Bouveret E, Rigal A et al (2001) Import of colicins across the outer membrane of *Escherichia coli* involves multiple protein interactions in the periplasm. Mol Microbiol 42:331–344

Karlsson M, Hannavy K, Higgins CF (1993) ExbB acts as a chaperone-like protein to stabilize TonB in the cytoplasm. Mol Microbiol 8:389–396

Kleanthous C, Hemmings AM, Moore GR et al (1998) Immunity proteins and their specificity for endonuclease colicins: telling right from wrong in protein–protein recognition. Mol Microbiol 28:227–233

Koronakis V, Li J, Koronakis E et al (1997) Structure of TolC, the outer membrane component of the bacterial type I efflux system, derived from two-dimensional crystals. Mol Microbiol 23:617–626

Koronakis V, Sharff A, Koronakis E et al (2000) Crystal structure of the bacterial membrane protein TolC central to multidrug efflux and protein export. Nature 405:914–919

Krone WJ, de Vries P, Koningstein G et al (1986) Uptake of cloacin DF13 by susceptible cells: removal of immunity protein and fragmentation of cloacin molecules. J Bacteriol 166:260–268

Kurisu G, Zakharov SD, Zhalnina MV et al (2003) The structure of BtuB with bound colicin E3 R-domain implies a translocon. Nat Struct Biol 10:948–954

Lakey JH, Slatin SL (2001) Pore-forming colicins and their relatives. Curr Top Microbiol Immunol 257:131–161

Larsen RA, Thomas MG, Postle K (1999) Protonmotive force, ExbB and ligand-bound FepA drive conformational changes in TonB. Mol Microbiol 31:1809–1824

Lazdunski C, Bouveret E, Rigal A et al (2000) Colicin import into *Escherichia coli* cells requires the proximity of the inner and outer membranes and other factors. Int J Med Microbiol 290:337–344

Lazdunski CJ, Bouveret E, Rigal A et al (1998) Colicin import into *Escherichia coli* cells. J Bacteriol 180:4993–5002

Lazzaroni JC, Portalier RC (1981) Genetic and biochemical characterization of periplasmic-leaky mutants of *Escherichia coli* K-12. J Bacteriol 145:1351–1358

Lazzaroni JC, Vianney A, Popot JL et al (1995) Transmembrane alpha-helix interactions are required for the functional assembly of the *Escherichia coli* Tol complex. J Mol Biol 246:1–7

Lazzaroni JC, Dubuisson JF, Vianney A (2002) The Tol proteins of *Escherichia coli* and their involvement in the translocation of group A colicins. Biochimie 84:391–397

Letain TE, Postle K (1997) TonB protein appears to transduce energy by shuttling between the cytoplasmic membrane and the outer membrane in *Escherichia coli*. Mol Microbiol 24:271–283

Liao CC, Hsiao KC, Liu YW et al (2001) Processing of DNase Domain during Translocation of Colicin E7 across the Membrane of *Escherichia coli*. Biochem Biophys Res Commun 284:556–562

Little JW (1982) Control of the SOS regulatory system by the level of RecA protease. Biochimie 64:585–589

Little JW, Mount DW (1982) The SOS regulatory system of *Escherichia coli*. Cell 29:11–22

Lloubes R, Granger-Schnarr M, Lazdunski C et al (1988) LexA repressor induces operator-dependent DNA bending. J Mol Biol 204:1049–1054

Locher KP, Rees B, Koebnik R et al (1998) Transmembrane signaling across the ligand-gated FhuA receptor: crystal structures of free and ferrichrome-bound states reveal allosteric changes. Cell 95:771–778

Loftus SR, Walker D, Mate MJ et al (2006) Competitive recruitment of the periplasmic translocation portal TolB by a natively disordered domain of colicin E9. Proc Natl Acad Sci USA 103:12353–12358

Masaki H, Toba M, Ohta T (1985) Structure and expression of the ColE2-P9 immunity gene. Nucleic Acids Res 13:1623–1635

Masaki H, Ogawa T (2002) The modes of action of colicins E5 and D, and related cytotoxic tRNases. Biochimie 84:433–438

Masi M, Vuong P, Humbard M et al (2007) Initial steps of colicin E1 import across the outer membrane of *Escherichia coli*. J Bacteriol 189:2667–2676

Matsuzawa H, Ushiyama S, Koyama Y et al (1984) *Escherichia coli* K-12 tolZ mutants tolerant to colicins E2, E3, D, Ia, and Ib: defect in generation of the electrochemical proton gradient. J Bacteriol 160:733–739

Mende J, Braun V (1990) Import-defective colicin B derivatives mutated in the TonB box. Mol Microbiol 4:1523–1533

Misra R, Reeves PR (1987) Role of micF in the *tolC*-mediated regulation of OmpF, a major outer membrane protein of *Escherichia coli* K-12. J Bacteriol 169:4722–4730
Moeck GS, Letellier L (2001) Characterization of in vitro interactions between a truncated TonB protein from *Escherichia coli* and the outer membrane receptors FhuA and FepA. J Bacteriol 183:2755–2764
Mora L, Diaz N, Buckingham RH et al (2005) Import of the transfer RNase colicin D requires site-specific interaction with the energy-transducing protein TonB. J Bacteriol 187:2693–2697
Mora L, Klepsch M, Buckingham RH et al (2008) Dual roles of the central domain of colicin D tRNase in TonB-mediated import and in immunity. J Biol Chem 283:4993–5003
Mosbahi K, Lemaitre C, Keeble AH et al (2002) The cytotoxic domain of colicin E9 is a channel-forming endonuclease. Nat Struct Biol 9:476–484
Mosbahi K, Walker D, Lea E et al (2004) Destabilization of the colicin E9 Endonuclease domain by interaction with negatively charged phospholipids: implications for colicin translocation into bacteria. J Biol Chem 279:22145–22151
Mosbahi K, Walker D, James R et al (2006) Global structural rearrangement of the cell penetrating ribonuclease colicin E3 on interaction with phospholipid membranes. Protein Sci 15:620–627
Ogawa T, Tomita K, Ueda T et al (1999) A cytotoxic ribonuclease targeting specific transfer RNA anticodons. Science 283:2097–2100
Ogierman M, Braun V (2003) Interactions between the outer membrane ferric citrate transporter FecA and TonB: studies of the FecA TonB box. J Bacteriol 185:1870–1885
Ogura T, Inoue K, Tatsuta T et al (1999) Balanced biosynthesis of major membrane components through regulated degradation of the committed enzyme of lipid A biosynthesis by the AAA protease FtsH (HflB) in *Escherichia coli.* Mol Microbiol 31:833–844
Oh KJ, Senzel L, Collier RJ et al (1999) Translocation of the catalytic domain of diphtheria toxin across planar phospholipid bilayers by its own T domain. Proc Natl Acad Sci USA 96:8467–8470
Ohno-Iwashita Y, Imahori K (1980) Assignment of the functional loci in colicin E2 and E3 molecules by the characterization of their proteolytic fragments. Biochemistry 19:652–659
Ohno-Iwashita Y, Imahori K (1982) Assignment of the functional loci in the colicin E1 molecule by characterization of its proteolytic fragments. J Biol Chem 257:6446–6451
Olschlager T, Turba A, Braun V (1991) Binding of the immunity protein inactivates colicin M. Mol Microbiol 5:1105–1111
Pawelek PD, Croteau N, Ng-Thow-Hing C et al (2006) Structure of TonB in complex with FhuA, E. coli outer membrane receptor. Science 312:1399–1402
Pilsl H, Glaser C, Gross P et al (1993) Domains of colicin M involved in uptake and activity. Mol Gen Genet 240:103–112
Ponting CP, Pallen MJ (1999) A beta-propeller domain within TolB. Mol Microbiol 31:739–740
Postle K, Good RF (1983) DNA sequence of the *Escherichia coli tonB* gene. Proc Natl Acad Sci USA 80:5235–5239
Postle K, Kadner RJ (2003) Touch and go: tying TonB to transport. Mol Microbiol 49:869–882
Prouty AM, Van Velkinburgh JC, Gunn JS (2002) *Salmonella enterica serovar typhimurium* resistance to bile: identification and characterization of the *tolQRA* cluster. J Bacteriol 184:1270–1276
Pugsley AP (1984) The ins and outs of colicins. Part I: Production, and translocation across membranes. Microbiol Sci 1:168–175
Qu JN, Makino SI, Adachi H et al (1996) The *tolZ* gene of *Escherichia coli* is identified as the ftsH gene. J Bacteriol 178:3457–3461
Ray MC, Germon P, Vianney A et al (2000) Identification by genetic suppression of *Escherichia coli* TolB residues important for TolB–Pal interaction. J Bacteriol 182:821–824
Riley MA, Wertz JE (2002) Bacteriocin diversity: ecological and evolutionary perspectives. Biochimie 84:357–364
Roos U, Harkness RE, Braun V (1989) Assembly of colicin genes from a few DNA fragments. Nucleotide sequence of colicin D. Mol Microbiol 3:891–902

Sauter A, Howard SP, Braun V (2003) *In vivo* evidence for TonB dimerization. J Bacteriol 185:5747–5754

Schaller K, Holtje JV, Braun V (1982) Colicin M is an inhibitor of murein biosynthesis. J Bacteriol 152:994–1000

Schendel SL, Click EM, Webster RE et al (1997) The TolA protein interacts with colicin E1 differently than with other group A colicins. J Bacteriol 179:3683–3690

Schumann W (1999) FtsH – a single-chain charonin? FEMS Microbiol Rev 23:1–11

Senior BW, Holland IB (1971) Effect of colicin E3 upon the 30S ribosomal subunit of *Escherichia coli*. Proc Natl Acad Sci USA 68:959–963

Sharma O, Cramer WA (2007) Minimum length requirement of the flexible N-terminal translocation subdomain of colicin E3. J Bacteriol 189:363–368

Sharma O, Yamashita E, Zhalnina MV et al (2007) Structure of the complex of the colicin E2 R-domain and its BtuB receptor. The outer membrane colicin translocon. J Biol Chem 282:23163–23170

Sharma O, Datsenko KA, Ess SC et al (2009) Genome-wide screens: novel mechanisms in colicin import and cytotoxicity. Mol Microbiol 73:571–585

Shi Z, Chak KF, Yuan HS (2005) Identification of an essential cleavage site in ColE7 required for import and killing of cells. J Biol Chem 280:24663–24668

Shultis DD, Purdy MD, Banchs CN et al (2006) Outer membrane active transport: structure of the BtuB:TonB complex. Science 312:1396–1399

Skare JT, Ahmer BM, Seachord CL et al (1993) Energy transduction between membranes. TonB, a cytoplasmic membrane protein, can be chemically cross-linked *in vivo* to the outer membrane receptor FepA. J Biol Chem 268:16302–16308

Smallwood CR, Marco AG, Xiao Q et al (2009) Fluoresceination of FepA during colicin B killing: effects of temperature, toxin and TonB. Mol Microbiol 72:1171–1180

Soelaiman S, Jakes K, Wu N et al (2001) Crystal structure of colicin E3: implications for cell entry and ribosome inactivation. Mol Cell 8:1053–1062

Striebel F, Kress W, Weber-Ban E (2009) Controlled destruction: AAA + ATPases in protein degradation from bacteria to eukaryotes. Curr Opin Struct Biol 19:209–217

Sturgis JN (2001) Organisation and evolution of the *tol-pal* gene cluster. J Mol Microbiol Biotechnol 3:113–122

Vakharia HL, Postle K (2002) FepA with globular domain deletions lacks activity. J Bacteriol 184:5508–5512

van der Wal FJ, Luirink J, Oudega B (1995) Bacteriocin release proteins: mode of action, structure, and biotechnological application. FEMS Microbiol Rev 17:381–399

Vandeputte-Rutten L, Kramer RA, Kroon J et al (2001) Crystal structure of the outer membrane protease OmpT from *Escherichia coli* suggests a novel catalytic site. EMBO J 20:5033–5039

Vankemmelbeke M, Zhang Y, Moore GR et al (2009) Energy-dependent immunity protein release during tol-dependent nuclease colicin translocation. J Biol Chem 284:18932–18941

Walburger A, Lazdunski C, Corda Y (2002) The Tol/Pal system function requires an interaction between the C-terminal domain of TolA and the N-terminal domain of TolB. Mol Microbiol 44:695–708

Walker D, Mosbahi K, Vankemmelbeke M et al (2007) The role of electrostatics in colicin nuclease domain translocation into bacterial cells. J Biol Chem 282:31389–31397

Wiener M, Freymann D, Ghosh P et al (1997) Crystal structure of colicin Ia. Nature 385:461–464

Wiener MC (2005) TonB-dependent outer membrane transport: going for Baroque? Curr Opin Struct Biol 15:394–400

Yamashita E, Zhalnina MV, Zakharov SD et al (2008) Crystal structures of the OmpF porin: function in a colicin translocon. EMBO J 27:2171–2180

Zakharov SD, Eroukova VY, Rokitskaya TI et al (2004) Colicin occlusion of OmpF and TolC channels: outer membrane translocons for colicin import. Biophys J 87:3901–3911

Zakharov SD, Zhalnina MV, Sharma O et al (2006) The colicin E3 outer membrane translocon: immunity protein release allows interaction of the cytotoxic domain with OmpF porin. Biochemistry 45:10199–10207

Zakharov SD, Sharma O, Zhalnina MV et al (2008) Primary events in the colicin translocon: FRET analysis of colicin unfolding initiated by binding to BtuB and OmpF. Biochemistry 47:12802–12809

Zeth K, Romer C, Patzer SI et al (2008) Crystal structure of colicin M, a novel phosphatase specifically imported by *Escherichia coli*. J Biol Chem 283:25324–25331

Zhang Y, Vankemmelbeke MN, Holland LE et al (2008) Investigating early events in receptor binding and translocation of colicin E9 using synchronized cell killing and proteolytic cleavage. J Bacteriol 190:4342–4350

Zhang Y, Li C, Vankemmelbeke MN et al (2010) The crystal structure of the TolB box of colicin A in complex with TolB reveals important differences in the recruitment of the common TolB translocation portal used by group A colicins. Mol Microbiol 75:623–636

Chapter 15
Class I Microcins: Their Structures, Activities, and Mechanisms of Resistance

Konstantin Severinov, Ekaterina Semenova, and Teymur Kazakov

Abstract Microcin J25, microcin B17, and microcin C7-C51 are the three known members of class I posttranslationally modified microcins (heavily posttranslationally modified antibacterial peptides produced by Enterobacteriaceae with molecular weights of less than 5 kDa). The three microcins are unrelated to each other; they have structures that are highly atypical for ribosomally synthesized peptides and target essential molecular machines that are validated drug targets. In this chapter, available data on mechanisms of action, structure–activity relationships, and immunity mechanisms for class I microcins and related compounds are discussed.

Introduction

Microcins are small ribosomally synthesized peptide antibiotics produced by Enterobacteriaceae and active against closely related bacterial species (Baquero et al. 1978). Microcins are extremely diverse, and most are not related to each other. Microcin synthesis is sharply activated when cells reach stationary phase or when they encounter conditions of nutrient limitation (Hernández-Chico et al. 1986; Connell et al. 1987; Chiuchiolo et al. 2001; Fomenko et al. 2001). Microcin-producing cells are resistant to the microcin they produce. The resistance appears to manifest itself even during exponential growth, i.e., resistance determinants are

K. Severinov (✉)
Waksman Institute for Microbiology, Rutgers, The State University of New Jersey, 190 Frelinghuysen Road, Piscataway, NJ 08854, USA
and
Department of Molecular Biology and Biochemistry, Rutgers, The State University of New Jersey, Piscataway, NJ 08854, USA
and
Institutes of Molecular Genetics and Gene Biology, Russian Academy of Sciences, Moscow 123182, Russia
e-mail: severik@waksman.rutgers.edu

D. Drider and S. Rebuffat (eds.), *Prokaryotic Antimicrobial Peptides: From Genes to Applications*, DOI 10.1007/978-1-4419-7692-5_15,

expressed even when microcins are not produced. Thus, gene clusters coding for microcin synthesis and immunity determinants may be adaptive and allow carrier cells to kill microcin-sensitive cells and to make more resources available. Unlike the case of cells producing bacteriocins, which have higher molecular weight than microcins, microcin-producing cells are capable of continuous toxin production without the loss of viability. The continuous production is ensured by dedicated systems that transport microcins through membranes of producing cells.

The diversity of microcins makes their rational classification difficult. Formerly, microcins were subdivided into two broad groups: those that were active directly as ribosomally synthesized peptides ("unmodified" microcins) and others that required extensive posttranslational modification of ribosomally synthesized precursor (promicrocin) for activity. A more refined classification of the latter group has recently been proposed (Duquesne et al. 2007a). Class I posttranslationally modified microcins are peptides with molecular weight of less than 5 kDa that are subject to very heavy posttranslational modifications. Class II microcins range in size from 5 to 10 kDa and contain disulfide bonds (class IIa) or C-terminal modifications (class IIb). Class I consists of three structurally unrelated compounds: microcin J25 (MccJ25), microcin B17 (MccB17), and microcin C7 (also known as microcin C51, MccC7-C51). Multiple occurrences of each of these three class I microcins around the world were reported. Thus, class I microcins genes appear to have a global spread, and the three basic compounds may likely exhaust the variety of this class of molecules in Enterobacteriaceae. In this chapter, available data on class I microcins' mechanisms of action, structure–activity relationships, and mechanisms contributing to microcin immunity are discussed.

Microcin J25

MccJ25 (Fig. 15.1) is a 21-amino-acid peptide produced from 58-amino-acid-long pro-MccJ25, a product of the *mcjA* gene (Solbiati et al. 1999). Pro-MccJ25 consists of a 37-amino-acid N-terminal leader and the C-terminal part that becomes MccJ25. MccJ25 is highly resistant to proteolysis and even withstands autoclaving without loss of activity (Salomón and Farías 1992; Blond et al. 1999). The structure of MccJ25 explains this exceptional stability. The amino group of MccJ25 Gly^1 is linked to the side-chain carboxyl of Glu^8 by a lactam bond, creating a ring. The C-terminal tail of MccJ25 forms a two-β-strand loop, the end of which passes through the ring to form a threaded-lasso structure (Bayro et al. 2003; Rosengren et al. 2003; Wilson et al. 2003). The side chains of Phe^{19} and Tyr^{20} at the end of the tail straddle the ring and prevent the escape of the tail from the ring. Extensive digestion of MccJ25 with thermolysin results in cleavage of the polypeptide chain between amino acids 10 and 11, while acid hydrolysis results in a derivative lacking loop amino acids 13–17 (Blond et al. 2002; Rosengren et al. 2004). Both of these compounds contain two separate polypeptide chains that remain stably associated with each other due to the lock formed by Phe^{19} and Tyr^{20}.

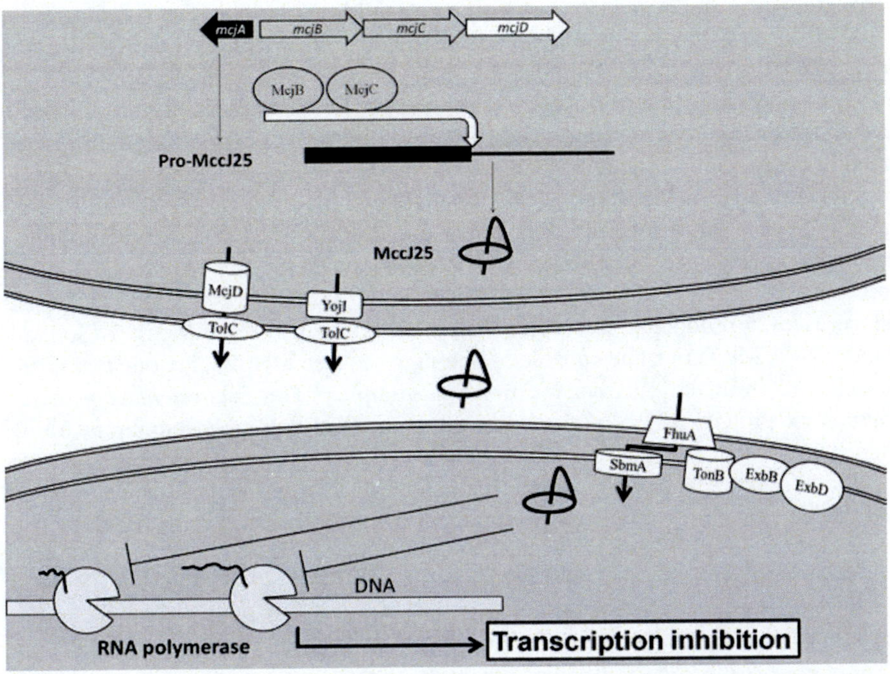

Fig. 15.1 MccJ25: maturation, export, import, and mechanism of action. The schematic summarizes available data on various aspects of MccJ25 discussed in the text. The *shaded areas* bordered by two curved *double lines* indicate MccJ25-producing cell (*top*) and MccJ25-sensitive cell (*bottom*). The structure of the *mcj* locus is shown with genes indicated by *arrows*. *Black arrow* denotes a gene coding for MccJ25 precursor; *light gray arrows* indicate genes whose products are involved in MccJ25 production/maturation, *white arrows* – a gene whose product is involved in MccJ25 export. The N-terminal MccJ25 leader in the linear pro-MccJ25 precursor is shown by a *thicker line*

The *mcjB* and *mcjC* gene products are required for production of MccJ25 from pro-MccJ25 in vivo and in vitro (Solbiati et al. 1996; Duquesne et al. 2007b). While precise biochemical functions of McjB and McjC remain unknown, sequence comparisons indicated that McjB is a protease and may therefore cleave the peptide bond joining McjA leader with the C-terminal sequence that becomes mature MccJ25. McjC is similar to class B asparagine synthetases, enzymes that convert aspartic acid to asparagine. Therefore, McjC may be responsible for the formation of an amide bond between the amino group of MccJ25 Gly^1 (generated upon proteolytic cleavage of pro-MccJ25) and the side-chain carboxyl of Glu^8 (Severinov et al. 2007).

The highly basic leader likely serves as a site of recognition by McjB and/or McjC and may also chaperone the MccJ25 part (1) helping it to adopt a conformation that favors the formation of the Gly^1–Glu^8 amide bond and (2) prefolding the MccJ25 part of pro-MccJ25 into essentially mature structure, since the tail with its bulky amino acids cannot be threaded through the ring once it is formed. Recent work has demonstrated that only seven leader amino acids of the leader immediately preceding the MccJ25 sequence are sufficient for MccJ25 production (Cheung et al. 2010). Though MccJ25 production from mutant pro-MccJ25 was significantly decreased, the result argues against the chaperone function of the leader.

The *mcjD* gene codes for an ABC transporter. The joined activity of McjB and McjC (measured by conversion of pro-MccJ25 into MccJ25) is detected in inner membrane fractions (Clarke and Campopiano 2007). Since no membrane-spanning motifs are found in McjB and McjC, it is possible that these proteins form a complex with each other and with the cytoplasmic domain of McjD. Such an arrangement would increase the efficiency of MccJ25 synthesis and export, helping to prevent accumulation of toxic levels of mature MccJ25 in the cytoplasm. However, since MccJ25-producing cells are resistant to externally added MccJ25 (Solbiati et al. 1996), McjD must be capable of pumping out free MccJ25, i.e. one that is not channelled from the putative McjB–McjC complex. The *Escherichia coli* outer membrane protein TolC may form an export complex with inner-membrane McjD and thus participate in MccJ25 secretion (Delgado et al. 1999). Another cellular protein, ABC transporter YojI, contributes to MccJ25 resistance (Delgado et al. 2005). YojI functions as an efflux pump and requires TolC to export MccJ25. The natural substrate of YojI is not known.

MccJ25 intake is mediated by the outer membrane receptor FhuA and the inner membrane proteins TonB, ExbB, ExbD, and SbmA (Salomón and Farías 1993, 1995; Destoumieux-Garzón et al. 2005). The TonB–ExbB–ExbD complex uses the proton-motive force from the cytoplasmic membrane to transduce energy to the outer membrane, thus providing energy for transport. To accomplish energy transfer, TonB contacts the outer membrane receptor FhuA. SbmA transports MccJ25 through the inner membrane (Fig. 15.1).

The intracellular target of MccJ25 is RNA polymerase (RNAP). In vitro, the activity of *E. coli* RNAP is inhibited in the presence of micromolar concentrations of MccJ25 (Delgado et al. 2001). Numerous MccJ25-resistant *rpoC* (RNAP β' subunit) and less frequent *rpoB* (RNAP β subunit) mutants were isolated (Yuzenkova et al. 2002; Mukhopadhyay et al. 2004). RNAP prepared from resistant cells is not inhibited by MccJ25 and unlike wild-type RNAP, MccJ25-resistant RNAP does not bind MccJ25 (Adelman et al. 2004). In the structural model of *Thermus aquaticus* RNAP, amino acids corresponding to *E. coli* RNAP positions that can lead to MccJ25 resistance when mutated are located around the circumference of the secondary channel, a narrow opening that leads from the surface to the active site of the enzyme (Zhang et al. 1999). The channel is thought to direct NTP substrates toward RNAP active site in active transcription elongation complexes and to accept the 3′ end-proximal portion of the nascent transcript in inactive backtracked elongation complexes. Transcript cleavage factors Gre reactivate backtracked complexes by reaching out to the RNAP active centre through the secondary channel and inducing endonucleolytic cleavage that creates a new nascent RNA 3′ end that is aligned with the active site (Stepanova et al. 2009).

The tight clustering of MccJ25-resistance mutations in the secondary channel suggests that MccJ25 binds in the channel and occludes it. Indeed, MccJ25 prevents formation of backtracked transcription complexes, presumably by preventing the threading of the transcript through the channel, and inhibits Gre factor-dependent cleavage, presumably by preventing factor access to RNAP active site (Adelman et al. 2004). MccJ25 also slows down transcription elongation, an effect consistent with its hindering NTP access to the RNAP active site. Single-molecule observations reveal that the presence of MccJ25 causes very long pauses in transcript elongation, interspersed by bursts of elongation at a normal rate (Adelman et al. 2004). Such an effect is consistent with complete blockage of NTP access by bound MccJ25. A cork-in-the-bottle is thus a good analogy of MccJ25 binding in the RNAP secondary channel (Mukhopadhyay et al. 2004).

A systematic structure–activity analysis of MccJ25 has been carried out (Pavlova et al. 2008). A collection of all possible point substitutions at every position of MccJ25, with the exception of Glu^{8} has been created. Each mutant has been characterized, viz., its ability (1) to mature (i.e., the ability of mutant pro-MccJ25 to interact with the McjBC maturation machinery and, upon maturation, to be exported by the McjD export pump), (2) to inhibit the growth of *E. coli* cells sensitive to wild-type MccJ25, and (3) to inhibit *E. coli* RNAP in vitro. The results of this effort led to the following conclusions. First, despite the complex structure of mature MccJ25, as many as 70% of point substitutions do not prevent maturation and export from the producing cells. Of those MccJ25 mutants that mature, about two thirds remain antibacterially active. The loop of the tail is important for cell uptake but is not involved in interactions with RNAP. Point mutants that increase the antibacterial activity of wild-type MccJ25 cluster in the loop. These mutants increase the antibacterial activity by facilitating MccJ25 uptake.

The results of mutational analysis are corroborated by functional studies of MccJ25 derivatives obtained by biochemical means. The previously mentioned MccJ25 derivatives obtained by thermolysin cleavage or acid hydrolysis inhibit RNAP in vitro but do not exhibit antibiotic activity (Destoumieux-Garzón et al. 2005; Semenova et al. 2005; Socias et al. 2009), confirming the importance of the loop for MccJ25 uptake and its dispensability for interactions with RNAP.

The collection of point substitutions in MccJ25 has proven to be a valuable resource for various follow-up studies, which are currently ongoing. Perhaps the most exciting result published to date is the work of Pomares et al. (2009) who showed that a mutant carrying a Gly^{12} to Tyr substitution in the loop holds significant promise as a food preservative in dairy products. The mutant retains the antibacterial activity of wild-type MccJ25. However, unlike the wild-type, the mutant is cleaved (and inactivated) by peptic proteases. Thus, the mutant, while active as a food protectant, is readily inactivated in the gastric tract and therefore has no adverse effect on normal gastric microflora.

Genome mining has identified clusters of genes whose products are homologous to MccJ25 maturation/immunity machinery in diverse bacteria (Severinov et al. 2007; Duquesne et al. 2007b). Adjacent short open reading frames coding for putative precursors of threaded-lasso peptides were also identified. In one case, heterologous coexpression in *E. coli* was used to prove that an *mcj*-like cluster *capABCD* present in *Burkholderia thailandensis* E264 genome is sufficient for production of capistruin, a 19-residue antibacterial threaded-lasso peptide active against *Burkholderia* and *Pseudomonas* (Knappe et al. 2008). Recent mutational analysis of *capA*, a homologue of *mcjA*, has identified CapA residues important for proteolytic cleavage and maturation (Knappe et al. 2009). Overall, the results of this structure–function analysis match very well those obtained during similar studies of MccJ25 and show that both threaded-lasso peptides are surprisingly resistant to mutational change and can therefore serve as scaffolds for additional rounds of derivatization. The target of capistruin action remains unknown at the time of this writing.

The structure of the MccJ25 ring and tail lock is virtually superimposable on the structure of RP 71955 (aborycin), an antibacterial threaded-lasso peptide produced by *Streptomyces* (Frechet et al. 1994; Potterat et al. 1994; Semenova et al. 2005). Unlike MccJ25, RP 71955 contains two intramolecular disulfide bonds that further stabilize the threaded-lasso fold. Several other pharmacologically active threaded-lasso peptides produced by various bacteria were isolated for their ability to inhibit processes as diverse as the HIV-1 infection, the function of endothelin 1, and the function of smooth muscle myosin kinase (Wyss et al. 1991; Chokekijchai et al. 1995; Katahira et al. 1995, 1996; Esumi et al. 2002; Iwatsuki et al. 2006). Clearly, these properties cannot reflect the biological function of these compounds. The structural similarity to MccJ25 suggests that threaded-lasso peptides might be functionally analogous to it as well. Preliminary data indicate that some non-MccJ25 threaded-lasso peptides indeed inhibit transcription by RNAPs purified from bacteria closely related to producer strains (ES and KS unpublished observations). Furthermore, at least some of these threaded-lasso peptides are encoded by genetic clusters that are homologous to the *mcj* and *capABCD* clusters (ES, Richard Ebright, and KS unpublished observations). Thus, threaded-lasso peptides, including peptides inhibiting RNAPs, may be commonly occurring throughout the bacterial world. These peptides as well as their mutational derivatives should serve as a rich source of antibacterial compounds for future studies and evaluation for possible practical use.

Microcin B17

No new data on MccB17, a Class I posttranslationally modified microcin whose structure and mechanism of action were intensely studied in the 1990s, have been published since several reviews on microcins appeared in 2007. MccB17 contains oxazole and thiazole rings that are posttranslationally synthesized by the McbBCD enzyme complex in subsequent reactions of cyclization, dehydration, and dehydrogenation from GlySer (oxazole) and GlyCys (thiazole) dipeptides present in pre-MccB17, a 69-amino-acid-long product of the *mcbA* gene (Li et al. 1996, Fig. 15.2). Modification of pre-MccB17 tripeptides $Gly^{39}Ser^{40}Cys^{41}$ and $Gly^{54}Cys^{55}Ser^{56}$ results in formation of, respectively, oxazole-thiazole and thiazole-oxazole 4,2-bis-heterocycles. Upon the removal of N-terminal 26-amino-acid-long modified pro-MccB17 leader peptide, mature MccB17 is exported outside the cell by a dedicated ABC transporter McbE–McbF. Modified pro-MccB17 accumulates inside cells harboring mutations in *tldD* or *tldE* genes (Allali et al. 2002). Thus, TldD/TldE may be a

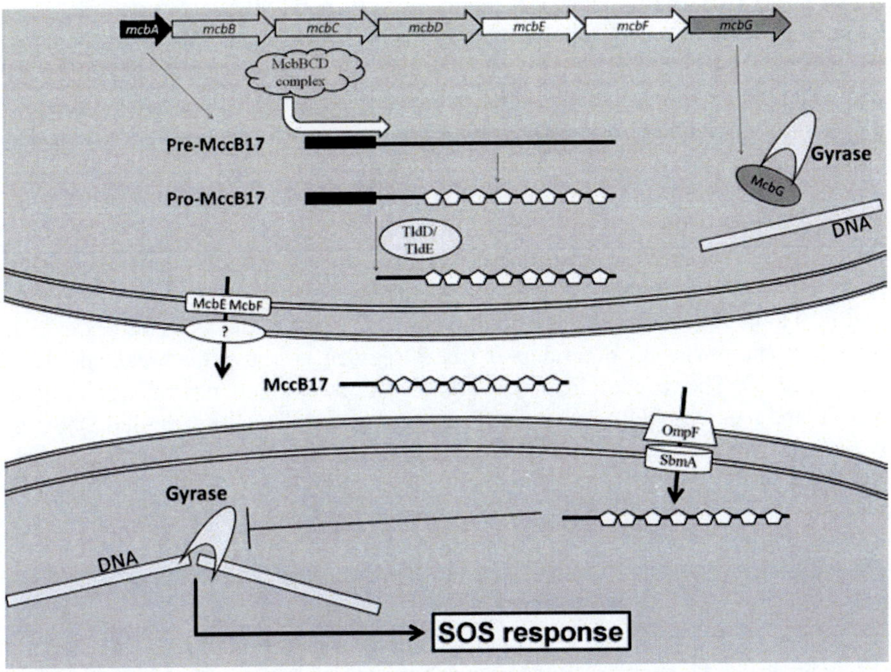

Fig. 15.2 MccB17: maturation, export, import, and mechanism of action. See Fig. 15.1 legend and text for details. *Dark gray arrow* denotes the *mcbG* gene whose product is involved in immunity. In modified precursor and mature MccB17 *pentagons* indicate heterocycles

protease that removes the leader peptide from pro-MccB17 and/or allows MccB17 export by McbE–McbF. The *tld* genes are evolutionarily conserved in *Eubacteria* and *Archaea*; however, their cellular function is not known.

The McbBCD enzyme complex recognizes pre-MccB17 leader peptide, which forms an amphipathic helix. Residues forming a hydrophobic patch on one side of the helix are essential for interaction with McbBCD (Roy et al. 1998). McbBCD acts in an N-to-C-terminus direction (Kelleher et al. 1999; Roy et al. 1999a). McbBCD action must be processive, since substitutions that affect the formation of the N-terminal 4,2-bis-heterocycle affect the degree of modification of C-terminal sites (Roy et al. 1999a). MccB17 species lacking one or more C-terminal heterocycles are secreted by wild-type MccB17-producing cells and must correspond to maturation intermediates resulting from dissociation of McbBCD. A species containing one extra oxazole ring, which becomes part of an additional extra bis-heterocycle, has also been described (Roy et al. 1999b). Antibacterial activity of such "overmodified" MccB17 is moderately increased compared to standard MccB17 activity, which was taken as evidence of importance of bis-heterocycles for activity.

MccB17 inhibits DNA replication in sensitive cells and causes a SOS response (Herrero et al. 1986). Modified pro-MccB17 that accumulates in *tld*⁻ cells has a similar effect (Allali et al. 2002). It therefore follows that the leader peptide does not interfere with MccB17 activity. MccB17 also induces accumulation of double-stranded breaks in plasmids incubated with extracts of sensitive cells (Vizan et al. 1991). Cells harboring a specific mutation in the *gyrB* gene that codes for a subunit of essential topoisomerase II DNA gyrase are resistant to MccB17 (Vizan et al. 1991). Moreover, DNA gyrase purified from the mutant cells is resistant to MccB17 in vitro, while the enzyme from the wild-type cells is sensitive (Heddle et al. 2001). Thus, the gyrase is the cellular target of MccB17.

DNA gyrase negatively supercoils DNA (see Roca et al. 1996 and references cited therein). Upon binding to DNA, gyrase wraps ~150 bp of DNA around itself, creating a loop. It then introduces a double-stranded break in the DNA duplex called the G segment, passes the other duplex (the T segment) through the break and then seals the break. The reaction involves formation of a covalent enzyme–DNA intermediate and requires ATP hydrolysis. The gyrase also relaxes supercoiled DNA in an ATP-independent reaction. The GyrA subunit is responsible for wrapping of DNA around the enzyme and G-segment breakage/ligation. GyrB is an ATPase that captures the T segment and passes it through the G segment hydrolyzing ATP in the process.

At conditions optimal for gyrase in vitro function, MccB17 has no effect on supercoiling/relaxation reactions (Heddle et al. 2001). However, modest (~threefold) inhibition of gyrase-catalyzed reactions is observed when reaction rates are attenuated by lowering the reaction temperature (Pierrat and Maxwell 2003). In the presence of MccB17 and ATP, complexes containing gyrase bound to cleaved DNA accumulate over time (Vizan et al. 1991; Heddle et al. 2001). These complexes impede the movement of DNA polymerase in vitro (Heddle et al. 2001) and might be responsible for triggering the cellular SOS response observed following MccB17 treatment.

The site of MccB17 interaction with the gyrase is not determined. MccB17 does not appear to interact with free gyrase (Pierrat and Maxwell 2005). The C-terminal DNA wrapping domain of GyrA is not needed for MccB17 inhibition and therefore is not part of the binding site (Pierrat and Maxwell 2005). In the presence of DNA and ATP, the C-terminal domain of GyrB is protected from limited proteolysis by MccB17 (Pierrat and Maxwell 2005). The only spontaneous mutation that leads to MccB17 resistance is Trp751Arg, which is located in the C-terminal domain of GyrB (Vizan et al. 1991). Limited site-specific mutagenesis of the corresponding codon of *gyrB* revealed that not every substitution at this position leads to MccB17 resistance (del Castillo et al. 2001). Clearly, systematic mutagenesis of gyrase genes is needed to better understand the details of MccB17 interaction with its target.

As of today, no systematic structure–function analysis of MccB17 has been reported. MccB17 derivatives containing bisoxazole or bisthiazole fused cycles (produced, respectively, by cells expressing mutant *mcbA* coding for GlySerSer or GlyCysCys tripeptides instead of GlySerCys and/or GlyCysSer tripeptides found in wild-type McbA) are active (Roy et al. 1999a, b). MccB17 mutant containing N-terminal bisoxazole is produced poorly, presumably because oxazole synthesis by McbBCD is slow and incomplete modification of leader-proximal site prevents efficient modification further downstream. A S39G substitution that created a single thiazole heterocycle instead of the N-terminal fused cycle also resulted in severe reduction of mature MccB17 production. However, no reduction in production of MccB17 with a S39N substitution was reported (Yorgey et al. 1994). The S39N compound was reported to be inactive, while activity of the S39G compound was found to be equal to that of the wild-type MccB17. MccB17 mutants in which the C-terminal bisheterocycle was replaced by a single cycle were obtained in normal yield (Roy et al. 1999a, b). However, their antibacterial and in vitro activities were severely reduced, indicating that this fused heterocycle is important for activity. Mutations destroying both cycles of C-terminal heterocycle caused complete loss of activity (Zamble et al. 2001).

Point mutations that substituted McbA glycines 48 or 49 for aspartates were reported to abolish antibacterial activity (Yorgey et al. 1993). Gly^{48} is not subject to posttranslational modifications, while Gly^{49} becomes part of a thiazole ring that is converted to the third fused heterocycle in "overmodified" MccB17 (Roy et al. 1999a, b). Parks et al. (2007) have shown that simultaneous chemical deamidation of two MccB17 asparagine residues at positions 53 and 59 to aspartates abolishes the in vivo and in vitro activity of MccB17 (Parks et al. 2007). Overall, it appears that the data on mutational analysis of MccB17 maturation and activity are incomplete and sometimes inconsistent. Further systematic structure–activity analyses are clearly required. Unfortunately, such studies are confounded by the complex nature of MccB17 maturation process (which leads to production of multiple intermediates containing different amounts of heterocycles) and peculiarities of the *mcbA* sequence (since MccB17 is very rich in glycines, the *mcbA* sequence is highly enriched in guanosine tracks, making site-specific mutagenesis difficult).

Significant amounts of MccB17 accumulate inside the producing cell. In addition to the McbE–McbF export pump, McbG protein contributes to the immunity of the producing cell to MccB17 (Fig. 15.2). McbG is a member of a burgeoning family of pentapeptide repeat proteins (Vetting et al. 2006). Proteins of this class are known to provide resistance to quinolones, clinically important antibiotics that also target gyrase and stabilize a covalent complex between the gyrase and cleaved DNA (Heddle et al. 2001 and references therein). Unlike MccB17, quinolones efficiently block the supercoiling reaction in vitro, and the cleaved complex accumulates even in the absence of ATP (Heddle et al. 2001). Cells harboring the *gyrB* mutation that leads to MccB17 resistance are sensitive to quinolones, suggesting that the binding sites are distinct. Nevertheless, McbG provides resistance to both MccB17 and quinolones when overexpressed (Garrido et al. 1988, Gilyarov and Severinov unpublished observations). Another pentapeptide repeat protein, Qnr, also provides resistance to quinolones, MccB17, and CcdB, an unrelated peptide inhibiting the gyrase (Tran and Jacoby 2002; Dao-Thi et al. 2005). The atomic structure of MfpA, a pentapeptide repeat protein responsible for low-level fluoroquinolone resistance in *Mycobacterium tuberculosis*, has been solved (Hegde et al. 2005). The 183-amino acid MfpA monomer is almost entirely composed of a right-handed eight-coil β helix. Two monomers interact through their C-termini to form a concave rod-shaped dimer with a length of 100 Å and diameters of 27 Å at the extremities and 18 Å in the middle, where the dimer interface is located. The dimer has an overall charge of −10, with negative charges forming two distinct patches on one side of the molecule. The right-handed helical nature of the MfpA fold, the dimensions and shape of the dimer, and the negative electrostatic surface potential led Hegde et al. (2005) to propose that MfpA might be mimicking a 30-bp segment of B-form duplex DNA. Both MfpA, Qnr, and McbG interact with the gyrase and relieve quinolone inhibition in vitro. In the absence of inhibitory compounds, MfpA inhibits the gyrase by preventing its interaction with DNA. It has been proposed that MfpA binds to the gyrase DNA-binding site, and a molecular model of such a complex was presented (Hegde et al. 2005). It was further argued that MfpA-induced inhibition of gyrase activity decreases the amount of DNA–gyrase complexes that are the targets of quinolones (and MccB17), which in turn should decrease the sensitivity of bacteria to the drug. Interestingly, unlike MfpA, Qnr does not inhibit gyrase in vitro (Mérens et al. 2009), while McbG actually stimulates it (our unpublished observations). It remains to be determined how homologous proteins interacting with the same enzyme can protect it from chemically unrelated inhibitors while having opposing effects on the enzyme activity.

E. coli GyrI, a protein unrelated to pentapeptide repeat proteins, has been also reported to protect cells from MccB17 (Chatterji and Nagaraja 2002). GyrI inhibits gyrase interactions with DNA and also protects cells from CcdB protein, inhibitor of gyrase. GyrI and chromosomally encoded Qnr homologues may exist in the cell specifically to counter the effects of diverse gyrase poisons, including MccB17. However, recent work has demonstrated that at least *qnr* genes can be induced during SOS response caused by double-stranded breaks in DNA (Da Re et al. 2009). Thus, the physiological function of "endogenous" gyrase inhibitors may be to protect the cell from potentially dangerous action of its own gyrase in condition of stress.

Genome mining revealed that clusters of genes whose products are similar to MccB17 synthase components are widespread in bacterial world (Lee et al. 2008). The small size of microcin precursors complicates bioinformatic identification of their genes. However, the unusual sequence composition of peptides into which thiozole and oxazole cycles can be introduced allowed confident identification of genes coding for precursor peptides immediately adjacent to predicted MccB17 synthase components gene homologues. In two cases, production of peptides containing oxazole and thiazole clusters was shown experimentally. The first case is that of trifolitoxin, a peptide antibiotic produced by some strains of *Rhizobium leguminosarum bv. trifolii* (Scupham and Triplett 2006). Like MccB17, trifolitoxin contains a fused thiazole–oxazole heterocycle and is produced from a longer ribosomally synthesized precursor. The second case is streptolysin S, a toxin produced by the human pathogen *Streptococcus pyogenes*. Interestingly, it was the bioinformatic identification of MccB17 synthase homologues in the locus responsible for streptolysin S synthesis that allowed insights into the structure of this compound, including evidence for existence of heterocycles. While the number of known or predicted heterocycle peptides with interesting antibacterial or pharmacological properties is certain to dramatically increase in the future, it remains to be seen how many of them will be targeting gyrase. While trifilotoxin target is unknown, streptolysin S is responsible for the hemolytic phenotype of *S. pyogenes*. To our knowledge, neither compound was ever tested for gyrase inhibition.

Microcin C7-C51

MccC7-C51 is a heptapeptide containing a modified adenosine monophosphate covalently attached to its C-terminus through an N-acyl phosphoramidate linkage (Guijarro et al. 1995, Fig. 15.3). A propylamine group is attached to the phosphate. The peptide moiety of MccC7-C51 is encoded by a 21-bp *mccA* gene, one of the shortest genes known (González-Pastor et al. 1994). The first residue of the MccC7-C51 peptide is a methionine. In MccC7-C51 produced by wild-type *E. coli* cells, this residue is formylated. Though the last residue of the peptide moiety of MccC7-C51 is an aspartic acid, the last codon of *mccA* codes for an asparagine (Fig. 15.3).

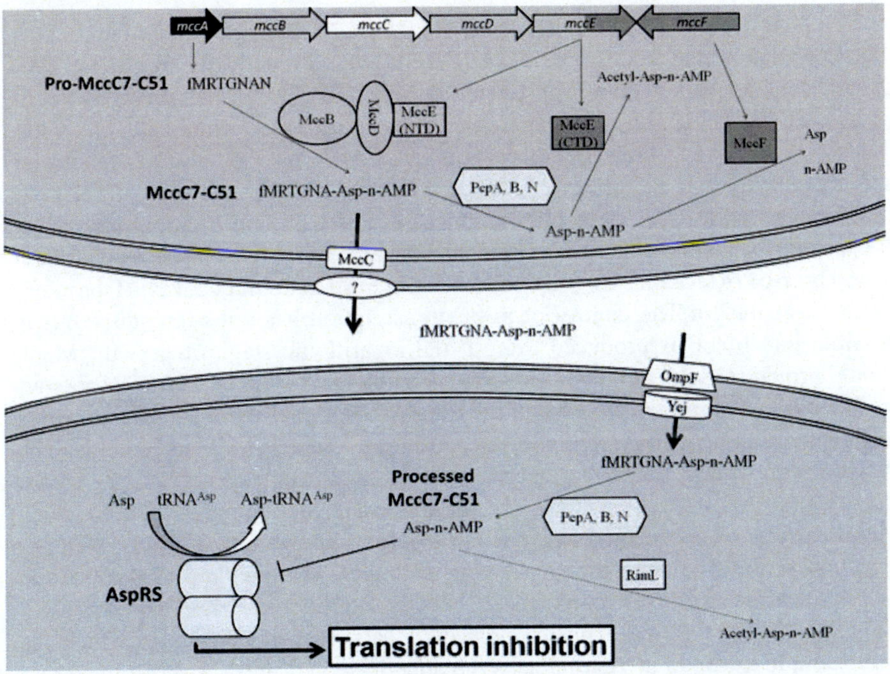

Fig. 15.3 MccC7-C51: maturation, export, import, and mechanism of action. See Fig. 15.1 legend and text for details. The dual shading of *gray* used for the *mccE* gene denotes the involvement of its N-terminal domain – MccE(NTD) – in MccC7-C51 in maturation, and C-terminal domain – MccE(CTD) – involvement in immunity. The two domains are shown as separate proteins on the scheme. *AspRS* aspartyl-tRNA synthetase

MccC7-C51 enters the *E. coli* cell through the outer membrane porin OmpF (Fig. 15.3, Metlitskaya, Novikova, Severinov unpublished data). It is actively transported through the inner membrane by the ABC transporter YejABEF (Novikova et al. 2007). Since *yej* mutants are fully resistant to MccC7-C51, YejABEF is the only transporter responsible for facilitated transport of MccC7-C51. The physiological function/substrate of YejABEF remains to be determined.

Once inside the cell, the N-terminal formyl group of MccC7-C51 is removed by peptide deformylase PDF (Kazakov et al. 2008). Deformylated MccC7-C51 is next processed by either one of the three broad-specificity aminopeptidases present in the cell – peptidase A, peptidase B, or peptidase N (Kazakov et al. 2008). Upon proteolytic cleavage of the ultimate peptide bond (between Ala^6 and Asp^7), processed MccC7-C51, a nonhydrolyzable analogue of aspartyl-adenylate is produced. Hydrolyzable aminoacyl adenylates are obligatory intermediates of tRNA aminoacylation reaction catalyzed by aminoacyl-tRNA synthetases, Processed MccC7-C51 prevents the synthesis of aminoacylated $tRNA^{Asp}$ by aspartyl-tRNA synthetase, which leads to translation inhibition (Metlitskaya et al. 2006). Unprocessed MccC7-C51 does not affect the $tRNA^{Asp}$ aminoacylation reaction. Conversely, processed MccC7-C51 does not affect the growth of MccC7-C51-sensitive cells. Thus, MccC7-C51 is a Trojan-horse inhibitor: the peptide moiety facilitates the entry of toxic part into sensitive cells.

The main cluster of *mcc* genes, *mccABCDE*, is transcribed from a single promoter (Fomenko et al. 2003). The *mccABCDE* promoter is activated by CAP–cAMP complex and appears to be recognized by both σ^{70} and σ^{S} RNA polymerase holoenzymes. The *mccABCDE* promoter transcription is activated when cells enter stationary phase. An additional gene, *mccF*, is located downstream of the main cluster and is transcribed in an opposite direction. The *mccABCDE* cluster is sufficient to endow a cell with capacity to both produce MccC7-C51 and become resistant to exogenously added MccC7-C51 (Fomenko et al. 2003). The *mccF* gene, when overexpressed from a plasmid, also makes cells resistant to MccC7-C51 (Gonzalez-Pastor et al. 1995). The control of *mccF* transcription has not been studied. It is possible that MccF is produced even at the logarithmic stage of growth, which should provide basal level of resistance to cells harboring the *mccABCDE* operon.

The product of the second gene of the *mccABCDE* operon, *mccB*, is homologous to E1 ubiquitin-activating enzymes that covalently attach AMP to proteins. The Walsh group demonstrated that MccB attaches AMP to pro-MccC7-C51 in vitro (Roush et al. 2008). These studies have established a reaction mechanism in which MccB carries out consecutive adenylation reactions consuming 2 mol of ATP per turnover. In the first step, the weakly nucleophilic α-carboxylate of Asn^7 of the MccA substrate attacks the α-phosphate of ATP to produce a labile peptidyl-AMP anhydride. Following this, an intramolecular rearrangement occurs in which the carboxamido nitrogen of Asn^7 attacks the anhydride and releases AMP to form a peptidyl-succinimide. The weakly nucleophilic succinimidyl nitrogen attacks the α-phosphate of a second molecule of ATP to form the N–P bond. A water-mediated ring opening generates the carboxy-terminal isoasparagine.

The Walsh group has also shown that MccB is able to utilize a synthetic substrate containing the first six amino acids of MccA coupled to succinimide, to yield the final product. An MccA heptapeptide in which the terminal Asn^7 is replaced with an $isoAsn^7$ is not a substrate, but rather an inhibitor of MccB. The structures of MccB with various substrates, including the MccA heptapeptide bearing an Iso-Asn at the 7th position have been reported (Regni et al. 2009). These structures reveal a core adenylating domain common to ubiquitin-conjugating enzymes, appended to a domain that holds the substrate peptide. There are considerable conformational differences in the orientation of the substrate peptide in different structures owing to local flexibility in the substrate peptide binding domain. This flexibility is consistent with results of systematic mutagenesis of the MccA peptide that indicate that the active site of MccB is flexible enough to tolerate multiple substitutions in almost all positions of the MccA peptide with the exceptions of the Thr^3 and carboxy-terminal Asn^7 (Kazakov et al. 2007). Structural analysis reveals that the terminal Asn^7 or succinimide of substrate peptides are too far away from the α-phosphate of ATP to allow catalysis to occur. Therefore, significant rearrangement of the bound ligand and/or the peptide-binding domain must occur in order to situate the two substrates (ATP and activated MccA) at a distance that would favor nucleophilic attack.

The products of the *mccD* and *mccE* genes are jointly required for the addition of the aminopropyl moiety to MccC7-C51 maturation intermediate that results from the action of MccB. The presence of aminopropyl group increases antibacterial activity several fold (Metlitskaya et al. 2009). Results of molecular modeling suggest that aminopropyl group increases processed MccC7-C51 binding to the target through electrostatic interactions between the positively charged amine and several negatively charged residues of the enzyme.

The product of *mccD* is homologous to S-AdoMet-dependent protein methylases. When *mccD* is disrupted, MccC7-C51 without the aminopropyl group is produced. MccE is a two-domain protein; the N-terminal domain is similar to decarboxylases, the C-terminal domain – to RimL, an acetylase of ribosomal protein L12 (Vetting et al. 2005). In several bioinformatically predicted MccC7-C51-like operons from bacteria other than *E. coli*, the two domains of the MccE homologue are encoded by separate genes (Severinov et al. 2007), suggesting that the two domains play distinct, and possibly unrelated, functions in MccC7-C51 production and/or immunity. Indeed, the N-terminal domain of MccE is required for production of MccC7-C51 with the propylamine group, while the C-terminal domain is not required. Since the aminopropyl increases the potency of MccC7-C51 by increasing the interaction of processed antibiotic with its target AspRS (Metlitskaya et al. 2009), it is of considerable interest to understand how this modification is accomplished. It is clear that both MccD and the N-terminal domain of MccE are required for propylamine synthesis and/or attachment to the product of the MccB-catalyzed reaction. Homoserine has been suggested as a source of the aminopropyl group (Fomenko et al. 2003). Homoserine is produced by aspartyl semialdehyde dehydrogenase (Asd), which uses aspartate phosphate as a substrate. Asd is the central enzyme of the pathway that converts aspartate to threonine (Shames et al. 1984). The pathway is also essential for biosynthesis of lysine, methionine, and isoleucine. Asd mutants are viable in the presence of diaminopimelic acid (DAP), which is used as a substrate for synthesis of amino acids whose synthesis is blocked by the absence of Asd. However, no homoserine is produced in this case. The *asd*⁻ *E. coli* cells grown in the presence of DAP and carrying the wild-type MccC7-C51 production plasmid produce MccC7-C51 containing the aminopropyl group (Metlitskaya and Severinov unpublished observations). Thus, homoserine cannot be the only source of the aminopropyl moiety. Another possible donor could be AdoMet, which could be decarboxylated by the N-terminal domain of MccE and then conjugated to the product of the MccB-catalyzed reaction by the MccD enzyme.

Plasmids harboring *mccABCDE* cluster with disrupted *mccE* (but not *mccD*) are rapidly lost from wild-type *E. coli* cells when they enter stationary growth phase, a condition when MccC7-C51 synthesis is induced. Such plasmids are stably maintained in cells lacking aminopeptidases A, B, and N (these cells are resistant to MccC7-C51, since no processed MccC7-C51 is produced, Kazakov et al. 2008). Thus, MccE contributes to immunity of the producing cell. When overproduced, both full-sized MccE and its C-terminal domain ($MccE^{CTD}$) confer resistance to MccC7-C51 (Novikova et al. 2010). $MccE^{CTD}$ is an acetylase. It recognizes processed MccC7-C51 as well as many other aminoacyl-adenylates and transfers an acetyl group from acetyl-CoA onto the amino group of the aminoacyl moiety. Acetylated processed MccC7-C51 no longer inhibits aspartyl-tRNA synthetase. The acetyltransferase activity of MccE is not needed for the addition of the propylamine group to MccC7-C51 maturation intermediate (Novikova et al. 2010). Overexpression of $MccE^{CTD}$ also makes cells resistant to albomycin, a Trojan-horse antibiotic unrelated to MccC7-C51, that upon intracellular processing gives rise to a serine coupled to thioribosyl pyrimidine that targets seryl-tRNA synthetase.

As already mentioned, $MccE^{CTD}$ is closely related to chromosomally encoded acetyltransferase RimL and more distantly to RimI and RimJ proteins, which acetylate N termini of ribosomal proteins S18, S5, correspondingly (Yoshikawa et al. 1987). Chromosomally encoded RimL, but not other Rim acetyltransferases, provides basal level of resistance to MccC7-C51 and, when overexpressed, also makes cells resistant to albomycin. Like $MccE^{CTD}$, RimL acts by acetylating processed MccC7-C51 and albomycin, which, along with ribosomal protein L12, should be considered as natural RimL substrates (Kazakov et al. in preparation).

As mentioned above, MccF provides resistance to exogenous MccC7-C51. Bioinformatic analysis indicates that MccF is homologous to LdcA, an L, D-carboxypeptidase whose substrate is monomeric murotetrapeptide L-Ala-D-Glu-*meso*-A_2pm-D-Ala or its UDP-activated murein precursor (Leguina et al. 1994; Korza and Bochtler 2005). The enzyme removes the C-terminal D-Ala residue from its substrates. MccF is also a peptidase, and it contributes to MccC7-C51 resistance through its peptidase activity. MccF hydrolyzes the terminal carboxamide bond connecting the nucleotide moiety with aspartate of the processed MccC7-C51 (Tikhonov et al., 2010). Intact MccC7-C51 is also a substrate (in this case, the entire MccA peptide is released upon cleavage).

The mechanism of the MccC7-C51 action makes it attractive to try to generate derivatives with altered C-terminal amino acids in the peptide part, for such compounds should target aminoacyl-tRNA synthetases other than the aspartyl-tRNA synthetase targeted by the wild-type MccC7-C51. However, saturating mutagenesis of the *mccA* gene failed to generate mature MccC7-C51 with substitutions in the last position (Kazakov et al. 2007). A synthetic strategy based on coupling a hexapeptide, corresponding to first six residues of MccA, to aminoacyl sulfamoyl adenylates, nonhydrolyzable analogs of aminoacyl adenylates, and strong inhibitors of aminoacyl tRNA synthetases was designed (Van de Vijver et al. 2009). The synthetic MccC7-C51-like compounds were shown to possess much higher antibacterial activities than parental aminoacyl sulfamoyl adenylates, which are poorly transported inside bacterial cells. The synthetic compounds retained the Trojan-horse mechanism of antibacterial action (i.e., they were taken up through the YejABEF transporter and required processing by cellular aminopeptidases). The target specificity of synthetic compounds was aminoacyl tRNA synthetases specified by the last (i.e., proximal to adenosine moiety) amino acid. Thus, in principle, the strategy allows to use the MccA hexapeptide as an effective transport vehicle to deliver inhibitors of up to 20 essential targets (the number of aminocyl tRNA synthetases present in most bacteria) into the cells.

Concluding Remarks

With the unfortunate exception of MccB17, the first class I posttranslationally modified microcin to be studied by modern biochemical methods, research conducted in recent years has tremendously increased our understanding of structure–function relationships, mechanisms of action, and immunity determinants of this exciting group of antibacterial compounds. Bioinformatic analyses indicate that peptides evolutionarily similar and structurally related to class I posttranslationally modified microcins first identified in *E. coli* are likely produced by a wide variety of phylogenetically diverse bacteria. Future research should illuminate physiological functions of these compounds and determine if their biological targets are the same as those of *E. coli* microcins. Bioinformatics will also be helpful in establishing structure–activity relationship of these interesting molecules. Ultimately, class I posttranslationally modified microcins and related molecules could become powerful platforms for rational design of pharmacologically active substances.

Acknowledgments This work was supported by NIH R01 grant GM64530, Russian Foundation for Basic Research grant 06-04-48865, and the Russian Academy of Sciences (program "Molecular and Cellular Biology") to KS.

References

Adelman K, Yuzenkova J, La Porta A, Zenkin N, Lee J, Lis JT, Borukhov S, Wang MD, Severinov K (2004) Molecular mechanism of transcription inhibition by peptide antibiotic Microcin J25. Mol Cell 14:753–762

Allali N, Afif H, Couturier M, Van Melderen L (2002) The highly conserved TldD and TldE proteins of Escherichia coli are involved in microcin B17 processing and in CcdA degradation. J Bacteriol 184:3224–3231

Baquero F, Bouanchaud D, Martinez-Perez MC, Fernandez C (1978) Microcin plasmids: a group of extrachromosomal elements coding for low-molecular-weight antibiotics in *Escherichia coli*. J Bacteriol 135:342–347

Bayro MJ, Mukhopadhyay J, Swapna GV, Huang JY, Ma LC, Sineva E, Dawson PE, Montelione GT, Ebright RH (2003) Structure of antibacterial peptide microcin J25: a 21-residue lariat protoknot. J Am Chem Soc 125:12382–12383

Blond A, Cheminant M, Destoumieux-Garzón D, Ségalas-Milazzo I, Peduzzi J, Goulard C, Rebuffat S (2002) Thermolysin-linearized microcin J25 retains the structured core of the native macrocyclic peptide and displays antimicrobial activity. Eur J Biochem 269:6212–6222

Blond A, Péduzzi J, Goulard C, Chiuchiolo MJ, Barthélémy M, Prigent Y, Salomón RA, Farías RN, Moreno F, Rebuffat S (1999) The cyclic structure of microcin J25, a 21-residue peptide antibiotic from *Escherichia coli*. Eur J Biochem 259:747–755

del Castillo FJ, del Castillo I, Moreno F (2001) Construction and characterization of mutations at codon 751 of the *Escherichia coli gyrB* gene that confer resistance to the antimicrobial peptide microcin B17 and alter the activity of DNA gyrase. J Bacteriol 183:2137–2140

Chatterji M, Nagaraja V (2002) GyrI: a counter-defensive strategy against proteinaceous inhibitors of DNA gyrase. EMBO Rep 3:261–267

Cheung WL, Pan SJ, Link AJ (2010) Much of the microcin J25 leader peptide is dispensable. J Am Chem Soc. doi:10.1021/ja910191u

Chiuchiolo MJ, Delgado MA, Farías RN, Salomón RA (2001) Growth-phase-dependent expression of the cyclopeptide antibiotic microcin J25. J Bacteriol 183:1755–1764

Chokekijchai S, Kojima E, Anderson S, Nomizu M, Tanaka M, Machida M, Date T, Toyota K, Ishida S, Watanabe K, Yoshioka H, Roller PP, Murakami K, Mitsuya H (1995) NP-06: a novel anti-human immunodeficiency virus polypeptide produced by a *Streptomyces* species. Antimicrob Agents Chemother 39:2345–2347

Clarke DJ, Campopiano DJ (2007) Maturation of McjA precursor peptide into active MccJ25. Org Biomol Chem 5:2564–2566

Connell N, Han Z, Moreno F, Kolter R (1987) An *E. coli* promoter induced by the cessation of growth. Mol Microbiol 1:195–201

Da Re S, Garnier F, Guérin E, Campoy S, Denis F, Ploy MC (2009) The SOS response promotes qnrB quinolone-resistance determinant expression. EMBO Rep 10:929–933

Dao-Thi MH, Van Melderen L, De Genst E, Afif H, Buts L, Wyns L, Loris R (2005) Molecular basis of gyrase poisoning by the addiction toxin CcdB. J Mol Biol 348:1091–1102

Delgado MA, Rintoul MR, Farías RN, Salomón RA (2001) *Escherichia coli* RNA polymerase is the target of the cyclopeptide antibiotic microcin J25. J Bacteriol 183:4543–4550

Delgado MA, Solbiati JO, Chiuchiolo MJ, Farías RN, Salomón RA (1999) Escherichia coli outer membrane protein TolC is involved in production of the peptide antibiotic microcin J25. J Bacteriol 181:1968–1970

Delgado MA, Vincent PA, Farías RN, Salomón RA (2005) YojI of Escherichia coli functions as a microcin J25 efflux pump. J Bacteriol 187:3465–3470

Destoumieux-Garzón D, Duquesne S, Peduzzi J, Goulard C, Desmadril M, Letellier L, Rebuffat S, Boulanger P (2005) The iron-siderophore transporter FhuA is the receptor for the antimicrobial peptide microcin J25: role of the microcin Val11-Pro16 beta-hairpin region in the recognition mechanism. Biochem J 389:869–876

Duquesne S, Destoumieux-Garzón D, Peduzzi J, Rebuffat S (2007a) Microcins, gene-encoded antibacterial peptides from enterobacteria. Nat Prod Rep 24:708–734

Duquesne S, Destoumieux-Garzón D, Zirah S, Goulard C, Peduzzi J, Rebuffat S (2007b) Two enzymes catalyze the maturation of a lasso peptide in *Escherichia coli*. Chem Biol 14:793–803

Esumi Y, Suzuki Y, Itoh Y, Uramoto M, Kimura K, Goto M, Yoshihama M, Ichikawa T (2002) Propeptin, a new inhibitor of prolyl endopeptidase produced by microbispora II. Determination of chemical structure. J Antibiot (Tokyo) 55:296–300

Fomenko D, Veselovskii A, Khmel I (2001) Regulation of microcin C51 operon expression: the role of global regulators of transcription. Res Microbiol 152:469–479

Fomenko DE, Metlitskaya AZ, Péduzzi J, Goulard C, Katrukha GS, Gening LV, Rebuffat S, Khmel IA (2003) Microcin C51 plasmid genes: possible source of horizontal gene transfer. Antimicrob Agents Chemother 47:2868–2874

Frechet D, Guitton JD, Herman F, Faucher D, Helynck G, Monegier du Sorbier B, Ridoux JP, James-Surcouf E, Vuilhorgne M (1994) Solution structure of RP 71955, a new 21 amino acid tricyclic peptide active against HIV-1 virus. Biochemistry 33:42–50

Garrido MC, Herrero M, Kolter R, Moreno F (1988) The export of the DNA replication inhibitor Microcin B17 provides immunity for the host cell. EMBO J 7:1853–1862

González-Pastor JE, San Millán JL, Moreno F (1994) The smallest known gene. Nature 369:281

Gonzalez-Pastor JE, San Millan JL, Castilla MA, Moreno F (1995) Structure and organization of plasmid genes required to produce the translation inhibitor microcin C7. J Bacteriol 177:7131–7140

Guijarro JI, González-Pastor JE, Baleux F, San Millán JL, Castilla MA, Rico M, Moreno F, Delepierre M (1995) Chemical structure and translation inhibition studies of the antibiotic microcin C7. J Biol Chem 270:23520–23532

Heddle JG, Blance SJ, Zamble DB, Hollfelder F, Miller DA, Wentzell LM, Walsh CT, Maxwell A (2001) The antibiotic microcin B17 is a DNA gyrase poison: characterisation of the mode of inhibition. J Mol Biol 307:1223–1234

Hegde SS, Vetting MW, Roderick SL, Mitchenall LA, Maxwell A, Takiff HE, Blanchard JS (2005) A fluoroquinolone resistance protein from Mycobacterium tuberculosis that mimics DNA. Science 308:1480–1483

Hernández-Chico C, San Millán JL, Kolter R, Moreno F (1986) Growth phase and *ompR* regulation of transcription of microcin B17 genes. J Bacteriol 167:1058–1065

Herrero M, Kolter R, Moreno F (1986) Effects of microcin B17 on microcin B17-immune cells. J Gen Microbiol 132:403–410

Iwatsuki M, Tomoda H, Uchida R, Gouda H, Hirono S, Omura S (2006) Lariatins, antimycobacterial peptides produced by *Rhodococcus sp.* K01-B0171, have a lasso structure. J Am Chem Soc 128:7486–7491

Katahira R, Shibata K, Yamasaki M, Matsuda Y, Yoshida M (1995) Solution structure of endothelin B receptor selective antagonist RES-701-1 determined by ^{1}H NMR spectroscopy. Bioorg Med Chem 3:1273–1280

Katahira R, Yamasaki M, Matsuda Y, Yoshida M (1996) MS-271, a novel inhibitor of calmodulin-activated myosin light chain kinase from *Streptomyces* sp. – II. Solution structure of MS-271: characteristic features of the "lasso" structure. Bioorg Med Chem 4:121–129

Kazakov T, Metlitskaya A, Severinov K (2007) Amino acid residues required for maturation, cell uptake, and processing of translation inhibitor microcin C. J Bacteriol 189:2114–2118

Kazakov T, Semenova E, Kuznedelov K, Mukhamedjarov D, Datsenko KA, Vondenhoff GH, Wanner B, Van Aerschot A, Severinov K. The RimL transacetylase provides resistance to translation inhibitors microcin C and albomycin (in preparation)

Kazakov T, Vondenhoff GH, Datsenko KA, Novikova M, Metlitskaya A, Wanner BL, Severinov K (2008) Escherichia coli peptidase A, B, or N can process translation inhibitor microcin C. J Bacteriol 190:2607–2610

Kelleher NL, Hendrickson CL, Walsh CT (1999) Posttranslational heterocyclization of cysteine and serine residues in the antibiotic microcin B17: distributivity and directionality. Biochemistry 38:15623–15630

Knappe TA, Linne U, Robbel L, Marahiel MA (2009) Insights into the biosynthesis and stability of the lasso peptide capistruin. Chem Biol 16:1290–1298

Knappe TA, Linne U, Zirah S, Rebuffat S, Xie X, Marahiel MA (2008) Isolation and structural characterization of capistruin, a lasso peptide predicted from the genome sequence of Burkholderia thailandensis E264. J Am Chem Soc 130:11446–11454

Korza HJ, Bochtler M (2005) Pseudomonas aeruginosa LD-carboxypeptidase, a serine peptidase with a Ser-His-Glu triad and a nucleophilic elbow. J Biol Chem 280:40802–40812

Lee SW, Mitchell DA, Markley AL, Hensler ME, Gonzalez D, Wohlrab A, Dorrestein PC, Nizet V, Dixon JE (2008) Discovery of a widely distributed toxin biosynthetic gene cluster. Proc Natl Acad Sci U S A 105:5879–5884

Leguina JI, Quintela JC, de Pedro MA (1994) Substrate specificity of Escherichia coli LD-carboxypeptidase on biosynthetically modified muropeptides. FEBS Lett 339:249–252

Li YM, Milne JC, Madison LL, Kolter R, Walsh CT (1996) From peptide precursors to oxazole and thiazole-containing peptide antibiotics: microcin B17 synthase. Science 274:1188–1193

Mérens A, Matrat S, Aubry A, Lascols C, Jarlier V, Soussy CJ, Cavallo JD, Cambau E (2009) The pentapeptide repeat proteins MfpAMt and QnrB4 exhibit opposite effects on DNA gyrase catalytic reactions and on the ternary gyrase-DNA-quinolone complex. J Bacteriol 191:1587–1594

Metlitskaya A, Kazakov T, Kommer A, Pavlova O, Praetorius-Ibba M, Ibba M, Krasheninnikov I, Kolb V, Khmel I, Severinov K (2006) Aspartyl-tRNA synthetase is the target of peptide nucleotide antibiotic Microcin C. J Biol Chem 281:18033–18042

Metlitskaya A, Kazakov T, Vondenhoff GH, Novikova M, Shashkov A, Zatsepin T, Semenova E, Zaitseva N, Ramensky V, Van Aerschot A, Severinov K (2009) Maturation of the translation inhibitor microcin C. J Bacteriol 191:2380–2387

Mukhopadhyay J, Sineva E, Knight J, Levy RM, Ebright RH (2004) Antibacterial peptide microcin J25 inhibits transcription by binding within and obstructing the RNA polymerase secondary channel. Mol Cell 14:739–751

Novikova M, Kazakov T, Vondenhoff GH, Semenova E, Rozenski J, Metlytskaya A, Zukher I, Tikhonov A, Van Aerschot A, Severinov K (2010) MccE provides resistance to protein synthesis inhibitor microcin C by acetylating the processed form of the antibiotic. J Biol Chem. doi:10.1074/jbc.M109.080192

Novikova M, Metlitskaya A, Datsenko K, Kazakov T, Kazakov A, Wanner B, Severinov K (2007) The Escherichia coli Yej transporter is required for the uptake of translation inhibitor microcin C. J Bacteriol 189:8361–8365

Parks WM, Botrill AR, Pierrat OA, Durrant MC, Maxwell A (2007) The action of the bacterial toxin, microcin B17, on DNA gyrase. Biochimie 89:500–507

Pavlova O, Mukhopadhyay J, Sineva E, Ebright RH, Severinov K (2008) Systematic structure-activity analysis of microcin J25. J Biol Chem 283:25589–25595

Pierrat OA, Maxwell A (2003) The action of the bacterial toxin microcin B17. Insight into the cleavage-religation reaction of DNA gyrase. J Biol Chem 278:35016–35023

Pierrat OA, Maxwell A (2005) Evidence for the role of DNA strand passage in the mechanism of action of microcin B17 on DNA gyrase. Biochemistry 44:4204–4215

Pomares MF, Salomón RA, Pavlova O, Severinov K, Farías R, Vincent PA (2009) Potential applicability of chymotrypsin-susceptible microcin J25 derivatives to food preservation. Appl Environ Microbiol 75:5734–5738

Potterat O, Stephan H, Metzger JW, Gnau V, Zähner H, Jung G (1994) Aborycin – a tricyclic 21-peptide antibiotic isolated from *Streptomyces griseoflavus*. Liebigs Ann Chem 1994:741–743

Regni CA, Roush RF, Miller DJ, Nourse A, Walsh CT, Schulman BA (2009) How the MccB bacterial ancestor of ubiquitin E1 initiates biosynthesis of the microcin C7 antibiotic. EMBO J 28:1953–1964

Roca J, Berger JM, Harrison SC, Wang JC (1996) DNA transport by a type II topoisomerase: direct evidence for a two-gate mechanism. Proc Natl Acad Sci U S A 93:4057–4062

Rosengren KJ, Blond A, Afonso C, Tabet JC, Rebuffat S, Craik DJ (2004) Structure of thermolysin cleaved microcin J25: extreme stability of a two-chain antimicrobial peptide devoid of covalent links. Biochemistry 43:4696–4702

Rosengren KJ, Clark RJ, Daly NL, Göransson U, Jones A, Craik DJ (2003) Microcin J25 has a threaded sidechain-to-backbone ring structure and not a head-to-tail cyclized backbone. J Am Chem Soc 125:12464–12474

Roush RF, Nolan EM, Löhr F, Walsh CT (2008) Maturation of an Escherichia coli ribosomal peptide antibiotic by ATP-consuming N-P bond formation in microcin C7. J Am Chem Soc 130:3603–3609

Roy RS, Allen O, Walsh CT (1999a) Expressed protein ligation to probe regiospecificity of heterocyclization in the peptide antibiotic microcin B17. Chem Biol 6:789–799

Roy RS, Kelleher NL, Milne JC, Walsh CT (1999b) In vivo processing and antibiotic activity of microcin B17 analogs with varying ring content and altered bisheterocyclic sites. Chem Biol 6:305–318

Roy RS, Kim S, Baleja JD, Walsh CT (1998) Role of the microcin B17 propeptide in substrate recognition: solution structure and mutational analysis of McbA1-26. Chem Biol 5:217–228

Salomón RA, Farías RN (1992) Microcin 25, a novel antimicrobial peptide produced by Escherichia coli. J Bacteriol 174:7428–7435

Salomón RA, Farías RN (1993) The FhuA protein is involved in microcin 25 uptake. J Bacteriol 175:7741–7742

Salomón RA, Farías RN (1995) The peptide antibiotic microcin 25 is imported through the TonB pathway and the SbmA protein. J Bacteriol 177:3323–3325

Scupham AJ, Triplett EW (2006) Determination of the amino acid residues required for the activity of the anti-rhizobial peptide antibiotic trifolitoxin. J Appl Microbiol 100:500–507

Semenova E, Yuzenkova Y, Péduzzi J, Rebuffat S, Severinov K (2005) Structure-activity analysis of microcin J25: distinct parts of the threaded lasso molecule are responsible for interaction with bacterial RNA polymerase. J Bacteriol 187:3859–3863

Severinov K, Semenova E, Kazakov A, Kazakov T, Gelfand MS (2007) Low-molecular-weight post-translationally modified microcins. Mol Microbiol 65:1380–1394

Shames SL, Ash DE, Wedler FC, Villafranca JJ (1984) Interaction of aspartate and aspartate-derived antimetabolites with the enzymes of the threonine biosynthetic pathway of *Escherichia coli*. J Biol Chem 259:15331–15339

Socias SB, Severinov K, Salomon RA (2009) The Ile13 residue of microcin J25 is essential for recognition by the receptor FhuA, but not by the inner membrane transporter SbmA. FEMS Microbiol Lett 301:124–129

Solbiati JO, Ciaccio M, Farías RN, Salomón RA (1996) Genetic analysis of plasmid determinants for microcin J25 production and immunity. J Bacteriol 178:3661–3663

Solbiati JO, Ciaccio M, Farías RN, González-Pastor JE, Moreno F, Salomón RA (1999) Sequence analysis of the four plasmid genes required to produce the circular peptide antibiotic microcin J25. J Bacteriol 181:2659–2662

Stepanova EV, Shevelev AB, Borukhov SI, Severinov KV (2009) Mechanisms of action of RNA polymerase-binding transcription factors that do not bind to DNA. Biofizika 54:773–790

Tikhonov A, Kazakov T, Semenova E, Vondenhoff GH, Van Aerschot V, and Severinov K (2010) Molecular mechanism of microcin C resistance due to the function of the MccF peptidase. J Biol Chem 285:37944–37952

Tran JH, Jacoby GA (2002) Mechanism of plasmid-mediated quinolone resistance. Proc Natl Acad Sci U S A 99:5638–5642

Van de Vijver P, Vondenhoff GHM, Kazakov T, Semenova E, Kuznedelov K, Metlitskaya A, Van Aerschot A, Severinov K (2009) Synthetic microcin C analogs targeting different aminoacyl-tRNA synthetases. J Bacteriol 191:6273–6280

Vetting MW, de Carvalho LP, Roderick SL, Blanchard JS (2005) A novel dimeric structure of the RimL Nalpha-acetyltransferase from Salmonella typhimurium. J Biol Chem 280:22108–22114

Vetting MW, Hegde SS, Fajardo JE, Fiser A, Roderick SL, Takiff HE, Blanchard JS (2006) Pentapeptide repeat proteins. Biochemistry 45:1–10

Vizan JL, Hernández-Chico C, del Castillo I, Moreno F (1991) The peptide antibiotic microcin B17 induces double-strand cleavage of DNA mediated by *E. coli* DNA gyrase. EMBO J 10:467–476

Wilson KA, Kalkum M, Ottesen J, Yuzenkova J, Chait BT, Landick R, Muir T, Severinov K, Darst SA (2003) Structure of microcin J25, a peptide inhibitor of bacterial RNA polymerase, is a lassoed tail. J Am Chem Soc 125:12475–12483

Wyss DF, Lahm HW, Manneberg M, Labhardt AM (1991) Anantin – a peptide antagonist of the atrial natriuretic factor (ANF). II. Determination of the primary sequence by NMR on the basis of proton assignments. J Antibiot (Tokyo) 44:172–180

Yorgey P, Davagnino J, Kolter R (1993) The maturation pathway of microcin B17, a peptide inhibitor of DNA gyrase. Mol Microbiol 9:897–905

Yorgey P, Lee J, Kordel J, Vivas E, Warner P, Jebaratnam D, Kolter R (1994) Posttranslational modifications in microcin B17 define an additional class of DNA gyrase inhibitor. Proc Natl Acad Sci U S A 91:4519–4523

Yoshikawa A, Isono S, Sheback A, Isono K (1987) Cloning and nucleotide sequencing of the genes *rimI* and *rimJ* which encode enzymes acetylating ribosomal proteins S18 and S5 of Escherichia coli K12. Mol Gen Genet 209:481–488

Yuzenkova J, Delgado M, Nechaev S, Savalia D, Epshtein V, Artsimovitch I, Mooney RA, Landick R, Farías RN, Salomón R, Severinov K (2002) Mutations of bacterial RNA polymerase leading to resistance to microcin J25. J Biol Chem 277:50867–50875

Zamble DB, Miller DA, Heddle JG, Maxwell A, Walsh CT, Hollfelder F (2001) In vitro characterization of DNA gyrase inhibition by microcin B17 analogs with altered bisheterocyclic sites. Proc Natl Acad Sci U S A 98:7712–7717

Zhang G, Campbell EA, Minakhin L, Richter C, Severinov K, Darst SA (1999) Crystal structure of *Thermus aquaticus* core RNA polymerase at 3.3 Å resolution. Cell 98:811–824

Chapter 16
Class II Microcins

Gaëlle Vassiliadis, Delphine Destoumieux-Garzón, and Jean Peduzzi

Abstract Class II microcins are 4.9- to 8.9-kDa polypeptides produced by and active against enterobacteria. They are classified into two subfamilies according to their structure and their gene cluster arrangement. While class IIa microcins undergo no posttranslational modification, class IIb microcins show a conserved C-terminal sequence that carries a salmochelin-like siderophore motif as a post-translational modification. Aside from this C-terminal end, which is the signature of class IIb microcins, some sequence similarities can be observed within and between class II subclasses, suggesting the existence of common ancestors. Their mechanisms of action are still under investigation, but several class II microcins use inner membrane proteins as cellular targets, and some of them are membrane-active. Like group B colicins, many, if not all, class II microcins are TonB- and energy-dependent and use catecholate siderophore receptors for recognition/translocation across the outer membrane. In that context, class IIb microcins are considered to have developed molecular mimicry to increase their affinity for their outer membrane receptors through their salmochelin-like posttranslational modification.

Introduction

Class II microcins are higher-molecular-mass microcins (4.9–8.9 kDa), and their peptide backbone does not undergo extensive posttranslational modifications (Duquesne et al. 2007). It has been recently shown that class II microcins are markers of virulence patterns in uropathogenic *Escherichia coli* (Azpiroz et al. 2009). In this review, we successively detail the structures, genetics, biosynthesis, and mechanisms of action of class II microcins. The organization of class II microcin

J. Peduzzi (✉)
CNRS, Muséum National d'Histoire Naturelle, UMR 7245,
CP 54, 57 rue Cuvier, 75005 Paris, France
e-mail: peduzzi@mnhn.fr

D. Drider and S. Rebuffat (eds.), *Prokaryotic Antimicrobial Peptides: From Genes to Applications*, DOI 10.1007/978-1-4419-7692-5_16,

gene clusters involves at least four clustered genes organized into one single or several transcription units. The minimal structure is composed of (1) the gene encoding the microcin precursor, (2) an adjacent self-immunity gene, which encodes the self-immunity protein that protects the producing strain from its own antibacterial substance, and (3) at least two genes encoding an inner membrane ABC (ATP-binding cassette) transporter and its accessory protein responsible for the secretion of the microcin. The nonspecialists should be aware that the name given to each gene is not standardized throughout the different microcin gene clusters. While class IIa microcins are not posttranslationally modified, class IIb microcins show a conserved C-terminal sequence that carries a salmochelin-like siderophore motif as a posttranslational modification. They are therefore considered as antibacterial siderophore-peptides (Thomas et al. 2004). Siderophores are small iron chelators synthesized by bacteria and fungi for the uptake of the very low amount of iron available in physiological media (Andrews et al. 2003). The enzymatic pathways leading to the class IIb microcin posttranslational modification have been recently studied (Nolan et al. 2007; Vassiliadis et al. 2007; Nolan and Walsh 2008; Mercado et al. 2008), and various steps of class IIb microcin biosynthesis could be achieved in vitro. The mechanisms of action of class II microcins are still under investigation, most of the knowledge being acquired on their recognition/uptake. Like group B colicins, many, if not all, class II microcins are TonB- and energy-dependent and use catecholate siderophore receptors for recognition/translocation across the outer membrane (Chehade and Braun 1988; Trujillo et al. 2001; Patzer et al. 2003; Thomas et al. 2004; Strahsburger et al. 2005). As a consequence, the salmochelin-like posttranslational modification of class IIb microcins increases their affinity for their outer membrane receptors (Thomas et al. 2004). Such a molecular mimicry is a unique case among antimicrobial peptides.

Class IIa Microcins

Microcins (Mcc) belonging to this subclass are characterized by the absence of posttranslational modification. MccV and MccL contain disulfide bond(s), whereas Mcc24 would be a linear unmodified peptide lacking such a bond.

Genetics and Structure

Class IIa microcin gene clusters (Fig. 16.1) are composed of only four plasmid-borne genes. MccV, formerly colicin V (ColV) (Fredericq et al. 1949) was the first antibiotic substance reported to be produced by *E. coli* (Gratia 1925). MccV is secreted by various *E. coli* strains harboring large (>80 kb) and low-copy-number pColV plasmids (Waters and Crosa 1991). MccL is produced by *E. coli* LR05 isolated from poultry intestine (Gaillard-Gendron et al. 2000), while Mcc24

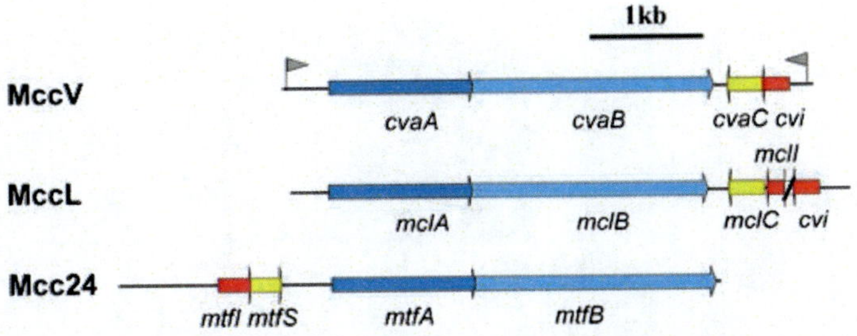

Fig. 16.1 *Genetic organization of class IIa microcin biosynthetic gene clusters.* A colour code identifies genes with similar functions. Genes encoding microcin precursors are shown in *yellow*. Genes required for self-immunity and microcin export are shown in *red* and *blue* respectively. Promoters are indicated by flags. The truncated gene in MccL gene cluster is crossed through

(formerly colicin 24) is secreted by the uropathogenic *E. coli* 2424, and its genetic determinants are located on the 43.5-kb conjugative plasmid p24-2 (O'Brien and Mahanty 1994). MccV and MccL gene clusters are composed of four genes organized in two converging transcription units (Gilson et al. 1987, 1990; Chehade and Braun 1988; Pons et al. 2004). The genes *cvaC and mclC*, encoding a 103-amino-acid MccV and a 105-amino-acid MccL precursor, respectively, as well as the self-immunity genes *cvi* and *mclI,* form the first operon. The second operon contains two genes (*cvaA/mclA* and *cvaB/mclB*) involved in the MccV and MccL export, respectively (Gilson et al. 1990; Pons et al. 2004). In MccL gene cluster, downstream of *mclI,* one open reading frame (ORF) exhibits 98% identity with *cvi*, which makes MccL-producing strains resistant to MccV (Sablé et al. 2003). Mcc24 gene cluster (GenBank accession number U47048) revealed four genes that would belong to one single operon (O'Brien and Mahanty 1996, unpublished work). The genes *mtfS* and *mtfI* encode the putative 88-amino-acid precursor and self-immunity protein of Mcc24, respectively, whereas *mtfA* and *mtfB* encode proteins similar to CvaA and CvaB, respectively.

The mature MccV is a 88-amino-acid peptide, originating from a 103-amino-acid precursor (CvaC) that possesses a single disulfide bond located in the C-terminal sequence (Table 16.1) (Håvarstein et al. 1994). MccL, which is generated from a 105-amino-acid precursor (MclC), is composed of 90 unmodified amino acids (Pons et al. 2004). MccL is an anionic and highly hydrophobic peptide characterized by two disulfide bonds (Pons et al. 2004). MccL and MccV share an identical 13-amino-acid C-terminal sequence that contains one disulfide bond (Table 16.1). Mcc24 has neither been isolated nor been characterized. Nevertheless, its precursor amino-acid sequence, which shares homology with MceA, a class IIb microcin precursor (Table 16.1), but lacks the 10-amino-acid C-terminal sequence typical of class IIb microcins (see "Structures"), and its gene cluster, which contains four genes only, suggest strongly that Mcc24 belongs to class IIa microcin. The putative mature Mcc24 corresponds to a 73-amino-acid peptide devoid of disulfide bond.

Table 16.1 Multiple amino acid sequence alignment of class II microcin precursors

Class	Precursor	Mature microcin	Leader peptide	¦	mature microcin
IIa	CvaC[a]	(MccV)[b]	MRTLTLNELDSVSGG	¦	ASGR--DIAMAIGTLSGQFVAGGIGAAAGGVAGGAIYDYASTHKPNPAMSPSGLGGTIKQKPEGIPSEAWNYAAGRLCNWSPNNLSDVCL
	McIC[a]	(MccL)[b]	MREITLNEMNNVSGA	¦	GDVNWVDVGKTVATNGAGVIGGAFGAGLCGPVCAGAFAVGSSAAVAALYDAAGNSNSAKQKPEGLPPEAWNYAEGRMCNWSPNNLSDVCL
	MtfS[a]	(Mcc24)[b]	MRELDREELNCVGGA	¦	GDPL-ADPNSQIVRQIMSNAAWGPPL-VPERFRGMAVGAAGGVTQTVLQGAAAHMPVNVPIPKVPMGPSWN-GS-KG
IIb	MceA[a]	(MccE492)[b]	MREISQKDLNLAFGA	¦	GE---TDPNTQLLNDLGNNMAWGAALGAPGGLGSAALGAAGGALQTVGQGLIDHGPVNVPIP-VLIGPSWN-GSGSGYNSATSSSGSGS
	McmA[a]	(MccM)[b]	MRKLSENEIKQISGG	¦	DG---NDGQAELIAIGS--LA-GT-FISPG-FGSIAGAYIGDKVHSWATT-ATVSPSMSPSGIGL-SSQF--GSGRGTSSASSSAGSGS
	MceL[a]	(MccG492)[b]	MRALTENDFFAVSGA	¦	DR---GDAAVAGAVAG--GTA-GAAAGGWAGAQMGATVGSLAGPVGTVVGFVAGAAAGA--YGGA----F---IYDSFSSPSNSSSSGS
	MchB[a]	(MccH47)[b]	MREITESQLRYISGA	¦	GG---APATSA--------NAA-GAAAIVGALAGIPGGPLGVVVGAVSAG--LTTA-----------------I--GSTVGSGSASSSAGGGS
	MchS2[a]	(MccI47)[b]	MREISDNMLDSVKGG	¦	MN---LNGLPA-------------STNVIDLRGKDMG--TYIDANGACWA----PDTPSIIMYPGGS-GPSY------SMSSSTSSANSGS

[a] Name of the microcin precursors.

[b] Name of the mature microcins (known or putative).

A vertical dash indicates the known or putative cleavage site of microcin precursors. Dashes in the amino acid sequence indicate gaps. Amino acids conserved in at least 50% of both class IIa and IIb sequences are highlighted in black. Amino acids conserved in at least 66% of class IIa sequences or 60% of class IIb sequences are highlighted in grey

Maturation and Export

Many bacterial proteins are translocated across the inner membrane by the Sec pathway (de Keyzer et al. 2003). Nevertheless, some bacteriocins, like class II microcins, use ABC transporters (Davidson and Maloney 2007; Hollenstein et al. 2007; Moussatova et al. 2008). For all class II microcins, maturation (cleavage of the leader peptide) and export appear to occur simultaneously. Except for MccV, whose mechanism of export has been partially characterized (Zhong et al. 1996), the class II microcin export machineries have been mainly identified based on genetics and sequence identity analysis. The class II microcin export machinery displays a canonical structure consisting of three components.

The first component is an ABC transporter. Encoded by the microcin gene cluster, this protein is an homodimer carrying a nucleotide fixation domain. CvaB would be responsible for the export of MccV and, based on their highly similar sequence (up to 70% identity), MclB and MtfB, for the export of MccL and Mcc24, respectively. All three proteins are similar to ABC transporters responsible for the export of other antibacterials such as class II bacteriocins from Gram positive bacteria (Håvarstein et al. 1995). Typically, ABC transporters exhibit three domains: (1) an N-terminal domain, supposed to be located in the cytoplasm (Franke et al. 1999; Wu et al. 2004), with a protease activity likely involved in the leader peptide cleavage during export, (2) a central transmembrane domain, poorly conserved, and (3) a C-terminal domain carrying a highly conserved nucleotide-binding cassette required for ATP binding (Benabdelhak et al. 2003). A model was proposed in which the binding of the MccV precursor to CvaB promotes a transition between an inactive dimer bound to nucleotide diphosphate and a high-energy dimer bound to nucleotide triphosphate (Guo et al. 2006).

The second component of the export machinery is the ABC transporter accessory protein. Based on bioinformatics studies, this protein, which is encoded by the microcin gene cluster, would be periplasmic with an N-terminal transmembrane helix that serves as an anchor at the inner membrane (Skvirsky et al. 1995; Franke et al. 1996). Like ABC transporters, accessory proteins are also highly conserved. CvaA, MclA, and MtfA are accessory proteins for MccV, MccL, and Mcc24, respectively. Their function in microcin secretion remains unclear. However, they have been proposed to serve as a connector between the ABC transporter and the third component of the class II microcin export machinery (Gilson et al. 1990; Pons et al. 2004).

The third component of the machinery is TolC, located at the outer membrane. Unlike genes encoding the ABC transporter and the accessory protein, *tolC* is located on the bacterial chromosome outside of the microcin gene cluster. This protein forms a trimeric channel with an α helix in the periplasmic space, and a β barrel channel spanning the outer membrane (Koronakis et al. 2000, 2004). The expression of TolC is required for the production of extracellular antimicrobials by MccV- and MccL-producing strains (Gilson et al. 1990; Pons et al. 2004). Therefore, TolC is considered to be required for the export of class II microcins.

Recognition/Uptake and Mechanism of Action

Class IIa microcins have been poorly studied in terms of mechanism of action compared to class IIb. Most of the data are speculative and are based on sequence similarities with microcins with known mechanism of action.

MccV (ColV) is active against Gram-negative bacteria belonging to the genera *Escherichia, Klebsiella, Salmonella*, and *Shigella* (Håvarstein et al. 1994). Like class IIb microcins, MccV is recognized by an outer membrane receptor for catecholate siderophores, and its uptake is TonB-dependent (Chehade and Braun 1988). It is therefore likely taken up into sensitive cells at the expense of energy that is provided by the proton-motive force of the inner membrane. However, while class IIb microcins are recognized by FepA, Cir, and Fiu, Cir alone appears to be involved in MccV uptake (Chehade and Braun 1988). MccV activity also depends on the inner membrane protein SdaC, also involved in serine uptake (Gerard et al. 2005). Although the mechanistic role of SdaC in MccV uptake and/or activity remains unknown, it could drive MccV to the inner membrane, thereby promoting channel formation and subsequent disruption of membrane potential (Yang and Konisky 1984).

MccL spectrum of activity covers a large number of Gram-negative species including *E. coli, Salmonella enterica, Klebsiella oxytoca, Shigella sonnei, Shigella flexneri, Pseudomonas aeruginosa,* and *Providencia stuartii* (Pons et al. 2004). However, to date, MccL has been poorly studied in terms of mechanism of action. Since it displays a 32-residue C-terminal sequence highly similar to that of MccV (87.5% identity), it could also have membrane-permeabilizing properties (Pons et al. 2004). This conserved region could also be required for recognition and uptake, since like MccV, MccL is recognized by Cir and imported in a TonB-dependent manner (Sablé, personal communication).

Mcc24 was reported to be active against the enterobacteria *E. coli* and *S. enterica* Typhimurium, but not against *Listeria monocytogenes* or *Campylobacter jejuni* (Wooley et al. 1999). To date, nothing is known regarding its uptake and mechanism of action. It is actually the only class II microcin for which nothing is known in terms of recognition by catecholate siderophore receptors and their associated TonB-ExbB-ExbD complex. Mcc24 displays striking sequence similarities with MccE492 (Lagos et al. 1999) but lacks the C-terminal region required for recognition by catecholate siderophore receptors. Based on their sequence identities, it was suggested that Mcc24 and MccE492 could have similar cellular targets and would both require ManYZ at the inner membrane (Bieler et al. 2006).

Class IIb Microcins: the Siderophore-Microcins

Class IIb microcins are linear polypeptides carrying a C-terminal siderophore as a posttranslational modification (MccE492, MccH47, MccM, and presumably MccG492 and MccI47). These microcins are characterized by a highly conserved

10-amino-acid C-terminal sequence, which is considered their signature (Vassiliadis et al. 2010).

Genetics

Contrary to class IIa microcins, which are all plasmid-encoded, class IIb microcins are chromosome-encoded and their gene clusters show a complex transcriptional organization (Fig. 16.2). In addition to the four genes described in class IIa microcin gene clusters, genes encoding posttranslational modification enzymes are encountered in class IIb microcin gene clusters.

The best characterized class IIb microcin is MccE492, which is secreted by *Klebsiella pneumoniae* RYC492 (de Lorenzo 1984). The entire MccE492 gene cluster is contained within a 13-kb DNA fragment that has been cloned in *E. coli* (Wilkens et al. 1997). Ten genes (*mceABCDEFGHIJ*) (Lagos et al. 2001), organized at least in five transcription units, are necessary for MccE492 biosynthesis. The gene *mceA* encodes the 99-amino-acid MccE492 precursor and *mceB* is involved in the self-immunity toward MccE492 (Lagos et al. 1999). Other genes are required for MccE492 posttranslational modification. Those are *mceC*, *mceD*, and *mceI*, which encode proteins homologous to a glycosyltransferase, an enterobactin esterase, and an acyltransferase, respectively (Nolan et al. 2007; Vassiliadis et al. 2007), as well as *mceJ*, which does not display evident homologies with known genes (Corsini et al. 2002; Nolan et al. 2007). Two genes, *mceG* and *mceH*, are necessary for the export of MccE492. They encode an ABC transporter and an accessory protein, respectively. The *mceF* gene would also be involved in export (Lagos et al. 2001), and the role of the last gene, *mceE*, remains unknown. The nucleotide sequence of this gene cluster has been recently corrected (GenBank accession number AF063590), showing that the orientation of *mceFGHIJ* is reverse to that previously described. The corrected sequence also unveiled six novel ORFs (*mceS2*/*mceS3*/*mceM*/*mceL*/*mceX*/*mceK*), among which *mceL* and *mceM* encode the 89-amino-acid precursor of a new putative microcin termed MccG492 and its self-immunity protein, respectively (Vassiliadis et al. 2010).

As with *K. pneumoniae* RYC492, several microcins can be expressed by one single *E. coli* isolate. Thus, *E. coli* H47 was reported to express both MccH47 and MccI47 (Laviña et al. 1990; Poey et al. 2006). Similarly, *E. coli* Nissle 1917 (also named Mutaflor), *E. coli* CA46, and *E. coli* CA58 strains express *MccM*, *MccH47*, and possibly MccI47 (Nissle 1925; Patzer et al. 2003; Duquesne et al. 2007; Vassiliadis et al. 2010). The genetic determinants required for MccH47, MccI47, and MccM biosynthesis are all located within a 10.5- to 14-kb DNA fragment (Laviña et al. 1990; Patzer et al. 2003; Vassiliadis et al. 2010). Analysis of the four microcin gene clusters showed that they share a common organization except for the 5′ region located upstream of *mchX* (Fig. 16.2). Each microcin gene cluster is characterized by common genes *(homologues)* involved in microcin export. Those are *mchE, mchF*, and *mcmM*, the latter being truncated in *E. coli* H47

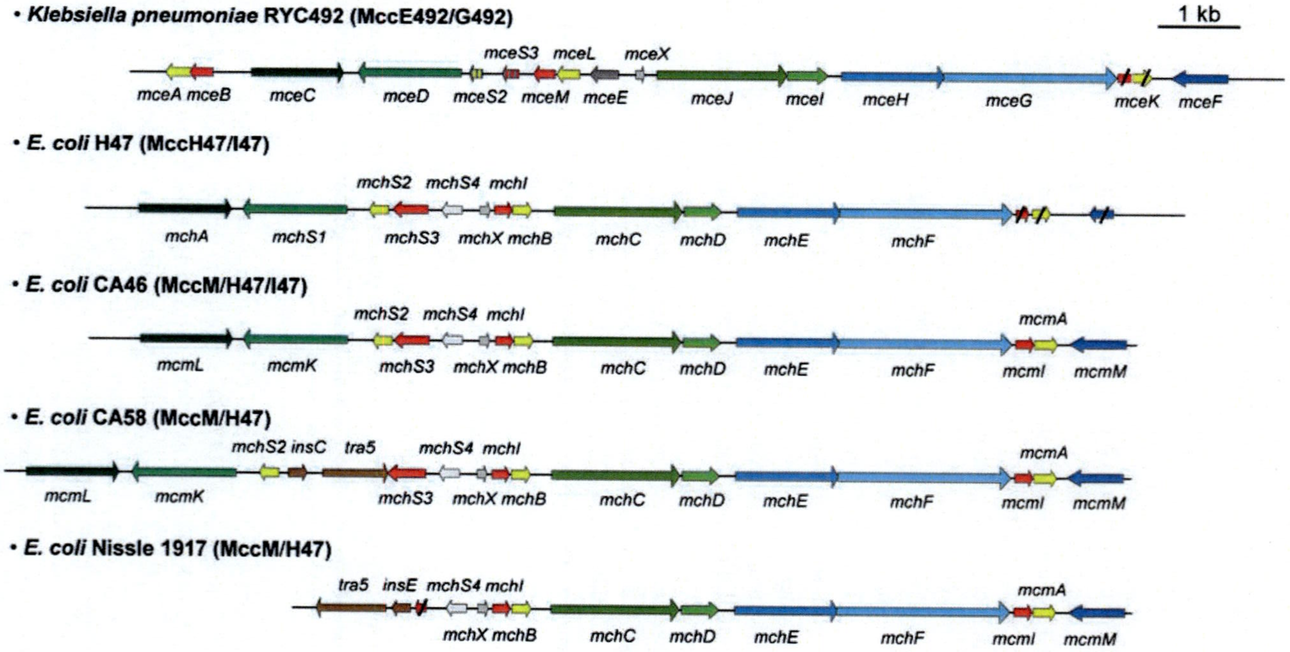

Fig. 16.2 *Genetic organization of class IIb microcin biosynthetic gene clusters*. A colour code identifies genes with similar functions. Genes encoding microcin precursors are shown in *yellow*. Genes required for self-immunity, microcin export, and posttranslational modifications are shown in *red*, *blue* and, *green*, respectively. Genes encoding transposases are shown in *brown*. Genes encoding proteins of unknown function are indicated either in *grey* or with vertical hatching. Genes encoding homologous or identical proteins in different clusters are coloured by different shades of the same colour. Truncated genes are crossed through

(Gaggero et al. 1993; Azpiroz et al. 2001; Patzer et al. 2003; Poey et al. 2006). Similarly, four genes required for the posttranslational modification of class IIb microcins are common to the microcin gene clusters. Those genes are *mchA*/*mcmL*, *mchS1*/*mcmK*, *mchC*, and *mchD*, which are homologous to *mceC*, *mceD*, *mceJ*, and *mceI* from MccE492 gene cluster, respectively (Vassiliadis et al. 2010). However, *mcmL* and *mcmK* are lacking in *E. coli* Nissle 1917. Finally, two specific genes encode each microcin precursor and self-immunity protein. Thus, *mchB* encodes the 75-amino-acid MccH47 precursor (Rodríguez et al. 1999), and *mchI* confers the self-immunity toward MccH47 (Rodríguez and Laviña 1998). The gene *mchS2* encodes the 77-amino-acid MccI47 precursor and *mchS3* confers the specific self-immunity toward MccI47 (Poey et al. 2006). The *mcmA* (formerly *mcmC*) (Braun et al. 2002) encodes the 92-amino-acid MccM precursor, and *mcmI* encodes the MccM self-immunity protein (Patzer et al. 2003). The role of *mchS4* and *mchX* located between *mchS3* and *mchI* remains unclear (Rodríguez and Laviña 1998; Azpiroz and Laviña 2004). In the 5′ region of microcin gene clusters from *E. coli* CA58 and Nissle 1917 are located genes encoding transposase and insertion sequences, which strongly supports the hypothesis of an horizontal gene transfer of microcin gene clusters. Complete or partial MccM gene cluster are also encountered in the pathogenicity island of the uropathogenic *E. coli* strains (Dobrindt et al. 2001; Welch et al. 2002; Dezfulian et al. 2004).

Structures

MccE492 was initially described as an 84-residue unmodified peptide (uMccE492) (Pons et al. 2002) deriving from the MccE492 precursor (MceA). Culture conditions were found to be critical to obtain mature MccE492, which carries an original posttranslational modification as revealed by mass spectrometry and nuclear magnetic resonance (Thomas et al. 2004). This modification consists of a C-glucosylated linear trimer of *N*-(2,3 dihydroxybenzoyl)-L-serine (DHBS) linked to the C-terminal serine carboxylate via an *O*-glycosidic bond (Fig. 16.3) (Thomas et al. 2004), a structure reminiscent of catecholate siderophores, and especially of salmochelin S4 (Hantke et al. 2003). Thus, MccE492 was the first example of a natural siderophore-peptide. With the characterization of novel microcins carrying a siderophore posttranslational modification, the name of siderophore-microcins was given to class IIb microcins. Indeed, MccM and MccH47 have been recently shown to be 77- and 60-residue peptides (Table 16.1) deriving from the microcin precursors McmA and MchB, and carrying a C-terminal posttranslational modification similar to that has previously been described for MccE492 (Vassiliadis et al. 2010). As such, they unambiguously belong to the siderophore-microcin family. However, when isolated from *E. coli* Nissle 1917, MccM and MccH47 did not carry any modification. The absence of posttranslational modification was correlated to the absence of *mcmL* and *mcmK* in microcin gene clusters, two genes involved in the biosynthesis of the

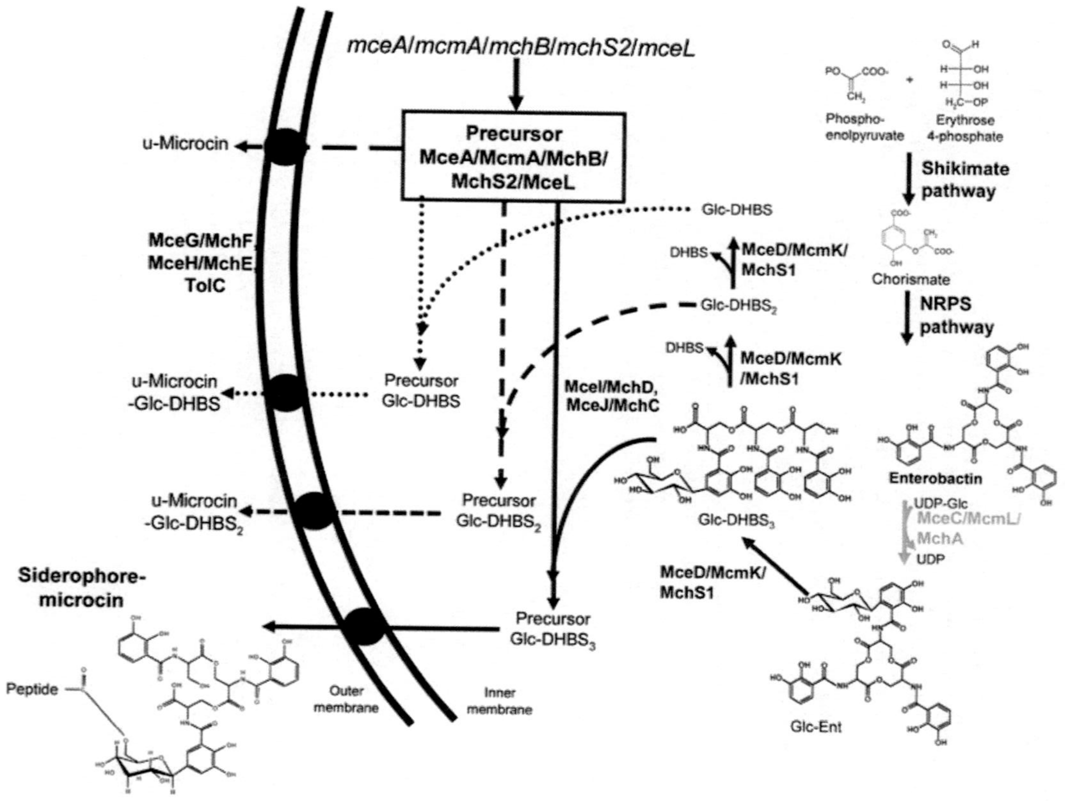

Fig. 16.3 *Model for the class IIb biosynthesis*. u-Microcin corresponds to unmodified microcin, while u-microcin-Glc-DHBS and u-microcin-Glc-DHBS$_2$ correspond to microcin intermediate forms. Ent, Glc, and DHBS stand for enterobactin, glucose, and *N*-(2,3 dihydroxybenzoyl)-L-serine, respectively

posttranslational modification (Vassiliadis et al. 2010). Two other putative members, MccG492 and MccI47, which remain to be fully characterized, would also belong to the siderophore-microcin family. After leader-peptide cleavage, putative MccG492 and MccI47 would contain 74 and 62 residues deriving from the microcin precursors MceL and MchS2, respectively (Table 16.1). Consequently, the class IIb microcins form the first homogeneous family among microcins, the siderophore-microcins.

Biosynthesis and Export

Several recent studies have shown that aside from the microcin gene clusters, genes involved in the biosynthesis of enterobactin and salmochelin are required for class IIb microcin biosynthesis and that both siderophores are precursors for the posttranslational modification (Azpiroz and Laviña 2004; Nolan et al. 2007; Vassiliadis et al. 2007; Mercado et al. 2008). This chapter describes the biosynthetic pathways that lead to siderophore-peptides, a unique combination of ribosomal and nonribosomal biosynthetic steps (Nolan and Walsh 2008).

Enterobactin and Salmochelin Biosynthesis

The maturation of class IIb microcins is dependent on catecholate siderophores. Catecholate-siderophores are named according to the chemical group that chelates iron and are one class of siderophores among the 500 siderophore structures described until now (Neilands 1995; Crosa and Walsh 2002; Miethke and Marahiel 2007). Enterobactin (also termed enterochelin) is a catecholate siderophore that was first isolated in *S. enterica* Typhimurium (Pollack and Neilands 1970) and *E. coli* (O'Brien et al. 1970), and later in *Streptomyces* (Fiedler et al. 2001), *Klebsiella*, and *Shigella* species (Payne et al. 1983; Podschun et al. 1992). It consists of a cyclic trimer of DHBS that chelates iron through its three dihydroxybenzoate (also termed catechol) groups. The coordination between enterobactin and iron is of the hexadentate type. Enterobactin was described as the best iron chelator isolated until now (Ecker et al. 1986). More recently, a new class of siderophore, named salmochelins, has been isolated from *S. enterica* Typhimurium (Hantke et al. 2003). Also produced by several uropathogenic *E. coli* and *K. pneumoniae* strains, salmochelins derive from enterobactin by the addition of one or two glucose on the catechol moiety.

Enterobactin Biosynthesis

Most of the known siderophores are synthesized by the multienzymatic pathway and particularly by the nonribosomal peptide synthetases (NRPS) pathway (Miethke and Marahiel 2007). Two main pathways lead to enterobactin biosynthesis.

The first pathway, which serves for the initial steps of enterobactin biosynthesis, is the shikimate pathway. From phosphoenol pyruvate and erythrose 4-phosphate, this pathway leads to chorismate, which is a common intermediate in enterobactin and aromatic-amino-acids biosynthesis. Genes involved in this pathway are *aroABCDEFGHKL*. Tyrosine, phenylalanine, and tryptophan have been described as inhibitors of the shikimate pathway (Cobbett and Delbridge 1987; Grove and Gunsalus 1987; Heatwole and Somerville 1992).

The second pathway, which serves for the last steps of enterobactin biosynthesis, involves an NRPS (Walsh et al. 1990; Crosa and Walsh 2002; Raymond et al. 2003). Extensive studies have been performed on specific NRPS that lead from chorismate to enterobactin. All genes involved in this pathway are located in the enterobactin gene cluster. This 20-kb fragment, organized in six operons, contains 16 genes. Seven (*entABCDEF* and *ybdB*) are involved in enterobactin synthesis, five (*fepABCDG*) in enterobactin uptake, one (*entS*) in enterobactin export, one (*fes*) in the release of iron from the siderophore, and two genes (*ybdz* and *fepE*) still have unknown functions. The enterobactin gene cluster is regulated by two *fur* (ferric uptake regulator) boxes, enabling the transcription regulation according to bacterial iron concentration (Brickman et al. 1990). Thus, to achieve enterobactin biosynthesis, chorismate is subjected to three reactions leading to dihydroxybenzoate (DHB) (Sakaitani et al. 1990; Gehring et al. 1998). Afterward, DHB is activated by adenylation and charged onto an aryl carrier protein. In parallel, a serine is also activated by adenylation and charged onto a peptidyl carrier protein (Reichert et al. 1992; Lambalot et al. 1996; Gehring et al. 1997). DHB and serine are finally ligated by a condensation domain and transferred onto another peptidyl carrier protein. These steps are repeated three times, and the final trimer of DHB-serine is transferred onto a thioesterification domain, which cyclizes and releases simultaneously the trimer to finally produce enterobactin (Shaw-Reid et al. 1999).

Salmochelin Biosynthesis

Conversion of enterobactin into salmochelins requires a 10-kb DNA fragment containing five genes, *iroBCDEN* (Bäumler et al. 1996, 1998), and two *fur* boxes (Bäumler et al. 1996). The gene *iroB* is involved in salmochelins biosynthesis, *iroC* in salmochelins export, *iroN* in salmochelins import, while *iroD* and *iroE* are involved in the release of iron chelated by salmochelins and degradation of these siderophores (Lin et al. 2005; Zhu et al. 2005). Enterobactin is di-glucosylated by IroB on two catechol groups, leading to salmochelin S4 formation (Bister et al. 2004; Fischbach et al. 2005). Salmochelin S2 is the linear form of salmochelin S4, while salmochelins S1, and SX are the monoglucosylated dimer and the monomer of DHBS, respectively. Salmochelins S2, S1, and SX are consequently degradation products of salmochelin S4.

Siderophore Microcin Biosynthesis

MccE492 was the first microcin found to carry a salmochelin-like siderophore as a posttranslational modification (Fig. 16.3). As a consequence, it has been the most extensively investigated of the siderophore-microcins. The genes *mceC* and *mceD* from MccE492 gene cluster display strong identities in sequences with *iroB* and *iroD*, which are involved in enterobactin glucosylation and linearization, respectively. In vivo and in vitro experiments showed that enterobatin and salmochelin are precursors for MccE492 posttranslational modification (Vassiliadis et al. 2007; Mercado et al. 2008). Thus, while microcins use siderophore biosynthetic pathways for the initial steps of their posttranslational modification, the last steps are encoded by the microcin gene clusters themselves. It has recently been shown for MccE492 that MceC and MceD monoglucosylate and linearize enterobactin, respectively (Nolan et al. 2007) and that MceI and MceJ are responsible for the *O*-glucosydic bond formation between the modified enterobactin and the peptide backbone (Nolan and Walsh 2008).

The biosynthesis of MccM and MccH47, the two recently characterized siderophore-microcins, was also investigated (Vassiliadis et al. 2010). Like MccE492, MccH47 posttranslational modification is dependent on the enterobactin synthesis pathway (Azpiroz and Laviña 2004). Interestingly, MchA and McmL display 85% identity with MceC, while MchS1 and McmK display 75% identity with MceD. By coupling gene complementation to mass spectrometry analysis, our group has recently showed that MchA, McmL, MceC, on one hand, and MchS1, McmK, MceD, on the other hand, are required for siderophore-microcin posttranslational modification and that each group of genes displays the same function, i.e., enterobactin *C*-glucosylation and linearization, respectively (Vassiliadis et al. 2007, 2010). Similarly, MchC and MchD display 71% and 80% identity with MceJ and MceI, respectively, and complement each other (Nolan and Walsh 2008). Therefore, MchC and MceJ, on one hand, as well as MchD and MceI, on the other hand, most likely display the same function in *O*-glucosylation of the class II microcin C-terminal serine. Together with earlier heterologous functional complementation assays (Poey et al. 2006), this indicates that all genes involved in siderophore-microcin posttranslational modification are probably interchangeable from one gene cluster to another.

MccI47 and the putative MccG492 have been very recently described, and as a consequence, little is known on their maturation. However, MccI47 structure and immunity genes are located within MccM/MccH47 gene cluster (Poey et al. 2006), while for MccG492, these genes are located within MccE492 gene cluster (Vassiliadis et al. 2010). Because MccM and MccH47 use the same genes for their posttranslational modification, one can speculate that these genes modify MccI47 so that it carries a similar siderophore posttranslational modification. Similarly, the genes required for MccE492 posttranslational modification could modify MccG492.

A model is proposed for siderophore-microcin biosynthesis (Fig. 16.3) in which after biosynthesis, enterobactin is *C*-glucosylated by MceC/McmL/MchA and linearized by MceD/McmK/MchS1. The linearized glucosylated enterobactin (salmochelin) is then linked to the microcin precursor MceA (for MccE492), McmA (for MccM), MchB (for MccH47), MchS2 (for MccI47), or MceL (for MccG492) through *O*-glycosylation. This biosynthetic step is catalyzed by MceIJ or MchCD. Finally, cleavage of the leader peptide occurs simultaneous to the export by a dedicated ABC transporter encoded by the microcin gene cluster (see "Genetics"). The existence of MccE492, MccM, and MccH47 intermediates carrying a dimer and a monomer of glucosylated DHBS instead of the classical trimer (Vassiliadis et al. 2007, 2010) could result from the degradation of glucosylated enterobactin by MceD/McmK/MchS1, followed by linkage to the microcin precursor and cleavage of the leader peptide, as described above.

Maturation and Export

Analyses of class IIb microcin gene clusters led to several hypotheses on the cleavage of class IIb microcin leader peptide and associated export. The mechanism by which the microcins are maturated and exported is supposed to be the same for all class II microcins (see "Maturation and Export"). The ABC transporters of MccE492/MccG492 and MccH47/MccM/MccI47 would be MceG and MchF, respectively. The accessory proteins of MccE492/MccG492 and MccH47/MccM/MccI47 would be MceH and MchE, respectively. Finally, TolC would be the outer membrane component of all class IIb microcin export machineries, as shown for MccH47, whose antimicrobial activity is dependent on *tolC* (Gaggero et al. 1993). An additional inner membrane protein could also be required for class IIb microcin export. Indeed, MccE492 export machinery seems to require MceF, a putative inner membrane protein that could interact with MceGH for processing or export (Lagos et al. 2001). Similarly, McmM, which displays 62% identity with MceF, could have the same function in MccM/MccH47/MccI47 export.

Recognition/Uptake and Mechanism of Action

Among class II microcins, the siderophore-microcins (class IIb microcins) have been the best studied in terms of mechanism of action. While the mechanisms responsible for class IIb microcin antibacterial activity are still incompletely understood, it is now established that their conserved C-terminal domain as well as their siderophore moiety are key determinants for their recognition/uptake at the outer membrane of enterobacteria.

Recognition/Uptake by the Receptors for Catecholate Siderophores

Spectra of antibacterial activity have been established very early for class IIb microcins. However, as presented in the Sect. Genetics, class IIb microcin-producing strains often express several class IIb microcins. The recent progresses in the purification of microcins have rendered possible the determination of reliable but nonexhaustive activity spectra. Thus, using pure microcin preparations or preparations that contain one single microcin, MccE492 was shown to be active against strains of *E. coli*, *Enterobacter cloacae*, *S. enterica*, and *K. pneumoniae,* MccM was active against *S. enterica* and *E. coli*, while MccH47 would to be active against *E. coli* only (Destoumieux-Garzón et al. 2003; Thomas et al. 2004; Vassiliadis et al. 2010).

- Determinants of class IIb microcin susceptibility in enterobacteria
  Two common features to all microcins are (1) an extremely potent activity, with minimum inhibitory concentrations (MICs) often below 0.1 μM (Thomas et al. 2004; Destoumieux-Garzón et al. 2005) and (2) a narrow spectrum of activity limited to few genera of enterobacteria, as illustrated above. The reason for that is a receptor-mediated recognition/uptake, which uses outer membrane proteins normally involved in nutrient uptake as receptors or docking molecules (Laviña et al. 1986; Pugsley et al. 1986; Salomón and Farias 1993, 1995; Trujillo et al. 2001). Those outer membrane proteins are also used by bacteriophages, antibiotics, and colicins for cell entry. They are considered an "Achilles' heel" for the bacterium.

  - The catecholate siderophore receptors FepA, Cir, Fiu, and IroN
    Over the past few years, substantial knowledge has been acquired on microcin recognition showing that only enterobacteria that express catecholate siderophore receptors are susceptible to class IIb microcins. In *E. coli,* class IIb microcins parasitize the FepA, Cir, and Fiu receptors involved in the uptake of catecholate siderophores (e.g. enterobactin) (Trujillo et al. 2001; Patzer et al. 2003; Thomas et al. 2004; Strahsburger et al. 2005). The requirement of MccE492 for catecholate siderophore receptors is illustrated by a shift from a potent activity against *E. coli* H1443 (MIC = 40 nM) to a lack of activity against the isogenic *fepA cir fiu* triple mutant (MIC > 10 μM) (Thomas et al. 2004). In *Salmonella*, class IIb microcins would use IroN, FepA, and Cir receptors (Patzer et al. 2003). FepA is by far the best studied of the four receptors. Its three-dimensional structure is composed of a β-barrel embedded in the outer membrane with an N-terminal globular domain, called either the plug or cork domain, folded inside the barrel (Buchanan et al. 1999). This domain spans most of the interior of the barrel and occludes it. The globular domain is connected to the β-barrel and to the external hydrophilic loops by numerous hydrogen bonds and salt bridges. The external loops contain the binding sites for iron-siderophore complexes (Annamalai et al. 2004). During the transport process, the globular domain of FepA would be expulsed from the β-barrel (Ma et al. 2007). Not only does FepA

transport catecholate siderophores but it also serves as a receptor/transporter for group B colicins and bacteriophages (Letellier and Santamaria 2002; Rabsch et al. 2007).

- The energy-transduction system TonB/ExbB/ExbD
  The catecholate siderophore receptors use the TonB-ExbB-ExbD inner membrane complex as an associated energy-transduction system for the active uptake of iron. This complex is responsible for the transduction of the proton-motive-force energy from the inner membrane, where it is generated, to the outer membrane, where it drives the active transport by high-affinity outer membrane transporters (Postle and Kadner 2003). ExbB and ExbD were proposed to influence the dimerization of TonB, although they would not be strictly required (Sauter et al. 2003), and TonB–ExbD interaction has been recently shown to be dependent on the proton-motive force (Ollis et al. 2009). TonB is required for the translocation across the outer membrane and the antibacterial activity of various microcins including class IIb microcins (Pugsley et al. 1986; Trujillo et al. 2001; Braun et al. 2002; Destoumieux-Garzón et al. 2003; Thomas et al. 2004; Strahsburger et al. 2005; Vassiliadis et al. 2010). However, ExbB and ExbD are often dispensable, as evidenced for MccE492, MccM, and MccH47 (Thomas et al. 2004; Vassiliadis et al. 2010). A similar ExbB-independent uptake has been reported for bacteriophage H8, which like class IIb microcins and group B colicins is a FepA ligand. Interestingly, when both ExbBD and TolQR systems are impaired, *E. coli* becomes resistant to bacteriophage H8 infection (Rabsch et al. 2007). Similar functional complementation of ExbBD by TolQR might occur for the uptake of class IIb microcins.

• Determinants of class IIb microcin recognition by catecholate siderophore receptors

  - The salmochelin posttranslational modification
    In a recent study, our group has showed that the three class IIb microcins on which structural data have been acquired (i.e., MccE492, MccM, and MccH47) possess a similar posttranslational modification (see "Structures") when expressed in a favorable genetic background (Vassiliadis et al. 2010). Since this modification mimics the catecholate siderophore salmochelin and behaves as a siderophore (Thomas et al. 2004), it was hypothesized that it could play a major role in class IIb microcin recognition. MccE492 is to date the only class IIb microcin for which modified and unmodified forms have been separated and purified in sufficient amounts to measure minimal inhibitory concentrations (MICs). Interestingly, the activity of the unmodified MccE492 (u-MccE492) against *E. coli* H1443 is four- to eightfold lower than that of the mature microcin (Thomas et al. 2004), indicating that the salmochelin posttranslational modification significantly enhances MccE492 recognition at the outer membrane of *E. coli*. As shown recently, the same posttranslational modification is observed in MccM, MccH47, and MccE492 (Vassiliadis et al. 2010) and is likely to occur in the recently

discovered MccI47 (Poey et al. 2006; Nolan and Walsh 2008) and MccG492 (Vassiliadis et al. 2010). Therefore, as shown for MccE492, the salmochelin posttran- slational modification is expected to increase the antibacterial activity of all class IIb microcins by mimicking the natural ligands of iron-siderophore receptors.

– The serine-rich C-terminal region
The unmodified MccE492 (u-MccE492) lacking the salmochelin posttranslational modification was shown to be active against *E. coli*, but, as the mature microcin, it was inactive against the isogenic *fepA cir fiu* triple mutant (Destoumieux-Garzón et al. 2006). This showed that molecular motives other than the sole posttranslational modification are involved in MccE492 recognition by the catecholate siderophore receptors. Later, Bieler et al. (2006) showed that the serine-rich C-terminal sequence of MccE492 was essential for the activity of extracellular, but not intracellular MccE492. Therefore, the C-terminal sequence is required for receptor recognition and/or translocation across the outer membrane. Because this C-terminal sequence is highly conserved among class IIb microcins, being considered as a signature for the family, it is very likely that, as demonstrated for MccE492, all class IIb microcins are recognized by catecholate-siderophore receptors through their serine-rich C-terminal sequence.

Antibacterial Activity of Class IIb Microcins

While class IIb microcins exhibit a conserved C-terminal domain required for recognition, the N-terminal domain is very diverse in terms of sequence (Table 16.1). Interestingly, the N-terminal domains of class IIb microcins display homologies with those of class IIa microcins. Therefore, class IIb N-terminal domains likely endow the antibacterial activity of the microcins. This is largely supported by the finding that a C-terminally truncated MccE492 remains active provided it is expressed intracellularly (Bieler et al. 2006). Among class IIb microcins, only MccE492 and MccH47 have been investigated in terms of mechanism of action, certainly due to the recent identification of MccI47, MccM, and MccG492. Data on those last three microcins are limited to their recognition at the outer membrane of *E. coli* (see the above section).

MccE492 membrane-disruption properties have been identified very early, well before the peptide was isolated and characterized. Thus, in vivo, culture supernatants containing MccE492 depolarised the inner membrane of *E. coli* (de Lorenzo and Pugsley 1985). This was later confirmed with homogeneous peptide preparations, showing that the MccE492 activity was energy- and TonB-dependent (Destoumieux-Garzón et al. 2003, 2006). Interestingly, the damages to the inner membrane were not responsible by themselves for the lethal effect of neither u-MccE492 nor MccE492 and did not result in cell lysis even at lethal concentrations (Destoumieux-Garzón et al. 2003, 2006). In vitro, the microcin was able to form ion channels in planar lipid bilayers (Lagos et al. 1993). The u-MccE492 pore-forming activity could be observed at concentrations as low as 2×10^{-10} M, and the insertion was shown to be voltage-independent (Destoumieux-Garzón et al. 2003). Altogether, this strongly suggests that the damages to *E. coli* inner membrane rely on a pore-forming activity.

Although the mechanistic details of MccE492 bactericidal activity are still incompletely understood, Bieler et al. (2006) evidenced that MccE492 targets the mannose permease, manY and manZ being critical genes for MccE492 antibacterial activity against *E. coli*. ManYZ is an inner membrane complex that functions together with the cytoplasmic ManX to form the mannose permease involved in the uptake of mannose and related hexoses (Williams et al. 1986; Erni et al. 1987). The *manYZ* mutants resistant to MccE492 were unable to metabolize mannose and became insensitive to the inner membrane depolarization mediated by periplasmic MccE492 (Bieler et al. 2006). Therefore, the mannose permease is required for MccE492 antibacterial activity but may not be the only intracellular target of the microcin.

Unlike MccE492, MccH47 has never been purified to homogeneity (Vassiliadis et al. 2010). Nonetheless, several studies by Laviña and collaborators have documented the mechanism of action of MccH47. The authors initially showed that the *atp* operon conferred susceptibility to MccH47 (Rodríguez et al. 1999). Later, they found that MccH47 targets the F_0F_1 ATP synthase, particularly its F_0 membrane component, which serves as a proton channel, while the F_1 catalytic unit was dispensable (Trujillo et al. 2001; Rodríguez and Laviña 2003). We have recently found that contrary to MccE492, MccM and MccH47 are active against *manXYZ* mutants (Peduzzi and Vandervennet unpublished results). Therefore, the mannose permease is not required for all siderophore-microcins antibacterial activity. Consequently, while the conserved C-terminal sequence of class IIb microcins is responsible for their recognition, this reinforces our hypothesis that their nonconserved N-terminal sequence is responsible for the interference with specific cellular functions.

Concluding Remarks

The recent progresses on class II microcins gathered in this review have shown that their mechanism of action relies on the recognition by iron-siderophore receptors and that one subclass (class IIb) uses molecular mimicry to improve recognition by iron receptors. Interestingly, the siderophores, which serve as precursors for the posttranslational modification of class IIb microcins, are recognized as important virulence factors involved in the battle for iron (Demir and Kaleli 2004; Lawlor et al. 2007; Caza et al. 2008). Moreover, the siderophore-encoding gene clusters are usually located in genomic islands, as also found for class IIb microcins in *E. coli* Nissle 1917 (Grozdanov et al. 2004). Together with the modular structure of their gene cluster, this strongly suggests that class II microcins are subject to extensive horizontal gene transfer, in agreement with an important role in microbial competition and/or virulence. The recent studies on colicins, large antimicrobial proteins expressed by enterobacteria that use similar iron receptors for recognition, indicate that their expression favors bacterial persistence in the gastrointestinal tract (Kirkup and Riley 2004; Gillor et al. 2009). Together with a recent study on uropathogenic class II microcins (Azpiroz et al. 2009), this strongly supports the role of class II microcins in enterobacterial virulence and in vivo fitness.

Acknowledgements We are grateful to Dominique Belin for the generous gifts of the *E. coli manXYZ* mutants.

References

Andrews SC, Robinson AK, Rodriguez-Quinones F (2003) Bacterial iron homeostasis. FEMS Microbiol Rev 27:215–237

Annamalai R, Jin B, Cao Z et al (2004) Recognition of ferric catecholates by FepA. J Bacteriol 186:3578–3589

Azpiroz MF, Laviña M (2004) Involvement of enterobactin synthesis pathway in production of microcin H47. Antimicrob Agents Chemother 48:1235–1241

Azpiroz MF, Rodríguez E, Laviña M (2001) The structure, function, and origin of the microcin H47 ATP-binding cassette exporter indicate its relatedness to that of colicin V. Antimicrob Agents Chemother 45:969–972

Azpiroz MF, Poey ME, Laviña M (2009) Microcins and urovirulence in *Escherichia coli*. Microb Pathog 47:274–280

Bäumler AJ, Tsolis RM, van der Velden AW et al (1996) Identification of a new iron regulated locus of *Salmonella typhi*. Gene 183:207–213

Bäumler AJ, Norris TL, Lasco T et al (1998) IroN, a novel outer membrane siderophore receptor characteristic of *Salmonella enterica*. J Bacteriol 180:1446–1453

Benabdelhak H, Kiontke S, Horn C et al (2003) A specific interaction between the NBD of the ABC-transporter HlyB and a *C*-terminal fragment of its transport substrate haemolysin A. J Mol Biol 327:1169–1179

Bieler S, Silva F, Soto C et al (2006) Bactericidal activity of both secreted and nonsecreted microcin E492 requires the mannose permease. J Bacteriol 188:7049–7061

Bister B, Bischoff D, Nicholson GJ et al (2004) The structure of salmochelins: C-glucosylated enterobactins of *Salmonella enterica*. Biometals 17:471–481

Braun V, Patzer SI, Hantke K (2002) Ton-dependent colicins and microcins: modular design and evolution. Biochimie 84:365–380

Brickman TJ, Ozenberger BA, McIntosh MA (1990) Regulation of divergent transcription from the iron-responsive fepB-entC promoter-operator regions in *Escherichia coli*. J Mol Biol 212:669–682

Buchanan SK, Smith BS, Venkatramani L et al (1999) Crystal structure of the outer membrane active transporter FepA from *Escherichia coli*. Nat Struct Biol 6:56–63

Caza M, Lepine F, Milot S et al (2008) Specific roles of the iroBCDEN genes in virulence of an avian pathogenic *Escherichia coli* O78 strain and in production of salmochelins. Infect Immun 76:3539–3549

Chehade H, Braun V (1988) Iron-regulated synthesis and uptake of colicin V. FEMS Microbiol Lett 52:177–182

Cobbett CS, Delbridge ML (1987) Regulatory mutants of the *aroF-tyrA* operon of *Escherichia coli* K-12. J Bacteriol 169:2500–2506

Corsini G, Baeza M, Monasterio O et al (2002) The expression of genes involved in microcin maturation regulates the production of active microcin E492. Biochimie 84:539–544

Crosa JH, Walsh CT (2002) Genetics and assembly line enzymology of siderophore biosynthesis in bacteria. Microbiol Mol Biol Rev 66:223–249

Davidson AL, Maloney PC (2007) ABC transporters: how small machines do a big job. Trends Microbiol 15:448–455

de Keyzer J, van der Does C, Driessen AJ (2003) The bacterial translocase: a dynamic protein channel complex. Cell Mol Life Sci 60:2034–2052

de Lorenzo V (1984) Isolation and characterization of microcin E492 from *Klebsiella pneumoniae*. Arch Microbiol 139:72–75

de Lorenzo V, Pugsley AP (1985) Microcin E492, a low-molecular-weight peptide antibiotic which causes depolarization of the *Escherichia coli* cytoplasmic membrane. Antimicrob Agents Chemother 27:666–669

Demir M, Kaleli I (2004) Production by *Escherichia coli* isolates of siderophore and other virulence factors and their pathogenic role in a cutaneous infection model. Clin Microbiol Infect 10:1011–1014

Destoumieux-Garzón D, Thomas X, Santamaria M et al (2003) Microcin E492 antibacterial activity: evidence for a TonB-dependent inner membrane permeabilization on *Escherichia coli*. Mol Microbiol 49:1031–1041

Destoumieux-Garzón D, Duquesne S, Peduzzi J et al (2005) The iron-siderophore transporter FhuA is the receptor for the antimicrobial peptide microcin J25: role of the microcin Val11-Pro16 beta-hairpin region in the recognition mechanism. Biochem J 389:869–876

Destoumieux-Garzón D, Peduzzi J, Thomas X et al (2006) Parasitism of iron-siderophore receptors of *Escherichia coli* by the siderophore-peptide microcin E492m and its unmodified counterpart. Biometals 19:181–191

Dezfulian H, Tremblay D, Harel J (2004) Molecular characterization of extraintestinal pathogenic *Escherichia coli* (ExPEC) pathogenicity islands in F165-positive *E. coli* strain from a diseased animal. FEMS Microbiol Lett 238:321–332

Dobrindt U, Blum-Oehler G, Hartsch T et al (2001) S-Fimbria-encoding determinant sfa(I) is located on pathogenicity island III(536) of uropathogenic *Escherichia coli* strain 536. Infect Immun 69:4248–4256

Duquesne S, Destoumieux-Garzón D, Peduzzi J et al (2007) Microcins, gene-encoded antibacterial peptides from enterobacteria. Nat Prod Rep 24:708–734

Ecker DJ, Matzanke BF, Raymond KN (1986) Recognition and transport of ferric enterobactin in *Escherichia coli*. J Bacteriol 167:666–673

Erni B, Zanolari B, Kocher HP (1987) The mannose permease of *Escherichia coli* consists of three different proteins. Amino acid sequence and function in sugar transport, sugar phosphorylation, and penetration of phage lambda DNA. J Biol Chem 262:5238–5247

Fiedler HP, Krastel P, Müller J et al (2001) Enterobactin: the characteristic catecholate siderophore of *Enterobacteriaceae* is produced by *Streptomyces* species. FEMS Microbiol Lett 196:147–151

Fischbach MA, Lin H, Liu DR et al (2005) In vitro characterization of IroB, a pathogen-associated *C*-glycosyltransferase. Proc Natl Acad Sci USA 102:571–576

Franke CM, Leenhouts KJ, Haandrikman AJ et al (1996) Topology of LcnD, a protein implicated in the transport of bacteriocins from *Lactococcus lactis*. J Bacteriol 178:1766–1769

Franke CM, Tiemersma J, Venema G et al (1999) Membrane topology of the lactococcal bacteriocin ATP-binding cassette transporter protein LcnC. Involvement of LcnC in lactococcin a maturation. J Biol Chem 274:8484–8490

Fredericq P, Joiris E, Betz-Barreau M et al (1949) Recherche des gènes producteurs de colicines dans les selles de malades atteints de fièvre paratyphoïde B. C R Soc Biol 143:556–559

Gaggero C, Moreno F, Laviña M (1993) Genetic analysis of microcin H47 antibiotic system. J Bacteriol 175:5420–5427

Gaillard-Gendron S, Vignon D, Cottenceau G et al (2000) Isolation, purification and partial amino acid sequence of a highly hydrophobic new microcin named microcin L produced by *Escherichia coli*. FEMS Microbiol Lett 193:95–98

Gehring AM, Bradley KA, Walsh CT (1997) Enterobactin biosynthesis in *Escherichia coli*: isochorismate lyase (EntB) is a bifunctional enzyme that is phosphopantetheinylated by EntD and then acylated by EntE using ATP and 2, 3-dihydroxybenzoate. Biochemistry 36:8495–8503

Gehring AM, Mori I, Walsh CT (1998) Reconstitution and characterization of the *Escherichia coli* enterobactin synthetase from EntB, EntE, and EntF. Biochemistry 37:2648–2659

Gerard F, Pradel N, Wu LF (2005) Bactericidal activity of colicin V is mediated by an inner membrane protein, SdaC, of *Escherichia coli.* J Bacteriol 187:1945–1950

Gillor O, Giladi I, Riley MA (2009) Persistence of colicinogenic *Escherichia coli* in the mouse gastrointestinal tract. BMC Microbiol 9:165

Gilson L, Mahanty HK, Kolter R (1987) Four plasmid genes are required for colicin V synthesis, export, and immunity. J Bacteriol 169:2466–2470

Gilson L, Mahanty HK, Kolter R (1990) Genetic analysis of an MDR-like export system: the secretion of colicin V. EMBO J 9:3875–3884

Gratia A (1925) Sur un remarquable exemple d'antagonisme entre deux souches de colibacille. C R Soc Biol 93:1041–1042

Grove CL, Gunsalus RP (1987) Regulation of the *aroH* operon of *Escherichia coli* by the tryptophan repressor. J Bacteriol 169:2158–2164

Grozdanov L, Raasch C, Schulze J et al (2004) Analysis of the genome structure of the nonpathogenic probiotic *Escherichia coli* strain Nissle 1917. J Bacteriol 186:5432–5441

Guo X, Harrison RW, Tai PC (2006) Nucleotide-dependent dimerization of the C-terminal domain of the ABC transporter CvaB in colicin V secretion. J Bacteriol 188:2383–2391

Hantke K, Nicholson G, Rabsch W et al (2003) Salmochelins, siderophores of *Salmonella enterica* and uropathogenic *Escherichia coli* strains, are recognized by the outer membrane receptor IroN. Proc Natl Acad Sci USA 100:3677–3682

Håvarstein LS, Holo H, Nes IF (1994) The leader peptide of colicin V shares consensus sequences with leader peptides that are common among peptide bacteriocins produced by gram-positive bacteria. Microbiology 140:2383–2389

Håvarstein LS, Diep DB, Nes IF (1995) A family of bacteriocin ABC transporters carry out proteolytic processing of their substrates concomitant with export. Mol Microbiol 16:229–240

Heatwole VM, Somerville RL (1992) Synergism between the Trp repressor and Tyr repressor in repression of the *aroL* promoter of *Escherichia coli* K-12. J Bacteriol 174:331–335

Hollenstein K, Dawson RJ, Locher KP (2007) Structure and mechanism of ABC transporter proteins. Curr Opin Struct Biol 17:412–418

Kirkup BC, Riley MA (2004) Antibiotic-mediated antagonism leads to a bacterial game of rock-paper-scissors in vivo. Nature 428:412–414

Koronakis V, Sharff A, Koronakis E et al (2000) Crystal structure of the bacterial membrane protein TolC central to multidrug efflux and protein export. Nature 405:914–919

Koronakis V, Eswaran J, Hughes C (2004) Structure and function of TolC: the bacterial exit duct for proteins and drugs. Annu Rev Biochem 73:467–489

Lagos R, Wilkens M, Vergara C et al (1993) Microcin E492 forms ion channels in phospholipid bilayer membrane. FEBS Lett 321:145–148

Lagos R, Villanueva JE, Monasterio O (1999) Identification and properties of the genes encoding microcin E492 and its immunity protein. J Bacteriol 181:212–217

Lagos R, Baeza M, Corsini G et al (2001) Structure, organization and characterization of the gene cluster involved in the production of microcin E492, a channel-forming bacteriocin. Mol Microbiol 42:229–243

Lambalot RH, Gehring AM, Flugel RS et al (1996) A new enzyme superfamily – the phosphopantetheinyl transferases. Chem Biol 3:923–936

Laviña M, Pugsley AP, Moreno F (1986) Identification, mapping, cloning and characterization of a gene (sbmA) required for microcin B17 action on *Escherichia coli* K12. J Gen Microbiol 132:1685–1693

Laviña M, Gaggero C, Moreno F (1990) Microcin H47, a chromosome-encoded microcin antibiotic of *Escherichia coli.* J Bacteriol 172:6585–6588

Lawlor MS, O'Connor C, Miller VL (2007) Yersiniabactin is a virulence factor for *Klebsiella pneumoniae* during pulmonary infection. Infect Immun 75:1463–1472

Letellier L, Santamaria M (2002) The biochemical and physiological characteristics of surface receptors of gram negative bacteria. Mini Rev Med Chem 2:343–351

Lin H, Fischbach MA, Liu DR et al (2005) In vitro characterization of salmochelin and enterobactin trilactone hydrolases IroD, IroE, and Fes. J Am Chem Soc 127:11075–11084

Ma L, Kaserer W, Annamalai R et al (2007) Evidence of ball-and-chain transport of ferric enterobactin through FepA. J Biol Chem 282:397–406

Mercado G, Tello M, Marin M et al (2008) The production in vivo of microcin E492 with antibacterial activity depends on salmochelin and EntF. J Bacteriol 190:5464–5471

Miethke M, Marahiel MA (2007) Siderophore-based iron acquisition and pathogen control. Microbiol Mol Biol Rev 71:413–451

Moussatova A, Kandt C, O'Mara ML et al (2008) ATP-binding cassette transporters in *Escherichia coli.* Biochim Biophys Acta 1778:1757–1771

Neilands JB (1995) Siderophores: structure and function of microbial iron transport compounds. J Biol Chem 270:26723–26726

Nissle A (1925) Weiteres über Grundlagen und Praxis der Mutaflorbehandlung. Dtsch Med Wochenschr 44:1809–1813

Nolan EM, Walsh CT (2008) Investigations of the MceIJ-catalyzed posttranslational modification of the microcin E492 C-terminus: linkage of ribosomal and nonribosomal peptides to form "trojan horse" antibiotics. Biochemistry 47:9289–9299

Nolan EM, Fischbach MA, Koglin A et al (2007) Biosynthetic tailoring of microcin E492m: post-translational modification affords an antibacterial siderophore-peptide conjugate. J Am Chem Soc 129:14336–14347

O'Brien GJ, Mahanty HK (1994) Colicin 24, a new plasmid-borne colicin from a uropathogenic strain of *Escherichia coli.* Plasmid 31:288–296

O'Brien IG, Cox GB, Gibson F (1970) Biologically active compounds containing 2, 3-dihydroxybenzoic acid and serine formed by Escherichia coli. Biochim Biophys Acta 201:453–460

Ollis AA, Manning M, Held KG et al (2009) Cytoplasmic membrane protonmotive force energizes periplasmic interactions between ExbD and TonB. Mol Microbiol 73:466–481

Patzer SI, Baquero MR, Bravo D et al (2003) The colicin G, H and X determinants encode microcins M and H47, which might utilize the catecholate siderophore receptors FepA, Cir, Fiu and IroN. Microbiology 149:2557–2570

Payne SM, Niesel DW, Peixotto SS et al (1983) Expression of hydroxamate and phenolate siderophores by *Shigella flexneri.* J Bacteriol 155:949–955

Podschun R, Fischer A, Ullmann U (1992) Siderophore production of *Klebsiella* species isolated from different sources. Zentralbl Bakteriol 276:481–486

Poey ME, Azpiroz MF, Laviña M (2006) Comparative analysis of chromosome-encoded microcins. Antimicrob Agents Chemother 50:1411–1418

Pollack JR, Neilands JB (1970) Enterobactin, an iron transport compound from *Salmonella Thyphimurium.* Biochem Biophys Res Commun 38:989–992

Pons AM, Zorn N, Vignon D et al (2002) Microcin E492 is an unmodified peptide related in structure to colicin V. Antimicrob Agents Chemother 46:229–230

Pons AM, Delalande F, Duarte M et al (2004) Genetic analysis and complete primary structure of microcin L. Antimicrob Agents Chemother 48:505–513

Postle K, Kadner RJ (2003) Touch and go: tying TonB to transport. Mol Microbiol 49:869–882

Pugsley AP, Moreno F, de Lorenzo V (1986) Microcin-E492-insensitive mutants of *Escherichia coli* K12. J Gen Microbiol 132:3253–3259

Rabsch W, Ma L, Wiley G et al (2007) FepA- and TonB-dependent bacteriophage H8: receptor binding and genomic sequence. J Bacteriol 189:5658–5674

Raymond KN, Dertz EA, Kim SS (2003) Enterobactin: an archetype for microbial iron transport. Proc Natl Acad Sci USA 100:3584–3588

Reichert J, Sakaitani M, Walsh CT (1992) Characterization of EntF as a serine-activating enzyme. Protein Sci 1:549–556

Rodríguez E, Laviña M (1998) Genetic analysis of microcin H47 immunity. Can J Microbiol 44:692–697
Rodríguez E, Laviña M (2003) The proton channel is the minimal structure of ATP synthase necessary and sufficient for microcin h47 antibiotic action. Antimicrob Agents Chemother 47:181–187
Rodríguez E, Gaggero C, Laviña M (1999) The structural gene for microcin H47 encodes a peptide precursor with antibiotic activity. Antimicrob Agents Chemother 43:2176–2182
Sablé S, Duarte M, Bravo D et al (2003) Wild-type *Escherichia coli* producing microcins B17, D93, J25, and L; cloning of genes for microcin L production and immunity. Can J Microbiol 49:357–361
Sakaitani M, Rusnak F, Quinn NR et al (1990) Mechanistic studies on trans-2, 3-dihydro-2, 3-dihydroxybenzoate dehydrogenase (Ent A) in the biosynthesis of the iron chelator enterobactin. Biochemistry 29:6789–6798
Salomón RA, Farias RN (1993) The FhuA protein is involved in microcin 25 uptake. J Bacteriol 175:7741–7742
Salomón RA, Farias RN (1995) The peptide antibiotic microcin 25 is imported through the TonB pathway and the SbmA protein. J Bacteriol 177:3323–3325
Sauter A, Howard SP, Braun V (2003) In vivo evidence for TonB dimerization. J Bacteriol 185:5747–5754
Shaw-Reid CA, Kelleher NL, Losey HC et al (1999) Assembly line enzymology by multimodular nonribosomal peptide synthetases: the thioesterase domain of *E. coli* EntF catalyzes both elongation and cyclolactonization. Chem Biol 6:385–400
Skvirsky RC, Reginald S, Shen X (1995) Topology analysis of the colicin V export protein CvaA in *Escherichia coli*. J Bacteriol 177:6153–6159
Strahsburger E, Baeza M, Monasterio O et al (2005) Cooperative uptake of microcin E492 by receptors FepA, Fiu, and Cir and inhibition by the siderophore enterochelin and its dimeric and trimeric hydrolysis products. Antimicrob Agents Chemother 49:3083–3086
Thomas X, Destoumieux-Garzón D, Péduzzi J et al (2004) Siderophore peptide, a new type of post-translationally modified antibacterial peptide with potent activity. J Biol Chem 279:28233–28242
Trujillo M, Rodríguez E, Laviña M (2001) ATP synthase is necessary for microcin H47 antibiotic action. Antimicrob Agents Chemother 45:3128–3131
Vassiliadis G, Peduzzi J, Zirah S et al (2007) Insight into siderophore-carrying peptide biosynthesis: enterobactin is a precursor for microcin E492 posttranslational modification. Antimicrob Agents Chemother 51:3546–3553
Vassiliadis G, Destoumieux-Garzón D, Lombard C et al (2010) Isolation and characterization of two members of the siderophore-microcin family, microcins M and H47. Antimicrob Agents Chemother 54:288–297
Walsh CT, Liu J, Rusnak F et al (1990) Molecular studies on enzymes in chorismate metabolism and the enterobactin biosynthetic pathway. Chem Rev 90:1105–1129
Waters VL, Crosa JH (1991) Colicin V virulence plasmids. Microbiol Rev 55:437–450
Welch RA, Burland V, Plunkett G 3rd et al (2002) Extensive mosaic structure revealed by the complete genome sequence of uropathogenic *Escherichia coli*. Proc Natl Acad Sci USA 99:17020–17024
Wilkens M, Villanueva JE, Cofre J et al (1997) Cloning and expression in *Escherichia coli* of genetic determinants for production of and immunity to microcin E492 from *Klebsiella pneumoniae*. J Bacteriol 179:4789–4794
Williams N, Fox DK, Shea C et al (1986) Pel, the protein that permits lambda DNA penetration of *Escherichia coli*, is encoded by a gene in *ptsM* and is required for mannose utilization by the phosphotransferase system. Proc Natl Acad Sci USA 83:8934–8938
Wooley RE, Gibbs PS, Shotts EB Jr (1999) Inhibition of *Salmonella typhimurium* in the chicken intestinal tract by a transformed avirulent avian *Escherichia coli*. Avian Dis 43:245–250
Wu TK, Huang CY, Ko CY et al (2004) Purification, tandem mass characterization, and inhibition studies of oxidosqualene-lanosterol cyclase enzyme from bovine liver. Arch Biochem Biophys 421:42–53

Yang CC, Konisky J (1984) Colicin V-treated *Escherichia coli* does not generate membrane potential. J Bacteriol 158:757–759

Zhong X, Kolter R, Tai PC (1996) Processing of colicin V-1, a secretable marker protein of a bacterial ATP binding cassette export system, requires membrane integrity, energy, and cytosolic factors. J Biol Chem 271:28057–28063

Zhu M, Valdebenito M, Winkelmann G et al (2005) Functions of the siderophore esterases IroD and IroE in iron-salmochelin utilization. Microbiology 151:2363–2372

Chapter 17
Microcins from Enterobacteria: On the Edge Between Gram-Positive Bacteriocins and Colicins

Sylvie Rebuffat

Abstract Most bacteria and archaea produce gene-encoded antimicrobial peptides/proteins called bacteriocins, which are secreted by the producing bacteria to compete against other microorganisms in a given niche. They are considered important mediators of intra- and interspecies interactions and therefore a factor in maintaining the microbial diversity and stability. They are ribosomally synthesized, and most of them are produced as inactive precursor proteins, which in some cases are further enzymatically modified. Bacteriocins generally exert potent antibacterial activities directed against bacterial species closely related to the producing bacteria. Bacteriocins are abundant and diverse in Gram-negative and Gram-positive bacteria. This chapter focuses on colicins and microcins from enterobacteria (mainly *Escherichia coli*) and on bacteriocins from lactic acid bacteria (LAB). Microcins are the lower-molecular-mass bacteriocins produced by Gram-negative bacteria with a repertoire of only 14 representatives. They form a very restricted family of bacteriocins, compared to the huge family of LAB bacteriocins that is constituted of several hundreds of peptides, with which microcins share common characteristics. Nevertheless, microcins also show similarities, particularly in their uptake mechanisms, with the higher-molecular-mass colicins, also produced by *E. coli* strains. On the edge between LAB bacteriocins and colicins, microcins appear to combine highly efficient strategies developed by both Gram-positive and Gram-negative bacteria at different levels, including uptake, translocation, killing of target cells, and immunity of the producing bacteria, making them important actors of bacterial competitions and fascinating models for novel concepts toward antimicrobial strategies and against resistance mechanisms.

S. Rebuffat (✉)
Muséum National d'Histoire Naturelle – Centre National de la Recherche Scientifique,
Laboratory of Communication Molecules and Adaptation of Microorganisms,
UMR 7245 CNRS-MNHN, CP 54, 57 rue Cuvier, 75005 Paris, France
e-mail: rebuffat@mnhn.fr

D. Drider and S. Rebuffat (eds.), *Prokaryotic Antimicrobial Peptides: From Genes to Applications*, DOI 10.1007/978-1-4419-7692-5_17,

Introduction

Most species of bacteria produce a broad array of chemical compounds acting as antimicrobials (Tagg et al. 1976). These include short-chain fatty acids, such as lactic acid, and hydrogen peroxide (H_2O_2) produced by LAB, protein toxins, and antimicrobial peptides that are termed bacteriocins. Bacteriocins are numerous and diverse in Gram-negative and Gram-positive bacteria; however, the production of bacteriocins is mainly encountered in enterobacteria and LAB. Bacteriocins act as important mediators of intra- and interspecies interactions, therefore representing a considerable factor in maintaining microbial diversity. Indeed, bacteriocins are of great interest in microbial ecology. To defend their ecological niche against competitors, bacteria develop original, complex, and subtle strategies for bacteriocins in terms of mechanisms of import into target microorganisms, mechanisms of killing the target microorganisms, and resistance of the producing strains to their own toxin(s). Therefore, bacteriocins constitute interesting models for the elaboration of novel strategies to fight against microorganisms, either for the conception of novel antibiotics with low levels of resistance or for the development of probiotic strains, since the production of bacteriocins is considered an important trait for a successful probiotic strain (Gillor et al. 2008). The strategies developed by Gram-negative and Gram-positive bacteria to act as potent competitors in their biotopes through the production of bacteriocins are compared and discussed in this chapter. Gram-negative bacteria produce both higher-molecular-mass (between 30 and 80 kDa) and lower-molecular-mass (<10 kDa) bacteriocins, termed colicins and microcins, respectively. On the other hand, Gram-positive bacteriocins range between 1 and 10 kDa. Table 17.1 sums up the main characteristics of these different classes of bacteriocins at different levels, in terms of structures, genetic systems and biosyntheses, mechanisms of action, and mechanisms of immunity of the producing strains. The chapter also provides a summary of the defense strategies associated with bacteriocins.

Structures and Features of Production

Structural diversity. Microcins and LAB bacteriocins share a number of common characteristics, in particular in terms of molecular masses (below 10 kDa), stability to high temperatures and extreme pHs, and to degradation by proteases in most cases (for reviews, see Drider et al. 2006; Cotter et al. 2005; Garneau et al. 2002; Jack et al. 1995). The three-dimensional structures of some bacteriocins have been determined by NMR/molecular modelling or by X-ray crystallography. They are extremely diverse and variable. For instance, circular bacteriocins adopt a four or five tightly packed α-helix structure (Sánchez-Barrena et al. 2003; Langdon et al. 1998; see Chap. 12 by John Vederas & coll.), while pediocin-like bacteriocins adopt structures assembling α-helix and β-sheet domains connected by hinges or unstructured regions (Drider et al. 2006). Other bacteriocins have completely

Table 17.1 Main characteristics of bacteriocins from Gram-positive and Gram-negative bacteria: colicins, microcins and LAB bacteriocins

Bacteriocins Characteristics	Colicins	Microcins	LAB bacteriocins
Structure			
Molecular mass	Proteins: 30–80 kDa	Peptides: 1–10 kDa Class I: 1–3 kDa Class II: 7–10 kDa	Peptides: 1–30 kDa Class I: <5 kDa Class II: <5 kDa Class III: >30 kDa
Sensitivity to proteases	Sensitive	± resistant according to the classes (class I more resistant)	± resistant according to the classes (class I more resistant)
Posttranslational modification	No	Yes: Class I Class IIb (siderophore)	Yes: Class I (lantibiotics) Class IIc (circular) No: other classes
Consensus sequence	No	Yes: Class IIb (C-terminal)	Yes: Class IIa (N-terminal) YGNGV sequence
Modular structure	Yes	Yes: Class IIb	Yes: Class IIb
Genetic system	Plasmid-encoded 3 genes	Plasmid-, chromosome-encoded 3–10 genes	Transposon-, plasmid-, chromosome-encoded 4–13 genes
Biosynthesis	No precursor, no leader	Precursor larger that the mature peptide and including a leader	Precursor larger that the mature peptide and including a leader
Mechanism of action	Trojan horse mechanisms	Trojan horse mechanisms	Dual mechanisms
Recognition/uptake	Outer membrane receptors (iron siderophore)/porine	Outer membrane receptors (iron siderophore): class I, class IIb Mannose permease (class IIb)	Mannose permease (class IIa, class IId)
Translocation	Tol/Pal or Ton B system	Ton B system	–
Killing mechanism	Membrane permeabilization Nuclease Peptidoglycan synthesis inhibition	Enzyme inhibition Membrane permeabilization Mannose permease requirement (class IIb: microcin E492)	Membrane permeabilization Mannose permease requirement (class IIb: mesentericin Y105)
Immunity	Immunity protein	Immunity protein (class IIb) Excretion through the export system (class I)	Immunity protein (class I; classes IIa, IIb, IId) ABC transporter (class I, class IIc)

atypical structures, as shown for lantibiotics (Chatterjee et al. 2005; see Chap. 9 by Gert Moll & coll.) and the lasso-peptide microcin J25 (Duquesne et al. 2007a, b).

Fourteen microcins only have been characterized until now. Despite this low number of representatives, they exhibit a large structural diversity that allows classifying them into two classes. One class encompasses microcins of lower molecular masses (class I; 1–3 kDa), which undergo extensive posttranslational modification (nucleotide-peptide, lasso-peptide…). The other class is composed of microcins of higher molecular masses (class II; 6–10 kDa) that can bear or not bear a C-terminal modification (siderophore-microcins) (Duquesne et al. 2007a; refer to Chap. 4 by S. Rebuffat for the detailed description of this classification). Forming a much larger group that includes several hundreds of representatives, LAB bacteriocins exhibit very high structural diversity, which conducted to several classifications (refer to Chap. 3 by Paul D. Cotter & coll. for an updated description of this classification). Among them, the following classes are considered in this chapter: (1) lantibiotics (class I of highly modified bacteriocins containing lanthionines and/or β-methyllanthionines; for reviews see Patton and van der Donk 2005; Chatterjee et al. 2005; for a detailed description refer to Chap. 9 by Gert Moll & coll.), (2) pediocin-like bacteriocins (class IIa of unmodified peptides that contain in the N-terminal region the YGNGV/LXC consensus motif, in which X represents any amino acid) and that display a specific anti-*Listeria* activity (for a detailed description refer to Chap. 10 by Djamel Drider), (3) the two-peptide bacteriocins (class IIb of bacteriocins that acquire their potent activity only when present in mixture, for a detailed description refer to Chap. 11 by Jon Nissen-Meyer & coll.), (4) circular bacteriocins (class IIc; for a detailed description refer to Chap. 12 by John Vederas & coll.), and (5) class IId of bacteriocins, which encompasses in particular the leaderless bacteriocins (enterocins L50A and L50 B) and the nonsubgrouped bacteriocins (for a detailed description refer to Chap. 13 by Kenji Sonomoto & coll.). Microcins and lantibiotics are both essentially hydrophobic peptides, which most often exhibit a high stability to extreme temperatures and pHs and a low sensitivity to proteases, as a result of their posttranslational modifications or/and their compact three-dimensional structures. The other classes of LAB bacteriocins also share part of these structural characteristics.

By contrast, colicins are unmodified proteins that are protease-sensitive and exhibit a typical modular structure organized in three functional domains: an N-terminal domain required for translocation across the outer membrane, a central domain necessary for binding to the target cell surface receptor, and a C-terminal domain supporting the killing function (Cascales et al. 2007; Braun et al. 2002). In microcins and LAB bacteriocins, a modular organization only occurs in some classes. The siderophore-microcins (class IIb) contain two domains, an N-terminal domain involved in toxicity and a C-terminal domain in uptake (Azpiroz and Laviña 2007; Duquesne et al. 2007a). This modular structure made possible the construction of chimeric active peptides (Azpiroz and Laviña 2007). Pediocin-like bacteriocins (class IIa of LAB bacteriocins) also contain two domains: a cationic and well-conserved N-terminal β-sheet domain that mediates binding of the bacteriocin to the target cell surface and a more diverse and hydrophobic C-terminal hairpin-like domain that penetrates into the hydrophobic part of the target cell membrane, acting as an important determinant of the target cell specificity (Fimland et al. 1996;

Johnsen et al. 2005a). The target specificity of pediocin-like bacteriocins and class IIb microcins seems to be determined by the C-terminal domain, which is recognized by a receptor. In pediocin-like bacteriocins, this domain has been shown to specifically interact with the C-terminal half of the immunity protein, directly or indirectly (Johnsen et al. 2005b).

Biosynthesis – roles of the leaders. All bacteriocins are synthesized using the ribosomal machinery. Microcins and LAB bacteriocins share a common maturation process. Both are produced as inactive precursor peptides, which typically contain leader peptides N-terminal to the mature peptide sequences. The mature bacteriocins are released upon cleavage of the leader peptides by dedicated ABC (ATP-binding cassette) transporters. Such a common maturation process is particularly exemplified by microcin V (also called colicin V) that shares more common points in this regard with pediocin-like LAB bacteriocins than with other microcins. Microcin V could be produced in *Lactococcus lactis* when fused C-terminally to the leader peptide of leucocin A (van Belkum et al. 1997). Prior to the proteolytic cleavage of the leader peptides, some precursors are posttranslationally modified: this is particularly the case with lantibiotics, class IIc circular LAB bacteriocins and class I and IIb microcins.

Leader peptides are highly variable in length. They are about 25–30 and 20–30 amino acid long in lantibiotics (class I) and pediocin-like bacteriocins (class IIa), respectively, while they range between 2 and 35 amino acids in circular bacteriocins (class IIc) and between 15 and 30 amino acids in two-peptide bacteriocins (class IIb). Leader peptides of class II microcins are ~15 amino acids in length, whereas posttranslationally processed microcins have longer leaders of 19–37 amino acids, microcin C7-C51 being the only one to be secreted without cleavage of a precursor peptide (García-Bustos et al. 1984).

Cleavage of the bacteriocin leaders most often occurs at a specific cleavage site. The leader peptides of some lantibiotics are typically rich in Asp and Glu, containing an ELXXBXG motif (B = V, L, I; X = any amino acid). They usually end in the so-called double-glycine (GG) motif where the cleavage occurs, as also found in pediocin-like bacteriocins. Similarly, class II microcin leader peptides contain a conserved sequence MRXI/LX(9)GA/G (X = any amino acid) that ends with the typical double-glycine or with the glycine–alanine motif. The latter is found in proteins exported through ABC transporters as an alternative to the GG motif (Duquesne et al. 2007a). However, leader peptides of class I microcins do not display common features.

In many cases, the role of the leader peptide still remains unclear. The roles most commonly ascribed are those of (1) a secretion signalling peptide that is recognized by the transport machinery, (2) recognition by the post-translational modification enzymes, (3) chaperones assisting in the folding of the precursor peptides, (4) protection of the precursor peptides against degradation, or (5) keeping the precursor peptides inactive inside the host cells during biosynthesis, until the appropriate time for secretion and proteolysis (Oman and van der Donk 2010; Duquesne et al. 2007a). Support for almost all of these roles has been reported (Oman and van der Donk 2010), but the function of the leader peptide differs for the different classes and subclasses of peptides considered.

Genetic systems. Although some are chromosomally encoded, most class II LAB bacteriocins are plasmid-encoded. The genetic determinants for class IIc of circular LAB bacteriocins can be located either on plasmids or on the chromosome. Similarly, only certain microcins have been found to be chromosomally encoded (class IIb), while all colicins are plasmid-encoded. In the case of lantibiotics, the gene clusters can be located on a transposon, on the chromosome, or on a plasmid (see Chap. 9 on lantibiotics by Gert Moll).

The large structural differences observed between the different bacteriocins reflect a variable organization of the encoding genetic systems. Colicins exhibit the simplest genetic systems among bacteriocins. A minimal gene cluster contains two genes encoding the toxin and an immunity protein that protects the producing strain from its own toxin. In some clusters, an extra gene is present, encoding a lysis protein responsible for toxin release (Cascales et al. 2007).

Although they share a similar gene cluster organization with colicins, microcins and LAB bacteriocins however have more complex genetic systems, with the presence of several genes encoding auxiliary proteins or modification enzymes. At least four genes are required for the synthesis of class II bacteriocins (Drider et al. 2006), each coding for a precursor peptide, an immunity protein, an ABC transporter required for secretion of the bacteriocin out of cells, and an accessory protein, whose function is not yet clear. In the case of two-peptide bacteriocins, a fifth gene is required, which encodes the second partner of the two-peptide bacteriocin. Circular LAB bacteriocin gene clusters contain five to ten genes (see Chap. 12 by John Vederas & coll.). For example, the production of and immunity to enterocin AS48, the most extensively studied circular bacteriocin, require the coordinated expression of ten genes (Martínez-Bueno et al. 1998). Lantibiotics, including two-peptide lantibiotics, which undergo extensive and complex posttranslational modifications, are encoded by the most complex genetic systems among bacteriocins that include 6–13 genes (Patton and van der Donk 2005; Chatterjee et al. 2005).

Similar to LAB bacteriocins, microcin gene clusters contain generally at least four genes: the structural gene encoding the precursor, the self-immunity gene typically adjacent to the former that encodes the self-immunity factors, and two genes encoding the export system. They may contain additional genes that code for maturation enzymes, as the cases of microcins J25 and E492 (Duquesne et al. 2007a). Besides the gene products required for the synthesis of the bacteriocins and for self-protection/immunity of the producer, proteins required for regulation are also encoded in the gene clusters of bacteriocins.

Regulation. Colicin gene clusters are regulated by the SOS response regulon, which in many prokaryotes plays a primary role in the response to DNA damages caused by an environmental stress, such as UV irradiation or exposure to DNA-damaging agents (Walker 1995). In contrast, the synthesis of microcins is not SOS-dependent, but rather induced directly by stress conditions, particularly by poor nutrient conditions. The biosynthesis of class II LAB bacteriocins is mainly regulated through the cell-density dependent three-component regulatory system that consists of an inducer peptide (peptide pheromone), a transmembrane histidine protein kinase (pheromone receptor), and cytosolic response regulators (Eijsink et al. 2002; Kleerebezem and Quadri 2001). In addition to this quorum-sensing mechanism, temperature has been shown to be a major factor of regulation. Moreover, some lantibiotics such as nisin, subtilin, and salivaricin have been shown to serve as sensing molecules that trigger the transcription of their prepeptides in an extracellular autoregulatory mechanism (Chatterjee et al. 2005). Microcins are produced under conditions of stress, such as nutrient depletion that occurs near the stationary phase, or oxygen starvation. However, they are not all regulated by identical stress stimuli. In the case of microcin J25, it has been shown that its production is not controlled by cell density (Chiuchiolo et al. 2001), contrary to class II LAB bacteriocins.

Export and release. Microcins and LAB bacteriocins share similar modes of secretion into the extracellular medium that use ABC transporters. The transporter cleaves off the typical GG-containing leader sequence of the precursor, concomitantly with translocation of the bacteriocins across the membrane. The dedicated membrane-associated ABC transporter for each bacteriocin typically contains an N-terminal cysteine protease domain (Håvarstein et al. 1995). In the case of some lantibiotics, the cleavage is rather ensured by a serine protease. For class II LAB bacteriocins, accessory proteins are supposed to facilitate membrane translocation and/or leader cleavage, but their precise roles remain to be firmly established. There is no standard export machinery for class I microcins: they utilize either ABC transporter-like systems (microcins B17, J25) or efflux pumps (microcin C7-C51), which are encoded by their genetic systems, in association with outer membrane components as a secretion mechanism. The machineries in charge of class II microcins secretion have a canonical organization consisting of two ABC transporter components and one accessory protein, all of which being encoded by the microcin gene cluster.

In contrast, the release of colicins results from the sole expression of lysis proteins. After induction by DNA-damaging agents, colicins are expressed and accumulate in the cytoplasm of the producing cells before they are released upon the action of colicin lysis proteins. These small lipoproteins, predominantly located in the outer membrane of colicinogenic bacteria and displaying high sequence homologies, are first synthesized as precursors and further acylated and processed in multiple steps. While the stable signal peptide accumulates in the inner membrane, the mature lysis protein would activate the outer membrane phospholipase A (OmpLA), both being responsible for the loss of membrane integrity that causes colicin release and simultaneously quasilysis of the producing strain (Cascales et al. 2007; van der Wal et al. 1995).

Mechanisms of Action

The mechanisms of action of Gram-negative bacteriocins take advantage of the specific organization of the Gram-negative bacterial membrane. It includes a succession of steps: (1) recognition of the bacteriocin at the target cell outer membrane,

(2) translocation into the periplasmic space, (3) lethal penetration into the inner membrane or the cytoplasm.

Colicins. Colicins penetrate target cells by exploiting existing systems in the bacteria, which are involved in transport, diffusion, or efflux. They have been classified into two classes, according to the translocation system they use to penetrate target bacteria: Tol/Pal system for group A colicins and TonB for group B (for review see Cascales et al. 2007; Braun et al. 2002). Both group A and B colicins parasitize specific nutrient receptors on the outer membrane. These are the vitamin B12 receptor Btub, the siderophore receptors FepA/Cir/Fiu or FhuA, and the nucleoside receptor Tsx; from these receptors the colicins can be translocated across the inner membrane via the Tol or TonB translocation systems. The major killing functions of colicins are either (1) a lethal pore-forming activity that affects the inner membrane or (2) an enzymatic activity, i.e., a nuclease activity (RNAse or DNase), which occurs in the cytoplasm, or a disruption of the peptidoglycan biosynthesis in the periplasm.

Colicins use complex mechanisms to kill target bacteria, where the first recognition step and the translocation step are tightly linked, but not compulsorily accomplished by the same partner (see Chap. 14 by Miklos de Zamaroczy & Mathieu Chauleau). Here are some examples: (1) colicin Ia (group B), which is a pore-former, uses the siderophore receptor Cir as both a receptor and translocator (Jakes and Finkelstein 2009); (2) colicin E3 (group A), which has a nuclease activity, binds to the receptor BtuB via its receptor domain, then elongates to locate the OmpF porin or the membrane protein TolC and finally forms a complex with both the porin and the high-affinity receptor (Zakharov and Cramer 2004; Zakharov et al. 2006); the final translocation of colicin E3 through OmpF, before it can exert its enzymatic activity, necessitates unfolding of the colicin, which would be induced by both the recognition of BtuB by the receptor domain and subsequent insertion of the translocation domain into OmpF (Zakharov et al. 2006, 2008); (3) colicin M (group B), which directs its toxic activity toward bacterial murein (Braun et al. 2002) is recognized by the iron siderophore receptor FhuA (Braun et al. 2002); the degradation of undecaprenyl phosphate-linked peptidoglycan precursors is then accomplished through a unique phosphatase activity (El Ghachi et al. 2006; Zeth et al. 2008).

LAB bacteriocins. Class IIa LAB bacteriocins target the inner membrane and induce its permeabilization, leading to dissipation of the proton-motive force, depletion of intracellular solutes, and cell death, similar to many antimicrobial peptides (for review, see Fimland et al. 2005; Drider et al. 2006). However, in addition to targeting the membrane bilayer, it has been shown that they interact with membrane-located proteins of the mannose-phosphotransferase system (Man-PTS) that serve as target receptors (Héchard et al. 2001; Kjos et al. 2010). In bacteria, transporters of the Man-PTS family are responsible for the concomitant import and phosphorylation of carbohydrates, such as mannose and glucose. They are composed of four subunits, two of which (IIC and IID) form a membrane-located complex through which sugars enter the cells. A recent bioinformatic approach has shown that only proteins of a defined Man-PTS phylogenetic sub-group are efficient receptors and confer sensitivity to class IIa bacteriocins (Kjos et al. 2009). Moreover, the level of sensitivity of bacteria to the bacteriocin is determined by the amino-acid sequence and the expression level of the Man-PTS (Kjos et al. 2009). It is proposed that lactococcin A, a member of the nonsubgrouped class IId LAB bacteriocins, would also require the Man-PTS system as receptor (Diep et al. 2007; refer to Chap. 13 on class IId bacteriocins by Kenji Sonomoto & coll.). In a very similar fashion, the toxicity of microcin E492 has been shown to be strictly dependent on the presence of the membrane-associated ManY and ManZ components of the *E. coli* Man-PTS, which are homologous to proteins IIC and IID of Gram-positive bacteria (Bieler et al. 2006). It is proposed that the ManYZ complex would play a transient or direct role in the formation of the microcin E492 toxic structures. Microcin 24, which has some structural similarity with microcin E492, could also share this mechanism (Bieler et al. 2006).

Another original and efficient strategy developed by LAB bacteriocins is afforded by the prototypic lantibiotic nisin, which has a dual mode of action. It uses Lipid II, a crucial precursor in peptidoglycan biosynthesis, as a docking molecule. The docking of nisin and lipid II not only prevents correct cell-wall synthesis, but also initiates the formation of heteromolecular stable pores made of eight nisin and four lipid II molecules within cell membranes (Breukink et al. 2003; for a detailed description of these aspects, refer to Chap. 9 by Gert Moll). Two-component lantibiotics that act synergistically to exert their potent activity use a similar strategy, each of the two peptides ensuring one of the two functions of nisin. The first peptide binds to Lipid II, blocking peptidoglycan synthesis, and the complex formed serves as a docking site for the second peptide, inducing pore formation. Therefore, such dual mechanisms developed by LAB bacteriocins lead to rapid and efficient cell death. Moreover, the membrane components of Man-PTS systems appear as an important target receptor for both Gram-positive and Gram-negative bacteriocins.

Microcins. Microcins use similar strategies of uptake by target bacteria as those used by colicins. The highly modified lasso-peptide microcin J25 Rebuffat et al. (2004) that belongs to class I microcins uses FhuA, the receptor for ferrichrome (a hydroxamate-type siderophore), to enter sensible bacteria. In a similar fashion, the siderophore-microcins that belong to class II microcins (principally microcins E492, M, H47) (Vassiliadis et al. 2010) use the receptors for catechol siderophores, FhuA/Cir/Fiu, to ensure an efficient penetration into target cells. Interestingly, the unmodified form of microcin E492 (without a siderophore anchored at the C-terminus) is also recognized by the same receptors, but with a lower affinity (Thomas et al. 2004).

If microcins use similar strategies as colicins for uptake into target bacteria, they differ from both colicins and bacteriocins as regards the strategies they use for killing sensible bacteria. While two major antimicrobial activities have been reported for colicins, either pore-forming or nuclease activity (hydrolysis of DNA or RNA strands), targets for class I microcins are intracellular enzymes responsible for the DNA or protein synthesis: DNA gyrase and RNA polymerase are inhibited by microcins B17 and J25, respectively (Duquesne et al. 2007a, b), while aspartyl tRNA synthetase is inhibited by the modified and nonhydrolyzable substrate that is generated upon processing of microcin C7-C51 inside target bacteria (Severinov et al. 2007). Class IIb microcins have been shown to target inner membrane proteins, either the F_0F_1 ATP-synthase for microcin H47 (Duquesne et al. 2007a) or the Man-PTS system for microcin E492 (see above). Finally, inhibition of peptidoglycan synthesis, which has been reported for colicin M (El Ghachi et al. 2006), has not been observed until now among microcins.

Therefore, despite their completely unrelated structures, the microcins for which the mechanisms of action have been studied in detail, microcins J25, E492, C7-C51, exhibit "Trojan horse" strategies for penetrating and killing target bacteria. The microcins are allowed by susceptible bacteria to use specific receptors designed for the uptake of essential elements to penetrate the cells, since they disguise themselves or mime useful molecules such as siderophores (microcin E492 and other related siderophore-peptides, the lasso-peptide microcin J25). Alternatively, the microcin is secreted as a harmless molecule (microcin C7-C51) that is friendly allowed to enter the bacteria, where the inactive microcin is further transformed by the host bacterium into a toxic entity, which is lethal to the misled bacterium. Although their target receptors can be similar (mannose permease Man-PTS systems), the subtle strategies that Gram-negative bacteria develop for microcins differ from those adopted by Gram-positive bacteria for bacteriocins, which rather combine two different efficient mechanisms to kill the competitors more efficiently in their microbial ecosystem.

Mechanisms of Immunity

The mechanisms by which colicin-producing strains are protected against their own toxins differ according to their killing mechanisms. Pore-forming colicins are inactivated by constitutive immunity proteins anchored in the inner membrane (Espesset et al. 1994), while colicins with nuclease activities are released bound to the cosynthesized immunity protein (Keeble and Kleanthous 2005), which renders the colicin inactive by either allosteric inhibition or steric/electrostatic hindrance. Immunity to colicin M, which displays a unique activity directed toward bacterial murein, is conferred during import by a specific immunity protein that is located in the periplasm, where it inhibits the action of the colicin (Gross and Braun 1996).

Immunity proteins that protect class IIa (pediocin-like) bacteriocin producers against their cognate bacteriocin (refer to Chap. 10 on pediocin-like bacteriocins by Djamel Drider) are in general highly charged. The three-dimensional (3D) structures of immunity proteins of enterocin A, carnobacteriocin B2, and pediocin PP1 (an analogue of pediocin PA-1 differing by a single amino acid) share a common conserved four-helix bundle structural motif (Sprules et al. 2004; Johnsen et al. 2005a; Kim et al. 2007). Despite significant similarities in their three-dimensional structures, immunity proteins of class IIa bacteriocins (pediocin-like) display a tight specificity with respect to the bacteriocin to which they confer resistance. It has been shown that the functionality of the immunity protein of this class of bacteriocins depends on a cellular component. In connection with their modes of action, which involve a binding to the IIC and/or IID subunits of the Man-PTS system (Diep et al. 2007), the immunity proteins that protect the producing cells from class IIa bacteriocins and from a one-peptide nonsubgrouped class IId bacteriocin (lactococcin A) bind strongly to the bacteriocin-Man-PTS complex, thereby preventing membrane leakage and cell killing (Diep et al. 2007; Johnsen et al. 2005b). Such a mechanism of immunity involving a binding to a membrane-associated target receptor could also take place for two-peptide class IIb bacteriocins, such as lactococcin G (Oppegård et al. 2010). However, such a receptor has yet to be identified for a two-peptide bacteriocin (refer to Chap. 11 by Jon Nissen-Meyer & coll.).

The immunity mechanisms used by microcin-producing strains remain elusive. Class I microcins would use their export systems as resistance mechanism, the newly synthesized microcin being actively pumped out of the producing cells by the export system. Similar to LAB bacteriocins, the producer strains of class II microcins express an immunity protein, which displays strong specificity to its cognate bacteriocin to protect itself from the newly synthesized antibacterial peptide. However, the structures of the immunity proteins and the mechanisms used to express the immunity remain unknown.

Physiological Roles and Applications

Cell–cell communication. Intercellular communication between bacteria (quorum-sensing) uses a chemical language through signalling molecules called autoinducers. This process regulates many bacterial behaviors, such as symbiosis, virulence, biofilm formation, and antibiotic production (Schauder and Bassler 2001), and plays a critical role in mediating interactions and competition in microbial communities (Riley and Wertz 2002; Kirkup and Riley 2004). Gram-positive bacteria employ peptides called peptide pheromones for cell–cell signalling, while Gram-negative bacteria essentially use *N*-acyl homoserine lactones. Peptide pheromones have similar characteristics to type II bacteriocins that in turn have been proved in some cases to play themselves the role of signal peptides (Kristiansen et al. 2005). Similar to bacteriocins, export of peptide pheromones requires ABC transporters that cleave off leader peptides containing the typical double-glycine motif, concomitant with translocation across the membrane. Therefore, it is postulated that class II microcins that require similar transporters (microcins V, 24, and E492), could similarly perform a signalling function in Gram-negative bacteria (Michiels et al. 2001). Recently, two DNA-degrading colicins (group A colicins E2 and E7) have been shown to induce the production of their counterparts. Such a cross-induction would presumably result from induction of the SOS response by the DNase colicins, which in turn would induce colicin production (Majeed et al. 2010).

Bacteriocin ecology; probiotic role. Recent studies on the ecological role played by bacteriocins have highlighted the importance of cell–cell interactions in mediating the outcome of competition (for a detailed description of these aspects, refer to Chap. 2 by Margaret Riley). Literature provides a large body of data, evidence, and theories that support the role played by bacteriocins in maintaining microbial diversity both at the population (within species) and community levels. Studies from Margaret Riley's group provided theoretical and empirical bases defining the conditions required for maintenance of bacteriocin-producing bacteria in microbial communities (Riley and Wertz 2002; Kirkup and Riley 2004). Colicins have been used as model systems for investigating and modelling the ecology of bacteriocins in general, leading to the rock-paper-scissors model (Kirkup and Riley 2004), which provides conclusive evidence of an antagonistic role played by colicins. Colicins, and potentially other bacteriocins and microcins, would promote rather than eliminate microbial diversity in the ecosystem (Kirkup and Riley 2004). Bacteriocin producers have been shown to competitively exclude sensitive, nonproducing strains. A recent study aimed at explaining the coexistence of competing species in microbial communities proposes models implying that colicin-mediated colicin induction enables producers to successfully compete and defend their niche against invaders (Majeed et al. 2010).

Bacteriocin production is an important means for probiotic bacteria to competitively exclude or inhibit invading bacteria (Gillor et al. 2008; Gordon 2009). A number of studies point the potential probiotic role of microcinogenic strains, as well as the involvement of microcins in this activity. Inhibition of *Shigella flexneri* by *E. coli* strain H22, which produces microcin C7-C51 together with two pore-forming colicins E1 and Ib, has been shown to be mediated by the microcin (Cursino et al. 2006). An *E. coli* strain producing microcin 24 has been shown to inhibit the growth of pathogenic *Salmonella* and *E. coli* O157:H7 in the intestinal tract of chickens (Wooley et al. 1999). The probiotic strain *E. coli* Nissle 1917 is currently commercialized against diarrhea under the name Mutaflor® (Grozdanov et al. 2004), although a clear role of the class II microcins in the inhibition of *Salmonella* invasion has not yet been established (Altenhoefer et al. 2004). However, it is likely that microcins play a significant role in homeostasis of the intestinal ecosystem; therefore, microcin-producing bacteria may have potentials as probiotics. To date, LAB and their bacteriocins are without contest the most popular bacteria for the development of probiotics. These bacteria are nonpathogenic, nontoxic and are used as food preservatives. As such, LAB-producing bacteriocins are used to avoid food spoilage by Gram-positive bacteria. Nisin produced by *L. lactis* is widely used since many years as food additive and therefore serves as a model for direct application of bacteriocins.

Concluding Remarks

This review has highlighted and compared the main differences between the bacteriocins from Gram-negative bacteria and those from Gram-positive bacteria (colicins, microcins, and LAB bacteriocins). This comparative analysis includes the structures, biosyntheses, mechanisms of action and of immunity of the producing bacteria.

Both colicins and microcins constitute the primary defense systems in *E. coli* and in some other enterobacteria. The brief survey above shows that these two classes of defense molecules produced by Gram-negative bacteria do not share much common characteristics. The essential common points are their target selectivity to enterobacteria closely related to the producer and, for some of them [the lasso-peptide microcin J25 and the siderophore-microcins (E492, M, and H47)], a similar strategy of uptake into susceptible bacteria. As such, they exploit key receptors that are normally used for the uptake of essential nutrients, such as iron-siderophore receptors, on the one hand. On the other hand, microcins share with LAB bacteriocins the ability to carry complex and atypical posttranslational modifications that are particularly exemplified in certain classes (class I microcins, class IIb of siderophore-microcins, class I of LAB bacteriocins [lantibiotics]) and to result from complex genetic systems containing several genes involved in the synthesis of modification enzymes. They share amazing, subtle, and efficient mechanisms of action that take advantage of the specificities of the target bacteria, such as specific outer membrane receptors of Gram-negative bacteria, or a thick and protective layer of peptidoglycan in Gram-positive bacteria. This rapid overview of the main characteristics displayed on the one hand by colicins, which are weapons complementary to microcins designed by Gram-negative bacteria, and on the other hand by LAB bacteriocins tends to point that microcins may be considered a blend of features and strategies exemplified by these two classes of antibacterial peptides, assembling some common traits of Gram-positive bacteriocins (leaders, maturation, immunity) with other traits specific of colicins, in particular their strategy of uptake into sensible bacteria. Such a natural engineering that results in an efficient assembly constitutes an amazing model for the design of novel antibacterial molecules.

With the worldwide emergence of antibiotic-resistant pathogens, there is a high and urgent demand for new antimicrobial agents. Owing to their potency and restricted spectra of activity, and also due to their efficient and subtle killing mechanisms, bacteriocins afford lessons for the design and engineering of new antimicrobials. Peptide hybrids and chimeras have already been designed and produced to target Gram-negative bacteria using LAB (McCormick et al. 1999), thus leading to possible applications for preventing food poisoning by Gram-negative pathogenic bacteria. Colicin-engineered antibiotics, fusion products of channel-forming colicins and Gram-positive bacteria pheromones have proved to be efficient and specific bactericidal agents against pathogenic Gram-positive strains, without toxicity to mammal cells (Qiu et al. 2003, 2005). Furthermore, siderophore-microcins could serve as a model for the conception of peptide conjugates with efficient penetration into target cells, and the typical knotted and compact structure of the lasso-peptide microcin J25 could constitute a platform for bioactive peptide engineering. The detailed knowledge of the mechanisms of immunity developed by Gram-positive and Gram-negative bacteria against their bacteriocins, which still remain incompletely deciphered, would help in finding novel strategies to fight against bacterial resistance to antibiotics. These different aspects and potential applications should stimulate continuing research on Gram-negative and Gram-positive bacteriocins.

References

Altenhoefer A, Oswald S, Sonnenborn U, Enders C, Schulze J, Hacker J, Oelschlaeger TA (2004) The probiotic *Escherichia coli* strain Nissle 1917 interferes with invasion of human intestinal epithelial cells by different enteroinvasive bacterial pathogens. FEMS Immunol Med Microbiol 40:223–229

Azpiroz MF, Laviña M (2007) Modular structure of microcin H47 and colicin V. Antimicrob Agents Chemother 51:2412–2419

Bieler S, Silva F, Soto C, Belin D (2006) Bactericidal activity of both secreted and nonsecreted microcin E492 requires the mannose permease. J Bacteriol 188:7049–7061

Braun V, Patzer SI, Hantke K (2002) Ton-dependent colicins and microcins: modular design and evolution (review). Biochimie 84:365–380

Breukink E, van Heusden HE, Vollmerhaus PJ, Swiezewska E, Brunner L, Walker S, Heck AJ, de Kruijff B (2003) Lipid II is an intrinsic component of the pore induced by nisin in bacterial membranes. J Biol Chem 278:19898–19903

Cascales E, Buchanan SK, Duché D, Kleanthous C, Lloubes R, Postle K, Riley M, Slatin S, Cavard D (2007) Colicin biology (review). Microbiol Mol Biol Rev 71:158–229

Chatterjee C, Paul M, Xie L, van der Donk WA (2005) Biosynthesis and mode of action of lantibiotics (review). Chem Rev 105:633–684

Chiuchiolo MJ, Delgado MA, Farías RN, Salomón RA (2001) Growth-phase-dependent expression of the cyclopeptide antibiotic microcin J25. J Bacteriol 183:1755–1764

Cotter PD, Hill C, Ross RP (2005) Bacteriocins: developing innate immunity for food (review). Nat Rev Microbiol 3:777–788

Cursino L, Smajs D, Smarda J, Nardi RM, Nicoli JR, Chartone-Souza E, Nascimento AM (2006) Exoproducts of the *Escherichia coli* strain H22 inhibiting some enteric pathogens both in vitro and in vivo. J Appl Microbiol 100:821–829

Diep DB, Skaugen M, Salehian Z, Holo H, Nes IF (2007) Common mechanisms of target cell recognition and immunity for class II bacteriocins. Proc Natl Acad Sci U S A 104: 2384–2389

Drider D, Fimland G, Héchard Y, McMullen LM, Prévost H (2006) The continuing story of class IIa bacteriocins (review). Microbiol Mol Biol Rev 70:564–582

Duquesne S, Destoumieux-Garzón D, Peduzzi J, Rebuffat S (2007a) Microcins, gene-encoded antibacterial peptides from enterobacteria (review). Nat Prod Rep 24:708–734

Duquesne S, Petit V, Peduzzi J, Rebuffat S (2007b) Structural and functional diversity of microcins, gene-encoded antibacterial peptides from enterobacteria (review). J Mol Microbiol Biotechnol 13:200–209

Eijsink VG, Axelsson L, Diep DB, Håvarstein LS, Holo H, Nes IF (2002) Production of class II bacteriocins by lactic acid bacteria; an example of biological warfare and communication (review). Antonie Leeuwenhoek 81:639–654

El Ghachi M, Bouhss A, Barreteau H, Touzé T, Auger G, Blanot D, Mengin-Lecreulx D (2006) Colicin M exerts its bacteriolytic effect via enzymatic degradation of undecaprenyl phosphate-linked peptidoglycan precursors. J Biol Chem 281:22761–22772

Espesset D, Piet P, Lazdunski C, Géli V (1994) Immunity proteins to pore-forming colicins: structure-function relationships. Mol Microbiol 13:1111–1120

Fimland G, Blingsmo OR, Sletten K, Jung G, Nes IF, Nissen-Meyer J (1996) New biologically active hybrid bacteriocins constructed by combining regions from various pediocin-like bacteriocins: the C-terminal region is important for determining specificity. Appl Environ Microbiol 62:3313–3318

Fimland G, Johnsen L, Dalhus B, Nissen-Meyer J (2005) Pediocin-like antimicrobial peptides (class IIa bacteriocins) and their immunity proteins: biosynthesis, structure, and mode of action (review). J Pept Sci 11:688–696

García-Bustos JF, Pezzi N, Asensio C (1984) Microcin 7: purification and properties. Biochem Biophys Res Commun 119:779–785

Garneau S, Martin NI, Vederas JC (2002) Two-peptide bacteriocins produced by lactic acid bacteria (review). Biochimie 84:577–592

Gillor O, Etzion A, Riley MA (2008) The dual role of bacteriocins as anti- and probiotics (review). Appl Microbiol Biotechnol 81:591–606

Gordon DM (2009) The potential of bacteriocin-producing probiotics and associated caveats. Future Microbiol 4:941–943

Gross P, Braun V (1996) Colicin M is inactivated during import by its immunity protein. Mol Gen Genet 251:388–396

Grozdanov L, Raasch C, Schulze J, Sonnenborn U, Gottschalk G, Hacker J, Dobrindt U (2004) Analysis of the genome structure of the nonpathogenic probiotic *Escherichia coli* strain Nissle 1917. J Bacteriol 186:5432–5441

Håvarstein LS, Diep DB, Nes IF (1995) A family of bacteriocin ABC transporters carry out proteolytic processing of their substrates concomitant with export. Mol Microbiol 16:229–240

Héchard Y, Pelletier C, Cenatiempo Y, Frère J (2001) Analysis of sigma(54)-dependent genes in *Enterococcus faecalis*: a mannose PTS permease (EII(Man)) is involved in sensitivity to a bacteriocin, mesentericin Y105. Microbiology 147:1575–1580

Jack RW, Tagg JR, Ray B (1995) Bacteriocins of gram positive bacteria (review). Microbiol Rev 59:171–200

Jakes KS, Finkelstein A (2009) The colicin Ia receptor, Cir, is also the translocator for colicin Ia. Mol Microbiol 75:567–578

Johnsen L, Dalhus B, Leiros I, Nissen-Meyer J (2005a) 1.6-Angstroms crystal structure of EntA-im. A bacterial immunity protein conferring immunity to the antimicrobial activity of the pediocin-like bacteriocin enterocin A. J Biol Chem 280:19045–19050

Johnsen L, Fimland G, Nissen-Meyer J (2005b) The C-terminal domain of pediocin-like antimicrobial peptides (class IIa bacteriocins) is involved in specific recognition of the C-terminal part of cognate immunity proteins and in determining the antimicrobial spectrum. J Biol Chem 280:9243–9250

Keeble AH, Kleanthous C (2005) The kinetic basis for dual recognition in colicin endonuclease-immunity protein complexes. J Mol Biol 352:656–671

Kim IK, Kim MK, Kim JH, Yim HS, Cha SS, Kang SO (2007) High resolution crystal structure of PedB: a structural basis for the classification of pediocin-like immunity proteins. BMC Struct Biol 7:35

Kirkup BC, Riley MA (2004) Antibiotic-mediated antagonism leads to a bacterial game of rock-paper-scissors in vivo. Nature 428:412–414

Kjos M, Nes IF, Diep DB (2009) Class II one-peptide bacteriocins target a phylogenetically defined subgroup of mannose phosphotransferase systems on sensitive cells. Microbiology 155:2949–2961

Kjos M, Salehian Z, Nes IF, Diep DB (2010). An extracellular loop of the mannose phosphotransferase 1 system component IIC is responsible for specific targeting by class IIa bacteriocins. J Bacteriol 192:5906–5913

Kleerebezem M, Quadri LE (2001) Peptide pheromone-dependent regulation of antimicrobial peptide production in gram-positive bacteria: a case of multicellular behavior (review). Peptides 22:1579–1596

Kristiansen PE, Fimland G, Mantzilas D, Nissen-Meyer J (2005) Structure and mode of action of the membrane-permeabilizing antimicrobial peptide pheromone plantaricin A. J Biol Chem 280:22945–22950

Langdon GM, Bruix M, Gálvez A, Valdivia E, Maqueda M, Rico M (1998) Sequence-specific ^{1}H assignment and secondary structure of the bacteriocin AS-48 cyclic peptide. J Biomol NMR 12:173–175

Majeed H, Gillor O, Kerr B, Riley MA (2011) Competitive interactions in *Escherichia coli* populations: the role of bacteriocins. ISME J 5:71–81

Martínez-Bueno M, Valdivia E, Gálvez A, Coyette J, Maqueda M (1998) Analysis of the gene cluster involved in production and immunity of the peptide antibiotic AS-48 in *Enterococcus faecalis*. Mol Microbiol 27:347–358

McCormick JK, Klaenhammer TR, Stiles ME (1999) Colicin V can be produced by lactic acid bacteria. Lett Appl Microbiol 29:37–41
Michiels J, Dirix G, Vanderleyden J, Xi C (2001) Processing and export of peptide pheromones and bacteriocins in gram-negative bacteria (review). Trends Microbiol 9:164–168
Oman TJ, van der Donk WA (2010) Follow the leader: the use of leader peptides to guide natural product biosynthesis (review). Nat Chem Biol 6:9–18
Oppegård C, Emanuelsen L, Thorbek L, Fimland G, Nissen-Meyer J (2010) The lactococcin G immunity protein recognizes specific regions in both peptides constituting the two-peptide bacteriocin lactococcin G. Appl Environ Microbiol 76:1267–1273
Patton GC, van der Donk WA (2005) New developments in lantibiotic biosynthesis and mode of action (review). Curr Opin Microbiol 8:543–551
Qiu XQ, Wang H, Lu XF, Zhang J, Li SF, Cheng G, Wan L, Yang L, Zuo JY, Zhou YQ, Wang HY, Cheng X, Zhang SH, Ou ZR, Zhong ZC, Cheng JQ, Li YP, Wu GY (2003) An engineered multidomain bactericidal peptide as a model for targeted antibiotics against specific bacteria. Nat Biotechnol 21:1480–1485, Erratum in: Nat Biotechnol (2004) 22:1590
Qiu XQ, Zhang J, Wang H, Wu GY (2005) A novel engineered peptide, a narrow-spectrum antibiotic, is effective against vancomycin-resistant *Enterococcus faecalis*. Antimicrob Agents Chemother 49:1184–1189
Rebuffat S, Blond A, Destoumieux-Garzón D, Goulard C, Peduzzi J (2004) Microcin J25, from the macrocyclic to the lasso structure: implications for biosynthetic, evolutionary and biotechnological perspectives (review). Curr Protein Pept Sci 5:383–391
Riley MA, Wertz JE (2002) Bacteriocins: evolution, ecology, and application (review). Annu Rev Microbiol 56:117–137
Sánchez-Barrena MJ, Martínez-Ripoll M, Gálvez A, Valdivia E, Maqueda M, Cruz V, Albert A (2003) Structure of bacteriocin AS-48: from soluble state to membrane bound state. J Mol Biol 334:541–549
Schauder S, Bassler BL (2001) The languages of bacteria (review). Genes Dev 15:1468–1480
Severinov K, Semenova E, Kazakov A, Kazakov T, Gelfand MS (2007) Low-molecular-weight post-translationally modified microcins (review). Mol Microbiol 65:1380–1394, Erratum in: Mol Microbiol (2007) 66:277
Sprules T, Kawulka KE, Vederas JC (2004) NMR solution structure of ImB2, a protein conferring immunity to antimicrobial activity of the type IIa bacteriocin, carnobacteriocin B2. Biochemistry 43:11740–11749
Tagg JR, Dajani AS, Wannamaker LW (1976) Bacteriocins of gram-positive bacteria (review). Bacteriol Rev 40:722–756
Thomas X, Destoumieux-Garzón D, Peduzzi J, Afonso C, Blond A, Birlirakis N, Goulard C, Dubost L, Thai R, Tabet JC, Rebuffat S (2004) Siderophore peptide, a new type of post-translationally modified antibacterial peptide with potent activity. J Biol Chem 279: 28233–28242
van der Wal FJ, Luirink J, Oudega B (1995) Bacteriocin release proteins: mode of action, structure, and biotechnological application. FEMS Microbiol Rev 17:381–399
van Belkum MJ, Worobo RW, Stiles ME (1997) Double-glycine-type leader peptides direct secretion of bacteriocins by ABC transporters: colicin V secretion in *Lactococcus lactis*. Mol Microbiol 23:1293–1301
Vassiliadis G, Destoumieux-Garzón D, Lombard C, Rebuffat S, Peduzzi J (2010) Siderophore microcins form the first family of structure-related antimicrobial peptides from Enterobacteriaceae: isolation and characterization of microcins M and H47. Antimicrob Agents Chemother 54:288–297
Walker GC (1995) SOS-regulated proteins in translesion DNA synthesis and mutagenesis (review). Trends Biochem Sci 20:416–420
Wooley RE, Gibbs PS, Shotts EB Jr (1999) Inhibition of *Salmonella typhimurium* in the chicken intestinal tract by a transformed avirulent avian *Escherichia coli*. Avian Dis 43:245–250
Zakharov SD, Cramer WA (2004) On the mechanism and pathway of colicin import across the *E. coli* outer membrane (review). Front Biosci 9:1311–1317

Zakharov SD, Zhalnina MV, Sharma O, Cramer WA (2006) The colicin E3 outer membrane translocon: immunity protein release allows interaction of the cytotoxic domain with OmpF porin. Biochemistry 45:10199–10207

Zakharov SD, Sharma O, Zhalnina MV, Cramer WA (2008) Primary events in the colicin translocon: FRET analysis of colicin unfolding initiated by binding to BtuB and OmpF. Biochemistry 47:12802–12809

Zeth K, Römer C, Patzer SI, Braun V (2008) Crystal structure of colicin M, a novel phosphatase specifically imported by *Escherichia coli*. J Biol Chem 283:25324–25331

Part VI
Applications and Perspectives

Chapter 18
Food Applications and Regulation

Antonio Gálvez, Hikmate Abriouel, Nabil Ben Omar, and Rosario Lucas

Abstract This chapter deals with food applications of bacteriocins. Regulatory issues on the different possibilities for incorporating bacteriocins as bioprotectants are discussed. Specific applications of bacteriocins or bacteriocin-producing strains are described for main food categories, including milk and dairy products, raw meats, ready-to-eat meat and poultry products, fermented meats, fish and fish products or fermented fish. The last section of the chapter deals with applications in foods and beverages derived from plant materials, such as raw vegetable foods, fruits and fruit juices, cooked food products, fermented vegetable foods and fermented beverages. Results obtained for application of bacteriocins in combination with other hurdles are also discussed for each specific case, with a special emphasis on novel food packaging and food-processing technologies, such as irradiation, pulsed electric field treatments or high hydrostatic pressure treatment.

Introduction

Microbes and/or their natural products have played key roles in the preservation of foods in mankind's history (Ross et al. 2002). The rational exploitation of microbial antagonism based on scientific knowledge has been possible after the discovery of the biochemical nature of the antimicrobial substances produced by micro-organisms. Bacteriocins produced by the lactic acid bacteria (LAB) have several features that still make them attractive for food preservation: (1) LAB have a long history of safe use in foods; (2) LAB and their cell products –including bacteriocins- are generally recognised as safe; (3) LAB bacteriocins are not active and non-toxic on eukaryotic cells and (4) owing to their proteinaceous nature,

A. Gálvez (✉)
Health Sciences Department, Microbiology Division, Faculty of Experimental Sciences, University of Jaen, Campus Las Lagunillas s/n, 23071 Jaen, Spain
e-mail: agalvez@ujaen.es

D. Drider and S. Rebuffat (eds.), *Prokaryotic Antimicrobial Peptides: From Genes to Applications*, DOI 10.1007/978-1-4419-7692-5_18,

bacteriocins are expected to become inactivated by digestive proteases and not exert significant effects on gut microbiota at the concentrations ingested with the food. In addition, LAB bacteriocins may be suitable as preservatives, given (1) their sometimes broad antimicrobial spectrum, including food poisoning and spoilage bacteria, (2) their synergistic activity with other antimicrobials, (3) a bactericidal mode of action exerted at membrane level, which avoids cross resistance with antibiotics of clinical use, (4) stability under the heat and pH conditions achieved during processing of many foods and (v) their genetic determinants are usually plasmid-encoded, which facilitates genetic manipulation and development of producer strains with improved technological properties. Bacteriocin-encoding plasmids may be transferred to other strains by natural processes, but at the same time there is a risk for loss of the plasmid along with the bacteriocin production capacity.

Application of bacteriocins in food preservation may be beneficial in several aspects (Thomas et al. 2000; Gálvez et al. 2007): (1) to decrease the risks of food poisoning, (2) decrease cross-contamination in the food chain, (3) improve the shelf life of food products, (4) protect food during temperature-abuse episodes, (5) decrease economic losses due to food spoilage, (6) reduce the levels of added chemical preservatives, (7) reduce the intensity of physical treatments, thereby achieving a better preservation of the food nutritional value and possibly decrease of processing cost, (8) possibly provide alternative preservation barriers for "novel" foods (less acidic, with a lower salt content, and with a higher water content) and (9) possibly satisfy the demands of consumers for foods that are fresh-tasting, lightly preserved, and ready to eat (RTE). There may also be a potential market for bacteriocins as natural substitutes for chemical preservatives, and in the preservation of functional foods and nutraceuticals (Robertson et al. 2004).

According to previous studies (Deegan et al. 2006; Gálvez et al. 2007), bacteriocins can be applied in foods in many different ways:

(a) Bacteriocin preparations, obtained by cultivation of the producer strain in a suitable growth medium. This is usually followed by bacterial heat inactivation, and concentration of the cultured cells (by lyophilisation or spray-drying) to obtain a bioactive powder which contains a mixture of the antimicrobial substances produced in broth (such as the bacteriocin and organic acids). Commercial preparations such as Nisaplin™, Alta™ products or Microgard™ are some examples. Other bacteriocins such as lacticin 3147, variacin from *Kokuria varians* or enterocin AS-48 have also been obtained as powder preparations (Morgan et al. 1999; O'Mahony et al. 2001; Ananou et al. 2010).
(b) Bacteriocin-producing cultures, ready for propagation in the food substrate and in situ bacteriocin production. Bacteriocin-producing strains can be applied as the main starter cultures in fermented foods provided that they offer the proper technological properties, or as an adjunct culture in combination with bacteriocin-resistant starter strains. They can also be applied as bioprotective cultures in non-fermented foods, provided that they do not have adverse effects on the food.

The purpose of the present chapter is to provide an overview of the possibilities for application of bacteriocins and their producer strains in food systems in view of the huge amount of research work that has been carried out in this field, and also to provide some insights into the regulatory aspects concerning the different approaches that may be used for incorporation of bacteriocins in food systems.

Regulations

Currently, there are many different regulations concerning the application of bacteriocins in foods depending on the food category, the mode of application (e.g. in the form of bacteriocin preparations or protective cultures) and the laws of each country. This includes not only direct food laws but also many others that deal with related aspects such as labelling, packaging, export or the use of genetically modified organisms. Even within the European Union, different member states may have particular regulations as to final food applications. Most EU laws are directives. Directives are framework laws that subsequently must be enacted by the parliament of each member state in a form that suits the member state best. However, there is a growing trend towards a harmonised legislation and replacement of directives by regulations, which are laws of direct application in all member states and override any member state legislation that might cover the same area (Wessels et al. 2004).

Nisin is the only bacteriocin currently approved as a food preservative (E234). Nisin was assessed to be safe for food use by the Joint Food and Agriculture Organization/World Health Organization Expert Committee on Food Additives in 1969 and was added to the European food additive list in 1983 (Directive 83/463/EEC; Directive 95/2/EC); (European Economic Community 1983; European Parliament and Council 1995). It was approved in 1988 by the US Food and Drug Agency (FDA) for use in pasteurised processed cheese spreads. The initial approval was followed by other licensed applications (e.g. FSIS and Inspection Service 2002). Nisin is legally used in over 80 countries (Adams 2003). However, there are major differences in national legislations concerning the presence and levels of nisin in food products.

Nisin (in the commercial forms Nisaplin™ and Chrisin™) is a lyophilised product obtained from a microbial fermentation. The industrially fermented products are regulated under general food laws. Concentrates or lyophilised powders obtained from fermentates may be added to foods as ingredients. Many commercial preparations currently on the market are sold as ingredients or shelf-life extenders. Alta™ 2351 and Fargo 23 are natural food ingredients with antilisterial activity produced by bacteriocinogenic strains through a fermentation process. Alta™ 2351 is labelled as "cultured dairy solids (Skim Milk, Dextrose, Whey, and Lactic Acid Culture)." It is "an ingredient with functionality against outgrowth of *Listeria* in dairy based products, and small spectrum lactic acid bacteria inhibition." Such products are approved in the USA and commercialised in several other countries as shelf-life extenders for application in a variety of food products.

According to Directive 2000/13/EC (European Parliament and Council 2000) on food labelling, "'ingredient' shall mean any substance, including additives, used in the manufacture or preparation of a foodstuff and still present in the finished product, even if in altered form." Fermented milk or whey concentrates or lyophilised preparations (regardless of whether they contain bacteriocins or not) can be added as ingredients in the preparation of dairy foods. In the European Union, approval of bacteriocins for application as food additives or preservatives must comply with specifications given in Regulation 1333/2008/EC (European Parliament and Council (2008b), which harmonises the use of food additives in foods in the European Community and updates Directive 89/107/EEC (European Commission 1988) concerning food additives authorised for use in foodstuffs intended for human consumption and Directive 95/2/EC (European Parliament and Council 1995) on food additives other than colours and sweeteners, and Regulation 258/97/EC (European Parliament and Council 1997) on novel foods and novel ingredients. Food additives are substances that are not normally consumed as food itself but are added to food intentionally for a technological purpose described in the above Regulation, such as the preservation of food. Preservatives are considered a functional class of food additives: "preservatives" are substances which prolong the shelf life of foods by protecting them against deterioration caused by micro-organisms and/or which protect against growth of pathogenic micro-organisms. Approved food additives must be listed in the Community lists and shall specify the following: (a) the name of the food additive and its E number, (b) the foods to which the food additive may be added, (c) the conditions under which the food additive may be used, (d) if appropriate, whether there are any restrictions on the sale of the food additive directly to the final consumer. Approved specifications should include information to adequately identify the food additive, including origin, and to describe the acceptable criteria of purity. Added preservatives must be listed in food labels with their specific name or EC number (Directive 2000/13/EC; European Parliament and Council 2000). The use and maximum levels of a food additive should take into account the intake of the food additive from other sources and the exposure to the food additive by special groups of consumers (e.g. allergic consumers). The risk assessment and approval of food additives should be carried out in accordance with the procedure laid down in Regulation (EC) No 1331/2008 (European Parliament and Council 2008a) establishing a common authorisation procedure for food additives, food enzymes and food flavourings. Food additives which were permitted before 20 January 2009 shall be subject to a new risk assessment carried out by the Authority.

Bacteriocin preparations could also be applied as processing aids in food manufacture. Directive 2000/13/EC (European Parliament and Council 2000) and Regulation 1333/2008/EC (European Parliament and Council 2008b) do not cover processing aids, but according to Regulation 1333/2008/EC "processing aid" shall mean any substance which (1) is not consumed as a food by itself, (2) is intentionally used in the processing of raw materials, foods or their ingredients, to fulfil a certain technological purpose during treatment or processing and (3) may result in the unintentional but technically unavoidable presence in the final product of residues of the substance or its derivatives provided they do not present any health risk and do not have any technological effect on the final product. According to this definition, bacteriocins could be applied as processing aids for the preservation of food ingredients, whereby the bacteriocin has no preservative or technological effect in the final food product.

Application of bacteriocins in activated packagings must follow specifications of Directive 2002/72/EC (European Parliament and Council 2002b) concerning plastic materials and articles intended to come into contact with foodstuffs and Regulation (EC) No 1935/2004 (European Parliament and Council 2004) on materials and articles intended to come into contact with food: Active food contact materials are designed to deliberately incorporate "active" components intended to be released into the food or to absorb substances from the food: "'active food contact materials and articles' (hereinafter referred to as active materials and articles) means materials and articles that are intended to extend the shelf life or to maintain or improve the condition of packaged food. They are designed to deliberately incorporate components that would release or absorb substances into or from the packaged food or the environment surrounding the food." Substances deliberately incorporated into active materials and articles to be released into the food or the environment surrounding the food shall be authorised and used in accordance with the relevant Community provisions applicable to food, and shall comply with the provisions of this Regulation and its implementing measures. These substances shall be considered as ingredients. Covering or coating materials forming part of the food and possibly being consumed with it (such as edible coatings) do not fall within the scope of this Regulation.

Bacteriocin-producing strains may be applied as starter or bioprotective cultures with the aim of contributing to microbiological safety (Aymerich et al. 2008). For example, Bactoferm F-Lc (Christian Hansen, Denmark) is an antilisterial mixed culture of *Pediococcus acidilactici* and *Lactobacillus curvatus* producing pediocin and sakacin A, respectively, for application in fermented sausages. The same company also sells bioprotective cultures containing *Lactobacillus sakei* and *Leuconostoc carnosum* 4010 for meat products packed under vacuum or modified atmosphere packaging (MAP), and a nisin-producing *Lactococcus lactis* preparation. Danisco (Copenhaguen, Denmark) markets a series of protective cultures (HOLDBAC™) for specific applications in meat and dairy foods based on their capacity to produce bacteriocins as well as other antimicrobial compounds and their competition in food systems. Such preparations include mainly strains of *Lactobacillus plantarum*, *Lactobacillus rhamnosus*, *Lb. sakei*, *Lb. paracasei* and *Propionibacterium freudenreichii* subsp. *shermanii*, whose primary functionalities are growth control of Gram-positive pathogens such as *Listeria*, spoilage micro-organisms such as yeasts and moulds, heterofermentative lactic bacteria and enterococci. Such strains have not been subjected to genetic modification, but the company advertises that local regulations should always be consulted concerning the status of these products as legislation regarding their use in food may vary from country to country.

From a regulatory point of view, bacteriocin-producing strains fall in the category of microbial cultures. In the USA, a new strain of micro-organism for use in food can be classified either as an additive or as a Generally Recognised as Safe (GRAS) substance (Wessels et al. 2004). Food additives are defined in a broad sense as "anything that might come into contact with food (excluding GRAS substances)," and require pre-market approval by the US FDA based on toxicological and efficacy data. The consideration of GRAS status is based on the availability of enough information relevant to the substance safe use for a given intended purpose: "generally recognized, among experts qualified by scientific training and experience to evaluate its safety, as having been adequately shown through scientific procedures (or, in the case as a substance used in food prior to January 1, 1958, through either scientific procedures or experience based on common use in food) to be safe under the conditions of its intended use" (US Food and Drug Administration 1999). The intended use is an essential part of the GRAS status concept. For instance, the GRAS status of a given strain for use in a yogurt product is not valid for the same strain in infant formulae (Wessels et al. 2004). The GRAS status is determined by qualified experts, not by the FDA. The food company that uses the bacterium assumes complete responsibility, regardless of its GRAS status.

Within the European Union, microbial food cultures with a long history of safe use are considered as traditional food ingredients and covered by general European food law (Regulation 178/2002/EC; European Parliament and Council 2002a). Microbial cultures must also be safe for their intended use. Novel use of microbial cultures is regulated by Regulation 258/97/EC (European Parliament and Council 1997) if a micro-organism has not been consumed to a significant degree before May 15, 1997. This may apply to selected bacteriocin producer strains isolated from a source different than the food where they will be applied. There is also an ongoing dispute in Europe regarding the food category of starter cultures with protective properties, since they may be considered as cultures with specific technological effects (preservatives). This may contradict current regulations on approval of new preservatives, given the long history of consumption of fermented foods and the fact that the original and primary purpose of fermenting food was to achieve a preservation effect. In this respect, guidance documents from the European Food Safety Authority (EFSA 2007, 2008) established a pre-market safety assessment of selected groups of micro-organisms leading to a "Qualified Presumption of Safety (QPS)"; if the taxonomic group did not raise safety concerns or if safety concerns existed, but could be defined and excluded (the qualification), the grouping could be granted QPS status. Thereafter, any strain of micro-organism the identity of which could be unambiguously established and assigned to a QPS group would be freed from the need for further safety assessment other than satisfying any qualifications specified. Micro-organisms not considered suitable for QPS would remain subject to a full safety assessment.

Application of genetically modified bacteriocin producer strains on foods may be affected by different EU regulations and Directives. Strains modified by naturally existing procedures (such as DNA transformation or plasmid conjugation) could be applied in foods with no other restrictions than those specified in the previous paragraph for naturally occurring strains. However, application of strains modified by procedures involving extensive DNA manipulation and artificial transfer to recipient cells is under much more strict control, including the specifications and limitations established by Directive 2001/118/EC (European Parliament and Council 2001) on the deliberate release of GMMs into the environment, Directive 2009/41/EC (European Parliament and Council 2009) on the contained use of genetically modified micro-organisms, and Regulation 1829/2003/EC (European Parliament and Council 2003) concerning the marketing of GMOs intended for food or feed and of food or feed products containing, consisting of, or produced from GMOs. Bacterial products such as fermented bioactive ingredients prepared from GMM should also need approval in accordance with specifications under Regulations 1829/2003/EC and 1333/2008/EC (European Parliament and Council 2003, 2008a, b).

Biopreservation of Milk and Dairy Products

Addition of Bacteriocin Preparations

Many different bacteriocins preparations have been tested for preservation of milks and dairy products, aimed at inactivation of food-borne pathogenic or spoilage bacteria. Nisin is widely used in the dairy industry for inhibition of gas blowing defect in cheeses caused by *Clostridium tyrobutyricum*, and also in processed cheeses and cheese products to inhibit *Clostridium botulinum*, and to prevent growth of post-process contaminating bacteria such as *Listeria monocytogenes* (Davies and Delves-Broughton 1999; Thomas et al. 2000; Thomas and Delves-Broughton 2001; Deegan et al. 2006; Sobrino-López and Martín-Belloso 2008) (Table 18.1). It is also used in many other pasteurised dairy products, such as chilled desserts, flavoured milk, clotted cream or canned evaporated milks (Thomas et al. 2000). Processing of dairy foods, such as slicing, can be a critical point for bacterial contamination. Packaging of sliced cheese with inserts containing immobilized nisin reduced the population of lactic acid bacteria, *Listeria innocua* and *Staphylococcus aureus* on the cheese slices.

Nisin addition in milk is permitted in certain countries to avoid shelf-life problems associated with hot weather temperature and/or long-distance transport and inadequate refrigeration systems (Davies and Delves-Broughton 1999; Thomas et al. 2000). The application of nisin in combination with heat treatments decreased the D values of bacteria such as *Bacillus cereus* and *Geobacillus stearothermophilus* and natural microbiota, making it possible to apply milder thermal treatments and at the same time extend the shelf life of milk even under poor refrigeration conditions.

Nisin activity in milk can increase in combination with other antimicrobials, such as monolaurin, the lactoperoxidase (LPS) system, lysozyme or reuterin (Gálvez et al. 2007). Application of nisin in combination with pulsed electric fields

Table 18.1 Examples of bacteriocin applications in dairy foods

Bacteriocin treatment	Effect(s)	Reference(s)
Nisin	Prevent proliferation of surviving endospore formers, mainly the gas-producing clostridia and *C. botulinum* in cheeses Prevent post-process contamination with *L. monocytogenes*	Thomas and Delves-Broughton (2001)
Nisin and PEF	Increased antimicrobial activity in milks against several bacteria such as *L. monocytogenes, S. aureus, B. cereus* and *E. coli*	Sobrino-López and Martín-Belloso (2008)
Nisin and HHP	Increased the inactivation of spoilage bacteria associated with milk	Black et al. (2005)
Lacticin 3147	Inactivation of *L. monocytogenes* in natural yogurt and in cottage cheese	Morgan et al. (2001)
Enterocin AS-48	Rapid inactivation of *L. monocytogenes* and slower inhibition of *S. aureus* in skim milk	Ananou et al. (2010)
Pediocin PA-1/AcH	Inhibition of *L. monocytogenes* in several dairy systems (dressed cottage cheese, half-and-half cream, cheese sauce and others)	Rodríguez et al. (2002)
Lacticin 3147-producing cultures	Inhibition *L. monocytogenes* in cottage cheese, and on the surface of a mould-ripened cheese and a smear-ripened cheese	O'Sullivan et al. (2006)
Enterocin AS-48 producer *E. faecalis* strain	Inhibition of *B. cereus* and *S. aureus* in cheeses and in skim milk	Muñoz et al. (2004, 2007)
Lacticin 3147-producing *L. lactis* IFPL 3593	Inhibition of gas formation by *C. tyrobutyricum* and heterofermentative lactobacilli in cheese	Martínez-Cuesta et al. (2010)
Bacteriocin producer *Lb. gasseri* K7	Reduced outgrowth of inoculated *C. tyrobutyricum* and butyric acid formation in the cheeses. Probiotic properties	Bogovič Matijašić et al. (2007)
S. macedonicus ACA-DC (producer of macedocin)	Inhibition of gas formation by *C. tyrobutyricum* in cheese	Anastasiou et al. (2009)
Lacticin 3147-producing cultures	Increased generation of 2-methylbutanal with the concomitant enhancement of the cheese aroma	Fernández de Palencia et al. (2004)
Lacticin 3147-producing cultures	Inhibition of adventitious non-starter LAB flora during ripening, enhancing the cheese quality	Ryan et al. (2001)

(PEF) or high hydrostatic pressure (HHP) has been investigated in recent years (Black et al. 2005; Sobrino-López and Martín-Belloso 2008). The combined application of nisin and PEF increased the microbial inactivation of *L. monocytogenes, S. aureus, B. cereus* and *Escherichia coli* in various substrates such as skim milk, whey or simulated milk ultrafiltrate media (Sobrino-López and Martín-Belloso 2008). Combined treatments of nisin and HHP (with or without lysozyme) increased the inactivation of bacteria associated with milk such as *E. coli, Pseudomonas fluorescens, L. innocua* and *Lactobacillus viridescens* and increased the inactivation of endospores and mesophilic bacteria in cheese (Gálvez et al. 2007, 2008). The combined treatment of nisin–HHP could improve the microbial stability and safety of cheeses, such as many traditional cheeses made from unpasteurised milk, decreasing the legal barriers to commercialisation.

Lacticin 3147 is another lactococcal bacteriocin with a high potential for application in the preservation of dairy foods (Ross et al. 1999; O'Sullivan et al. 2002). Added lacticin 3147 powder rapidly inactivated *L. monocytogenes* and reduced *S. aureus* viable cell counts in an infant milk formulation and was highly effective against *L. monocytogenes* in natural yogurt and in cottage cheese (Morgan et al. 2001). However, optimisation of lacticin 3147 powder to increase specific activity may be necessary to decrease the amount of added powder required for an effective microbial inhibition. Antibacterial activity of this lacticin against *L. monocytogenes* and *S. aureus* increased in combination with HHP treatment when tested in milk and in whey, thereby reducing the amounts of bacteriocin required.

Several enterococal bacteriocins have been tested for preservation of dairy foods (Giraffa 1995; Foulquié Moreno et al. 2006; Galvez et al. 2008). Enterocins CCM 4231, CRL35 or AS-48 can reduce the levels of *L. monocytogenes, S. aureus* or *B. cereus* in milk and dairy products. Enterocin AS-48 has been widely investigated among enterocins, because of its cyclic structure and broad antibacterial spectrum. In skim milk, combinations of enterocin AS-48 with or without nisin and PEF treatment increased the inactivation of *S. aureus* (Sobrino et al. 2009). Addition of a whey-derived bioactive powder containing enterocin AS-48 to skim milk rapidly inactivated *L. monocytogenes* cells and progressively reduced the viable counts of *S. aureus* (Ananou et al. 2010).

Pediocin PA-1/AcH preparations are interesting for application in dairy products because of the bacteriocin antilisterial activity, stability in aqueous solutions at ambient temperature and also during freezing and heating and a wide pH range for activity (Nes et al. 1996; Rodríguez et al. 2002). The commercial preparations containing pediocin in the form of Alta™ products can be used as ingredients in dairy foods. Added pediocin PA-1/AcH was effective in reducing the levels of *L. monocytogenes* in several types of dairy products such as dressed cottage cheese, half-and-half cream and cheese sauce (Rodríguez et al. 2002).

Other bacteriocins of interest in preservation of dairy foods are the propionicins. Propionibacteria are used in some dairy fermentations and may produce bacteriocins with broad inhibitory spectra (Holo et al. 2002). Microgard™ is a commercial preparation containing an antimicrobial peptide produced by *P. freudenreichii* ssp. *shermanii* (Weber and Broich 1986), which is approved in certain countries for commercial use as an ingredient mainly in dairy products such as cottage cheese and yogurt. Bacteriocins produced by *Propionibacterium jensenii* P126 and P1264 strains have been patented as antibacterial agents for controlling the growth of certain lactic acid bacteria. These bacteriocins could be particularly useful in controlling the over-acidification of yogurt to decrease the sour taste often found in this product.

Some bacteriocins from bacteria not associated with milk fermentations have been investigated for application in dairy foods. Variacin in the form of a dry milk-based ingredient inhibited the proliferation of *B. cereus* in chilled dairy products, vanilla and chocolate desserts (Mollet et al. 2004). There is also a growing interest in exploitation of bacteriocins from bacilli. Cerein 8A is an antimicrobial peptide produced by the soil isolate *B. cereus* 8A, with bactericidal activity towards *L. monocytogenes* and *B. cereus* (Bizani et al. 2005). Cerein addition inhibited growth of *L. monocytogenes* in milk and on the surface of Minas-type cheese during refrigeration storage, suggesting its potential use as biopreservative in dairy products.

Application of Bacteriocin-Producing Strains for Inhibition of L. monocytogenes and Other Food-Borne Pathogens

L. monocytogenes is considered the main food-borne pathogen of concern in cheese and dairy products. Therefore, many different studies have focused on the application of antilisterial starter or adjunct cultures for inhibition of this bacterium (Table 18.1). Nisin-producing lactococcal strains inhibit *L. monocytogenes* in several types of cheeses such as cottage, Camembert or Manchego cheese made from raw milk. They can also reduce *S. aureus* viable counts (Deegan et al. 2006; Galvez et al. 2008), but often lack the technological properties required for cheese making such as fast acidification capacity and proteolytic activity. For this reason, they should be recommended as adjunct cultures in combination with suitable nisin-resistant strains as the primary starters.

The lactococci can also produce other bacteriocins (such as lacticin 3147) in fermented dairy products (Guinane et al. 2005). In order to develop suitable starters, the plasmid coding for lacticin 3147 production was transferred to suitable recipient lactococci. Lacticin production by those modified starters successfully inhibited *L. monocytogenes* in cottage cheese, in semi-hard raw-milk cheeses and on the surface of a mould-ripened cheese and a smear-ripened cheese (O'Sullivan et al. 2006). The lactococci have also been tested for heterologous production of other bacteriocins such as enterocin A. The resulting starter derivative successfully controlled the levels of *L. monocytogenes* during cottage cheese fermentation (Liu et al. 2008).

Bacteriocin-producing enterococci have been investigated as adjunct cultures because of their robustness, natural presence in cheeses, and production of several bacteriocins with marked antilisterial activity (Giraffa 1995; Foulquié Moreno et al. 2003; Franz et al. 2007; Galvez et al. 2008). When tested in cheeses, strains producing enterocin AS-48 showed strong inhibition of *L. monocytogenes*, as well as *B. cereus* and *S. aureus* (Muñoz et al. 2004, 2007). *Enterococcus faecium* RZS C5 (a natural cheese isolate carrying the structural genes for enterocins A, B and P) was reported to be effective as an antilisterial bacteriocin-producing co-culture in Cheddar cheese manufacture. Strains of enterococci and lactococci producing bacteriocins (such as enterocins I, TAB 7, TAB 57, AS-48, nisin A, nisin Z and lacticin 481) have been tested in combination with HHP treatments as a way to improve the safety of cheeses made from raw milk. Inoculation of milk with bacteriocinogenic strains before cheese making followed by application of HHP treatment to the cheeses was reported to increase the bactericidal activity against *L. monocytogenes, S. aureus* and *E. coli* O157:H7.

Pediococci are not well adapted to dairy substrates due to their lack or very slow lactose fermentation (Papagianni and Anastasiadou 2009). However, some strains such as *P. acidilactici* NRRL-B-18925 are particularly effective in producing bacteriocin in milk-based media. Since pediocin PA-1/AcH is not inhibitory to bacterial species employed as yogurt starters (Gonzalez and Kunka 1987), there is a great interest for application of producer strains in developing more naturally preserved yogurts and to avoid proliferation of cross-contaminating pathogens during yogurt processing. Vedamuthu Ebenezer (1995) patented a method for producing a yogurt product which contains bacteriocin active against undesirable flora. The yogurt product can be dried, either by lyophilisation or spray-drying or other means, preferably to a powder for use in various foods. Recently, pediocin production in milk has been reported in coculture with yogurt starter cultures, at the expense of the excess sugar released from lactose hydrolysis by the starters (Somkuti and Steinberg 2010). A different approach has been to develop pediocin-producing genetically engineered *L. lactis* strains, which have shown significant potential for inhibition of food-borne pathogens in cheeses (Galvez et al. 2008).

Bacteriocin-producing lactobacilli have also been suggested for preservation of dairy foods, such as the human isolate probiotic strain *Lactobacillus gasseri* K7 producer of bacteriocins with wide range of inhibition (Čanžek Majhenič et al. 2003). Application of bacteriocin-producing probiotic strains could be exploited with the purpose of improving food safety and quality and at the same time providing health benefits.

Application of Bacteriocin-Producing Strains for Inhibition of Bacteria Producing Gas-Blowing Defects in Cheeses

Gas production is an undesirable defect in most cheeses and may be caused by outgrowth of *C. tyrobutyricum* (and other clostridia) spores surviving heat treatments applied to milk before processing (Cocolin et al. 2004; Le Bourhis et al. 2007), and also by some heterofermentative LAB. Application of nisin-producing starter cultures to prevent gas-blowing defects in cheese was proposed as early as 1951. Strains producing the natural variant nisin Z have equally shown to reduce the levels of *C. tyrobutyricum* in cheese. Recently, lacticin 3147-producing lactococci have been shown to inhibit *C. tyrobutyricum* spores and prevent late blowing in semi-hard cheeses and also demonstrated a considerable inhibition of heterofermentative lactobacilli and their associated blowing defects in cheese (Martínez-Cuesta et al. 2010). The authors suggested that application of lacticin-producers in cheese manufacture is a promising alternative to the addition of lysozyme, given the increasing concerns about the potential allergenicity of this additive in egg allergic consumers.

Thermophilic streptococci could be applied to inhibit *C. tyrobutyricum* in some cheeses. Thermophilin from *Streptococcus thermophilus* ST580 is active against *C. tyrobutyricum,* but not against the thermophilic lactobacilli used as starters. This strain inhibits gas production in cheese by *C. tyrobutyricum* and could be included in thermophilic starters for hard-cheese making. *Streptococcus macedonicus* ACA-DC 198, which produces the food-grade lantibiotic macedocin, could also be employed to inhibit gas formation in cheese (De Vuyst and Tsakalidou 2008). When tested as an adjunct culture in Kasseri cheese production, this strain produced macedocin in the cheese and inhibited outgrowth of *C. tyrobutyricum* spores during the cheese production and ripening (Anastasiou et al. 2009).

Among lactobacilli, the bacteriocin producer strain *L. gasseri* K7 was able to survive during semi-hard-type cheese manufacturing and reduced outgrowth of *C. tyrobutyricum* and butyric acid formation in the cheeses (Bogovič Matijašić et al. 2007).

Application of Bacteriocin-Producing Strains to Improve Cheese Quality, Accelerate Ripening and/or Improve Cheese Flavour

With some exceptions, the primary target of bacteriocins is the bacterial cytoplasmic membrane, where they insert and modify membrane permeability. A secondary effect of bacteriocins is the induction of cell lysis, as a result of deregulation of cell wall autolysins (Gálvez et al. 1990). This effect was further observed on dairy starter cultures and led to more detailed studies on the potential applications for the release of bacterial intracellular enzymes (such as lipases, proteases, peptidases and amino-acid-converting enzymes) of technological relevance in cheese ripening (Lortal and Chapot-Chartier 2005; Peláez and Requena 2005; Deegan et al. 2006).

Lacticin 3147 production in cheese induced bacterial lysis of cheese starters, increased cheese proteolysis and facilitated the access of bacterial amino-acid-converting enzymes to amino acids (Deegan et al. 2006). The accelerated starter cell lysis enhanced reactions such as isoleucine transamination, increasing the formation of alpha-keto-beta-methyl-n-valeric acid and 2-hydroxy-3-methyl-valeric acid and cheese aroma intensity due to the higher 2-methylbutanal formation. Optimisation of aroma production could be achieved by selective combination of starters, such as a lacticin 3147-producing *L. lactis* transformant in combination with adjunct cultures producing aminotransferase and a-keto acid decarboxylase activities (Fernández de Palencia et al. 2004). LAB strains producing other bacteriocins such as lactococcins A, B and M, lacticin 481, enterocin AS-48 or nisin) provided similar benefits in cheeses, such as increased lipolysis and proteolysis, formation of hydrophilic and hydrophobic peptides and overall improvement of sensory characteristics (Galvez et al. 2008).

The application of bacteriocin-producing adjunct cultures to accelerate cheese ripening can be a cheaper approach compared to the addition of exogenous lytic enzymes (Papagianni and Anastasiadou 2009). Application of bacteriocins and/or their producer strains in the development of stabilised cheese flavouring systems has been patented (Dias et al. 2009). Using microbial cells with high aminopeptidase activity in combination with an antimicrobial that can permeabilise the cells can decrease the levels of aminopeptidases that need to be added to the matrix, thereby increasing efficiency. Cheeses produced at the expense of enzymes released form the bacterial cells have a more rounded flavour (Dias et al. 2009). Recently, an adjunct *P. acidilactici* culture that accelerates and enhances flavour formation in Cheddar and semi-hard cheeses due to the production of bacteriocins has been marketed by Danisco (CHOOZIT™ Lyo. Flav 43).

Another suggested application of bacteriocin-producing cultures is the inactivation of adventitious non-starter LAB (NSLAB) microbiota during cheese ripening. Control of NSLAB is still a pending issue in dairy industries. Growth of NSLAB may induce batch-to-batch variations in the sensory quality of cheese and cause defects such as the formation of calcium lactate crystals (due to racemation of L-lactate to D-lactate), slit formation and off-flavour development, but they may also exert beneficial effects on the cheeses. Application of lacticin 3147-producing starters has been proposed as a way to enhance cheese quality through inhibition of adventitious NSLAB microbiota during ripening (Ryan et al. 2001; Deegan et al. 2006). During Cheddar cheese manufacture, ripening can be accelerated by increasing the temperature from 7 to 12°C, but this also results in a higher risk of spoilage due to a more rapid proliferation of NSLAB. Inoculation with a lacticin 3147-producing strain allowed a better control of NSLAB during cheese ripening at elevated temperature. Lacticin 481 production in cheese has also shown to reduce the concentrations of NSLAB in cheese by 4 to 2 orders of magnitude during ripening. Bacteriocin production in the cheese resulted in selection of NSLAB that were much more resistant to the bacteriocin than isolates from control cheeses. Therefore, it would be possible to select bacteriocin-resistant strains that do not have negative effects on cheese ripening as the predominant NSLAB.

Biopreservation of Meats and Meat Products

Control of Pathogenic and Spoilage Bacteria in Raw Meats

Bacteriocins have been tested for preservation of raw meats alone or in combination with other hurdles for carcass decontamination and/or to inhibit bacterial growth on stored fresh meats (Table 18.2). Washing, spraying or dipping with bacteriocin solutions have been tested alone or in combination with other antimicrobials to potentiate bacteriocin activity. In order to increase the efficacy of treatments and/or avoid cross contamination, raw meats are chilled, packaged under different atmospheric conditions such as vacuum packaging, MAP or active packaging with O_2 scavengers or CO_2 generating systems (Coma 2008; McMillin 2008). Additional combinations such as low-dose irradiation, UV surface decontamination or HHP

Table 18.2 Examples of applications of bacteriocin preparations in meat and poultry products

Bacteriocin preparations	Effect(s)	Reference(s)
Raw meats		
Nisin combinations (organic acids, chelators, lysozyme, vacuum packaging, MAP)	Decontamination of raw meat surfaces before processing	Thomas et al. (2000)
Nisin activated film with EDTA	Inhibition of LAB, carnobacteria and *B. thermosphacta* and reduction of *Enterobacteriaceae* load on beef cuts	Ercolini et al. (2010)
Pediocins	Anti-*Listeria* protection by pediocins in raw meats	Rodríguez et al. (2002)
Pentocin 31-1	Reduction of growth of *Listeria* and *Pseudomonas* and total volatile basic nitrogen production in chill-stored tray-packaged pork meat	Zhang et al. (2010)
RTE meats		
Nisin activated films	Increased inactivation of *L. monocytogenes* in several vacuum-packaged products	Aymerich et al. (2008)
Nisin in combination with HHP	Increased inactivation of *E. coli* and staphylococci in cooked ham, avoiding regrowth of *E. coli* and slime-forming bacteria	Garriga et al. (2002)
Nisin and pulsed light	Application of a Nisaplin dip followed by exposure to pulsed light reduced the population of *L. innocua* on sausages	Uesugi and Moraru (2009)
Nisin-pectin film, in combination with low-dose irradiation	Increased microbial inactivation of *L. monocytogenes* on RTE turkey meat and inhibition of survivor proliferation during storage	Jin et al. (2009)
Pediocin in combination with post-packaging irradiation or thermal treatment	Effective combination with to control *L. monocytogenes* on frankfurters	Chen et al. (2004a, b)
Enterocin alginate film, in combination with HHP	Prevention of *L. monocytogenes* regrowth in the treated cooked ham during cold storage as well as cold chain break	Marcos et al. (2008)

have been proposed (Aymerich et al. 2008). All these processing treatments have selective effects on the initial microbiota and may act in synergy with bacteriocins to increase the product safety and shelf life. Although raw meat products are further processed prior to consumption by treatments that usually destroy pathogenic bacteria, they can be a considerable source of cross-contamination and in some cases may also carry heat-stable microbial toxins.

Nisin has been widely tested for preservation of raw meats (Thomas et al. 2000). However, application of nisin in meats faces some limitations derived from its poor solubility, interaction with phospholipids and inactivation by glutathione (Thomas et al. 2000; Stergiou et al. 2006). Nevertheless, positive results have been reported for surface decontamination of raw meats before processing and packaging for combinations of nisin with other antimicrobials or hurdles such as organic acids, chelators, lysozyme, vacuum packaging or MAP, increasing the microbial inactivation of *L. monocytogenes, Brochothrix thermosphacta* and *E. coli* O157:H7. Treatment of raw meats with pediocins (especially pediocin PA-1/Ach) can also retard growth of spoilage Gram-positive bacteria (such as *B. thermosphacta*) and/or reduce *L. monocytogenes* and *Clostridium perfringens* populations (Rodríguez et al. 2002; Nieto-Lozano et al. 2006). Other bacteriocins such as sakacins, carnobacteriocins, bifidocins, lactocins, lactococcins or pentocins have shown variable inhibitory effects against spoilage or pathogenic bacteria in raw meats or poultry (Aymerich et al. 2008; Galvez et al. 2008). In minced meats, the combination of bacteriocins with plant essential oils at levels where they would not impart undesirable flavour is being considered as a way to increase inactivation of *L. monocytogenes* and inhibition of *Salmonella* Enteritidis (Solomakos et al. 2008; Govaris et al. 2010).

One attractive approach to increase the activity of bacteriocins in raw meats has been immobilisation in substrates (such as beads, liposomes, coatings or films). Nisin (alone or in combinations with citric acid, EDTA and Tween 80) incorporated in a variety of substrates (such as calcium alginate gels, agar coatings, palmitoylated alginate-based films, polyvinyl chloride, LDPE or nylon) showed strong inhibition of bacteria such as *L. monocytogenes, B. thermosphacta, S. aureus* or *S. typhimurium* on refrigerated raw meats (Chen and Hoover 2003; Aymerich et al. 2008; Gálvez et al. 2007, 2008). The combination of chill temperature and antimicrobial packagings has proven to be effective in enhancing the microbiological quality of beef cuts by inhibiting LAB, carnobacteria and *B. thermosphacta* (Ercolini et al. 2010).

Since many LAB naturally associated with meats can grow at refrigeration temperatures, bacteriocin-producing strains that do not have adverse effects on meats can be selected as protective cultures for raw meat preservation (Table 18.3). Previous works have demonstrated the effectiveness of bacteriocin-producing *Lb. sakei* and *Lb. curvatus* strains in inhibiting *L. monocytogenes* or *B. thermosphacta* in raw meat products (Castellano et al. 2008). BLIS-producing *Lb. sakei* strains delayed blown-pack spoilage caused by *Clostridium estertheticum* and reduced the survival of *Campylobacter jejuni* on beef meat (Jones et al. 2009). In chicken ground meat, growth of *S. enteritidis* was adversely affected by the protective culture *L. fermentum* ACA-DC179 (Maragkoudakis et al. 2009). Fresh beef meat inoculated with *Lb. curvatus* CRL705 as protective culture showed a net increase of free amino acids due to the complementary activity of the bacterial and meat proteases on meat sarcoplasmic proteins, and it was proposed that this protective culture could contribute to meat ageing while improving shelf life (Fadda et al. 2008).

Control of Pathogenic and Spoilage Bacteria in Semi-Processed and Cooked Meats

Cooked meat products are widely consumed RTE foods. They may consist of whole primary meat pieces, but usually they are made by grinding and mixing secondary meats, fat, animal organs or blood with other ingredients, followed by stuffing/molding and cooking. The cooking process inactivates natural microbiota, paving the way for

Table 18.3 Examples of applications of bacteriocin-producing cultures in meat and poultry products

Starter or protective cultures		
Raw meats		
Bacteriocin producer *Lb. curvatus* CRL705	Effective inhibition of *L. innocua* and *B. thermosphacta* and indigenous contaminant LAB in fresh beef; contribution to meat ageing by limited proteolysis	Fadda et al. (2008)
BLIS-producing *Lb. sakei*	Delayed blown-pack spoilage caused by *C. estertheticum* and reduced survival of *C. jejuni* on meat	Jones et al. (2009)
BLIS-producing *Lb. fermentum* ACA-DC179	Growth inhibition of *S. enteritidis* in refrigerated chicken ground meat	Maragkoudakis et al. (2009)
RTE meats		
Bacteriocin-producing *P. acidilactici* strains	Inhibition of *L. monocytogenes* in cooked meats	Rodríguez et al. (2002)
Sakacin K-producing *Lb. sakei* CTC494	Inhibition of *L. monocytogenes* in cooked meat products	Hugas et al. (1998)
Bacteriocin-producing *Lb. sakei*	Growth inhibition of *L. monocytogenes* and *Escherichia coli* O157.H7 in cooked, sliced, vacuum-packaged meats	Bredholt et al. (2001)
Fermented meats		
Bacteriocin-producing *Lb. sakei* starter cultures	Reduction of *Listeria* populations in fermented sausages	Ravyts et al. (2008)
Curvacin-producing *Lb. curvatus*	Antilisterial effects in meat fermentation	Dicks et al. (2004)
Pediocin-producing *P. acidilactici*	Commercial starter cultures for fermentation of meat products to reduce the numbers of *L. monocytogenes* in the final product	Amezquita and Brashears (2002)
E. faecalis CECT7121 (producer of enterocin MR99)	Reduction of viable counts of *Enterobacteriaceae*, *S. aureus* and other Gram-positive cocci in craft dry-fermented sausages	Sparo et al. (2008)

growth of post-process contaminants. The pH values of most cooked meat products are compatible with growth of pathogenic and spoilage bacteria, which can proliferate at refrigeration temperatures during the product's shelf life. Some of these products may also be submitted to further processing such as slicing, peeling and packaging, which increase the risks for cross-contamination (Murphy et al. 2005). Bacteriocin preparations (mainly pediocin and nisin) have great market opportunities as hurdles against pathogenic and spoilage bacteria in cooked meat products. The main approaches tested are based on addition of bacteriocin preparations to the meat slurries before the heating process, surface application of the bacteriocins before packaging or application of films or coatings containing bacteriocins.

Strains of LAB (mainly *Lactobacillus* and *Leuconostoc*) are the major group of spoilage bacteria developing on various types of vacuum-packed meats, where they produce typical sensory changes such as souring, gas, SH_2 and slime (Korkeala et al. 1988; Björkroth and Korkeala 1997). These LAB rapidly recontaminate the products during handling and slicing (Lücke 2000). Enterocins A and B inhibited the production of slime in vacuum-packaged sliced cooked pork ham by *Lb. sakei*, and nisin prevented ropiness by *L. carnosum* (Hugas et al. 2003; Aymerich et al. 2008). When the above bacteriocins and pediocin PA-1/AcH were incorporated in blended cooked ham in combination with HHP treatment, only nisin avoided regrowth of *L. carnosum*. The combined HHP treatment with nisin increased the microbial inactivation of *E. coli* and avoided regrowth during storage (Garriga et al. 2002). Nisin was also the bacteriocin causing a greater decline of staphylococci, but did not prevent regrowth of *L. monocytogenes* (while enterocins, sakacin and pediocin did). These results illustrate the variable efficacy of different bacteriocins depending on the target bacteria and food-processing conditions.

The efficacy of nisin and lacticin 3147 against Gram-positive and Gram-negative bacteria in sausages (such as fresh pork, ham or bologna sausages) increased in combination with organic acids, lysozyme and EDTA. Similarly, greater inhibitory effects were reported for pediocin combinations with sodium diacetate or sodium lactate against *L. monocytogenes* on frankfurters or *L. monocytogenes* and *Yersinia enterocolitica* on cooked poultry cuts stored under MAP at 3.5°C (O'Sullivan et al. 2002; Chen and Hoover 2003; Aymerich et al. 2008; Galvez et al. 2008). These results paved the way for application of activated packagings containing cockatils of antimicrobial substances on RTE meats (Coma 2008). The bacteriocin-activated films may be quite appropriate for cooked meat products, acting as barriers against external contamination of the processed product. Among the various kinds of edible coatings tested on vacuum-packaged products (hot dogs, frankfurters or ham), best results have been reported for coatings containing nisin in combination with other antimicrobials under refrigeration storage. A recent study has shown that cellulose films containing pediocin PA-1/AcH retarded growth of *Listeria* in vacuum-packaged sliced ham stored at 12°C simulating abusive temperatures that can occur in supermarkets (Santiago-Silva et al. 2009).

Pre-surface application of bacteriocins in combination with post-packaging treatments is another approach of recent interest. Bacteriocin application followed by in-package thermal treatments can provide an effective combination to control *L. monocytogenes* on products such as frankfurters or turkey bologna, as shown for pediocin, nisin, nisin–lysozyme or combinations of these bacteriocins with sodium lactate/sodium diacetate (Chen et al. 2004b; Mangalassary et al. 2008). Application of a Nisaplin dip followed by exposure to pulsed light (PL; 9.4 J/cm^2) reduced the population of *L. innocua* in sausages and inhibited the microbial growth during refrigeration storage (Uesugi and Moraru 2009). Since application of PL is approved for decontamination of food and food surfaces, the combined treatment could be applied as a post-processing step to reduce surface contamination and increase the safety of RTE meat products. Low-dose irradiation of pediocin-treated frankfurters or RTE turkey meat vacuum-packaged with a pectin–nisin film increased the inactivation of *L. monocytogenes* greatly and reduced the proliferation of survivors during storage (Chen et al. 2004a; Jin et al. 2009). The authors concluded that the combined treatment could serve to prevent listeriosis due to post-processing contamination while reducing radiation doses and impact on product quality, or to prevent *L. monocytogenes* growth in accidentally recontaminated packages of irradiated RTE meats. Application of bacteriocin-activated films in combination with HHP could increase the microbial inactivation in RTE meat products as well (Aymerich et al. 2008). In cooked ham packaged in films activated with enterocins, the combined treatment with HHP reduced the numbers of *L. monocytogenes* and inhibited the recovery of pressure-injured *Listeria* during temperature abuse (Marcos et al. 2008). The application of nisin through interleavers combined with an HHP treatment was an effective treatment to achieve absence of *Salmonella* in 25 g samples of sliced cooked ham during refrigeration storage (Jofré et al. 2008).

Bacteriocin-producing LAB strains could be used as protective cultures for semi-processed and cooked meats provided that they cause only a minimal change in the desired sensory properties of the products while inhibiting *Listeria* and lactic acid bacteria involved in spoilage (Hugas et al. 1998; Lücke 2000; Chen and Hoover 2003; Aymerich et al. 2008; Galvez et al. 2008). Inoculation of strains producing sakacins, pediocins, leucocins, plantaricins, enterocins, bavaricins or curvaticins in processed meat products was shown to inhibit growth of *Listeria* and in some cases to inhibit slime formation as well (Aymerich et al. 1998; Hugas et al. 1998). There are already several LAB cultures in the market introduced as starter or bioprotective culture with the aim of contributing to microbiological safety of semi-processed and cooked meats (Aymerich et al. 2008).

Applications in Fermented Meats

Bacteriocin preparations can be added to meat batters for inactivation of microbial pathogens in fermented meat products. The lower pH attained in sausages compared to fresh meats may increase the solubility of some bacteriocins such as nisin and probably their antimicrobial activity as well. Bacteriocins such as nisin, enterocins (CCM 4231, A, B and AS-48) or leucocins improved the reduction of *L. monocytogenes* or *S. aureus* populations in fermented meats (Rodríguez et al. 2002; Chen and Hoover 2003; Aymerich et al. 2008; Galvez et al. 2008). Microbial inactivation by bacteriocin addition may be an attractive hurdle for slightly fermented sausages, in which the higher pH and water content may facilitate survival and proliferation of certain pathogenic bacteria.

LAB play key roles in meat fermentations. Therefore, bacteriocin-producing strains have been proposed as starter cultures to combat pathogens such as *L. monocytogenes* (Työppönen et al. 2003; Leroy et al. 2006; Aymerich et al. 2008). Bacteriocin-producing lactobacilli (mainly *Lb. sakei* and *Lb. curvatus*, and also *Lb.*

rhamnosus and *Lb. plantarum*) have demonstrated anti-listerial effects in sausage or salami fermentations, depending to a great extent on strain and type of meat (Erkkilä et al. 2001; Leroy et al. 2005; Dicks et al. 2004; Benkerroum et al. 2005; Todorov et al. 2007) (Table 18.3). *Lb. sakei* CTC 494 (producing sakacin K) is a promising functional starter culture with antilisterial activity, being capable to successfully suppress *L. monocytogenes* in Spanish-style and German-style fermented sausages (Aymerich et al. 2008) or to reduce *Listeria* populations in Belgian-style sausages, Italian salami and Cacciatore salami (Ravyts et al. 2008). The efficacy of *Lb. sakei* is influenced by environmental factors such as sausage ingredients, salt, fat and nitrite content, acidification level and temperature (Leroy et al. 2006). Since *Lb. sakei* and *Lb. curvatus* can hydrolyse muscle sarcoplasmic proteins and, to a lesser extent, myofibrillar proteins, they can contribute to the generation of small peptides and amino acids which contribute as direct flavour enhancers or as precursors of other flavour compounds during the ripening of dry-fermented sausages (Leroy et al. 2006). Exploitation of these activities may lead to the use of a new generation starter cultures with industrially or nutritionally important functionalities (Leroy et al. 2006).

Bacteriocin-producing pediococci can reduce *L. monocytogenes* populations in fermented meats (Amezquita and Brashears 2002; Rodríguez et al. 2002; Aymerich et al. 2008). Pediococci are preferred as starters in certain products (rather than lactobacilli), e.g. in American-style sausages fermented at higher temperatures. Pediocin PA-1 producers do not inhibit bacteria relevant to the fermentation, such as staphylococci and micrococci (Gonzalez and Kunka 1987). Nisin-producing lactococcal strains isolated from fermented sausages have also been suggested as adjunct cultures for improving the food safety of meat fermented products manufactured under poor hygienic conditions such as indigenous fermentations (Noonpakdee et al. 2003).

Enterococci are often part of the normal microbiota in meat fermentations and have demonstrated robustness as antilisterial agents in fermented meats, and also to inhibit *S. aureus* (Foulquié Moreno et al. 2003; Aymerich et al. 2008; Galvez et al. 2008). However, their application in foods is controversial because of their potential virulence as opportunistic pathogens. Strain *Enterococcus faecalis* CECT7121 (producer of the broad-spectrum enterocin MR99) is interesting because it is devoid of the genes for haemolysin and gelatinase production and does not produce biogenic amines (Sparo et al. 2008). Craft dry-fermented sausages inoculated with strain CECT7121 had lower viable counts of *Enterobacteriaceae*, *S. aureus* and other Gram-positive cocci at the end of fermentation, with no detectable enterobacteria and *S. aureus* at the end of drying.

Staphylococci and micrococci may also be exploited as sources for antibacterial substances applicable in sausage fermentations. The introduction of the lysostaphin gene (an endopeptidase that specifically cleaves the glycine–glycine bonds unique to the interpeptide cross-bridge of the *S. aureus* cell wall) into meat starter lactobacilli (Cavadini et al. 1998) can be used to prevent the growth of *S. aureus*. One *Staphylococcus xylosus* sausage isolate that produces an antilisterial substance increased the microbial inactivation of *L. monocytogenes* in Naples-type sausage

(Villani et al. 1997). *K. varians* produces the lantibiotic variacin (Pridmore et al. 1996), but there are still no reports on the effects of strains producing this antimicrobial peptide in meat fermentations.

Biopreservation of Seafoods

Treatment of Seafoods with Bacteriocin Preparations

The most popular commercial sources of nisin (Nisaplin™) and pediocin (Alta™ 2341) are based on a milk formulation. These preparations may not be the best formulations for fish application because of the inappropriate taste they can impart when used at high concentrations and also because of the lower solubility of bacteriocins complexed with the milk solids. The use of fishery by-products as peptide sources for the production of nisin and pediocin has been proposed recently (Vázquez et al. 2004, 2006). In a similar way, Tahiri et al. (2009a) described an extract of snow crab hepatopancreas, a food-grade by-product of crustacean processing, as useful substrate for stimulating growth of *Carnobacterium divergens* M35 and divergicin M35 production. This approach could be exploited for the industrial production of bacteriocin preparations intended for preservation of seafood products.

Several approaches based on bacteriocin treatment have been proposed for inactivation of unwanted bacteria (mainly *L. monocytogenes*) in seafood products, although most tests have been carried out on cold-smoked salmon (CSS) where listerial contamination is of greatest concern (O'Sullivan et al. 2002; Chen and Hoover 2003; Drider et al. 2006; Calo-Mata et al. 2008; Galvez et al. 2008). These include addition of bacteriocin preparations (either by immersion or by spraying or by mixing with food matrix), bacteriocin injection or immobilisation on plastic films or coatings or even on the bacteriocin producer cells (Table 18.4).

Washing crabmeat with antimicrobials (PerLac 1911, Microgard™, Alta™ 2341, nisin or *E. faecium* 1083 culture supernatant containing the bacteriocin-like substance (BLIS) enterocin 1083 reduced the viable counts of *L. monocytogenes* during storage at 4°C only in the samples treated with 20,000 AU of Alta™ 2341, nisin or enterocin 1083. In brined shrimp, addition of nisin Z and bavaricin A preparations extended the product shelf-life, although the efficacy of bacteriocin treatments was much more limited compared to brined shrimp stored in a benzoate–sorbate solution. In ready-to-eat seafoods, nisin has been tested in combination with heat treatments with the purpose of reducing the processing cost and/or the intensity of treatments. The reported effects include increased killing of *L. monocytogenes* in cold-pack lobster, and synergistic activity against *Listeria* and total mesophilic bacteria in caviars (Al-Holy et al. 2004).

Proliferation of *L. monocytogenes* in slightly processed products which are consumed without further cooking (such as cold-smoked seafood products) is a matter

Table 18.4 Examples of bacteriocin applications in seafoods

Bacteriocin treatment	Effect(s)	Reference(s)
Nisin- radio frequency heating at 65°C	Complete inactivation of *L. innocua* in sturgeon caviar or ikura	Al-Holy et al. (2004)
Nisin-coated plastic films	Inactivation of *L. monocytogenes* in CSS during refrigeration storage Inhibition of background spoilage microbiota	Neetoo et al. (2008)
C. divergens V41 or its culture supernatant	Inhibitory effect on *L. innocua* 2030c growth cold-smoked salmon trout	Vaz-Velho et al. (2005)
C. divergens M35, or divergicin M35	Suggested as bioingredient for application to the inactivation of *L. monocytogenes* in ready-to-eat seafood	Tahiri et al. (2009b)
Lb. curvatus CWBI-B28 culture, spraying with bacteriocin, packaging in bacteriocin-coated plastic film, cell-adsorbed bacteriocin	Variable inactivation of *L. monocytogenes* in CSS. Best results were reported for bacteriocin adsorbed on heat-inactivated cells.	Ghalfi et al. (2006)
Lb. curvatus ET30	Reduction of *L. innocua* counts on salmon fillets before and after cold-smoking and during vacuum pack storage	Tomé et al. (2008)
Leuconostoc spp., *Lb. fuchuensis, C. alterfunditum*	Broad-spectrum bioprotective cultures	Matamoros et al. (2009)
Bifidobacterium-thymol combination	Extended shelf life of fresh plaice fillets; inhibition of fresh packaged fish spoilers	Altieri et al. (2005)
S. xylosus	Use as protective culture to decrease biogenic amine formation in salted and fermented anchovy	Mah and Hwang (2009)
L. lactis (nisin-producer)	Use as starter culture to improve Senegalese guedj fish fermentation	Diop et al. (2009)

of concern, and extensive work has been carried out on application of bacteriocins in this field. Purified sakacin P completely eliminated *L. monocytogenes* on CSS, despite the proteolytic degradation of the bacteriocin by salmon tissue proteases. Treatments of cold-smoked salmon trout with *C. divergens* V41 culture supernatant showed a significant inhibitory effect on *L. innocua* growth (Vaz-Velho et al. 2005; Rihakova et al. 2009). In vacuum-packaged CSS, growth of *L. monocytogenes* could be prevented by a combination of carbon dioxide, nisin, NaCl and low temperature. On smoked salmon slices inoculated with *L. monocytogenes* and surface-treated with nisin or Alta™ 2341, both antimicrobials reduced the growth of listeria to some extent. When the treated smoked salmon was packaged in 100% CO_2, counts of *L. monocytogenes* were reduced below detectable levels (2 logs) in both cases.

Inhibition of aerobic bacteria is important to prevent seafood spoilage. The combination of nisin and Microgard™ reduced the total aerobic bacteria populations and delayed growth of *L. monocytogenes* in fresh-chilled salmon, increasing the product shelf life (Calo-Mata et al. 2008; Galvez et al. 2008). Similarly, when nisin was tested in combination with other antimicrobials (such as the lactoperoxidase system or with headspace CO_2 levels and EDTA), increased inactivation or growth delay of spoilage microbiota was observed in sardines and in fish muscle extract. In fish spreads, enterocins 1071A and 1071B inhibited the growth of aerobic mesophilic bacteria during cold storage.

Antimicrobial activity of bacteriocins in seafoods can improve with immobilisation in packaging materials (Galvez et al. 2008). In CSS vacuum-packed inside nisin-coated plastic films, nisin (2,000 IU/cm^2) inhibited the proliferation of *L. monocytogenes* strains and background microbiota (aerobic, anaerobic and LAB) during storage at 4°C in a concentration-dependent manner (Neetoo et al. 2008). Similarly, chitosan-coated plastic films dosed with sodium lactate (2.3 mg/cm^2) in combination with nisin (500 IU/cm^2) inhibited the proliferation of *L. monocytogenes* as well as aerobic and anaerobic background microbiota in the vacuum-packaged CSS during storage at 4°C (Ye et al. 2008). Packaging CSS in plastic film coated with bacteriocin from *Lb. curvatus* CWBI-B28 caused *L. monocytogenes* inactivation late during refrigeration storage (Ghalfi et al. 2006). However, best results (complete inactivation of *L. monocytogenes* during storage for 22 days) were reported for CSS treated with bacteriocin adsorbed to its heat-inactivated producer cells.

Application of Bacteriocin-Producing Strains as Protective Cultures in Seafood Products

Bacteriocin production by fish-acclimatised bacterial species is of great interest for inhibition of pathogenic micro-organisms in seafood products (Table 18.4). Antagonistic bacterial strains (such as those isolated from cold-smoked seafood products) could be applied for the competitive exclusion of *L. monocytogenes* in the processed food products. Many LAB strains are able to grow at refrigeration temperatures. They tolerate atmosphere packaging, low pH, high salt concentrations and the presence of additives such as lactic acid, ethanol or acetic acid. The selected antagonistic strains should meet several criteria: (1) to be able to grow on the fish product during cold storage and produce antimicrobials to inactivate *L. monocytogenes*, or at least inhibit growth of the pathogen, (2) do not cause adverse effects on the food product (such as off flavours, colour changes), (3) do not have adverse effects on health (e.g. production of biogenic amines) or carry antibiotic resistance or virulence traits. Inoculated strains could have probiotic properties, but this approach for administration of probiotics through seafood products has not been exploited yet.

Since LAB comprise the dominant microflora in CSS (González-Rodríguez et al. 2002; Cardinal et al. 2004), research has focused on selection of antagonistic LAB strains from the processed products. *L. monocytogenes* can be inhibited by carnobacteria cultures that do not produce bacteriocins, partly due to glucose depletion (Nilsson et al. 2005). However, LAB strains producing bacteriocins (mainly *Carnobacterium* and *Lactobacillus* species) may be superior for biopreservation. Trials carried out with antagonistic bacteria in CS foods (such as CSS, cold-smoked salmon trout, or cold-smoked surubim) included antagonistic strains of *C. piscicola*, *C. divergens*, *Lb. sakei*, *Lb. casei*, *Lb. curvatus*, *Lactobacillus delbrueckii*, *Lb. plantarum, P. acidilactici* or *E. faecium* (Leisner et al. 2007; Calo-Mata et al. 2008; Galvez et al. 2008; Tomé et al. 2008; Tahiri et al. 2009b). Bacteriocin-producing enterococci from seafoods are gaining interest for biocontrol of *L. monocytogenes* in the processed products. Several strains have been previously isolated from seafoods such as *E. faecium* strains and *Enterococcus mundtii*, producers of unknown bacteriocins (Campos et al. 2006; Tomé et al. 2008; Hosseini et al. 2009), or *E. faecium* strains, producers of enterocin P (Arlindo et al. 2006) or enterocin B (Pinto et al. 2009). Antilisterial effects are increased by using combinations of bacteriocin-producing strains (such as *Lb. casei–Lb. plantarum*) and also by exogenously added bacteriocin in combination with the producer strain.

While most studies have focused on inhibition of *L. monocytogenes* in seafood products, other pathogenic bacteria (such as *Vibrio parahaemolyticus*, *Vibrio vulnificus* and *Vibrio cholerae*, *C. botulinum*, histamine-producing bacteria and post-contaminating bacteria, such as *S. aureus* or *Salmonella* spp.) or spoilage bacteria (such as *Shewanella putrefaciens*, *Photobacterium phosphoreum*, *Aeromonas* spp. and *Pseudomonas* spp.) are still a matter of concern (Gram and Dalgaard 2002; Calo-Mata et al. 2008). For this reason, there is a growing interest to extend the spectrum of inhibition of bacteriocins in combination with other hurdles and also on isolation of LAB strains with broader spectrum of inhibitory activity. It was reported that treatment of whole shrimp with potassium sorbate with *Bifidobacterium breve* cells extended the product's microbiological shelf life (Al-Dagal and Bazaraa 1999), and treatment of plaice fillets with a preparation of *Bifidobacterium bifidum* cells and thymol combined with low storage temperature and anoxia/hypoxia showed a great efficacy against the main fresh-packaged fish spoilage species (Altieri et al. 2005). Selected strains of *Leuconostoc gelidum*, *Lactococcus piscium*, *Lactobacillus fuchuensis* and *Carnobacterium alterfunditum* (psychrotrophs, lacking antibiotic resistance traits and unable to produce histamine or tyramine) are currently being investigated as broad-spectrum bioprotective cultures in fish preservation (Matamoros et al. 2009).

Bacteriocin Applications in Fermented Fish

Fermented fish products are very popular in the Asiatic and Pacific regions, but their microbiological aspects are not known so well as other fermented foods. Salted fermented foods contain abundant amino acids, which can generate relatively large amounts of biogenic amines (Mah et al. 2003). Inoculation with

S. xylosus, producer of a BLIS, has been proposed as a protective culture to decrease biogenic amine formation in salted and fermented anchovy (Mah and Hwang 2009).

L. lactis subsp. *lactis* strain CWBI B1410 (which produces various antibacterial compounds including organic acids and nisin) was tested to improve the traditional Senegalese fish fermentation into guedj (Diop et al. 2009). The inoculated starter (in combination with glucose addition) released nisin onto the fish fermentate, produced a faster acidification and reduced the counts of enteric bacteria in the fermented fish. The authors proposed a new fish fermentation strategy based on inoculation with this strain as starter, combined with salting and drying, to enhance the safety of guedj.

Biopreservation of Vegetable Foods

Inhibition of Pathogenic Bacteria on Fresh Produce

Bacteriocin treatments (with nisin, pediocin or enterocin AS-48) have been proposed to inactivate food-borne pathogenic or toxinogenic bacteria (such as *L. monocytogenes, B. cereus,* and *Bacillus weihenstephanensis, E. coli, Salmonella* and other enterobacteria) on the surfaces of fresh-cut vegetables and on sprouted seeds (Galvez et al. 2008; Randazzo et al. 2009; Abriouel et al. 2010) (Table 18.5). Bacteriocin treatments have also been proposed for decontamination of whole fruit surfaces and to avoid transmission of pathogenic bacteria from fruit surfaces to processed fruits (Ukuku et al. 2005; Silveira et al. 2008), and to decrease bacterial survival on sliced fruit surfaces during storage. In general, the single treatments with bacteriocins significantly reduced the microbial loads of Gram-positive pathogens but were ineffective against Gram-negatives unless they were applied in combination with other antimicrobials (such as organic acids, inorganic acids or phenolic compounds). The combined treatments not only increased microbial inactivation on the fruit and vegetable food surfaces but also avoided proliferation of survivors during storage (Cobo Molinos et al. 2008). Some of the combined treatments proposed could find industrial applications, especially in added-value food products such as those intended for consumption by the elderly, immunocompromised people or debilitated hospital patients.

Inoculation with live cultures is another proposed alternative for inhibition of pathogenic bacteria on fresh produce surfaces. Bacterial strains (including species of genera such as *Bacillus, Pseudomonas, Enterococcus, Lactococcus, Leuconostoc, Weissella* or *Lactobacillus*) isolated from raw vegetables may produce antagonistic substances to food-borne pathogens (Galvez et al. 2008; Trias et al. 2008a, b). These strains may be better adapted to vegetable substrates and growth under cold or moderate temperatures. The efficacy of such treatments greatly depends on ecological factors such as the capacity to grow and produce antimicrobials in situ by the protective cultures in competition with resident microbiota.

Table 18.5 Examples of bacteriocin applications in vegetable foods and beverages

Bacteriocin treatment	Effect(s)	Reference(s)
Nisin Z, coagulin, nisin:coagulin cocktail	Reduced viable cell counts of *L. monocytogenes* on fresh-cut iceberg lettuce stored in microperforated plastic bags	Allende et al. (2007)
Enterocin AS48 washing treatments alone or in combination with other antimicrobials	Inactivation of *L. monocytogenes, B. cereus, B. weihenstephanensis* and *Enterobacteria* on sprouts	Abriouel et al. (2010)
Nisin in combination with of hydrogen peroxide, sodium lactate and citric acid as a sanitizer	Decontamination of whole cantaloupe and honeydew melon surfaces. Prevented transfer of *L. monocytogenes* and *E. coli* to fresh-cut pieces	Ukuku et al. (2005)
Nisin combination with lysozyme and PEF	Inactivation of *Salmonella* Typhimurium in orange juice	Liang et al. (2002)
Enterocin AS-48 alone or in combination with PEF, chelators or heat	Inactivation of pathogenic and spoilage bacteria in fruit juices	Abriouel et al. (2010)
Nisin	Prevented spoilage caused by non-aciduric and aciduric spore formers in canned foods and in other foods	Thomas et al. (2000, 2001)
Enterocin AS-48	Inactivation of endospore formers in boiled rice, purees and canned vegetables	Abriouel et al. (2010)
Nisin	Inactivation of wine LAB at lower sulphite concentrations	Rojo-Bezares et al. (2007)
Pediocin PD-1	Control of *O. oeni* in wines	Bauer et al. (2003)
Enterocins L50A and L50B	Inhibition of beer-spoilage LAB in worts and lager beers	Basanta et al. (2008)
Plantaricin-producing starter culture	Improved microbiological control of table olives fermentation	Vega Leal-Sánchez et al. (2003)
Bacteriocin-producing strains	Inhibition of rope-forming bacilli in breads	Settanni and Corsetti (2008)
	Enhanced competitiveness of strains in fermented doughs	
Bacteriocin-producing *L. plantarum*	Decreased survival of *B. cereus*, *E. coli* O157:H7 and *S. enterica* in millet gruels	Sánchez Valenzuela et al. (2008)

Inactivation of Spoilage and Pathogenic Bacteria in Fruit Juices, Drinks, and Fermented Beverages

In fruit juices and drinks, bacteriocin addition (nisin, enterocins) has been proposed for inactivation of endospore-forming bacteria causing spoilage, such as *Alicyclobacillus acidoterrestris* and thermophilic spore formers such as *G. stearothermophilus*. Bacteriocin addition may also be useful for microbial inactivation of bacteria causing ropiness (such as EPS-producing *Bacillus licheniformis*, pediococci and lactobacilli) as well as acrolein-producing bacteria. While fruit juices and drinks usually have a pH that is too low for proliferation of food-borne pathogenic bacteria, some less acidic juices and drinks can support bacterial growth. Inactivation of food-borne pathogens (*L. monocytogenes, B. cereus, S. aureus*) by enterocins has been reported in lettuce juices, soy milk and sport and energy drinks with lower acidity (Galvez et al. 2008; Abriouel et al. 2010). Since freshly made fruit juices have been implicated in transmission of enteric pathogens, bacteriocins (such as nisin and enterocin AS-48) have been tested in combination with other agents to increase the bacterial outer membrane permeability (such as chelators or PEF). The combined treatments of bacteriocins and PEF greatly increased the bactericidal effects and decreased the risks of survivor proliferation in the treated samples (Liang et al. 2002; Mosqueda-Melgar et al. 2008).

In fermented beverages, bacteriocin preparations can be applied against spoilage LAB. Several applications have been proposed for nisin in the production of beer: (1) cleaning of the equipment and final cleansing rinse, (2) addition to fermenters to control contamination, (3) increasing the shelf life of uncontaminated beers, (4) reduction of pasteurization regimes, (5) washing pitching yeasts to eliminate contaminating bacteria and (6) development of wort bioacidifiying LAB and/or yeast starter cultures genetically modified to produce bacteriocins. Other bacteriocins such as lacticin M30, and enterocins L50A and L50B have been shown to inhibit spoilage LAB in beers (Basanta et al. 2008).

Bacteriocin-producing strains can be isolated from raw as well as malted barley. Fermented worts containing bacteriocins could be used to prevent beer spoilage LAB (Vaughan et al. 2005). Bacteriocin production may be a desirable trait for wort bioacidifying LAB starter cultures, enhancing the implantation and proliferation of such strains over spoiling LAB. Development of yeast starter cultures genetically modified to produce bacteriocins has also been suggested, and heterologous production of bacteriocins such as pediocins, leucocins, plantaricins and enterocins by yeasts has been reported (Schoeman et al. 1999; Du Toit and Pretorius 2000; Van Reenen et al. 2002; Gutiérrez et al. 2005; Sánchez et al. 2008; Basanta et al. 2009). The bactericidal yeast strains could be used as starters or protective cultures in the fermentations of brewing, wine and baking processes as biological control agents to inhibit growth of spoilage bacteria.

Addition of bacteriocins could prevent spoilage of other alcoholic beverages, such as fermented apple ciders (Galvez et al. 2008; Abriouel et al. 2010) or wines (Du Toit and Pretorius 2000; Bartowsky 2009). Nisin addition is permitted in beer in certain countries, but not in wine. However, nisin has been reported to act synergistically with sulphites against wine LAB. Nisin addition could aid to reduce the sulphite content in musts before fermentation. Application of pediocin PD-1 has been proposed in removal of *Oenococcus oeni* biofilms from stainless steel surfaces and also to control growth of *O. oeni* in wine (Bauer et al. 2003).

Bacteriocin producers can be isolated from wine and vineyards, including species of *Lb. plantarum, O. oeni* or *Pediococcus pentosaceus* (Navarro et al. 2000; Rojo-Bezares et al. 2007; Knoll et al. 2008; Yanagida et al. 2008). Selected strains could be useful against undesired LAB in vinification (such as spoilers or histamine producers), and for proper control of the wine malolactic fermentation.

Improvement of Vegetable Fermentations, and Inactivation of Food Poisoning and Spoilage Microbes in Fermented Vegetable Foods

Addition of bacteriocins (such as nisin) has been proposed as a way to direct the microbiota of vegetable fermentations towards selection of desirable bacteriocin-tolerant or bacteriocin-resistant strains with desirable effects while at the same time inhibiting strains causing defects such as overripening (as in the case of kimchi fermentation).

Bacteriocin-producing strains can be used as starter cultures in table olive fermentation to improve the microbiological control of the process, increase the lactic acid yield and provide a consistently high quality of the product (Vega Leal-Sánchez et al. 2003). Such starters can also be applied in newly operating plants that still lack the appropriate resident LAB microbiota to accelerate the fermentation process and avoid stuck fermentations. Alternatively, starter cultures could be applied to control food-borne pathogens such as *L. monocytogenes* in certain fermented foods such as kimchi (Galvez et al. 2008).

There is a growing interest in bacteriocin-producing strains from ethnic fermented vegetable foods (Kostinek et al. 2007; Yoon et al. 2008; Ge et al. 2009; Hata et al. 2009; Huang et al. 2009; Tamang et al. 2009; Gao et al. 2010). The nisin-producer *L. lactis* subsp. *lactis* IFO12007 isolated from miso was used as starter for fermentation of cooked rice and rice koji supplemented with soybean extract (Kato et al. 2001). The producer strain proliferated in the cooked rice and produced enough nisin activity to inhibit *Bacillus subtilis* without any adverse effect on growth of *Aspergillus oryzae* during the koji fermentation.

Fermented millet flours are very popular in the African continent, but only a limited number of studies have been carried out on their LAB microbiota and their bacteriocins (Ben Omar et al. 2006, 2008). The plantaricin-producing strain *Lb. plantarum* 2.9 (isolated from ben-saalga, a traditional pearl millet fermented food from Burkina Faso) produced strong inhibitory activity in malted millet flour, decreasing the survival of *B. cereus*, *E. coli* O157:H7 and *S. enterica* (Sánchez Valenzuela et al. 2008). This strain could be applied as a starter culture to improve the safety of cereal gruels.

LAB strains isolated from cereals and fermented doughs may produce several bacteriocins and BLIS (Messens and De Vuyst 2002; Narbutaite et al. 2007; Settanni and Corsetti 2008), as well as antifungal compounds (Valerio et al. 2009; Dalié et al. 2010). The production of antimicrobial substances, such as reutericyclin and bacteriocins, may enhance the competitiveness of strains in fermented doughs (Gänzle and Vogel 2003; Leroy et al. 2007) and is considered a desirable trait for starter culture development (Corsetti and Settanni 2007; De Vuyst et al. 2009).

It has been suggested that sourdoughs or cultured broths fermented with bacteriocin-producing lactobacilli could be applied to inhibit rope formation by bacilli in yeast-leavened breads (Menteş et al. 2007; Settanni and Corsetti 2008; Valerio et al. 2008).

Inactivation of Food-Borne Pathogens and Endospore-Forming Bacteria in Ready-to-Eat, Processed, and Canned Vegetable Foods

Addition of bacteriocins such as nisin or enterocin AS-48 has been reported to reduce viable counts of *L. monocytogenes* in RTE vegetable foods such as tofu or deli-type salads (Thomas et al. 2000; Abriouel et al. 2010). However, the strong interaction of bacteriocin molecules with food components required addition of high bacteriocin concentrations or the combined addition of other antimicrobials. This approach was also effective in the control of *S. enterica* in salads and *S. aureus* in vegetable sauces (Abriouel et al. 2010).

In cooked vegetables (such as cooked potato products, sous-vide mushrooms, "home-made"-type soups, purees or cooked rice foods) and in canned vegetables (such as canned tomato, peas, corn, etc.), addition of bacteriocins (such as nisin, nisin–pediocin combination or enterocin AS-48) has been proposed as a way to inhibit endospore outgrowth and production of enterotoxins (such as *B. cereus* or *C. botulinum* toxins) during storage and/or to increase the efficacy of thermal treatments against endospores (Thomas et al. 2000; Galvez et al. 2008; Cabo et al. 2009; Abriouel et al. 2010). Incorporation of bacteriocins in canned vegetables can be an effective hurdle to prevent spoilage caused by non-aciduric as well as aciduric spore formers.

Concluding Remarks

There is a great body of knowledge on the antimicrobial effects of bacteriocins and bacteriocin-producing strains in many different types of foods. However, only a few bacteriocin preparations are on the market, which are being used as ingredients in some types of foods, and only one bacteriocin preparation is licensed as a food preservative so far. The great efforts put in by many research groups in using Nature's weapons to improve food preservation are not being exploited to their full potential by the food industries. Nevertheless, novel food-processing technologies such as UV irradiation, low-dose radiation, PEF or HHP treatments may enhance the value of bacteriocins as additional hurdles to bypass some of the limitations of the novel treatments, improve microbial inactivation, decrease the intensity of treatments (and at the same time the impact on food quality) and afford protection against pathogenic and spoilage bacteria during the product storage. Application of bacteriocin preparations in activated packaging technology also offers new possibilities, especially for RTE foods, decreasing the amounts of bacteriocin necessary for inhibition of bacteria.

While application of bacteriocin preparations faces several limitations derived from legal constraints, costly toxicological studies, low yields of industrial-scale production processes and higher processing cost, application of bacteriocin-producing strains seems a much cheaper and feasible alternative. Since most LAB are considered to have a QPS status (EFSA 2008), their application in foods is widely accepted. However, in addition to not having adverse effects on the food and not being able to produce biogenic amines, LAB strains used in food production should also be free of transferable resistance to antibiotics of clinical use. Therefore, bacteriocin-producing strains to be used in foods must be selected carefully, and specific assessment of their QPS status may be required in future. Commercial cultures based on bacteriocin producers are already available for application as bioprotective or adjunct cultures in foods. In addition to inhibition of food-borne pathogens, bacteriocin-producing strains may also play other functions such as accelerated ripening of fermented foods, improvement of fermentation processes, or inhibition of NSLAB. The bacteriocin trait may also serve to enhance the proliferation and predominance of LAB strains with functional properties in food systems and in animal systems as well. The linkage between bacteriocin production and probiosis is increasingly becoming stronger (Rodgers 2008), and functional foods based on bacteriocin-producing probiotic LAB are to be expected in market.

References

Abriouel H, Lucas R, Ben Omar N et al (2010) Potential applications of the cyclic peptide enterocin AS-48 in the preservation of vegetable foods and beverages. Probiot Antimicrob Prot 2:77–89

Adams M (2003) Nisin in multifactorial food preservation. In: Roller S (ed) Natural antimicrobials for the minimal processing of foods. CRC Press LLC, Boca Raton, FL, pp 11–33

Al-Dagal MM, Bazaraa WA (1999) Extension of shelf life of whole and peeled shrimp with organic acid salts and bifidobacteria. J Food Prot 62:51–56

Al-Holy M, Ruiter J, Lin M et al (2004) Inactivation of *Listeria innocua* in nisin-treated salmon (*Oncorhynchus keta*) and sturgeon (*Acipenser transmontanus*) caviar heated by radio frequency. J Food Prot 67:1848–1854

Allende A, Martínez B, Selma V et al (2007) Growth and bacteriocin production by lactic acid bacteria in vegetable broth and their effectiveness at reducing *Listeria monocytogenes* in vitro and in fresh-cut lettuce. Food Microbiol 24:759–766

Altieri C, Speranza B, Del Nobile MA et al (2005) Suitability of bifidobacteria and thymol as biopreservatives in extending the shelf life of fresh packed plaice fillets. J Appl Microbiol 99:1294–1302

Amezquita A, Brashears MM (2002) Competitive inhibition of *Listeria monocytogenes* in ready-to-eat meat products by lactic acid bacteria. J Food Prot 65:316–325

Ananou S, Muñoz A, Martínez-Bueno M et al (2010) Evaluation of an enterocin AS-48 enriched bioactive powder obtained by spray drying. Food Microbiol 27:58–63

Anastasiou R, Aktypis A, Georgalaki M et al (2009) Inhibition of *Clostridium tyrobutyricum* by *Streptococcus macedonicus* ACA-DC 198 under conditions mimicking Kasseri cheese production and ripening. Int Dairy J 19:330–335

Arlindo S, Calo P, Franco C et al (2006) Single nucleotide polymorphism analysis of the enterocin P structural gene of *Enterococcus faecium* strains isolated from nonfermented animal foods. Mol Nutr Food Res 50:1229–1238

Aymerich MT, Hugas M, Monfort JM (1998) Review: bacteriocinogenic lactic acid bacteria associated with meat products. Food Sci Technol Int 4:141–158

Aymerich T, Picouet PA, Monfort JM (2008) Decontamination technologies for meat products. Meat Sci 78:114–129

Bartowsky EJ (2009) Bacterial spoilage of wine and approaches to minimize it. Lett Appl Microbiol 48:149–56

Basanta A, Sánchez J, Gómez-Sala B et al (2008) Antimicrobial activity of *Enterococcus faecium* L50, a strain producing enterocins L50 (L50A and L50B), P and Q, against beer-spoilage lactic acid bacteria in broth, wort (hopped and unhopped), and alcoholic and non-alcoholic lager beers. Int J Food Microbiol 125:293–307

Basanta A, Herranz C, Gutiérrez J et al (2009) Development of bacteriocinogenic strains of *Saccharomyces cerevisiae* heterologously expressing and secreting the leaderless enterocin L50 peptides L50A and L50B from *Enterococcus faecium* L50. Appl Environ Microbiol 75:2382–2392

Bauer R, Nel HA, Dicks LMT (2003) Pediocin PD-1 as a method to control growth of *Oenococcus oeni* in wine. Am J Enol Vitic 54:86–91

Benkerroum N, Daoudi A, Hamraoui T et al (2005) Lyophilized preparations of bacteriocinogenic *Lactobacillus curvatus* and *Lactococcus lactis* subsp. *lactis* as potential protective adjuncts to control *Listeria monocytogenes* in dry-fermented sausages. J Appl Microbiol 98:56–63

Ben Omar N, Abriouel H, Lucas R et al (2006) Isolation of bacteriocinogenic *Lactobacillus plantarum* strains from ben saalga, a traditional fermented gruel from Burkina Faso. Int J Food Microbiol 112:44–50

Ben Omar N, Abriouel H, Keleke S et al (2008) Bacteriocin producing *Lactobacillus* strains isolated from poto poto, a Congolese fermented maize product, and genetic fingerprinting of their plantaricin operons. Int J Food Microbiol 127:18–25

Bizani D, Motta AS, Morrissy JAC et al (2005) Antibacterial activity of cerein 8A, a bacteriocin-like peptide produced by *Bacillus cereus*. Int Microbiol 8:125–131

Black EP, Kelly AL, Fitzgerald GF (2005) The combined effect of high pressure and nisin on inactivation of microorganisms in milk. Inn Food Sci Emerg Technol 6:286–292

Bogovič Matijašić B, Koman Rajšp M, Perko B et al (2007) Inhibition of *Clostridium tyrobutyricum* in cheese by *Lactobacillus gasseri*. Int Dairy J 17:157–166

Björkroth J, Korkeala H (1997) Ropy slime-producing *Lactobacillus sake* strains possess a strong competitive ability against a commercial biopreservative. Int J Food Microbiol 38:117–123

Bredholt S, Nesbakken T, Holck A (2001) Industrial application of an antilisterial strain of *Lactobacillus sakei* as a protective culture and its effect on the sensory acceptability of cooked, sliced, vacuum-packaged meats. Int J Food Microbiol 66:191–196

Cabo ML, Torres B, Herrera JJ et al (2009) Application of nisin and pediocin against resistance and germination of *Bacillus* spores in sous vide products. J Food Prot 72:515–623

Calo-Mata P, Arlindo S, Boehme K et al (2008) Current applications and future trends of lactic acid bacteria and their bacteriocins for the biopreservation of aquatic food products. Food Bioprocess Technol 1:43–63

Campos CA, Rodriguez O, Calo-Mata P et al (2006) Preliminary characterization of bacteriocins *from Lactococcus lactis*, *Enterococcus faecium* and *Enterococcus mundtii* strains isolated from turbot (*Psetta maxima*). Food Res Int 39:356–364

anžek Majhenič A, Bogovič Matijašić B, Rogelj I (2003) Chromosomal location of genetic determinants for bacteriocins produced by *Lactobacillus gasseri* K7. J Dairy Res 70:199–203

Cardinal M, Gunnlaugsdottir H, Bjoernevik M et al (2004) Sensory characteristics of cold-smoked Atlantic salmon (*Salmo salar*) from European market and relationships with chemical, physical and microbiological measurements. Food Res Int 37:181–193

Castellano P, Belfiore C, Fadda S et al (2008) A review of bacteriocinogenic lactic acid bacteria used as bioprotective cultures in fresh meat produced in Argentina. Meat Sci 79:483–499

Cavadini C, Hertel C, Hammes WP (1998) Application of lysostaphin-producing lactobacilli to control staphylococcal food poisoning in meat products. J Food Prot 61:419–424

Chen H, Hoover DG (2003) Bacteriocins and their food applications. Comp Rev Food Sci Food Safety 2:82–100

Chen CM, Sebranek JG, Dickson JS et al (2004a) Combining pediocin with postpackaging irradiation for control of *Listeria monocytogenes* on frankfurters. J Food Prot 67:1866–1875

Chen CM, Sebranek JG, Dickson JS et al (2004b) Combining pediocin (ALTA 2341) with post-packaging thermal pasteurization for control of *Listeria monocytogenes* on frankfurters. J Food Prot 67:1855–1865

Cobo Molinos A, Abriouel H, Lucas López R et al (2008) Combined physico-chemical treatments based on enterocin AS-48 for inactivation of Gram-negative bacteria in soybean sprouts. Food Chem Toxicol 46:2912–2921

Cocolin L, Innocente N, Biasutti M et al (2004) The late blowing in cheese: a new molecular approach based on PCR and DGGE to study the microbial ecology of the alteration process. Int J Food Microbiol 90:83–91

Coma V (2008) Bioactive packaging technologies for extended shelf life of meat-based products. Meat Sci 78:90–103

Corsetti A, Settanni L (2007) Lactobacilli in sourdough fermentation. Food Res Int 40:539–558

Dalié DKD, Deschamps AM, Richard-Forget F (2010) Lactic acid bacteria – Potential for control of mould growth and mycotoxins: A review. Food Control 21:370–380

Davies EA, Delves-Broughton J (1999) Nisin. In: Robinson R, Batt C, Patel P (eds) Encyclopedia of food microbiology. Acad. Press, London, pp 191–198

Deegan LH, Cotter PD, Hill C et al (2006) Bacteriocins: biological tools for bio-preservation and shelf-life extension. Int Dairy J 16:1058–1071

De Vuyst L, Tsakalidou E (2008) *Streptococcus macedonicus*, a multi-functional and promising species for dairy fermentations. Int Dairy J 18:476–485

De Vuyst L, Vrancken G, Ravyts F et al (2009) Biodiversity, ecological determinants, and metabolic exploitation of sourdough microbiota. Food Microbiol 26:666–675

Dias BE, Galer CD, Moran JW et al (2009) Cheese flavoring systems prepared with bacteriocins. US Patent 7,556,833 (Appl. No.: 10/723,257)

Dicks LMT, Mellett FD, Hoffman LC (2004) Use of bacteriocin-producing starter cultures of *Lactobacillus plantarum* and *Lactobacillus curvatus* in production of ostrich meat salami. Meat Sci 66:703–708

Diop MB, Dubois-Dauphin R, Destain J et al (2009) Use of a nisin-producing starter culture of *Lactococcus lactis* subsp, *lactis* to improve traditional fish fermentation in Senegal. J Food Prot 72:1930–1934

Drider D, Fimland G, Héchard Y et al (2006) The continuing story of class IIa bacteriocins. Microbiol Molec Biol Rev 70:564–582

Du Toit M, Pretorius IS (2000) Microbial spoilage and preservation of wine: using weapons from nature's own arsenal. South Afr J Enol Vit 21:74–96

EFSA (2007) Introduction of a Qualified Presumption of Safety (QPS) approach for assessment of selected microorganisms referred to EFSA. EFSA J 587:1–16

EFSA (2008) The maintenance of the list of QPS microorganisms intentionally added to foods or feeds. Scientific opinion of the Panel on Biological Hazards. EFSA J 923:1–48

Ercolini D, Ferrocino I, La Storia A et al (2010) Development of spoilage microbiota in beef stored in nisin activated packaging. Food Microbiol 27:137–143

Erkkilä S, Suihko ML, Eerola S et al (2001) Dry sausage fermented by *Lactobacillus rhamnosus* strains. Int J Food Microbiol 64:205–210

European Commission (1988) Council Directive of 21 December 1988 on the approximation of the laws of the Member States concerning food additives authorized for use in foodstuffs intended for human consumption (89/107/EEC). O J L40: 27–37

European Economic Community (1983) European Economic Community Commission Directive 83/463/EEC. O J 255:1–6

European Parliament and Council (1995) Directive No 95/2/EC of 20 February 1995 on food additives other than colours and sweeteners. O J L61:1–53
European Parliament and Council (1997) Regulation (EC) No 258/97 of 27 January 1997 concerning novel foods and novel food ingredients. O J L043:1–6
European Parliament and Council (2000) Directive 2000/13/EC of 20 March 2000 on the approximation of the laws of the Member States relating to the labelling, presentation and advertising of foodstuffs. O J L109:29–56
European Parliament and Council (2001) Directive 2001/18/EC of 12 March 2001 on the deliberate release into the environment of genetically modified organisms and repealing Council Directive 90/220/EEC. O J L106:1–38
European Parliament and Council (2002a) Regulation (EC) No 178/2002 of 28 January 2002 laying down the general principles and requirements of food law, establishing the European Food Safety Authority and laying down procedures in matters of food safety. O J L31:1–37
European Parliament and Council (2002b) Commission Directive 2002/72/EC of 6 August 2002 relating to plastic materials and articles intended to come into contact with foodstuffs. O J L220:18–58
European Parliament and Council (2003) Regulation (EC) No 1829/2003 of 22 September 2003 on genetically modified food and feed. O J L268:1–22
European Parliament and Council (2004) Regulation (EC) No 1935/2004 of 27 October 2004 on materials and articles intended to come into contact with food and repealing Directives 80/590/EEC and 89/109/EEC. O J L338:4–17
European Parliament and Council (2008a). Regulation (EC) No 1331/2008 of 16 December 2008 establishing a common authorisation procedure for food additives, food enzymes and food flavourings. O J L354:1–6
European Parliament and Council (2008b). Regulation (EC) No 1333/2008 of 16 December 2008 on food additives. O J L354:16–33
European Parliament and Council (2009) Directive 2009/41/EC of 6 May 2009 on the contained use of genetically modified micro-organisms. O J L125:75–97
Fadda S, Chambon C, Champomier-Vergès MC et al (2008) *Lactobacillus* role during conditioning of refrigerated and vacuum-packaged Argentinean meat. Meat Sci 79:603–610
Fernández de Palencia P, de la Plaza M, Mohedano ML et al (2004) Enhancement of 2-methylbutanal formation in cheese by using a fluorescently tagged Lacticin 3147 producing *Lactococcus lactis* strain. Int J Food Microbiol 93:335–347
Foulquié Moreno MR, Rea MC, Cogan TM et al (2003) Applicability of a bacteriocin-producing *Enterococcus faecium* as a co-culture in Cheddar cheese manufacture. Int J Food Microbiol 81:73–84
Foulquié Moreno MR, Sarantinopoulos P, Tsakalidou E et al (2006) The role and application of enterococci in food and health. Int J Food Microbiol 106:1–24
Franz CMAP, van Belkum MJ, Holzapfel WH et al (2007) Diversity of enterococcal bacteriocins and their grouping into a new classification scheme. FEMS Microbiol Rev 31:293–310
FSIS, Food Safety and Inspection Service (2002) Safe and suitable ingredients used in the production of meat and poultry products. Directive 7120.1, Washington, DC.
Gálvez A, Valdivia E, Martínez-Bueno M et al (1990) Induction of autolysis in *Enterococcus faecalis* by peptide AS-48. J Appl Bacteriol 69:406–413
Gálvez A, Abriouel H, López RL et al (2007) Bacteriocin-based strategies for food biopreservation. Int J Food Microbiol 120:51–70
Galvez A, Lopez RL, Abriouel H et al (2008) Application of bacteriocins in the control of foodborne pathogenic and spoilage bacteria. Crit Rev Biotechnol 28:125–152
Gänzle MG, Vogel RF (2003) Contribution of reutericyclin production to the stable persistence of *Lactobacillus reuteri* in an industrial sourdough fermentation. Int J Food Microbiol 80:31–45
Gao Y, Jia S, Gao Q et al (2010) A novel bacteriocin with a broad inhibitory spectrum produced by *Lactobacillus sake* C2, isolated from traditional Chinese fermented cabbage. Food Control 21:76–81

Garriga M, Aymerich MT, Costa S et al (2002) Bactericidal synergism through bacteriocins and high pressure in a meat model system during storage. Food Microbiol 19:509–518

Ge J, Ping W, Song G et al (2009) Paracin 1.7, a bacteriocin produced by *Lactobacillus paracasei* HD1.7 isolated from Chinese cabbage sauerkraut, a traditional Chinese fermented vegetable food. Wei Sheng Wu Xue Bao 49:609–616

Ghalfi H, Allaoui A, Destain J et al (2006) Bacteriocin activity by *Lactobacillus curvatus* CWBI-B28 to inactivate *Listeria monocytogenes* in cold-smoked salmon during 4°C storage. J Food Prot 69:1066–1071

Giraffa G (1995) Enterococcal bacteriocins: their potential as *anti-Listeria* factors in dairy technology. Food Microbiol 12:291–299

Gonzalez CF, Kunka BS (1987) Plasmid-Associated Bacteriocin Production and Sucrose Fermentation in *Pediococcus acidilactici.* Appl Environ Microbiol 53:2534–2538

González-Rodríguez MN, Sanz JJ, Santos JA et al (2002) Numbers and types of microorganisms in vacuum-packed cold smoked freshwater fish at the retail level. Int J Food Microbiol 77:161–168

Govaris A, Solomakos N, Pexara A et al (2010) The antimicrobial effect of oregano essential oil, nisin and their combination against *Salmonella* Enteritidis in minced sheep meat during refrigerated storage. Int J Food Microbiol 137:175–180

Gram L, Dalgaard P (2002) Fish spoilage bacteria – problems and solutions. Curr Opin Biotechnol 13:262–266

Guinane CM, Cotter PD, Hill C et al (2005) Microbial solutions to microbial problems: Lactococcal bacteriocins for the control of undesirable biota in food. J Appl Microbiol 98:1316–1325

Gutiérrez J, Criado R, Martín M et al (2005) Production of enterocin P, an antilisterial pediocin-like bacteriocin from *Enterococcus faecium* P13, in *Pichia pastoris.* Antimicrob Agents Chemother 49:3004–3008

Hata T, Alemu M, Kobayashi M et al (2009) Characterization of a bacteriocin produced by *Enterococcus faecalis* N1-33 and its application as a food preservative. J Food Prot 72:524–30

Holo H, Faye T, Brede DA et al (2002) Bacteriocins of propionic acid bacteria. Lait 82:59–68

Hosseini SV, Arlindo S, Böhme K et al (2009) Molecular and probiotic characterization of bacteriocin-producing *Enterococcus faecium* strains isolated from nonfermented animal foods. J Appl Microbiol 107:1392–403

Huang Y, Luo Y, Zhai Z et al (2009) Characterization and application of an anti-*Listeria* bacteriocin produced by *Pediococcus pentosaceus* 05-10 isolated from Sichuan Pickle, a traditionally fermented vegetable product from China. Food Contr 20:1030–1035

Hugas M, Pagés F, Garriga M et al (1998) Application of the bacteriocinogenic *Lactobacillus sakei* CTC494 to prevent growth of *Listeria* in fresh and cooked meat products packaged with different atmospheres. Food Microbiol 15:639–650

Hugas M, Garriga M, Aymerich MT (2003) Functionalty of enterococci in meat products. Int J Food Microbiol 88:223–233

Jin T, Liu L, Sommers CH et al (2009) Radiation sensitization and postirradiation proliferation of *Listeria monocytogenes* on ready-to-eat deli meat in the presence of pectin-nisin films. J Food Prot 72:644–649

Jofré A, Aymerich T, Garriga M et al (2008) Assessment of the effectiveness of antimicrobial packaging combined with high pressure to control *Salmonella* sp. in cooked ham. Food Control 19:634–638

Jones RJ, Zagorec M, Brightwell G et al (2009) Inhibition by *Lactobacillus sakei* of other species in the flora of vacuum packaged raw meats during prolonged storage. Food Microbiol 26:876–881

Kato T, Inuzuka L, Kondo M et al (2001) Growth of nisin-producing lactococci in cooked rice supplemented with soybean extract and its application to inhibition of *Bacillus subtilis* in rice miso. Biosci Biotechnol Biochem 65:330–337

Knoll C, Divol B, du Toit M (2008) Genetic screening of lactic acid bacteria of oenological origin for bacteriocin-encoding genes. Food Microbiol 25:983–991

Korkeala H, Suortti T, Mäkelä P (1988) Ropy slime formation in vacuum-packed cooked meat products caused by homofermentative lactobacilli and a *Leuconostoc* species. Int J Food Microbiol 7:339–347

Kostinek M, Specht I, Edward VA et al (2007) Characterisation and biochemical properties of predominant lactic acid bacteria from fermenting cassava for selection as starter cultures. Int J Food Microbiol 114:342–351

Le Bourhis AG, Doré J, Carlier JP et al (2007) Contribution of *C. beijerinckii* and *C. sporogenes* in association with *C. tyrobutyricum* to the butyric fermentation in Emmental type cheese. Int J Food Microbiol 113:154–163

Leisner JJ, Laursen BG, Prévost H et al (2007) *Carnobacterium*: positive and negative effects in the environment and in foods. FEMS Microbiol Rev 31:592–613

Leroy F, Lievens K, De Vuyst L (2005) Interactions of meat-associated bacteriocin-producing Lactobacilli with *Listeria innocua* under stringent sausage fermentation conditions. J Food Prot 68:2078–2084

Leroy F, Verluyten J, De Vuyst L (2006) Functional meat starter cultures for improved sausage fermentation. Int J Food Microbiol 106:270–285

Leroy F, De Winter T, Moreno MRF et al (2007) The bacteriocin producer *Lactobacillus amylovorus* DCE 471 is a competitive starter culture for type II sourdough fermentations. J Sci Food Agric 87:1726–1736

Liang Z, Mittal GS, Griffiths MW (2002) Inactivation of *Salmonella* Typhimurium in orange juice containing antimicrobial agents by pulsed electric field. J Food Prot 65:1081–1087

Liu L, O'Conner P, Cotter PD et al (2008) Controlling *Listeria monocytogenes* in Cottage cheese through heterologous production of enterocin A by *Lactococcus lactis*. J Appl Microbiol 104:1059–1066

Lortal S, Chapot-Chartier MP (2005) Role, mechanisms and control of lactic acid bacteria lysis in cheese. Int Dairy J 15:857–871

Lücke FK (2000) Utilization of microbes to process and preserve meat. Meat Sci 56:105–115

Mah JH, Ahn JB, Park JH et al (2003) Characterization of biogenic amine-producing microorganisms isolated from Myeolchi-Jeot, Korean salted and fermented anchovy. J Microbiol Biotechnol 13:692–699

Mah JH, Hwang HJ (2009) Inhibition of biogenic amine formation in a salted and fermented anchovy by *Staphylococcus xylosus* as a protective culture. Food Control 20:796–801

Mangalassary S, Han I, Rieck J et al (2008) Effect of combining nisin and/or lysozyme with in-package pasteurization for control of *Listeria monocytogenes* in ready-to-eat turkey bologna during refrigerated storage. Food Microbiol 25:866–870

Maragkoudakis PA, Mountzouris KC, Psyrras D et al (2009) Functional properties of novel protective lactic acid bacteria and application in raw chicken meat against *Listeria monocytogenes* and *Salmonella enteritidis*. Int J Food Microbiol 130:219–226

Marcos B, Aymerich T, Monfort JM et al (2008) High-pressure processing and antimicrobial biodegradable packaging to control *Listeria monocytogenes* during storage of cooked ham. Food Microbiol 25:177–182

Martínez-Cuesta M, Bengoechea J, Bustos I et al (2010) Control of late blowing in cheese by adding lacticin 3147-producing *Lactococcus lactis* IFPL 3593 to the starter. Int Dairy J 20:18–24

Matamoros S, Pilet MF, Gigout F et al (2009) Selection and evaluation of seafood-borne psychrotrophic lactic acid bacteria as inhibitors of pathogenic and spoilage bacteria. Food Microbiol 26:638–644

McMillin KW (2008) Where is MAP Going? A review and future potential of modified atmosphere packaging for meat. Meat Sci 80:43–65

Menteş Ö, Ercan R, Akçelik M (2007) Inhibitor activities of two *Lactobacillus* strains, isolated from sourdough, against rope-forming *Bacillus* strains. Food Control 18:359–363

Messens W, De Vuyst L (2002) Inhibitory substances produced by lactobacilli isolated from sourdoughs – a review. Int J Food Microbiol 72:31–43

Mollet B, Peel J, Pridmore D et al (2004) Bactericide compositions prepared and obtained from micrococcus varians. US Patent 6,689,750 (Appl. No.: 08/693,353)

Morgan SM, Galvin M, Kelly J et al (1999) Development of a lacticin 3147-enriched whey powder with inhibitory activity against foodborne pathogens. J Food Prot 62:1011–1016

Morgan SM, Garvin M, Ross RP et al (2001) Evaluation of a spray-dried lacticin 3147 powder for the control of *Listeria monocytogenes* and *Bacillus cereus* in a range of food systems. Lett Appl Microbiol 33:387–391

Mosqueda-Melgar J, Elez-Martínez P, Raybaudi-Massilia RM et al (2008) Effects of pulsed electric fields on pathogenic microorganisms of major concern in fluid foods: a review. Crit Rev Food Sci Nutr 48:747–759

Muñoz A, Maqueda M, Gálvez A et al (2004) Biocontrol of psychrotrophic enterotoxigenic *Bacillus cereus* in a non fat hard type cheese by an enterococcal strain-producing enterocin AS-48. J Food Prot 67:1517–1521

Muñoz A, Ananou S, Gálvez A et al (2007) Inhibition of *Staphylococcus aureus* in dairy products by enterocin AS-48 produced in situ and ex situ: Bactericidal synergism through heat and AS-48. Int Dairy J 17:760–769

Murphy RY, Hanson RE, Feze N et al (2005) Eradicating *Listeria monocytogenes* from fully cooked franks by using an integrated pasteurization-packaging system. J Food Prot 68:507–511

Narbutaite V, Fernandez A, Horn N et al (2007) Influence of baking enzymes on antimicrobial activity of five bacteriocin-like inhibitory substances produced by lactic acid bacteria isolated from Lithuanian sourdoughs. Lett Appl Microbiol 47:555–560

Navarro L, Zarazaga M, Sáenz J et al (2000) Bacteriocin production by lactic acid bacteria isolated from Rioja red wines. J Appl Microbiol 88:44–51

Neetoo H, Ye M, Chen H et al (2008) Use of nisin-coated plastic films to control *Listeria monocytogenes* on vacuum-packaged cold-smoked salmon. Int J Food Microbiol 122:8–15

Nes IF, Diep DB, Havarstein LS et al (1996) Biosynthesis of bacteriocins in lactic acid bacteria. Antonie van Leeuwenhoek 70:113–128

Nieto-Lozano JC, Reguera-Useros JI, Peláez-Martínez MC et al (2006) Effect of a bacteriocin produced by *Pediococcus acidilactici* against *Listeria monocytogenes* and *Clostridium perfringens* on Spanish raw meat. Meat Sci 72:57–61

Nilsson L, Hansen TB, Garrido P et al (2005) Growth inhibition of *Listeria monocytogenes* by a nonbacteriocinogenic *Carnobacterium piscicola*. J Appl Microbiol 98:172–183

Noonpakdee W, Santivarangkna C, Jumriangrit P et al (2003) Isolation of nisin-producing *Lactococcus lactis* WNC 20 strain from nham, a traditional Thai fermented sausage. Int J Food Microbiol 81:137–145

O'Mahony T, Rekhif N, Cavadini C et al (2001) The application of a fermented food ingredient containing 'variacin', a novel antimicrobial produced by *Kocuria varians*, to control the growth of *Bacillus cereus* in chilled dairy products. J Appl Microbiol 90:106–114

O'Sullivan L, Ross RP, Hill C (2002) Potential of bacteriocin-producing lactic acid bacteria for improvements in food safety and quality. Biochimie 84:593–604

O'Sullivan L, O'Connor EB, Ross RP et al (2006) Evaluation of live-culture-producing lacticin 3147 as a treatment for the control of *Listeria monocytogenes* on the surface of smear-ripened cheese. J Appl Microbiol 100:135–143

Papagianni M, Anastasiadou S (2009) Pediocins: The bacteriocins of Pediococci. Sources, production, properties and applications. Microbial Cell Factories 8:3

Peláez C, Requena T (2005) Exploiting the potential of bacteria in the cheese ecosystem. Int Dairy J 15:831–844

Pinto AL, Fernandes M, Pinto C et al (2009) Characterization of anti-*Listeria* bacteriocins isolated from shellfish: potential antimicrobials to control non-fermented seafood. Int J Food Microbiol 129:50–58

Pridmore D, Rekhif N, Pittet AC et al (1996) Variacin, a new lanthionine-containing bacteriocin produced by *Micrococcus varians*: comparison to lacticin 481 of *Lactococcus lactis*. Appl Environ Microbiol 62:1799–1802

Randazzo CL, Pitino I, Scifo GO et al (2009) Biopreservation of minimally processed iceberg lettuces using a bacteriocin produced by *Lactococcus lactis* wild strain. Food Control 20:756–763

Ravyts F, Barbuti S, Frustoli MA et al (2008) Competitiveness and antibacterial potential of bacteriocin-producing starter cultures in different types of fermented sausages. J Food Prot 71:1817–1827

Rihakova J, Belguesmia Y, Petit VW et al (2009) Divercin V41 from gene characterization to food applications: 1998-2008, a decade of solved and unsolved questions. Lett Appl Microbiol 48:1–7

Robertson A, Tirado C, Lobstein T et al (eds), (2004) Food and health in Europe: a new basis for action. WHO Regional Publications, European Series, No. 96, Geneva.

Rodgers S (2008) New applications of live bacteria in food services: probiotics and protective cultures. Trends Food Sci Technol 19:188–197

Rodríguez JM, Martinez MI, Kok J (2002) Pediocin PA-1, a wide-spectrum bacteriocin from lactic acid bacteria. Crit Rev Food Sci Nutr 42:91–121

Rojo-Bezares B, Sáenz Y, Zarazaga M et al (2007) Antimicrobial activity of nisin against *Oenococcus oeni* and other wine bacteria. Int J Food Microbiol 116:32–36

Ross RP, Galvin M, McAuliffe O et al (1999) Developing applications for lactococcal bacteriocins. Antonie van Leeuwenhoek 76:337–346

Ross RP, Morgan S, Hill C (2002) Preservation and fermentation: past, present and future. Int J Food Microbiol 79:3–16

Ryan MP, Ross RP, Hill C (2001) Strategy for manipulation of cheese flora using combinations of lacticin 3147-producing and -resistant cultures. Appl Environ Microbiol 67:2699–2704

Sánchez J, Borrero J, Gómez-Sala B et al (2008) Cloning and heterologous production of Hiracin JM79, a Sec-dependent bacteriocin produced by *Enterococcus hirae* DCH5, in lactic acid bacteria and *Pichia pastoris*. Appl Environ Microbiol 74:2471–2479

Sánchez Valenzuela A, Díaz Ruiz G, Ben Omar N et al (2008) Inhibition of food poisoning and pathogenic bacteria by *Lactobacillus plantarum* strain 2.9 isolated from ben saalga, both in a culture medium and in food. Food Control 19:842–848

Santiago-Silva P, Nilda FF, Soares Juliana E et al (2009) Antimicrobial efficiency of film incorporated with pediocin (ALTA® 2351) on preservation of sliced ham. Food Control 20:85–89

Settanni L, Corsetti A (2008) Application of bacteriocins in vegetable food biopreservation. Int J Food Microbiol 121:123–138

Silveira AC, Conesa A, Aguayo E et al (2008) Alternative sanitizers to chlorine for use on fresh-cut "Galia" (*Cucumis melo* var. *catalupensis*) melon. J Food Sci 73:405–411

Schoeman H, Vivier MA, du Toit M et al (1999) The development of bactericidal yeast strains by expressing the *Pediococcus acidilactici* pediocin gene (*pedA*) in *Saccharomyces cerevisiae*. Yeast 15:647–656

Sobrino-López A, Martín-Belloso O (2008) Use of nisin and other bacteriocins for preservation of dairy products. Int Dairy J 18:329–343

Sobrino A, Martínez Viedma P, Abriouel H et al (2009) The impact of adding antimicrobial peptides to milk inoculated with *Staphylococcus aureus* and processed by High-intensity pulsed electric field. J Dairy Sci 92:2514–2523

Solomakos N, Govaris A, Koidis P et al (2008) The antimicrobial effect of thyme essential oil, nisin, and their combination against *Listeria monocytogenes* in minced beef during refrigerated storage. Food Microbiol 25:120–127

Somkuti GA, Steinberg DH (2010) Pediocin production in milk by *Pediococcus acidilactici* in co-culture with *Streptococcus thermophilus* and *Lactobacillus delbrueckii* subsp. *bulgaricus*. J Ind Microbiol Biotechnol 37:65–69

Sparo M, Nuñez GG, Castro M et al (2008) Characteristics of an environmental strain, *Enterococcus faecalis* CECT7121, and its effects as additive on craft dry-fermented sausages. Food Microbiol 25:607–615

Stergiou VA, Thomas LV, Adams MR (2006) Interactions of nisin with glutathione in a model protein system and meat. J Food Prot 69:951–956

Tahiri I, Desbiens M, Kheadr E et al (2009a) Growth of *Carnobacterium divergens* M35 and production of divergicin M35 in snow crab by-product, a natural-grade medium. LWT-Food Sci Technol 42:624–632

Tahiri I, Desbiens M, Kheadr E et al (2009b) Comparison of different application strategies of divergicin M35 for inactivation of *Listeria monocytogenes* in cold-smoked wild salmon. Food Microbiol 26:783–793

Tamang JP, Tamang B, Schillinger U et al (2009) Functional properties of lactic acid bacteria isolated from ethnic fermented vegetables of the Himalayas. Int J Food Microbiol 135:28–33

Thomas LV, Delves-Broughton J (2001) New advances in the application of the food preservative nisin. Adv Food Sci 2:11–22

Thomas LV, Clarkson MR, Delves-Broughton J (2000) Nisin. In: Naidu AS (ed) Natural food antimicrobial systems. CRC-Press, FL, pp 463–524

Todorov SD, Koep KSC, Van Reenen CA et al (2007) Production of salami from beef, horse, mutton, Blesbok (Damaliscus dorcas phillipsi) and Springbok (Antidorcas marsupialis) with bacteriocinogenic strains of *Lactobacillus plantarum* and *Lactobacillus curvatus*. Meat Sci 77:405–412

Tomé E, Gibbs PA, Teixeira PC (2008) Growth control of *Listeria innocua* 2030c on vacuum-packaged cold-smoked salmon by lactic acid bacteria. Int J Food Microbiol 121:285–294

Trias R, Bañeras L, Badosa E et al (2008a) Bioprotection of Golden Delicious apples and Iceberg lettuce against foodborne bacterial pathogens by lactic acid bacteria. Int J Food Microbiol 123:50–60

Trias R, Badosa E, Montesinos E et al (2008b) Bioprotective *Leuconostoc* strains against *Listeria monocytogenes* in fresh fruits and vegetables. Int J Food Microbiol 127:91–98

Ukuku DO, Bari ML, Kawamoto S et al (2005) Use of hydrogen peroxide in combination with nisin, sodium lactate and citric acid for reducing transfer of bacterial pathogens from whole melon surfaces to fresh-cut pieces. Int J Food Microbiol 104:225–233

Työppönen S, Petäjä E, Mattila-Sandholm T (2003) Bioprotectives and probiotics for dry sausages. Int J Food Microbiol 83:233–244

Uesugi AR, Moraru CI (2009) Reduction of *Listeria* on Ready-to-Eat sausages after exposure to a combination of pulsed light and Nisin. J Food Prot 72:347–353

US Food and Drug Administration (1999) Federal food, drug, and cosmetic act. Washington, DC

Valerio F, De Bellis P, Lonigro SL et al (2008) Use of *Lactobacillus plantarum* fermentation products in bread-making to prevent *Bacillus subtilis* ropy spoilage. Int J Food Microbiol 122:328–332

Valerio F, Favilla M, De Bellis P et al (2009) Antifungal activity of strains of lactic acid bacteria isolated from a semolina ecosystem against *Penicillium roqueforti*, *Aspergillus niger* and *Endomyces fibuliger* contaminating bakery products. Syst Appl Microbiol 32:438–448

Van Reenen CA, Chikindas ML, van Zyl WH et al (2002) Characterisation and heterologous expression of a class IIa bacteriocin, plantaricin 423 from *Lactobacillus plantarum* 423, in *Saccharomyces cerevisiae*. Int J Food Microbiol 81:29–40

Vaughan A, O'Sullivan T, van Sinderen D (2005) Enhancing the microbiological stability of malt and beer—a review. J Inst Brewing 111:355–371

Vázquez JA, González MP, Murado MA (2004) Peptones from autohydrolysed fish viscera for nisin and pediocin production. J Biotechnol 112:299–311

Vázquez JA, González MP, Murado MA (2006) Preliminary tests on nisin and pediocin production using waste protein sources. Factorial and kinetic studies. Bioresour Technol 97:605–613

Vaz-Velho M, Todorov S, Ribeiro J et al (2005) Growth control of *Listeria innocua* 2030c during processing and storage of cold smoked salmon-trout by *Carnobacterium divergens* V41 culture and supernatant. Food Control 16:541–549

Vedamuthu Ebenezer R (1995) Method of producing a yogurt product containing bacteriocin PA-1. US Patent 5,445,835 (Appl. No.: 08/192,960).

Vega Leal-Sánchez M, Ruiz-Barba JL, Sánchez AH et al (2003) Fermentation profile and optimization of green olive fermentation using *Lactobacillus plantarum* LPCO10 as a starter culture. Food Microbiol 20:421–430

Villani F, Sannino L, Moschetti G et al (1997) Partial characterization of an antagonistic substance produced by *Staphylococcus xylosus* 1E and determination of the effectiveness of the producer strain to inhibit *Listeria monocytogenes* in Italian sausages. Food Microbiol 14:555–566

Weber GH, Broich WA (1986) Shelf-life extension of cultured dairy foods. C Dairy Prod J 21:19
Wessels S, Axelsson L, Bech Hansen E et al (2004) The lactic acid bacteria, the food chain, and their regulation. Trends Food Sci Technol 15:498–505
Yanagida F, Srionnual S, Chen YS (2008) Isolation and characteristics of lactic acid bacteria from koshu vineyards in Japan. Lett Appl Microbiol 47:134–139
Ye M, Neetoo H, Chen H (2008) Effectiveness of chitosan-coated plastic films incorporating antimicrobials in inhibition of *Listeria monocytogenes* on cold-smoked salmon. Int J Food Microbiol 127:235–240
Yoon MY, Kim YJ, Hwang HJ (2008) Properties and safety aspects of *Enterococcus faecium* strains isolated from *Chungkukjang*, a fermented soy product. LWT Food Sci Technol 41:925–933
Zhang J, Liu G, Li P et al (2010) Pentocin 31-1, a novel meat-borne bacteriocin and its application as biopreservative in chill-stored tray-packaged pork meat. Food Control 21:198–202

Chapter 19
Medical and Personal Care Applications of Bacteriocins Produced by Lactic Acid Bacteria

L.M.T. Dicks, T.D.J. Heunis, D.A. van Staden, A. Brand, K. Sutyak Noll, and M.L. Chikindas

Abstract The frequent use of antibiotics has led to a crisis in the antibiotic resistance of pathogens associated with humans and animals. Antibiotic resistance and the emergence of multiresistant bacterial pathogens have led to the investigation of alternative antimicrobial agents to treat and prevent infections in both humans and animals. Research on antimicrobial peptides, with a special interest on bacteriocins of lactic acid bacteria, is entering a new era with novel applications other than food preservation. Many scientists are now focusing on the application of these peptides in medicinal and personal care products. However, it is difficult to assess the success of such ventures due to the dearth of information that has been published and the lack of clinical trials.

Introduction

Bacteriocins of lactic acid bacteria are ribosomally synthesized, are generally cationic, have less than 100 amino-acid residues (Marcus et al. 1999; Jenssen et al. 2006), and contain a substantial portion (30% or more) of hydrophobic residues (Dijkshoorn et al. 2004). The cationic charge of these peptides ensures electrostatic affinity with the negatively charged bacterial outer membrane, whereas the hydrophobic section of the peptide interacts with the cell membrane and enters the double lipid membrane. Most bacteriocins have a relatively narrow spectrum of antimicrobial activity, i.e., the growth of only certain species, usually those phylogenetically related to the producer strain, are affected. However, some bacteriocins exhibit a much broader spectrum of antimicrobial activity and may extend beyond the borders of bacteria to include protozoa, yeast, fungi, and viruses (Reddy et al. 2004). A few bacteriocins are cytotoxic, with activity against sperm and tumor cells (Reddy et al. 2004).

L.M.T. Dicks (✉)
Department of Microbiology, Stellenbosch University, 7600 Stellenbosch, South Africa
e-mail: lmtd@sun.ac.za

D. Drider and S. Rebuffat (eds.), *Prokaryotic Antimicrobial Peptides: From Genes to Applications*, DOI 10.1007/978-1-4419-7692-5_19,

Bacteriocins may be seen as defense peptides and may thus be grouped into the same category as killer toxins of yeast, defensins of mammals, cecropins of insects, tachyplesins of crabs, magainins of amphibia, pandanins of scorpions, and thionins of plants (Boheim 1995; Corzo et al. 2001; de Vuyst and Vandamme 1994; Reddy et al. 2004). Mature peptides are produced after the cleavage of inactive prepeptides (Nes and Holo 2000; Sahl and Bierbaum 1998).

Colicins, produced by *Escherichia coli*, were the first described bacteriocins. The mode of action of these peptides varies from pore formation to nuclease activity (Braun et al. 1994; Riley and Wertz 2002; Smarda and Smajs 1998). Colicins E1 and K inhibit macromolecular synthesis without arrest of respiration. Colicin E2 causes DNA breakdown, and colicin E3 stops protein synthesis (Nomura 1967). The genes encoding colicin production are usually located on plasmids (Pugsley and Oudega 1987). Type A plasmids are small (6–10 kb), usually present as multiple copies, and are conjugative. Type B plasmids are approximately 40 kb in size, occur as single copies, and are also conjugative.

Microcins, also produced by *E. coli*, are smaller than colicins and share more properties with bacteriocins produced by Gram-positive bacteria (Baquero and Mreno 1984; Gillor et al. 2004; Pons et al. 2002). Fourteen microcins have been reported to date, of which only seven have been isolated and fully characterized (Duquesne et al. 2007a, b; Severinov et al. 2007).

Bacteriocins of Gram-positive bacteria are more diverse than bacteriocins described for Gram-negative bacteria (Riley and Wertz 2002). Over the years many classification methods have been proposed for bacteriocins. Fredericq (1957) classified bacteriocins based on specificity of absorption and proposed subclasses according to immune responses. Reeves (1965) implemented a system consisting of 16 classes of bacteriocins based on the species that produce them. Ten years later Bradley (1967) classified bacteriocins based on molecular weight. Klaenhammer (1993) classified bacteriocins into four classes based on structure, mechanism of action, genetics, and biochemical properties. Finally, Cotter et al. (2005) proposed two major classes, each divided into subclasses (Table 19.1).

Bacteriocins of class I contain lanthionine or β-methyllanthionine and are classified as lantibiotics. They undergo posttranslational modifications to produce

Table 19.1 Classification of bacteriocins (Cotter et al. 2005)

Classes	Characteristics
Class I	Lantibiotics
Class Ia	Small (19–38 amino acids), elongated, positively charged peptides that form pores
Class Ib	Globular peptides that interfere with essential enzymes
Class II	Nonlanthionine containing bacteriocins
Class IIa	Pediocin-like peptides that contain the YGNGVXCXXXXVXV consensus sequence in their N-terminal
Class IIb	Two-peptide bacteriocins, require both peptides for activity
Class IIc	Cyclic peptides, N- and C-terminal are covalently linked
Class IId	Single nonpediocin-like peptides

peptides of less than 5 kDa in size. Type A lantibiotics such as nisin, epidermin, and subtilin are screw-shaped, elongated, flexible, and amphipathic peptides with pore-forming activities (Kordel et al. 1989; Ruhr and Sahl 1985), whereas type B lantibiotics, of which mersacidin is a typical example, are small and compact peptides that target specific components of the bacterial membrane (Brötz et al. 1998). Type B lantibiotics kill by interfering with cellular enzymatic reactions, such as cell wall synthesis (Pag and Sahl 2002; Sahl and Bierbaum 1998; Sahl et al. 1995).

Nisin A and mutacin B-Ny266 are active against a range of organisms, including species of *Actinomyces, Bacillus, Clostridium, Corynebacterium, Enterococcus, Gardnerella, Lactococcus, Listeria, Micrococcus, Mycobacterium, Propionibacterium, Streptococcus,* and *Staphylococcus* (Mota-Meira et al. 2000, 2005). Activity against a number of Gram-negative pathogens such as *Campylobacter, Haemophilus, Helicobacter,* and *Neisseria* spp. has also been reported (Morency et al. 2001).

Nonlanthionine bacteriocins are grouped in class II. These peptides are heat-stable and do not undergo extensive posttranslational modifications (Cotter et al. 2005). The majority of bacteriocins in class II kill by inducing membrane permeabilization and the subsequent leakage of molecules from target bacteria. Two subgroups are differentiated. Class IIa peptides form the largest subgroup. They are active against *Listeria* and have a conserved amino-terminal sequence (YGNGVXaaC) that facilitates nonspecific binding to the target surface (Drider et al. 2006; Oppegård et al. 2007). Similar to type A lantibiotics, class IIa bacteriocins form pores in the cytoplasmic membrane. Typical examples include pediocin and pediocin-like bacteriocins, sakacin A, and leucocin A (Drider et al. 2006; Héchard and Sahl 2002; Oppegård et al. 2007). Class IIb bacteriocins such as lacticin F and lactococcin G form pores and have two different proteins (Garneau et al. 2002; Héchard and Sahl 2002). In the case of lacticin 3147, one of the peptides depolarizes the membrane, while the other forms pores (Martin et al. 2004). Sec-dependent bacteriocins, such as acidocin 1B, are placed in subgroup IIc (Han et al. 2007). Nonpediocin-like bacteriocins are classified as subgroup IId (Cotter et al. 2005).

Genes encoding the biosynthesis of bacteriocins are organized in clusters located on either the chromosome or plasmids or on possibly both if located on a transposon (McAuliffe et al. 2001; Nes et al. 1996; van Reenen et al. 1998). The class I (lantibiotic) gene cluster consists of a structural gene that codes for a prebacteriocin, genes encoding accessory proteins involved in structure modifications such as proteolytic processing of the leader peptide, transport genes that code for an ABC-superfamily of transport proteins, regulation genes, and immunity genes that confer resistance to the producer strain (de Vos et al. 1995; Jack et al. 1995; Kolter and Moreno 1992; Sahl et al. 1995). The structural gene encodes an inactive prepeptide with an N-terminal extension or leader peptide connected to the C-terminal of the propeptide (McAuliffe et al. 2001; Sahl and Bierbaum 1998). The propeptide is modified to become a mature active peptide (McAuliffe et al. 2001). At the end of biosynthesis the leader peptide is cleaved by a protease, before or after the peptide is translocated by ABC (ATP-binding) transport proteins (Klaenhammer 1993; Sahl 1998). The bacteriocin is then released to execute its antibacterial activity (McAuliffe et al. 2001).

Genes involved in the production of class II bacteriocins are somewhat functionally similar to the genes for class I, i.e., they include a structural gene, an immunity gene, and two genes encoding a membrane-associated ATP-dependent binding cassette (ABC) transporter and an accessory protein (Eijsink et al. 1998; Ennahar et al. 2006; Klaenhammer 1993; Nes et al. 1996). The structural genes also encode a prepeptide with a leader peptide, which becomes active once the leader peptide is cleaved by a protease (Eijsink et al. 2002; Nes and Holo 2000; Nissen-Meyer and Nes 1997). Class IIa bacteriocins have a leader peptide with double-glycine residues, which acts as the processing site and is secreted once the leader peptide has been cleaved (Klaenhammer 1993; Michiels et al. 2001).

Mode of Action of Bacteriocins

Class Ia lantibiotics, especially nisin, have been extensively studied for their mode of action because of various industrial applications. Nisin, a typical example of type A lantibiotics, binds to lipid II on the cell wall surface and prevents cell wall synthesis. Insertion of the peptide into the phospholipid bilayer of the cell membrane leads to drastic changes in permeability and may cause cell death (Wiedemann et al. 2001). Nisin kills sensitive organisms by disruption of the proton motive force (PMF) (Abee et al. 1995; Chung and Hancock 2000; de Vuyst and Vandamme 1994; Kraaij et al. 1999; Ruhr and Sahl 1985). The PMF, composed of a chemical component (pH gradient) and an electrical component (membrane potential), plays an important role in the synthesis of adenosine triphosphate (ATP) and the influx of molecules by PMF-driven transport systems (de Vuyst and Vandamme 1994; McAuliffe et al. 2001). The amphiphilic nature of nisin allows it to interact with hydrophilic heads and hydrophobic regions of the plasma membrane and cause an efflux of ions, solutes, and small molecules, forcing biosynthetic processes in the cell to a halt (Sahl 1998). Although the formation of pores or channels with the depletion of the PMF is the primary activity of nisin, the peptide also forms a complex with lipid II (undecaprenyl-pyrophosphoryl-MurNAc-(pentapeptide)-GlcNAc), thereby inhibiting the synthesis of lipids and peptidoglycan (de Vuyst and Vandamme 1994; Sahl 1998). Compared to pore formation, inhibition of cell wall biosynthesis is a relatively slow process (McAuliffe et al. 2001). Class Ib lantibiotics have a different mode of action. They hinder the activities of essential enzymes (McAuliffe et al. 2001).

Class II bacteriocins also disrupt the PMF. They recognize specific protein receptors and form voltage independent pores or channels in the plasma membrane of sensitive cells (Héchard and Sahl 2002). This has been shown for pediocin PA-1, lactococcin, and sakacin A and B (Chikindas et al. 1993). Two-peptide bacteriocins such as lactococcin G have no effect on the pH gradient, but dissipate the membrane potential, which leads to the efflux of cations (Moll et al. 1998, 1999).

Various models for the formation of pores or channels by bacteriocins have been proposed, such as the "barrel-stave" model, the "carpet" model, and the "wedge" model

(Driessen et al. 1995; Héchard and Sahl 2002; Moll et al. 1999; Sahl 1991; van den Hooven et al. 1996). The "wedge" model is proposed for lantibiotics, while the "barrel-stave" and "carpet" models are proposed for the class II bacteriocins (Moll et al. 1999). In the "barrel-stave" model, the bacteriocin forms a transmembrane barrel of α-helices with a hydrophilic interior and hydrophobic exterior (Ojcius 1994; Sahl 1991). Electrostatic interactions form between the positively charged amino acids of the peptide and the negatively charged heads of the phospholipid bilayer. The hydrophobic part of the peptide forms interactions with the hydrophobic acyl chains of the membrane lipids. Once inserted in the plasma membrane, the bacteriocin molecules form pores that lead to disruption of the PMF (Ennahar et al. 2006; Moll et al. 1999). In the "carpet" model, single peptide molecules interfere with plasma membrane organization and result in pore formation (Moll et al. 1999). In the "wedge" model the hydrophilic, positively charged amino acids of the peptide interact with the negatively charged heads of the phospholipid bilayer. The hydrophobic region of the peptide inserts itself into the outer leaflet of the bilayer (Driessen et al. 1995; Sahl 1998; van den Hooven et al. 1996).

Medical Applications of Bacteriocins

While nowadays everyone expects to see bacteriocins related to food applications, their many possible uses for the control of nondesired microorganisms in the human environment are greatly underappreciated. With the advent of multidrug resistant bacteria, it has become a priority to develop alternative medicinal treatments/preventive measures against these pathogens. Since the mode of action of bacteriocins is remarkably different from conventional antibiotics, they may be considered as a novel source for the control of microbial pathogens. This is discussed in greater detail in a later section of this review.

Staphylococcus aureus and methicillin-resistant *S. aureus* (MRSA) are the most prevalent organisms in skin infections and have become a serious problem, especially in hospitals (Guggenheim et al. 2009; Lesseval and Hadjiiski 1996; Taylor et al. 1992). MRSA was first reported in the 1960s after methicillin was introduced to treat *S. aureus* infections (Hackbarth and Chambers 1989; Kim 2009). Resistant strains have acquired the *mecA* gene that codes for a low-affinity penicillin-binding protein, PBP2a (Fraise et al. 1997; Hiramatsu 1995). Oxacillin, nafcillin, quinopristin–dalfopristin, rifampicin, ciprofloxacillin, teichoplanin, cefazolin, and cephalothin A are also used to treat *S. aureus* infection (Gould and Chamberlain 1995; Lowy 1998) but with limited success. In many cases, vancomycin is used as the last resort but also with limited success (Dicks et al. 2009). Mersacidin, a lantibiotic produced by *Bacillus* sp. strain HIL Y-85,54728 (Sass et al. 2008), inhibits the growth of MRSA strains in vivo in mice (Kruszewska et al. 2004) and may be considered an alternative treatment. The peptide inhibits cell wall synthesis of MRSA strains with efficiency equal to that reported for vancomycin (Chatterjee et al. 1992; Limbert et al. 1991). Mersacidin is also active against *Propionibacterium acnes* and may thus be used in the treatment of acne (Jung 1991a, b; Kellner et al. 1988; Niu and Neu 1991). The mode of action of mersacidin differs from vancomycin, which leaves the option of using the two substances in combination (Brötz et al. 1995).

In a study conducted by Kruszewska et al. (2004), 2-month-old female BALB/cA mice were immune suppressed, intranasally infected with *S. aureus* 99308 and then treated with mersacidin. Mersacidin effectively inhibited the growth of *S. aureus* in the mice.

Lacticin 3147, a two-peptide lantibiotic produced by *Lactococcus lactis* subsp. *lactis*, inhibits the growth of *S. aureus*, MRSA, and vancomycin-resistant strains of *Enterococcus faecalis* (Galvin et al. 1999). Epidermin and gallidermin, also classified as lantibiotics but produced by *Staphylococcus gallinarum* and *Staphylococcus epidermidis*, proved effective in the treatment of skin infections (Kellner et al. 1988).

Epidermin and gallidermin are also active against *P. acnes* (Jung 1991a, b; Kellner et al. 1988; Niu and Neu 1991). Bonelli et al. (2006) measured the antimicrobial activity of epidermin and gallidermin by determining the release of potassium (K^+) from membrane models they have constructed. Treatment with gallidermin resulted in K^+ release from *Staphylococcus simulans* and *Micrococcus flavus*. Epidermin was active against *M. flavus,* whereas nisin inactivated both pathogens. Nisin and IB-367, a protegrin-like cationic peptide produced by Intrabiotics (Mountain View, California), have recently entered phase I clinical trials for acne infections.

A number of studies claim that lantibiotics are effective in the prevention of tooth decay and gingivitis (Blackburn and Goldstein 1995; Howell et al. 1993; McConville 1995; Peek et al. 1995). Nisin exhibits antimicrobial activity against plaque and gingivitis-causing bacteria and has been included in mouthwashes (van Kraaij et al. 1999). In vivo experiments were done on beagle dogs (Howell et al. 1993). Lacticin 3147 prevented the growth of *Streptococcus mutans* associated with dental decay (Galvin et al. 1999). BLIS K12, a commercial product that contains a strain of *Streptococcus salivarius* that produces salivaricin A2 and B, inhibits bacteria associated with bad breath (Tagg 2004).

Mastitis, a bacterial infection of the mammary glands, causes huge economic losses in the dairy industry (Bradley 2002; Riffon et al. 2001; Soltys and Quinn 1999; Sordelli et al. 2000; Twomey et al. 2000). Trials carried out by Taylor et al. (1949) indicated that a single intramammary infusion of nisin was effective in treating both streptococcal and staphylococcal infections. However, infusions of the nisin preparation into the udder produced an adverse cellular response, although there was no correlation between the nisin concentration and local intolerance produced in the udder. Broadbent et al. (1989) showed that nisin inhibited growth of several Gram-positive, mastitis-causing pathogens in vivo. Nisin A in combination with lysostaphin was also administered through intramammary infusions (Sears et al. 1992). During these experiments, cure rates of 66% for *Staphylococcus aureus*, 95% for *Streptococcus agalactiae*, and 100% for *Streptococcus uberis* were observed. Nisin A has also been incorporated in teat wipes (Broadbent et al. 1989; Cotter et al. 2005; Ross et al. 1999).

Lacticin 3147 proved effective in the treatment of bovine mastitis and showed activity against mastitic staphylococci and streptococci (Ryan et al. 1999). A bacteriocin produced by *Bacillus subtilis* LFB112 inhibited the growth of *S. aureus* associated with mastitis (Xie et al. 2009). *Streptococcus gallolyticus* subsp. *macedonicus* ST91KM produces a bacteriocin (macedocin ST91KM) active against *Streptococcus agalactiae, Streptococcus dysgalactiae* subsp. *dysgalactiae, Streptococcus uberis, Staphylococcus aureus*, and *Staphylococcus epidermidis*, including strains resistant to methicillin and oxacillin (Pieterse et al. 2008, 2010). Macedocin ST91KM may thus be used as an alternative in the treatment of mastitis in dairy cows.

Several cases of infections caused by contaminated biomedical implant devices have been reported (Campoccia et al. 2005). Nisin, adsorbed to silanized surfaces, prevented the growth of *Listeria monocytogenes* (Bower et al. 1995). In a separate study, the authors coated Teflon® FEP intravenous catheters with nisin and inserted the devices in the jugular veins of sheep. In a similar experiment PVC tracheotomy tubes were coated and placed in the upper airways of ponies. The endotracheal tubes coated with nisin prevented colonization of *S. aureus*, *S. epidermidis*, and *Streptococcus faecalis* (Bower et al. 2002). The authors, however, concluded that the antimicrobial activity of nisin may be short lived in vivo, as intravascular catheters introduced in sheep controlled bacterial infection for only 5 h. Tracheotomy tubes treated with nisin controlled infection to some degree (Bower et al. 2002).

Bacteriocins for the Control of Upper Respiratory Tract Infections

Pseudomonas aeruginosa is the main causative agent of nosocomial pneumonia in cystic fibrosis (CF) patients (Linden et al. 2003) and has been diagnosed in more than 50% of CF patients with lung infections (CF patient registry, 2008). *P. aeruginosa* is also one of the major pathogens responsible for chronic/acute otitis (Roland and Stroman 2002; Wright et al. 2009). Otitis media is one of the most common diseases diagnosed in children under the age of 2 years (Segal et al. 2005). Long-term side effects of persistent otitis media include impaired hearing and delayed speech (Ryding et al. 1997; Teele et al. 1990). In addition to lung infection in CF patients and ear infections, *P. aeruginosa* is also associated with burn wound infections, especially among immunodeficient individuals (Deretic 2000; Hachem et al. 2007). Treatment of *P. aeruginosa* depends on several factors, including age of patient and severity of disease (Doring et al. 2000). Antibiotic treatments include Cefepime, Ceftazidime, Ciprofloxacin, Colistin, Piperacillin–Tazobactam, Gatifloxacin, Amikacin, and Gentamicin (Doring et al. 2000; Osih et al. 2007).

Nisin inhibited the growth of *S. pneumonia* associated with otitis media in in vivo trials (Goldstein et al. 1998). Peptide ST4SA, a class II bacteriocin, showed better activity toward Gram-positive middle-ear pathogens compared to other antimicrobial agents (Knoetze et al. 2008). Peptide ST4SA remained active when incubated in blood and middle-ear fluid and thus has the potential to be used in the treatment of otitis media.

Peptide IB-367 has recently undergone phase I safety trials on humans with the objective of using the peptide against chronic *P. aeruginosa* lung infections, specifically on patients suffering from cystic fibrosis. Toward the end of the last century, approximately one third of the world's population had been infected with *Mycobacterium tuberculosis* and was at risk of acquiring tuberculosis (Bloom and Murray 1992). In 2007 alone, more than 13 million people were diagnosed with tuberculosis, and more than nine million new cases were reported. The World Health Organization reported more than 1.8 million deaths caused by *M. tuberculosis*, of whom almost half a million were HIV positive (WHO 2009). Class II bacteriocins are active against *M. tuberculosis*. According to in vivo experiments, a bacteriocin–liposome complex increased the survival rate of animals challenged with the pathogen (Sosunov et al. 2007). The authors created an acute TB model in C57BL/6JCit (B6) mice by injecting the animals with a *M. tuberculosis* strain into the tail vein. Six hours after infection bacteriocin Bcn5 was injected into the tail vein and repeated every day for the next 5 days. Treatment with this bacteriocin showed less mortality than the negative control group, but greater mortality than the positive control group, which was treated with the clinically used antibiotic, rifampicin.

Nisin F, described by de Kwaadsteniet et al. (2009), inhibited the growth of *S. aureus* in the respiratory tract of rats when administered intranasally. The trachea and lungs of immunosuppressed rats that were infected with *S. aureus* and then treated with nisin F remained healthy. No significant differences were recorded in blood cell indices. The antimicrobial activity of low concentrations of nisin F (80–320 AU/ml) was slightly stimulated by lysozyme and lactoferrin. The authors concluded that Nisin F is nontoxic and may be used to control respiratory tract infections caused by *S. aureus*. This is, however, a preliminary study with an animal model and needs to be confirmed with human studies.

A strain of *Streptococcus mutans* that produces mutacin 1140 is active against tooth decay bacteria (Hillman et al. 1998). The strain has since been genetically modified to produce less lactic acid. The genetically modified strain, referred to as strain SMaRT, has been evaluated as a replacement therapy to eliminate decay-causing strains of *S. mutans*. Preclinical tests on laboratory animals have shown that the SMaRT strain eliminates disease-causing *S. mutans* strains, but not other types of bacteria commonly found in the oral cavity.

Use of Bacteriocins in Systemic Infections

S. aureus, *Listeria monocytogenes*, and *P. aeruginosa* are often associated with systemic infections (Czuprynski et al. 2002; Drake and Montie 1988; Harbarth et al. 1998; Klug et al. 1997). *Clostridium perfringens*, *Salmonella* spp., *S. aureus*, *Helicobacter* (*Campylobacter*) spp., *E. coli*, and *L. monocytogenes* are the most prominent bacteria causing gastrointestinal disorders and food poisoning (Tyopponen et al. 2003).

L. monocytogenes is especially dangerous to young children, pregnant women, immunocompromised individuals, and the elderly (Tyopponen et al. 2003). The pathogen usually enters through the intestine and attaches to the epithelial cells where it is taken up via phagocytosis (Portnoy et al. 2002; Ramaswamy et al. 2007). Hemolysin secreted by the bacteria then induces cytolysis of the phagosome membrane, which initiates intracytoplasmic replication (Portnoy et al. 1988, 2002; Ramaswamy et al. 2007). *Listeria* survives in monocytes/macrophages and crosses the blood–brain barrier to cause meningitis (Yildiz et al. 2007).

Listeria is resistant to β-lactams, monolactams, and cephalosporins, including cefotaxime and ceftazidime (Guinane et al. 2006). *Listeria* infection is usually treated with penicillin and ampicillin, which may be used in combination with an aminoglycoside, e.g., amikacin (Charpentier and Courvalin 1999; Yildiz et al. 2007). A trimethoprim–sulfamethoxazole combination is often used in patients allergic to penicillin (Charpentier and Courvalin 1999; Yildiz et al. 2007). Class II bacteriocins are all active against *L. monocytogenes* and may be used to treat listeriosis (Nes and Holo 2000). Pediocin PA-1, produced by *Pediococcus acidilactici*, revealed in vivo antimicrobial activity against various *L. monocytogenes* strains (Naghmouchi et al. 2006, 2007). Additionally, pediocin PA-1 does not inhibit other intestinal bacteria when administered intragastrically, in contrast to nisin A and nisin Z (le Blay et al. 2007).

Extensive in vitro studies have been done on antilisterial bacteriocins including pediocin PA-1, divergicin 35, and nisin, while only a few were done in vivo. When injected intravenously into the tail vein of BALB/c mice, piscicolin 126 relieved *Listeria* infection in various tissues (Ingham et al. 2003). Abp118, a bacteriocin produced by *Lactobacillus salivarius* UCC118, also showed good antilisterial activity in infected mice (Corr et al. 2007). Pediocin PA-1 offered protection against *L. monocytogenes* infection when administered onto the gastrointestinal tract of ICR mice (Dabour et al. 2009). Rihakova et al. (2010) have shown in vivo activity of divercin V41 against *L. monocytogenes* EGDe.

Mersacidin inhibited the growth of MRSA in vivo, and lacticin 3147 acted against *S. aureus* and MRSA strains (Galvin et al. 1999; Kruszewska et al. 2004, Limbert et al. 1991). Nisin inhibited the growth of *P. aeruginosa* when used in combination with polymyxin E and clarithromycin (Giacometti et al. 1999). Intrabiotics claims that the peptide protegrin PG-1 confers up to 100% systemic protection against intraperitoneal infections caused by *S. aureus,* MRSA, and *P. aeruginosa.* Another derivative of protegrin-1, iseganan, is undergoing phase II/III clinical testing for ventilator-associated pneumonia.

Clostridium botulinum causes a neuroparalytic disease known as botulism (Dolly et al. 2009; Franciosa et al. 2009; Kalb et al. 2009). The neurotoxins inhibit release of acetylcholine from the neuromuscular junction, bind to the cell surface of cholinergic nerve endings, and inhibit acetylcholine release, causing flaccid paralysis (Seyler et al. 2008). To eliminate the toxin from the body, an antitoxin is administered that binds to the toxin, rendering it inactive (Domingo et al. 2008; Shukla and Sharma 2005).

Most strains of *C. botulinum* are susceptible to antibiotics such as tetracycline, metronidazole, and chloramphenicol. Antibiotics that may worsen the paralysis, such as aminoglycoside and clindamycin, should be avoided (Dolly et al. 2009). Nisin proved effective in the inhibition of *C. botulinum* (Delves-Broughton et al. 1996) and *Clostridium tyrobutyricum* (de Carvalho et al. 2007). Nisin also proved effective against *Clostridium difficile* (Bartoloni et al. 2004), leading the authors to conclude that nisin may be used in the treatment of diarrhea.

Salmonella typhi causes typhoid fever and is prevalent in underdeveloped countries (Hoffner et al. 2000; Ling et al. 1996; Perera et al. 2007). *S. typhi* attaches to host epithelial cells with the help of fimbriae (Baumler et al. 1996) and invades cells of the intestinal epithelium (House et al. 2001; Sukhan 2000).

Salmonella infection is usually treated with fluoroquinolones (ciprofloxacin, levofloxacin, ofloxacin and ciprofloxacin) and chloramphenicol. However, the increase in antibiotic resistance is a cause for major concern (Kadhiravan et al. 2005; Perera et al. 2007). Two mainstream vaccines are currently available for the prevention of typhoid fever; a live-attenuated oral vaccine (Ty2Ia) and a vaccine based on purified capsular polysaccharide of *S. typhi* Vi antigen (Vi polysaccharide vaccine; Fraser et al. 2007a, b).

Bacteriocins of LAB are usually not active against Gram-negative bacteria, and only a few papers referred to activity against *Salmonella* spp. Bacteriocin AS-48, produced by *Enterococcus faecalis*, inhibited the growth of *Salmonella choleraesuis* at pH 4.0 (Abriouel et al. 1998). A bacteriocin-like substance produced by a strain of *L. plantarum* inhibited the growth of *Salmonella* spp. isolated from mango (Ragazzo-Sanchez et al. 2009). Enhanced antimicrobial activity was observed when NaCl (40 mg/mL) was used in combination with the bacteriocin-like substance. *Enterococcus gallinarum* strain 012, isolated from the duodenum of ostrich, produced enterocin 012 (3.4 kDa in size) which is active against *Salmonella typhimurium* (Jennes et al. 2000).

Gastric colonization of *Helicobacter pylori* causes upper GIT disorders such as chronic gastritis, peptic ulcer disease, tissue lymphoma, and gastric cancer (Correa 1992; Isreal and Peek 2001; Kusters et al. 2006). *H. pylori* is not only able to survive the acidic gastric conditions of the GIT but is also able to colonize these areas (Salama et al. 2001). Gastritis may protect the bacterium from host defenses (Isreal and Peek 2001). Proteins CagA and VacA secreted by the bacterium increase its virulence by increasing the inflammatory responses, causing vacuolization in epithelium cells, inducing apoptosis, and promoting activation and proliferation of T-cells (Cover and Blaser 1992, 2005; Kuipers et al. 1995; Peek et al. 1995; Salama et al. 2001).

Nisin and Lacticins A164 and BH5 inhibited the growth of *H. pylori* in vitro and may thus be used in the treatment of peptic ulcers (Delves-Broughton et al. 1996; Kim et al. 2003). Nisin, produced by AMBI (Purchase, New York), and IB-367, a protegrin-like cationic peptide from Intrabiotics (Mountain View, California), have successfully undergone phase I (safety) clinical trials. Both these peptides are being considered for treatment of stomach ulcers caused by *H. pylori* and oral mucositis. The companies Astra and Merck have commercialized nisin for treatment of gastric

Helicobacter infections and ulcers, while other nisin variants (nisin A and Z) have entered preclinical trials for treating vancomycin-resistant enterococci.

Phospholipase A2 plays a regulatory role in the arachidonic acid cascade, leading to the formation of potent mediators of inflammation and allergy, including the prostaglandins, leukotrienes, and hydroxyeicosatetraenoic acids (Braquet et al. 1987; Irvine 1982; Johnson et al. 1983; Page et al. 1984; Zipser and Laffi 1985). The lantibiotics duramycin, duramycin B and C, and cinnamycin inhibit phospholipase A2 indirectly by sequestering the substrate phosphatidylethanolamine (Fredenhagen et al. 1990; Marki et al. 1991) and may thus be used as anti-inflammatory drugs (van Kraaij et al. 1999).

The angiotensin-converting enzyme catalyzes the conversion of angiotensin I to angiotensin II and degrades bradykinin, thereby regulating blood pressure and fluid balance (Cushman et al. 1982; Imig 2004; Skeggs et al. 1956; Zhang et al. 2000). Cinnamycin-like lantibiotics and ancovenin, a type B lantibiotic, inhibit the activity of the angiotensin-converting enzyme (Jung 1991a, b; Shiba et al. 1991). Lantibiotics may thus have potential for treating high blood pressure (Kido et al. 1983).

Potential Applications of Bacteriocins in Infections of the Urogenital Tract

High numbers of *Lactobacillus* spp. form part of the normal bacterial flora in the vagina and ensure a reduced risk of bacterial vaginosis and urinary tract infections (Nomoto 2005). Although bacteriocin production by lactobacilli is one of their major protective mechanisms, these antimicrobial peptides are not typically active against yeasts. As such, this review is focused only on those microorganisms sensitive to the antimicrobials in question, with bacterial vaginosis (BV) as the primary focus. However, it should be noted that when bacteriocins are combined with other natural antimicrobials, there may be increased activity against pathogenic fungi in the vaginal ecosystem.

In the past, antimicrobially active lactobacilli were commonly used to develop products for the prevention and treatment of genital infections (Barbes and Boris 1999; Famularo et al. 2001; Maggi et al. 2000). The most consumer-accepted product is based on an application of lactobacilli that produce H_2O_2 (McLean and Rosenstein 2000; Ocaña et al. 1999). Based on the health-promoting action of healthy vaginal *Lactobacillus crispatus*, The Medicines Company (Cambridge, MA) and University of Pittsburgh (PA) initiated an NIH-sponsored study of the *L. crispatus* strain CTV-05 as an adjunct to standard antibiotic treatment of bacterial vaginosis. It was hypothesized that *L. crispatus* could replace vaginal pathogens and promote the reestablishment of normal vaginal microflora due to high-level production of hydrogen peroxide, which is an effective antimicrobial agent. However, strain CTV-05 did not improve clinical cure rates at 30 days, the primary end point of the trial. This failure was not an accidental one. To succeed in a strain replacement therapy approach as described above, the healthy microorganism has to (1) migrate to the surface of the epithelial cells, (2) adhere onto the epithelial cells, (3) successfully colonize the epithelial cells and develop a functional biofilm, and finally, having survived all these initial phases, (4) start producing antimicrobials such as hydrogen peroxide, bacteriocins, lactic acid, etc. that will force the invading pathogens away from the ecological niche conquered back by the healthy microorganism. Failure in any of the major outlined phases (and other undescribed circumstances) will lead to failure in the replacement of nondesired pathogens by healthy bacteria. Recently published data have shown that a daily oral and vaginal intake of *L. rhamnosus* GG (1,010 cells per dose, Culturelle, USA) was unsuccessful in reaching the goal of colonizing the vagina and correcting the disturbed vaginal ecology (Colodner et al. 2003; Devillard et al. 2005).

Lactobacilli inhibited the growth and attachment of uropathogenic *E. coli* in vitro (McGroarty and Reid 1988; Reid et al. 1987; Velraeds et al. 1998). *L. rhamnosus* GR-1, *L. fermentum* B-54, and *Lactobacillus* strains with high adhesion and hydrogen peroxide production were weekly administered as a vaginal suppository (Reid and Bruce 2001; Reid et al. 1987; 1995). Urinary tract infection was significantly decreased. Orally administered *L. rhamnosus* GR-1 and *L. fermentum* RC-14 decreased *E. coli* and fungi in the vagina. Future studies should include the inhibition of STBs by probiotics taken orally or as vaginal suppositories (Sewankambo et al. 1997). A proposed mechanism is the induction of an immune response via the urethra or vagina, which is functional in the bladder (Reid and Burton 2002). Probiotic colonization in the vagina might prevent infection by competition for nutrients and mannose and hydrophobic adhesion to receptors (Braun 1999; Masuoka and Hazen 1999) and possibly bacteriocin production with fungistatic activity (Okkers et al. 1999).

An overall low vaginal pH is essential for the prevention of vaginal infections. Intravaginal products such as Acidgel, BufferGel, etc. are based on the acid-producing ability of the lactobacilli, which help maintain a vaginal pH lower than 4.5 (Amaral et al. 1999; Andersch et al. 1990; Garg et al. 2001). However, low vaginal pH alone is not sufficient to inhibit vaginal pathogens and to prevent infection.

Bacterial vaginosis (BV), a common condition found in up to 30% of women in North America (Schwebke 2003), is typically associated with a multispecies infection, where *Gardnerella vaginalis* is often the major contributor to BV development. In addition to *G. vaginalis, Prevotella bivia* and *Peptostreptococcus* spp. contribute significantly to the development of BV (Dover et al. 2008; Nikolaithcouk et al. 2008), which is associated with a relatively high pH, a decrease in antimicrobial activity of the vaginal fluid compared to healthy women, and local impairment of the multiple innate immune pathways (Pybus and Onderdonk 1999; Valore et al. 2006).

Almost one third of childbearing-age women in the world are diagnosed with BV (Schwebke 1997). The harmful effects of BV range from complications with pregnancies to the development of pelvic inflammatory diseases (Hillier et al. 1995; Ness et al. 2005). Studies have also associated BV with a higher risk of acquisition of HIV infection (Cohen et al. 1995; Martin et al. 1999; Sha et al. 2005; Taha et al. 1998) and herpes simplex virus type 2 infections (Cherpes et al. 2003). It is estimated that nearly 60% of HIV+ women have concurrent BV infections (Mascellino et al. 1991), a statistic made especially disturbing by the fact that BV directly causes an increase in the rate of HIV replication and disease progression (Al-Harthi et al. 1999; Hashemi et al. 2000). Toxins from BV-associated microorganisms (such as lipopolysaccharides) may also cross the placenta and cause brain injuries in fetuses. These toxins may cause permanent neurological brain damages such as cerebral palsy, a risk of developing Parkinson's disease and schizophrenia (Grether and Nelson 2000; Ling et al. 2004; Urakubo et al. 2001).

While 60% of BV cases can be successfully treated with metronidazole and clindamycin, as recommended by the Centers for Disease Control and Prevention (Paavonen et al. 2000; Sobel et al. 2001), about 20% of these cases return with highly-developed antibiotic resistance (Boris et al. 1997; Bannatyne and Smith 1998; Ferris et al. 1995; Liebetrau et al. 2003; Lubbe et al. 1999). The risk of developing antimicrobial drug resistance increases dramatically with overall increased use of antimicrobial (feminine hygiene and treatment) preparations (Uehara et al. 2006). In addition, in vitro studies showed that clindamycin and metronidazole inhibit healthy vaginal *Lactobacillus* spp. at concentrations lower than doses topically applied for treatment (Aroutcheva et al. 2001; Simoes et al. 2001). Therefore, there is an interest in developing alternative treatments against BV, such as selective antimicrobials that will inhibit BV-associated bacteria without killing healthy *Lactobacillus* spp.

One promising alternative is the bacteriocin subtilosin A. Originally isolated from the wild-type *B. subtilis* 168 by Babasaki et al. (1985), subtilosin has been recently found to be concurrently produced by *B. amyloliquefaciens*, a similar but divergent *Bacillus* species isolated from a fermented dairy beverage (Sutyak et al. 2008a). It is a circular molecule of 35 amino acids, with the distinctive posttranslational modification of three sulfur cross-links between cysteine and the α-carbon of two phenylalanines and one threonine residue. This structure is unique among bacteriocins, suggesting it may belong in a new, undefined class of antimicrobial peptides (Kawulka et al. 2004).

In contrast to many bacteriocins, which have an overall positive charge at physiological pH, subtilosin A is anionic, having only one lysine and a total of three aspartate and glutamate residues (Kawulka et al. 2004; Thennarasu et al. 2005). Because bacterial membranes also have a net anionic charge, it has been hypothesized that subtilosin may not interact solely with the cell membrane, but may first bind a surface receptor prior to insertion into target membranes (Thennarasu et al. 2005). It has been shown that at high concentrations (much greater than its MIC values), subtilosin A interacts with the lipid head group region of bilayer membranes of target cells, causing membrane perturbation, the extent of which is dependent on lipid composition (Thennarasu et al. 2005). At these high concentrations, subtilosin aggregates into multimeric units; therefore, its primary mode of action may be by interaction with a membrane component or receptor (Thennarasu et al. 2005).

Subtilosin has proven antimicrobial activity against a wide variety of human pathogens, including *L. monocytogenes*, *G. vaginalis*, *S. agalactiae*, and *Micrococcus luteus* (Sutyak et al. 2008a). Its activity against *G. vaginalis*, combined with its lack of effect on probiotic vaginal *Lactobacillus* isolates (Sutyak et al. 2008a), indicates that subtilosin could target the vaginal pathogen while leaving the healthy vaginal microflora intact. Cytotoxicity testing conducted in vivo using the EpiVaginal

(VEC-100) human ectocervical tissue model (MatTek, Ashland, MA) demonstrated that subtilosin caused only a 5% decrease in cell viability after 24 h of continuous exposure (20% after 48 h) (Sutyak et al. 2008b). Its safety for human tissue as well as the probiotic microflora of the specific ecological niche gives subtilosin great potential as a future inclusion in personal care products.

Spermicidal Activity and Potential Contraceptive Usage of Bacteriocins

Some of the bacteriocins that are active against vaginal pathogens are also reported as having spermicidal activity. This feature makes them attractive for formulation in feminine health care and contraceptive products.

In order to evaluate nisin's spermicidal activity, Aranha et al. (2004) developed a contraceptive model in rats. Nisin, dissolved in saline, was administered into the vagina of the animals for 14 consecutive days during the proestrus–estrous transition phase. Animals were then immediately allowed to mate, and none of the nisin-treated animals became pregnant. No histopathological lesions were observed in the vaginal epithelium, and liver and kidney function remained normal. Fertility was also restored after experiments. According to the authors, 1 mg of nisin was able to completely halter sperm motility. This is an interesting finding, since many commonly used contraceptive products contain Nonoxynol-9 (N-9), a compound harmful to epithelium.

Subtilosin, the previously described bacteriocin produced by *B. amyloliquefaciens*, was also shown to have potent spermicidal activity. When tested against human spermatozoa, it was able to decrease motility in a dose-dependent manner and has an IC_{50} value of 64.5 μg/mL (Sutyak et al. 2008b). Interestingly, subtilosin also has demonstrated significant spermicidal activity when tested on various animals, including boar, horse/pony, rat, and bovine models (Silkin et al. 2008). Nisin inhibited sperm motility and caused no abnormalities when applied intravaginally in rats (Aranha et al. 2004). Intravaginal application of nisin also prevented conception in rabbits and showed no inflammation or damage to the vaginal epithelium when applied for 2 weeks (Reddy et al. 2004). However, if the concentrations of nisin used in the animal model are extrapolated for human usage, they are well above the limits of what the healthy vaginal microflora can survive. Thus, nisin cannot be practically considered for use in human products.

Recombinant Antimicrobial Peptides

Because of their low resistance to plasma and serum proteolytic activity (Bracci et al. 2003; Pini et al. 2005), cationic peptides show high in vitro activity and limited in vivo activity. It is thus necessary to resort to different strategies to increase peptide stability for therapeutic application. Multimeric peptides have a remarkably increased half-life in vivo and enhanced antimicrobial activity with respect to linear homologues (Pini et al. 2005; Tam et al. 2002). This topic is covered by another chapter in the book.

Bacteriocins, Immunity, and Resistance: Issues and Concerns

Genes involved in the biosynthesis of bacteriocins are organized in a cluster on the genome, a plasmid, or a transposon (McAuliffe et al. 2001; Nes et al. 1996; Van Reenen et al. 1998). In some instances the immunity gene is located in the same operon as the structural gene that codes for the bacteriocin (Gasson 1984). However, in some cases immunity genes are located on plasmids or transposons and thus not linked to bacteriocin production (Froseth et al. 1988; Klaenhammer and Sanozky 1985; McKay and Baldwin 1984). Mobile genetic elements can, however, be exchanged with other organisms of the same or different species, rendering sensitive strains insensitive to some bacteriocins.

Resistance in spontaneous mutants can be ascribed to changes in the membrane charge and fluidity, cell wall thickness, cell wall charge, and combinations of the aforementioned changes (Abachin et al. 2002; Bierbaum and Sahl 1987; Crandall and Montville 1998; Li et al. 2002; Maisnier-Patin and Richard 1996; Mantovani and Russell 2001; Vadyvaloo et al. 2002, 2004; Verheul et al. 1997). Spontaneous nisin resistance frequency in *L. monocytogenes* varied from 10^{-2} to 10^{-7} in a strain-dependent manner (Gravesen et al. 2002a, b; Davies and Adams 1994), and *L. monocytogenes* mutants were detected at a frequency of 10^{6}–10^{8} that were resistant to 50 g/ml nisin (Harris et al. 1991). Nisin-resistant *S. pneumoniae* has also been found when this organism is exposed to nisin over long periods of time, with the minimum inhibitory concentration (MIC) increasing from 0.4 to 6.4 mg/ml (Severina et al. 1998). Spontaneous mutants of *P. acidilactici* resistant to pediocin AcH are lost when grown in the absence of the bacteriocin. Pediocin-resistant mutants had a reduction of up to 44% of the maximum specific growth rate as compared to the wild-type strain (Gravesen et al. 2002a, b). Nisin resistance has also been reported in *C. botulinum* spores and vegetative cells (Mazzotta and Montville 1997).

Most bacteriocin resistance studies have been carried out using resistant *L. monocytogenes* strains. Resistance of *L. monocytogenes* to nisin can be ascribed to changes in fatty acid and phospholipid composition, a lower ratio of C-15 to C-17 fatty acids, more zwitterionic phosphatidylethanolamine, less anionic phosphatidylglycerol and cardiolipin, and a requirement for divalent cations (Crandall and Montville 1998; Mazzotta and Montville 1997; Ming and Daeschel 1993, 1995; Verheul et al. 1997). An enzyme, nisinase, can also confer resistance to nisin. Nisinase activity, which inactivates nisin, has been reported in several *Bacillus* spp. (Jarvis 1967).

L. monocytogenes resistance toward class II bacteriocins correlates with a reduction in the expression of a mannose permease of the phosphotransferase system (man-PTS) (Cotter et al. 2005). Because class II bacteriocin-resistant mutants display this reduction in the expression of man-PTS, it was speculated that the man-PTS serves as a target site for some class II bacteriocins (Ramnath et al. 2000). This was confirmed by Diep et al. (2007), who reported that class II bacteriocins use the IIC and IID components of the man-PTS as receptors. They also reported that the immunity proteins form a complex with the receptor proteins, thus rendering the producer strain resistant to its own bacteriocin. Gravesen et al. (2002b) reported that *L. monocytogenes* resistance toward class II bacteriocins is characterized by the upregulation of EII^{Bgl} and phospho-β-glucosidase, as well as the prevention of EII_t^{Man} synthesis. Inhibited *mpt* expression confers the resisitance, while upregulated EII^{Bgl} and phospho-β-glucosidase expression are merely a result of inhibited *mpt* expression.

Cross-resistance has been reported between different bacteriocins. Nisin resistance conferred cross-resistance to pediocin PA-1 and leuconocin S in *L. monocytogenes* (Crandall and Montville 1998). Leucocin A-resistant strains of *L. monocytogenes* showed no significant cross-resistance toward other bacteriocins, including nisaplin and ESF1-7GR, although they were shown to be resistant to pediocin PA-2 (51, 200 AU/ml) (Ramnath et al. 2000). Studies have demonstrated that resistance to bacteriocins is still relatively weak because low levels of resistant strains are being isolated and in some cases bacteriocin resistance is unstable. Resistance can easily be lost if the strains are cultured without the bacteriocins (Breuer and Radler 1996; Dykes and Hastings 1998; Ming and Daeschel 1993; Rasch and Knøchel 1998; Rekhif et al. 1994; Song and Richard 1997). Resistant mutants have also been shown to have a lower growth rate than wild-type strains and were unable to outgrow them (Gravesen et al. 2002a, b; Dykes and Hastings 1998; Maisnier-Patin et al. 1995). This indicates that a resistance mechanism toward bacteriocins has a negative influence on the strain, and when grown without bacteriocin this organism is outcompeted by other organisms (Dykes 1995; Noerlis and Ray 1994). No resistance has been reported to lacticin 3147 in sensitive gram-positive organisms (Ross et al. 1999), which is especially desirable with the increase in antibiotic resistance seen in numerous human pathogens.

Although as yet there are no reports in the literature on the acquisition of bacteriocin immunity gene(s) by pathogenic bacteria, there are examples where these genes have been cloned and expressed by Venema et al. (1995) and other groups (Fimland et al. 2002), in a bacteriocin-sensitive host, resulting in resistance to the antimicrobial protein by the bacteria. Also, a number of examples of the production of identical bacteriocins (and their cognate immunity proteins) by bacteria of different species (Ennahar et al. 1999; Sutyak et al. 2008a) raise additional concern regarding the emergence of bacteriocin resistance via interspecies gene transfer. In addition, it was reported for *Streptococcus mutans* that the microorganism's increased sensitivity to antimicrobials such as tetracycline, penicillin, and triclosan is triggered by the repression of the bacteriocin immunity gene (Matsumoto-Nakano and Kuramitsu 2006). All of these render the elucidation of the mechanism of immunity even more important, since it will contribute to our general understanding of how antimicrobial resistance may be prevented. Furthermore, strains of

Enterococcus faecium and *Bacillus licheniformis* were found to carry gene homologues that provide protection against a bacteriocin. This phenomenon was named "resistance through immune mimicry" (Draper et al. 2009).

Almost no or very low levels of cross-resistance between bacteriocins and antibiotics have been found (Bower et al. 2001; Mantovani and Russell 2001). Bacteriocins and antibiotics have different modes of action. Antibiotics can inhibit cell wall synthesis, protein synthesis, DNA synthesis, RNA synthesis and can cause the competitive inhibition of folic acid synthesis, while bacteriocins forms pores in the membrane of sensitive cells, leading to leakage of intracellular material and cell death (Ennahar et al. 1999; Kraaij et al. 1999; Levy and Marshall 2004; Nes and Holo 2000; Neu 1992; Sahl and Bierbaum 1998). Bacteriocins and antibiotics can potentially be used together to prevent or hamper the emergence of resistant pathogens because they have different modes of action and acquiring resistance toward two different antimicrobials is very unlikely (Diep et al. 2007).

Intelligent Delivery Systems for Antimicrobial Peptides

Bacteriocins show the potential to have various applications in the biomedical industry. However, the stability of bacteriocins can become a problem in a complex in vivo environment. Polymeric delivery systems can help overcome the stability problems of peptides in an in vivo system and can release the bacteriocins in a bioactive form to a specific site of interest. A variety of biodegradable biomedical polymers are available that can be used as delivery systems. Polylactic acid is one such polymer that has received FDA approval for use in humans (Nair and Laurencin 2007). Most studies aimed at the delivery of bacteriocins have dealt with antimicrobial packaging films or materials aimed at food preservation, and very few studies have been conducted to generate delivery systems for biomedical applications of bacteriocins (Cutter et al. 2001; Malheiros et al. 2010; Marcos et al. 2007; Scannell et al. 2000). For the most recent comprehensive review see Balasubramanian et al. (2009).

Nisin has been encapsulated into poly(L-lactide) (PLA) nanoparticles by semi-continuous compressed CO_2 antisolvent precipitation (Salmaso et al. 2004). Nisin was released in the active form for up to 1,000 h, and the release was dependent on the salt concentration and the pH of the release medium. Nisin released from the PLA nanoparticles was able to inhibit the growth of *Lactobacillus delbrueckeii* subsp. *bulgaricus* when nisin-loaded PLA nanoparticles were incubacted in MRS containing the sensitive strain. These nanoparticles could potentially be used in antimicrobial pharmaceutical products.

Encapsulation of bacteriocins in electrospun nanofibers has recently been reported (Heunis et al. 2010). Electrospinning is the process where a high voltage is applied to a polymer solution, which forms a Taylor cone when charged (Taylor 1969; Yarin et al. 2001). When the electric forces overcome the surface tension of the solution, a charged polymer jet is ejected from the Taylor cone and will start to accelerate toward the collector. The solvent will evaporate during this process, and very thin fibers are formed (Agarwal et al. 2008; Liang et al. 2007). Plantaricin 423, produced by *Lactobacillus plantarum* 423, was electrospun into polyethylene oxide (PEO) nanofibers. A slight decrease in bacteriocin activity was seen in the fibers; however, enough activity was retained to inhibit the growth of *Lactobacillus sakei* DSM 20017 and *Enterococcus faecium* HKLHS. The activity decreased from 51,200 to 25,600 AU/ml and from 204,800 to 51,200 AU/ml after electrospinning, as determined against *L. sakei* DSM 20017 and *E. faecium* HKLHS, respectively. Thus, nanofibers could also be used in various biomedical applications.

Drug delivery systems will play an integral role in the use of bacteriocins in the biomedical industry. These systems will help to protect and keep the peptides active and will release them in a controlled manner to exert activity. Drug delivery systems will have added potential with regard to the localized release of bacteriocins, which would be desired if the bacteriocins are to be used as a topical treatment for skin infections or as coatings for biomedical devices to combat device-related infections. More studies need to be conducted on the controlled release of bacteriocins from drug delivery systems. These studies will not only help to increase the already huge potential biomedical applications of bacteriocins, but will also bring bacteriocins into a new era of biomedicine.

References

Abachin E, Poyart C, Pellegrini E, Milohanic E, Fiedler F, Berche P, Triu-Cot P (2002) Formation of D-alanyl-lipoteichoic acid is required for adhesion and virulence of *Listeria monocytogenes*. Mol Microbiol 43(1):1–14

Abee T, Krockel L, Hill C (1995) Bacteriocins: modes of action and potentials in food preservation and control of food poisoning. Int J Food Microbiol 28:169–185

Abriouel H, Valdivia E, Gálvez A, Maqueda M (1998) Response of *Salmonella choleraesuis* LT2 spheroplasts and permeabilized cells to the bacteriocin AS-48. Appl Environ Microbiol 64:4623–4626

Agarwal S, Wendorff JH, Greiner A (2008) Use of electrospinning technique for biomedical applications. Polymer 49:5603–5621

Al-Harthi L, Roebuck KA, Olinger GG, Landay A, Sha BE, Hashemi FB, Spear GT (1999) Bacterial vaginosis-associated microflora isolated from the female genital tract activates HIV-1 expression. J Acquir Immune Defic Syndr 21:194–202

Amaral E, Faundes A, Zaneveld L, Waller D, Garg S (1999) Study of the vaginal tolerance to Acidform, an acid-buffering, bioadhesive gel. Contraception 60:361–366

Andersch B, Lindell D, Dahlen I, Brandberg A (1990) Bacterial vaginosis and the effect of intermittent prophylactic treatment with an acid lactate. Gynecol Obstet Invest 30:114–119

Aranha C, Gupta S, Reddy KV (2004) Contraceptive efficacy of antimicrobial peptide nisin: in vitro and in vivo studies. Contraception 69(4):333–338

Aroutcheva A, Simoes J, Shott S, Faro S (2001) The inhibitory effect of clindamycin on *Lactobacillus* in vitro. Infect Dis Obstet Gynecol 9:239–242

Babasaki K, Takao T, Shimonishi Y, Kurahashi K (1985) Subtilosin A, a new antibiotic peptide produced by *Bacillus subtilis* 168: isolation, structural analysis, and biogenesis. J Biochem 98:585–603

Balasubramanian A, Rosenberg LE, Yam K, Chikindas ML (2009) Antimicrobial packaging: potential VS reality. J Appl Pack Res 3:193–221

Bannatyne RM, Smith AM (1998) Recurrent bacterial vaginosis and metronidazole resistance in *Gardnerella vaginalis*. Sex Transm Infect 74:455–456

Baquero F, Mreno F (1984) The microcins. FEMS Microbiol Lett 23(2–3):117–124

Barbes C, Boris S (1999) Potential role of lactobacilli as prophylactic agents against genital pathogens. AIDS Patient Care Stds 13:747–751

Bartoloni A, Mantella A, Goldstein BP, Dei R, Benedetti M, Sbaragli S, Paradisi F (2004) *In-vitro* activity of nisin against clinical isolates of *Clostridium difficile*. J Chemother 16:119–121

Baumler A, Tsolis R, Heffron F (1996) Contribution of fimbrial operons to attachment to and invasion of epithelial cell lines by *Salmonella typhimurium*. Infect Immun 64:1862–1865

Bierbaum G, Sahl HG (1987) Autolytic system of *Staphylococcus simulans* 22: influence of cationic peptides on activity of N-acetylmuramoyl-L-alanine amidase. J Bacteriol 169:5452–5458

Blackburn, P. and B.P. Goldstein (1995) Applied Microbiology Inc. International Patent Application WO 97/10801

Bloom BR, Murray CJL (1992) Tuberculosis: Commentary on a reemergant killer. Science 257:1055–1063

Boheim HG (1995) Peptide antibiotics and their role in innate immunity. Annu Rev Immunol 13:61–92

Bonelli R, Wiedemann RI, Sahl HG (2006) Lantibiotics. In: Kastin A (ed) Handbook of biologically active peptides. Elsevier, New York, NY

Boris J, Påhlson C, Larsson P (1997) Six years observation after successful treatment of bacterial vaginosis. Infect Dis Obstet Gynecol 5:297–302

Bower CK, Bothwell MK, McGuire J (2001) Lantibiotics as surface active agents for biomedical applications. Colloids Surf B Biointerfaces 22:259–265

Bower CK, McGuire J, Daeschel MA (1995) Suppression of *Listeria monocytogenes* colonization following adsorption of nisin onto silica surfaces. Appl Environ Microbiol 61(3):992–997

Bower CK, Parker JE, Higgins AZ, Oest ME, Wilson JT, Valentine BA, Bothwell MK, McGuire J (2002) Protein antimicrobial barriers to bacterial adhesion: in vitro and in vivo evaluation of nisin-treated implantable materials. Surf B: Biointerfaces 25:81–90

Bracci L, Falciani C, Lelli B, Lozzi L, Runci Y, Pini A (2003) Synthetic peptides in the form of dendrimers become resistant to protease activity. J Biol Chem 278:46590–46595

Bradley AJ (2002) Bovine mastitis: an evolving disease. Vet J 164:116–128

Bradley D (1967) Ultrastructure of bacteriophage and bacteriocins. Bacteriol Rev 31:230–314

Braquet P, Touqui L, Shen TS, Vargafttig BB (1987) Perspectives in platelet-activating factor research. Pharmacol Rev 39:97–145

Braun PC (1999) Nutrient uptake by *Candida albicans*: the influence of cell surface mannoproteins. Can J Microbiol 45:353–359

Braun V, Pilsl H, Grob (1994) Colicins: Structures, mode of action, transfer through membranes, and evolution. Arch Microbiol. 161:199–206

Breuer B, Radler F (1996) Inducible resistance against nisin in *Lactobacillus casei*. Arch Microbiol 165:114–118

Broadbent JR, Chou YC, Gillies K, Kondo JK (1989) Nisin inhibits several Gram-positive, mastitis-causing pathogens. J Dairy Sci 72:3342–3345

Brötz H, Bierbaum G, Leopold K, Reynolds PE, Sahl HG (1998) The lantibiotic mersacidin inhibits peptidoglycan synthesis by targeting lipid II. Antimicrob Agents Chemother 42:154–160

Brötz H, Bierbaum G, Markus A, Molitor E, Sahl HG (1995) Mode of action of the lantibiotic mersacidin: inhibition of peptidoglycan biosynthesis via a novel mechanism? Antimicrob Agents Chemother 39:714–719

Campoccia D, Montanaro L, Baldassarri L, An YH, Arciola CR (2005) Antibiotic resistance in *Staphylococcus aureus* and *Staphylococcus epidermidis* clinical isolates from implant orthopedic infections. Int J Artif Organs 28:1186–1191

Charpentier E, Courvalin P (1999) Antibiotic resistance in *Listeria* spp. Antimicrob Agents Chemother 43:2103–2108

Chatterjee S, Chatterjee DK, Jani RH, Blumbach J, Ganguli BN, Klesel N, Limbert M, Seibert G (1992) Mersacidin, a new antibiotic from *Bacillus*, in vitro and in vivo antibacterial activity. J Antibiot 45:839–845

Cherpes TL, Meyn LA, Krohn MA, Hillier SL (2003) Risk factors for infection with herpes simplex virus type 2: Role of smoking, douching, uncircumcised males, and vaginal flora. Sex Trans Dis 30:405–410

Chikindas ML, Garcia-Garcera MJ, Driessen AJ, Ledeboer AM, Nissen-Meyer J, Nes IF, Abee T, Konings WN, Venema G (1993) Pediocin PA-1, a bacteriocin from *Pediococcus acidilactici* PAC1.0, forms hydrophilic pores in the cytoplasmic membrane of target cells. Appl Environ Microbiol 59:3577

Chung W, Hancock REW (2000) Action of lysozyme and nisin mixtures against lactic acid bacteria. Int J Food Microbiol 60:25–32

Cohen CR, Duerr A, Pruithithada N, Rugpao S, Hillier S, Garcia P, Nelson K (1995) Bacterial vaginosis and HIV seroprevalence among female commercial sex workers in Chiang Mai, Thailand. AIDS 9:1093–1097

Colodner R, Edelstein H, Chazan B, Raz R (2003) Vaginal colonization by orally administered *Lactobacillus rhamnosus* GG. Isr Med Assoc J 5:767–769

Corr S, Li Y, Riedel CU, O'Tooles PW, Hill C, Gahan CG (2007) Bacteriocin production as a mechanism for the antiinfective activity of *Lactobacillus salivarius* UCC118. Proc Nat Acad Sci 104:7617–7621

Correa P (1992) Human gastric carcinogenesis: a multistep and multifactorial process First American Cancer Society Award Lecture on cancer epidemiology and prevention. Cancer Res 52:6735–6740

Corzo G, Escoubas P, Villegas E, Barnham KJ, He W, Norton RS, Nakajima T (2001) Characterization of unique amphipathic antimicrobial peptides from venom of the scorpion *Pandinus imperator*. Biochem J 359:35–45

Cotter PD, Hill C, Ross RP (2005) Bacteriocins: developing innate immunity for food. Nat Rev Microbiol 3:777–788

Cover T, Blaser M (1992) Purification and characterization of the vacuolating toxin from *Helicobacter pylori*. J Biol Chem 267:10570–10575

Cover TL, Blanke SR (2005) Helicobacter pylori VacA, a paradigm for toxin multifunctionality. Nat Rev Microbiol 3:320–332

Crandall AD, Montville TJ (1998) Nisin resistance in *Listeria monocytogenes* ATCC 700302 is a complex phenotype. Appl Environ Microbiol 64(1):231–237

Cushman DW, Cheung HS, Sabo EF, Ondetti MA (1982) Development and design of specific inhibitors of angiotensin-converting enzyme. Am J Cardiol 49:1390–1394

Cutter CN, Willett JL, Siragusa GR (2001) Improved antimicrobial activity of nisin-incorporated polymer films by formulation change and addition of food grade chelator. Lett Appl Microbiol 33:325–328

Czuprynski CJ, Faith NG, Steinberg H (2002) Ability of the *Listeria monocytogenes* strain Scott A to cause systemic infection in mice infected by the intragastric route. Appl Environ Microbiol 68:2893–2900

Dabour N, Zihler A, Kheadr E, Lacroix C, Fliss I (2009) In vivo study on the effectiveness of pediocin PA-1 and *Pediococcus acidilactici* UL5 at inhibiting *Listeria monocytogenes*. Int J Food Microbiol 133:225–233

Davies EA, Adams MR (1994) Resistance of *Listeria monocytogenes* to the bacteriocin nisin. Int J Food Microbiol 21:341–347

de Carvalho AA, Mantovani HC, Vanetti MC (2007) Bactericidal effect of bovicin HC5 and nisin against *Clostridium tyrobutyricum* isolated from spoiled mango pulp. Lett Appl Microbiol 45:68–74

de Kwaadsteniet M, ten Doeschate KT, Dicks LMT (2009) Nisin F in the treatment of respiratory tract infections caused by *Staphylococcus aureus*. Lett Appl Microbiol 48:65–70

de Vos WM, Kuipers OP, van der Meer JR, Siezen RJ (1995) Maturation pathway of nisin and other lantibiotics: post-translationally modified antimicrobial peptides exported by Gram-positive bacteria. Mol Microbiol 17(3):427–437

de Vuyst L, Vandamme EJ (eds) (1994) Bacteriocins of lactic acid bacteria. Chapman and Hall, London

Delves-Broughton J, Blackburn RJ, Evans RJ, Hugenholtz J (1996) Applications of the bacteriocin nisin. Antonie van Leeuwenhoek. Int J Gen Microbiol 69:193–202

Deretic V (2000) *Pseudomonas aeruginosa* infections. In: Nataro JP, Blaser MJ, Cunningham-Rundles S (eds), Persistent bacterial infections. Press Am Soc Microbiol, Washington DC

Devillard E, Burton J, Reid G (2005) Complexity of vaginal microflora as analyzed by PCR denaturing gradient gel electrophoresis in a patient with recurrent bacterial vaginosis. Infect Dis Obstet Gynecol 13:25–30

Dicks LMT, Todorov SD, Franco BDGM (2009) Current status of antibiotic resistance in lactic acid bacteria. In: Antibiotic Resistance: Causes and Risk Factors, Mechanisms and Alternatives (Eds: Bonilla AR, Muniz KP), Nova Publications, pp. 379–425

Diep DB, Skaugen M, Salehian Z, Holo H, Nes IF (2007) Common mechanisms of target cell recognition and immunity for class II bacteriocins. Proc Natl Acad Sci 104:2384

Dijkshoorn L, Brouwer CPJM, Bogaards SJP, Nemec A, van den Broek PJ, Nibbering PH (2004) The synthetic N-terminal peptide of human lactoferrin, hLF(1-11), is highly effective against experimental infection caused by multidrug-resistant *Acinetobacter baumanni*. Antimicrob Agents Chemother 48:4919–4921

Dolly JO, Lawrence GW, Meng J, Wang J (2009) Neuro-exocytosis: botulinum toxins as inhibitory probes and versatile therapeutics. Curr Opin Pharmacol 9:326–335

Domingo RM, Haller JS, Gruenthal M (2008) Infant botulism: two recent cases and literature review. J Child Neurol 23:1336–1346

Doring G, Conway S, Heijerman H, Hodson M, Hoiby N, Smyth A, Touw D (2000) Antibiotic therapy against *Pseudomonas aeruginosa* in cystic fibrosis: a European consensus. Eur Resp J 16:749–767

Dover SE, Aroutcheva AA, Faro S, Chikindas ML (2008) Natural antimicrobials and their role in vaginal health: a short review. Int J Probiotics and Prebiotics 3:219–230

Drake D, Montie TC (1988) Flagella, Motility and Invasive Virulence of *Pseudomonas aeruginosa*. J Gen Microbiol 134:43–52

Draper LA, Grainger K, Deegan LH, Cotter PD, Hill C, Ross RP (2009) Cross-immunity and immune mimicry as mechanisms of resistance to the lantibiotic lacticin 3147. Mol Microbiol 71:1043–1054

Drider D, Fimland G, Hechard Y, McMullen LM, Prevost H (2006) The continuing story of class IIa bacteriocins. Microbiol Mol Biol Rev 70:564–582

Driessen AJM, van den Hooven HW, Kuiper W, van de Kamp M, Sahl HG, Konings RNH, Konings WN (1995) Mechanistic studies of lantibiotic-induced permeabilization of phospholipid vesicles. Biochem 34:1606–1614

Duquesne S, Destoumieux-Garzon D, Peduzzi J, Rebuffat S (2007a) Microcins, gene-encoded antibacterial peptides from enterobacteria. Nat Prod Rep 24:708–734

Duquesne S, Petit V, Peduzzi J, Rebuffatd S (2007b) Structural and functional diversity of microcins, gene-encoded antibacterial peptides from enterobacteria. J Mol Microbiol Biotechnol 13:200–209

Dykes GA (1995) Bacteriocins: ecological and evolutionary significance. Trends Ecol Evol 10(5):186–189

Dykes GA, Hastings JW (1998) Fitness costs associated with class IIa bacteriocin resistance in *Listeria monocytogenes* B73. Lett Appl Microbiol 26(1):5–8

Eijsink VGH, Axelsson L, Diep DB, Håvarstein LS, Holo H, Nes IF (2002) Production of class II bacteriocins by lactic acid bacteria; an example of biological warfare and communication. Antonie Van Leeuwenhoek 81:639–654

Eijsink VGH, Skeie M, Middelhoven PH, Brurberg MB, Nes IF (1998) Comparative studies of class IIa bacteriocins of lactic acid bacteria. Appl Environ Microbiol 64:3275

Ennahar S, Sashihara T, Sonomoto K, Ishizaki A (2006) Class IIa bacteriocins: biosynthesis, structure and activity. FEMS Microbiol Rev 24:85–106

Ennahar S, Sonomoto K, Ishizaki A (1999) Class IIa bacteriocins from lactic acid bacteria: antibacterial activity and food preservation. J Biosci Bioeng 87:705–716

Famularo G, Pieluigi M, Coccia R, Mastroiacovo P, De Simone C (2001) Microecology, bacterial vaginosis and probiotics: perspectives for bacteriotherapy. Med Hypotheses 56:421–430

Ferris DG, Litaker MS, Woodward L, Mathis D, Hendrich J (1995) Treatment of bacterial vaginosis: a comparison of oral metronidazole, metronidazole vaginal gel, and clindamycin vaginal cream. J Fam Pract 41:443–449

Fimland G, Eijsink VGH, Nissen-Meyer J (2002) Comparative studies of immunity proteins of pediocin-like bacteriocins. Microbiol 148:3661–3670

Fraise AP, Mitchell K, O'Brien SJ, Oldfield K, Wise R (1997) Methicillin-resistant *Staphylococcus aureus* (MRSA) in nursing homes in a major UK city: an anonymized point prevalence survey. Epidemiol Infect 118(1):1–5

Franciosa G, Maugliani A, Scalfaro C, Aureli P (2009) Evidence that plasmid-borne Botulinum neurotoxin Type B genes are widespread among *Clostridium botulinum* serotype B strains. PLoS One 4(3):e4829. doi:10.1371/journal.pone.0004829

Fraser A, Goldberg E, Acosta CJ, Paul M, Leibovici L (2007a) Vaccines for preventing typhoid fever. Cochrane Database Systematic Reviews, Issue 3, Art. No.: CD001261. DOI: 10.1002/14651858.CD001261.pub2

Fraser A, Paul M, Goldberg E, Acosta CJ, Leibovici L (2007b) Typhoid fever vaccines: systematic review and meta-analysis of randomised controlled trials. Vaccine 25:7848–7857

Fredenhagen A, Fendrich G, Märki F, Märki W, Gruner J, Raschdorf F, Peter HH (1990) Duramycin B and C, two new lanthionine containing antibiotics as inhibitors of phospholipase A2. Structural revision of duramycin and cinnamycin. J Antibiot (Tokyo) 43(11):1403–1412

Fredericq P (1957) Colicins. Ann Rev Microbiol 11:7–22

Froseth BR, Herman RE, McKay LL (1988) Cloning of nisin resistance determinant and replication origin on 7.6-kilobase *EcoRI* fragment of pNP40 from *Streptococcus lactis* subsp. *diacetylactis* DRC3. Appl Environ Microbiol 54(8):2136–2139

Galvin M, Hill C, Ross RP (1999) Lacticin 3147 displays activity in buffer against gram-positive bacterial pathogens which appear insensitive in standard plate assays. Lett Appl Microbiol 28:355–358

Garg S, Anderson RA, Chany CJ 2nd, Waller DP, Diao XH, Vermani K, Zaneveld LJD (2001) Properties of a new acid-buffering bioadhesive vaginal formulation (ACIDFORM). Contraception 64:67–75

Garneau S, Martin NI, Vederas JC (2002) Two-peptide bacteriocins produced by lactic acid bacteria. Biochem 84:577–592

Gasson MJ (1984) Transfer of sucrose fermenting ability, nisin resistance and nisin production into *Streptococcus lactis* 712. FEMS Microbiol Lett 21(1):7–10

Giacometti A, Cirioni O, Barchiesi F, Fortuna M, Sealise G (1999) In vitro activity of cationic peptides alone and in combination with clinically used antimicrobial agents against *Pseudomonas aeruginosa.* J Antimicrob Chemother 44:641–645

Gillor O, Kirkup BC, Riley MA (2004) Colicins and microcins: the next generation antimicrobials. Adv Appl Microbiol 54:129–146

Goldstein BP, Wei J, Greenberg K, Novick R (1998) Activity of nisin against *Streptococcus pneumoniae,* in vitro, and in a mouse infection model. J Antimicrob Chemother 42(2):277–278

Gould DJ, Chamberlain A (1995) *Staphylococcus aureus:* a review of the literature. J Clin Nurs 4:5–12

Gravesen A, Jydegaard Axelsen AM, Mendes da Silva J, Hansen TB, Knochel S (2002a) Frequency of bacteriocin resistance development and associated fitness costs in *Listeria monocytogenes.* Appl Environ Microbiol 68:756–764

Gravesen A, Ramnath M, Rechinger KB, Andersen N, Jansch L, Hechard Y, Hastings JW, Knochel S (2002b) High-level resistance to class IIa bacteriocins is associated with one general mechanism in *Listeria monocytogenes.* Microbiol 148:2361

Grether JK, Nelson KB (2000) Possible decrease in prevalence of cerebral palsy in premature infants? Lett T J Pediatr 136:133

Guggenheim M, Zbinden R, Handschin A, Gohritz A, Altintas MA, Giovanoli P (2009) Changes in bacterial isolates from burn wounds and their antibiograms: A 20-year study (1986–2005) Burns 35:553–560

Guinane CM, Cotter PD, Ross RP, Hill C (2006) Contribution of penicillin-binding protein homologs to antibiotic resistance, cell morphology, and virulence of *Listeria monocytogenes* EGDe. Antimicrob Agents Chemother 50:2824–2828

Hachem RY, Chemaly RF, Ahmar CA, Jiang Y, Boktour MR, Rjaili GA, Bodey GP, Raad II (2007) Colistin is effective in treatment of infections caused by multidrug-resistant *Pseudomonas aeruginosa* in cancer patients. Antimicrob Agents Chemother 51:1905–1911

Hackbarth CJ, Chambers HF (1989) Methicillin-resistant staphylococci: genetics and mechanisms of resistance. Antimicrob Agents Chemother 33(7):991–994

Han KS, Kim Y, Kim SH, Oh S (2007) Characterization and purification of acidocin 1B, a bacteriocin produced by *Lactobacillus acidophilus* GP1B. J Microbiol Biotechnol 17:774–783

Harbarth S, Rutschmann O, Sudre P, Pittet D (1998) Impact of methicillin resistance on the outcome of patients with bacteremia caused by *Staphylococcus aureus*. Arch Iintern Med 158:182–189

Harris LJ, Fleming HP, Klaenhammer TR (1991) Sensitivy and resistance of *Listeria monocytogenes* ATCC 19115, Scott A and UAL 500 to nisin. J Food Prot 54:836–840

Hashemi FB, Ghassemi M, Faro S, Aroutcheva A, Spear GT (2000) Induction of human immunodeficiency virus type 1 expression by anaerobes associated with bacterial vaginosis. J Infect Dis 181:1574–1580

Héchard Y, Sahl HG (2002) Mode of action of modified and unmodified bacteriocins from Gram-positive bacteria. Biochimie 84:545–557

Heunis TDJ, Botes M, Dicks LMT (2010) Encapsulation of *Lactobacillus plantarum* 423 and its bacteriocin in nanofibers. Probiotics Antimicrob Proteins 2:46–51

Hillier SL, Nugent RP, Eschenbach DA, Krohn MA, Gibbs RS, Martin DH, Cotch MF, Edelman R, Pastorek JG 2nd, Rao AV, McNellis D, Regan JA, Carey JC, Klebanoff MA (1995) Association between bacterial vaginosis and preterm delivery of a low-birth-weight infant. The Vaginal Infections and Prematurity Study Group. N Engl J Med 333:1737–1742

Hillman JD, Novák J, Sagura E, Gutierrez JA, Brooks TA, Crowley PJ, Hess M, Azizi A, Leung K, Cvitkovitch D, Bleiweis AS (1998) Genetic and biochemical analysis of mutacin 1140, a lantibiotic from *Streptococcus mutans*. Infect Immun 66:2743–2749

Hiramatsu K (1995) Molecular evolution of MRSA. Microbiol Immunol 39:531–543

Hoffner RJ, Slaven E, Perez J, Magana RN, Henderson SO (2000) Emergency department presentations of typhoid fever. J Emerg Med 19:317–321

House D, Bishop A, Parry C, Dougan G, Wain J (2001) Typhoid fever: pathogenesis and disease. Curr Opin Infect Dis 14:573–578

Howell TH, Fiorellini JP, Blackburn P, Projan SJ, de la Harpe J, Williams ARC (1993) The effect of a mouthrinse based on nisin, a bacteriocin, on developing plaque and gingivitis in beagle dogs. J Clin Periodontol 20:335–339

Imig JD (2004) ACE inhibition and bradykinin-mediated renal vascular responses: EDHF Involvement. Hypertension 43:533–535

Ingham A, Ford M, Moore RJ, Tizard M (2003) The bacteriocin piscicolin 126 retains antilisterial activity in vivo. J Antimicrob Chemother 51:1365–1371

Irvine RF (1982) How is the level of free arachidonic acid controlled in mammalian cells? Biochem J 204:3–16

Isreal D, Peek R (2001) Review article: pathogenesis of *Helicobacter pylori*-iduced gastric inflammation. Alim Pharmacol Therap 15:1271–1290

Jack RW, Tagg JR, Ray B (1995) Bacteriocins of gram-positive bacteria. Microbiol Rev 59:171–200

Jarvis B (1967) Resistance to nisin and production of nisin-inactivating enzymes by several *Bacillus* species. J Gen Microbiol 47:33–48

Jennes W, Dicks LMT, Verwoerd DJ (2000) Enterocin 012, a bacteriocin produced by *Enterococcus gallinarum*, isolated from the intestinal tract of ostrich. J Appl Microbiol 88:349–357

Jenssen H, Hamill P, Hancock REW (2006) Peptide antimicrobial agents. Clin Microbiol Rev 19:491–511

Johnson M, Carey F, McMillan RM (1983) Alternative pathways of arachidonate metabolism: prostaglandins, thromboxane and leukotrienes. In: Campbell PN, Marshall RD (eds), Essays in Biochemistry vol. 19, pp 40–141

Jung G (1991a) Lantibiotics-ribosomally synthesized biologically active polypeptides containing sulfide bridges and a, B-didehydroamino acids. Angew Chem Int Ed Engl 30:1051–1068

Jung G (1991b) Lantibiotics: a survey. In: Jung G, Sahl HG (eds) Nisin and novel lantibiotics. ESCOM Science Publishers, Leiden, The Netherlands, pp 1–34

Kadhiravan T, Wig N, Kapil A, Kabra SK, Renuka K, Misra A (2005) Clinical outcomes in typhoid fever: adverse impact of infection with nalidixic acid-resistant Salmonella typhi. BMC Infect Dis. 5 (37): doi:10.1186/1471-2334-5-37

Kalb SR, Lou J, Garcia-Rodriguez C, Geren IN, Smith TJ, Moura H, Marks JD, Smith LA, Pirkle JL, Barr (2009). Extraction and inhibition of enzymatic activity of botulinum neurotoxins/A1,/ A2, and/A3 by a panel of monoclonal anti-BoNT/A antibodies. PLoS ONE 4(4): e5355. doi:10.1371/journal.pone.0005355

Kawulka KE, Sprules T, Diaper CM, Whittal RM, McKay RT, Mercier P, Zuber P, Vederas JC (2004) Structure of subtilosin A, a cyclic antimicrobial peptide from *Bacillus subtilis* with unusual sulfur to [alpha]-carbon cross-links: formation and reduction of [alpha]-thio-[alpha]-amino acid derivatives. Biochem 43:3385–3395

Kellner R, Jung G, Horner T, Zahner H, Schnell N, Entian KD, Götz F (1988) Gallidermin: a new lanthionine-containing polypeptide antibiotic. Eur J Biochem 177:53–59

Kido Y, Hamakado T, Yoshida T, Anno M, Motoki Y, Wakamiya T, Shiba T (1983) Isolation and characterization of ancovenin, a new inhibitor of angiotensin I converting enzyme, produced by actinomycetes. J Antibiot 36:1295–1299

Kim JY (2009) Understanding the evolution of methicillin-resistant *Staphylococcus aureus*. Clin Microbiol Newsl 31:17–23

Kim TS, Hur JW, Yu MA, Cheigh CI, Kim KN, Hwang JK, Pyun YR (2003) Antagonism of *Helicobacter pylori* by bacteriocins of lactic acid bacteria. J Food Prot 66:3–12

Klaenhammer TR (1993) Genetics of bacteriocins produced by lactic acid bacteria. FEMS Microbiol Rev 12:39–85

Klaenhammer TR, Sanozky RB (1985) Conjugal transfer from *Streptococcus lactis* ME2 of plasmids encoding phage resisitance, nisin resistacne and lactrose-fermenting ability: evedince for a high-frequency conjugative plasmid responsible for abortive infection of virulent bacteriophage. J Gen Microbiol 131:1531–1541

Klug D, Lacroix D, Savoye C, Goullard L, Grandmougin D, Hennquin JL, Kacet S, Lekieffre J (1997) Systemic infection related to endocarditis on pacemaker leads clinical presentation and management. Circulation 95:2098–2107

Knoetze H, Todorov SD, Dicks LMT (2008) A class IIa peptide from *Enterococcus mundtii* inhibits bacteria associated with otitis media. Int J Antimicrob Agents 31:228–234

Kolter R, Moreno F (1992) Genetics of ribosomally synthesized peptide antibiotics. Annu Rev Microbiol 46:141–161

Kordel M, Schüller F, Sahl HG (1989) Interaction of the pore forming-peptide antibiotics Pep 5, nisin and subtilin with non-energized liposomes. FEBS Lett 244:99–102

Kraaij C, Vos WM, Siezen RJ, Kuipers OP (1999) Lantibiotics: biosynthesis, mode of action and applications. Nat Prod Rep 16:575–587

Kruszewska D, Sahl HG, Bierbaum G, Pag U, Hynes SO, Ljungh A (2004) Mersacidin eradicates methicillin-resistant *Staphylococcus aureus* (MRSA) in a mouse rhinitis model. J Antimicrob Chemother 54:648–653

Kuipers EJ, Perez-Perez GI, Meuwissen SGM, Blaser MJ (1995) *Helicobacter pylori* and atrophic gastritis: importance of the cagA status. JNCI J Natl Cancer Inst 87:1777–1780

Kusters JG, van Vliet AHM, Kuipers EJ (2006) Pathogenesis of *Helicobacter pylori* infection. Clin Microbiol Rev 19:449–490

Le Blay G, Lacroix C, Zihler A, Fliss I (2007) In vitro inhibition activity of nisin A, nisin Z, pediocin PA-1 and antibiotics against common intestinal bacteria. Lett Appl Microbiol 45:252–257

Lesseval MI, Hadjiiski G (1996) Staphylococcal infections in the Sofia Burn Centre, Bulgaria. Burns Vol. 2. pp. 279–282

Levy SB, Marshall B (2004) Antibacterial resistance worldwide: causes, challenges and responses. Nat Med Sup 10:122–129

Li J, Chikindas ML, Ludescher RD, Montville TJ (2002) Temperature- and surfactant-induced membrane modifications that alter *Listeria monocytogenes* nisin sensitivity by different mechanisms. Appl Environ Microbiol 68:5904–5910

Liang D, Hsiao BJ, Chu B (2007) Functional electrospun nanofibrous scaffolds for biomedical applications. Adv Drug Deliv Revs 59:1392–1412

Liebetrau A, Rodloff AC, Behra-Miellet J, Dubreuil L (2003) In vitro activities of a new desfluoro(6)quinolone, garenoxacin, against clinical anaerobic bacteria. Antimicrob Agents Chemother 47:3667–3671

Limbert M, Isert D, Klesel N, Markus A, Seibert G, Chatterjee S, Chatterjee DK, Jani RH, Ganguli BN (1991) Chemotherapeutic properties of mersacidin in vitro and in vivo. In: Jung G, Sahl HG (eds) Nisin and novel lantibiotics. ESCOM, Leiden, The Netherlands

Linden PK, Kusne S, Coley K, Fontes P, Kramer DJ, Paterson D (2003) Use of parenteral colistin for the treatment of serious infection due to antimicrobial-resistant *Pseudomonas aeruginosa*. Clin Infect Dis 37:e154–e160

Ling J, Lo N, Ho Y, Kam K, Ma C, Wong S, Cheng A (1996) Emerging resistance in *Salmonella enterica* serotype Typhi in Hong Kong. Int J Antimicrob Agents 7:161–166

Ling ZD, Chang Q, Lipton JW, Tong CW, Landers TM, Carvey PM (2004) Combined toxicity of prenatal bacterial endotoxin exposing and postnatal 6-hydroxydopamine in the adult rat midbrain. Neuroscience 124:619–628

Lowy FD (1998) *Staphylococcus aureus* infections. N Engl J Med 339:520–532

Lubbe MM, Botha PL, Chalkley LJ (1999) Comparative activity of eighteen antimicrobial agents against anaerobic bacteria isolated in South Africa. Eur J Clin Microbiol Infec Dis 18:46–54

Maggi L, Mastromarino P, Macchia S, Brigidi P, Pirovano F, Matteuzzi D, Conte U (2000) Technological and biological evaluation of tablets containing different strains of lactobacilli for vaginal administration. Eur J Pharm and Biopharm 50:389–395

Maisnier-Patin S, Richard J (1996) Cell wall changes in nisin-resistant variants of *Listeria* innocua grown in the presence of high nisin concentrations. FEMS Microbiol Lett 140:29–35

Maisnier-Patin S, Tatini SR, Richard J (1995) Combined effect of nisin and moderate heat on destruction of *Listeria monocytogenes* in milk. Lait 75:81–91

Malheiros PS, Daroit DJ, da Silveira NP, Brandelli A (2010) Effect of nanovesicle-encapsulated nisin on growth of *Listeria monocytogens* in milk. Food Microbiol 27:175–178

Mantovani HC, Russell JB (2001) Nisin resistance of *Streptococcus bovis*. Appl Environ Microbiol 67:808–813

Marcos B, Aymerich T, Monfort JM, Garriga M (2007) Use of antimicrobial biodegradable packaging to control *Listeria monocytogenes* during storage of cooked ham. Int J Food Microbiol 120:152–158

Marcus JP, Green JL, Goulter KC, Manners JM (1999) A family of antimicrobial peptides is produced by processing of a 7S globulin protein in *Macadamia integrifolia* kernels. J Plant 19:699–710

Marki F, Hanni E, Fredenhagen A, van Oostrum J (1991) Mode of action of the Lanthionine-containing peptide antibiotics duramycin, duramycin B and C, and cinnamycin as indirect inhibitors of phospholipase A2. Biochem Pharmacol 42:2027–2035

Martin HL Jr, Richardson BA, Nyange PM, Lavreys L, Hillier SL, Chohan B, Mandaliya K, Ndinya-Achola JO, Bwayo J, Kreiss J (1999) Vaginal lactobacilli, microbial flora, and risk of human immunodeficiency virus type 1 and sexually transmitted disease acquisition. J Infect Dis 180:1863–1868

Martin NI, Sprules T, Carpenter MR, Cotter PD, Hill C, Ross RP, Vederas JC (2004) Structural characterization of lacticin 3147, a two-peptide lantibiotic with synergistic activity. Biochem 43:3049–3056

Mascellino MT, Iona E, Iegri F, Catania S, Trinchieri V, Oliva P, Amenta L, Reverberi L, Sorice F (1991) Evaluation of vaginal microflora in patients infected with HIV. Microbiologica 14:343–349

Masuoka JC, Hazen K (1999) Differences in the acid-labile component of *Candida albicans* mannan from hydrophobic and hydrophilic yeast. Glycobiology 9:1281–1286

Matsumoto-Nakano M, Kuramitsu HK (2006) The role of bacteriocin immunity proteins in the antimicrobial sensitivity of *Streptococcus mutans*. J Bacteriol 188:8092–8102

Mazzotta AS, Montville TJ (1997) Nisin induces changes in membrane fatty acid composition of *Listeria monocytogenes* nisin-resistant strains at 10°C and 30°C. J Appl Bacteriol 82:32–38

McAuliffe O, Ross RP, Hill C (2001) Lantibiotics: structure, biosynthesis and mode of action. FEMS Microbiol Rev 25:285–308

McConville P (1995) SmithKline Beecham Plc. International Patent Application WO 97 : 06772

McGroarty JA, Reid G (1988) Detection of a lactobacillus substance which inhibits *Escherichia coli*. Can J Microbiol 34:974–978

McKay LL, Baldwin KA (1984) Conjugative 40-megadalton plasmid in *Streptococcus lactis* subsp. *diacetylactis* DRC3 is associated with resistance to nisin and bacteriophage. Appl Environ Microbiol 47:68–74

McLean NW, Rosenstein IJ (2000) Characterization and selection of a *Lactobacillus* species to recolonise the vagina of women with recurrent bacterial vaginosis. J Med Microbiol 49:543–552

Michiels J, Dirix G, Vanderleyden J, Xi C (2001) Processing and export of peptide pheromones and bacteriocins in Gram-negative bacteria. Trends in Microbiol 9:164–168

Ming X, Daeschel MA (1993) Nisin resistance of foodborne bacteria and the specific resistance response of *Listeria monocytogenes* Scott A. J Food Prot 56:944–948

Ming X, Daeschel MA (1995) Correlation of cellular phospholipid content with nisin resistance of *Listeria monocytogenes* Scott A. J Food Prot 58:416–420

Moll G, Hildeng-Hauge H, Nissen-Meyer J, Nes IF, Konings WL, Driessen AJM (1998) Mechanistic properties of the two-component bacteriocin Lactococcin G. J Bacteriol 180:(1)96–99

Moll GN, Konings WN, Driessen AJM (1999) Bacteriocins: mechanism of membrane insertion and pore formation. Antonie van Leeuwenhoek 76:185–198

Morency H, Mota-Meira M, LaPointe G, Lacroix C, Lavoie MC (2001) Comparison of the activity spectra against pathogens of bacterial strains producing a mutacin or a lantibiotic. Can J Microbiol 47:322–331

Mota-Meira M, LaPointe G, Lacroix C, Lavoie MC (2000) MICs of mutacin B-Ny266, nisin A, vancomycin, and oxacillin against bacterial pathogens. Antimicrob Agents Chemother 44:24–29

Mota-Meira M, Morency H, Lavoie MC (2005) In vivo activity of mutacin B-Ny266. J Antimicrob Chemother 56:869–871

Naghmouchi K, Drider D, Kheadr E, Lacroix C, Prévost H, Fliss I (2006) Multiple characterizations of *Listeria monocytogenes* sensitive and insensitive variants to divergicin M35, a new pediocin-like bacteriocin. J App Microbiol 10:29–39

Naghmouchi K, Kheadr E, Lacroix C, Fliss I (2007) Class I/Class IIa bacteriocin crossresistance phenomenon in *Listeria monocytogenes*. J Microbiol 24:718–727

Nair LS, Laurencin CT (2007) Biodegradable polymers as biomaterials. Prog Polymer Sci 32:762–798

Nes IF, Holo H (2000) Class II antimicrobial peptides from lactic acid bacteria. Biopolymers, Peptide Sci 55:50–61

Nes IF, Diep DB, Håvarstein LS, Brurberg MB, Eijsink V, Holo H (1996) Biosynthesis of bacteriocins in lactic acid bacteria. Antonie van Leeuwenhoek 70:113–128

Ness RB, Kip KE, Hillier SL, Soper DE, Stamm CA, Sweet RL, Rice P, Richter HE (2005) A cluster analysis of bacterial vaginosis-associated microflora and pelvic inflammatory disease. Am J Epidemiol 162:585–590

Neu HC (1992) The crisis in antibiotic resistance. Science 257:1064–1073

Nikolaithcouk N, Andersch B, Falsen E, Strömbeck L, Mattsby-Baltzer I (2008) The lower genital tract microbiota in relation to cytokine-, SLPI- and endotoxin levels: application of checker-board DNA-DNA hybridization (CDH). APMIS 116:263–277

Nissen-Meyer J, Nes IF (1997) Ribosomally synthesized antimicrobial peptides: their function, structure, biogenesis, and mechanism of action. Arch Microbiol 167:67–77

Niu WW, Neu HC (1991) Activity of merscidin, a novel peptide, compared with that of vancomycin, tecoplanin, and datomycin. Antimicrob Agents Chemother 35:988–1000

Noerlis Y, Ray B (1994) Factors influencing immunity and resistance of *Pediococcus acidilactici* to the bacteriocin, pediocin AcH. Lett Appl Microbiol 18:138–143

Nomoto K (2005) Prevention of infections by probiotics. J Biosci Bioeng 100:583–592

Nomura M (1967) Colicins and related bacteriocins. Annu Rev Microbiol 21:257–284

Ocaña VS, de Ruiz P, Holgado AA, Nader-Macías ME (1999) Selection of vaginal H2O2-generating *Lactobacillus* species for probiotic use. Curr Microbiol 38:279–284

Okkers D, Dicks LMT, Silvester M, Joubert JJ, Odendaal HJ (1999) Characterization of pentocin TV35b, a bacteriocin-like peptide isolated from *Lactobacillus pentosus* with a fungistatic effect on *Candida albicans*. J Appl Microbiol 87:726–734

Oppegård C, Rogne P, Emanuelsen L, Kristiansen PE, Fimland G, Nissen-Meyer J (2007) The two-peptide class II bacteriocins:structure, production, and mode of action. J Mol Microbiol Biotechnol 13:210–219

Osih R, McGregor J, Rich S, Moore A, Furuno J, Perencevich E, Harris A (2007) Impact of empiric antibiotic therapy on outcomes in patients with *Pseudomonas aeruginosa* bacteremia. Antimicrob Agents Chemother 51:839–844

Paavonen J, Mangioni C, Martin MA, Wajszczuk CP (2000) Vaginal clindamycin and oral metronidazole for bacterial vaginosis: a randomized trial. Obstet Gynecol 96:256–260

Pag U, Sahl HG (2002) Multiple activities in lantibiotics-models for the design of novel antibiotics? Curr Pharm Des 8:815–833

Page CP, Archer CB, Paul W, Morley J (1984) Paf-acether: a mediator of inflammation and asthma. Trends Pharm Sci 5:239–241

Peek R, Miller G, Tham K, Perez-Perez G, Zhao X, Atherton J, Blaser M (1995) Heightened inflammatory response and cytokine expression in vivo to cagA+ *Helicobacter pylori* strains. Lab Invest 73:760–770

Perera N, Geary C, Wiselka M, Rajakumar K, Andrewswann K (2007) Mixed Salmonella infection: Case report and review of the literature. J Travel Med 14:134–135

Pieterse R, Todorov S, Dicks LMT (2008) Bacteriocin ST91KM, produced by *Streptococcus gallolyticus* subsp. *macedonicus* ST91KM, is a narrow-spectrum peptide active against bacteria associated with mastitis in dairy cattle. Can J Microbiol 54:525–531

Pieterse R, Todorov S, Dicks LMT (2010) Mode of action and in vitro susceptibility of mastitis pathogens to macedocin ST91KM and preparation of a teat seal containing the bacteriocins. Braz J Microbiol 41:133–145

Pini A, Giuliani A, Falciani C, Runci Y, Ricci C, Lelli B (2005) Antimicrobial activity of novel dendrimeric peptides obtained by phage display selection and rational modification. Antimicrob Agents Chemother 49:2665–2672

Pons AM, Lanneluc I, Cottenceau G, Sable S (2002) New developments in non-post translationally modified microcins. Biochimie 84:531–537

Portnoy D, Jacks P, Hinrichs D (1988) Role of hemolysin for the intracellular growth of *Listeria monocytogenes*. J Exper Med 167:1459–1471

Portnoy DA, Auerbuch V, Glomski IJ (2002) The cell biology of *Listeria monocytogenes* infection: the intersection of bacterial pathogenesis and cell-mediated immunity. J Cell Biol 158:409–414

Pugsley AP, Oudega B (1987) Methods for studying colicins and their plasmids. In: Hardy KG (ed) Plasmids: a practical approach. IRL Press, Oxford, Washington DC

Pybus V, Onderdonk AB (1999) Microbial interactions in the vaginal ecosystem, with emphasis on the pathogenesis of bacterial vaginosis. Microbes Infect 1:285–292

Ramaswamy V, Cresence V, Rejitha J, Lekshmi M, Dharsana K, Prasad, Vijila S (2007) *Listeria*-review of epidemiology and pathogenesis. J Microbiol Imm Infect 40:4–13
Ragazzo-Sanchez JA, Sanchez-Prado L, Gutie´rrez-Martı´nez P, Luna-Solano G, Gomez-Gil B, Calderon-Santoyo M (2009) Inhibition of *Salmonella* spp. isolated from mango using bacteriocin-like produced by lactobacilli. CyTA – J Food 7:181–187
Ramnath M, Beukes M, Tamura K, Hastings JW (2000) Absence of a putative mannose-specific phosphotransferase system enzyme IIAB component in a leucocin a-resistant strain of *Listeria monocytogenes*, as shown by two-dimensional sodium dodecyl sulfate-polyacrylamide gel electrophoresis. Appl Environ Microbiol 66:3098–3101
Rasch M, Knøchel S (1998) Variations in tolerance of *Listeria monocytogenes* to nisin, pediocin PA-1 and bavaricin A. Lett Appl Microbiol 27:275–278
Reddy KV, Aranha C, Gupta SM, Yedery RD (2004) Evaluation of antimicrobial peptide nisin as a safe vaginal contraceptive agent in rabbits: in vitro and in vivo studies. Reproduction 128:117–126
Reeves P (1965) The Bacteriocins. Bacteriol Rev 29:24–45
Reid G, Bruce AW (2001) Selection of *Lactobacillus* for urogenital probiotic applications. J Infect Dis 183:77–80
Reid G, Bruce AW, Taylor M (1995) Instillation of *Lactobacillus* and stimulation of indigenous organisms to prevent recurrence of urinary tract infections. Microecol Ther 23:32–45
Reid G, Burton J (2002) Use of lactobacillus to prevent infection by pathogenic bacteria. Microbes Infect 4:319–324
Reid G, Cook RL, Bruce AW (1987) Examination of strains of lactobacilli for properties which may influence bacterial interference in the urinary tract. J Urol 138:330–335
Rekhif N, Atrih A, Lefebvre G (1994) Selection and properties of spontaneous mutants of *Listeria monocytogenes* ATCC 15313 resistant to different bacteriocins produced by lactic acid bacteria strains. Curr Microbiol 28:237–241
Riffon R, Sayasith K, Khalil H, Dubreuil P, Drolet M, Lagace J (2001) Development of a rapid and sensitive test for identification of major pathogens in bovine mastitis by PCR. J Clin Microbiol 39:2584–2589
Rihakova J, Cappelier JM, Hue I, Demnerova K, Fédérighi M, Prévost H, Drider D (2010) In vivo activities of recombinant divercin V41 and its structural variants against *Listeria monocytogenes*. Antimicrob Agents Chemother 54:563–564
Riley MA, Wertz JE (2002) Bacteriocins: Evolution, ecology, and application. Ann Rev Microbiol 56:117–137
Roland PS, Stroman DW (2002) Microbiology of acute otitis externa. Laryngoscope 112:1166–1177
Ross RP, Galvin MG, McAuliffe O, Morgan SM, Ryan MP, Twomey DP, Meaney WJ, Hill C (1999) Developing applications for lactococcal bacteriocins. Antonie van Leeuwenhoek 76:337–346
Ruhr E, Sahl HG (1985) Mode of action of the peptide antibiotic nisin and influence on the membrane potential of whole cells and on cytoplasmic and artificial membrane vesicles. Antimicrob Agents Chemother 27:841–845
Ryan MP, Flynn J, Hill C, Ross RP, Meaney WJ (1999) The natural food grade inhibitor, lacticin 3147, reduced the incidence of mastitis after experimental challenge with *Streptococcus dysgalactiae* in nonlactating dairy cows. J Dairy Sci 82:2625–2631
Ryding M, Konradsson K, Kalm O, Prellner K (1997) Sequelae of recurrent acute otitis media: ten-year follow-up of a prospectively studied cohort of children. Acta Paediatrica 86:1208–1213
Sahl HG (1991) Pore formation in bacterial membranes by cationic lantibiotics. In: Jung G, Sahl HG (eds) Nisin and novel lantibiotics. Escom, Leiden, The Netherlands
Sahl HG, Bierbaum G (1998) Lantibiotics: biosynthesis and biological activities of uniquely modified peptides from gram-positive bacteria. Ann Rev Microbiol 52:41–79
Sahl HG, Jack RW, Bierbaum G (1995) Biosynthesis and biological activities of lantibiotics with unique post-translational modifications. Eur J Biochem 230:827–853
Salama NR, Otto G, Tompkins L, Falkow S (2001) Vacuolating cytotoxin of *Helicobacter pylori* plays a role during colonization in a mouse model of infection. Infect Immun 69:730–736

Salmaso S, Elvassore N, Bertucco A, Lante A, Caliceti P (2004) Nisin-loaded poly-L-lactide nano-particles produced by CO_2 anti-solvent precipitation for sustained antimicrobial activity. Int J Pharm 287:163–173.

Sass P, Jansen A, Szekat C, Sass V, Sahl HG, Bierbaum G (2008) BMC Microbiol 2008, 8:186doi:10.1186/1471-2180-8-186

Scannell AGM, Hill C, Ross RP, Marx S, Hartmeier W, Arendt EK (2000) Development of bioactive food packaging materials using immobilised bacteriocins Lacticin 3147 and Nisaplin®. Int J Food Microbiol 60:241–249

Schwebke JR (1997) Bacterial vaginosis-more questions than answers. Genitourinary Med 73:333–334

Schwebke JR (2003) Gynecologic consequences of bacterial vaginosis. Obstet Gynecol Clin North Am 30:685–694

Sears PM, Smith BS, Stewart WK, Gonzalez RN (1992) Evaluation of a nisin based germicidal formulation on teat skin of live cows. J Dairy Sci 75:3185–3190

Segal N, Leibovitz E, Dagan R, Lieberman A (2005) Acute otitis media-diagnosis and treatment in the era of antibiotic resistant organisms: updated clinical practice guidelines. Int J Pediatr Otorhinolaryngol 69:1311–1319

Severina E, Severin A, Tomasz A (1998) Antibacterial efficacy of nisin against multidrug-resistant Gram-positive pathogens. J AntimicrobChemother 41:341–347

Severinov K, Semenova E, Kazakov A, Teymur Kazakov T, Gelfand MS (2007) Low-molecular-weight post-transcriptionally modified microcins. Mol Microbiol 65:1380–1394

Sewankambo N, Gray RH, Wawer MJ (1997) HIV-1 infection associated with abdominal vaginal flora morphology and bacterial vaginosis. Lancet 350:530–531

Seyler TM, Smith BP, Marker DR, Ma J, Shen J, Smith TL, Mont MA, Kolaski K, Koman LA (2008) Botulinum neurotoxin as a therapeutic modality in orthopaedic surgery: More than twenty years of experience. J Bone Joint Surg 90:133–145

Sha BE, Zariffard MR, Wang QJ, Chen HY, Bremer J, Cohen MH, Spear GT (2005) Female genital-tract HIV load correlates inversely with *Lactobacillus* species but positively with bacterial vaginosis and Mycoplasma hominis. J Infect Dis 91:25–32

Shiba T, Wakamiya T, Fukase K, Ueki Y, Teshima T, Nishikawa M (1991) Structure of the lanthionine peptides nisin, ancovenin and lanthiopeptin. In: Jung G, Sahl HG (eds) Nisin and novel lantibiotics. Escom, Leiden, The Netherlands

Shukla H, Sharma S (2005) *Clostridium botulinum*: a bug with beauty and weapon. Crit Revs Microbiol 31:11–18

Silkin L, Hamza S, Kaufman S, Cobb SL, Vederas JC (2008) Spermicidal bacteriocins: lacticin 3147 and subtilosin A. Bioo Medic Chem Lett 18:3103–3106

Simoes J, Aroutcheva A, Shott S, Faro S (2001) Effect of metronidazole on the growth of vaginal lactobacilli in vitro. Infect Dis Obstet Gynecol 9:41–46

Skeggs LT, Kahn JR, Shumway NP (1956) The preparation and function of the hypertension-converting enzume. J Exp Med 103:295–299

Smarda J, Smajs D (1998) Colicins- exocellular lethal proteins of *Escherichia coli.* Folia microbiologica 43:563–582

Sobel J, Peipert JF, McGregor JA, Livengood C, Martin M, Robbins J (2001) Efficacy of clindamycin vaginal ovule (3-day treatment) in bacterial vaginosis. Infect Dis Obstet Gynecol 9:9–17

Soltys J, Quinn MT (1999) Selective recruitment of T-cell subsets to the udder during staphylococcal and streptococcal mastitis: analysis of lymphocyte subsets and adhesion molecule expression. Infect Immun 67:6293–6302

Song H-J, Richard J (1997) Antilisterial activity of three bacteriocins used at sub minimal inhibitory concentrations and cross-resistance of the survivors. Int J Food Microbiol 36:155–161

Sordelli DO, Buzzola FR, Gomez MI, Steele-Moore L, Berg D, Gentilini E, Catalano M, Reitz AJ, Tollersrud T, Denamiel G, Jeric P, Lee JC (2000) Capsule expression by bovine isolates of *Staphylococcus aureus* from Argentina: genetic and epidemiologic analyses. J Clin Microbiol 38:846–850

Sosunov V, Mischenko V, Eruslanov B, Svetoch E, Shakina Y, Stern N, Majorov K, Sorokoumova G, Selishcheva A, Apt A (2007) Antymicobacterial activity of bacteriocins and their complexes with liposomes. J Antimicrob Chemother 59:919–925

Sukhan A (2000) The invasion-associated type III secretion system of *Salmonella typhimurium*: common and unique features. Cell Mol Life Sci 57:1033–1049

Sutyak KE, Wirawan RE, Aroutcheva AA, Chikindas ML (2008a) Isolation of the *Bacillus subtilis* antimicrobial peptide subtilosin from the dairy product-derived *Bacillus amyloliquefaciens*. J Appl Microbiol 104:1067–1074

Sutyak KE, Anderson RA, Dover SE, Feathergill KA, Aroutcheva AA, Faro S, Chikindas ML (2008b) Spermicidal activity of the safe natural antimicrobial peptide subtilosin. Infect Diseases Obstet Gynecol Article ID 540758, 6 pages

Tagg JR (2004) Prevention of streptococcal pharyngitis by anti-*Streptococcus pyogenes* bacteriocin-like inhibitory substances (BLIS) produced by *Streptococcus salivarius*. Indian J Med Res 119:13–16

Taha TE, Hoover DR, Dallabetta GA, Kumwenda NI, Mtimavalye LA, Yang LP, Liomba GN, Broadhead RL, Chiphangwi JD, Miotti PG (1998) Bacterial vaginosis and disturbances of vaginal flora: association with increased acquisition of HIV. AIDS 12:1699–1706

Tam PJ, Lu YA, Yang JL (2002) Antimicrobial dendrimeric peptides. Eur J Biochem 269:923–932

Taylor GI (1969) Electrically driven jets. Proc Royal Soc Lond A 313:453–475

Taylor JI, Hirsch A, Mattick ATR (1949) The treatment of bovine streptococcal and staphylococcal mastitis with nisin. Vet Rec 61:197–198

Taylor GD, Kibsey P, Kirkland T, Burroughs E, Tredget E (1992) Predominance of staphylococcal organisms in infections occurring in a burns intensive care unit. Burns 18:332–335

Teele DW, Klein JO, Chase C, Menyuk P, Rosner BA (1990) Otitis media in infancy and intellectual ability, school achievement, speech, and language at age 7 years. Greater Boston Otitis Media Study Group. J Infect Dis 162:685–694

Thennarasu S, Lee DK, Poon A, Kawulka KE, Vederas JC, Ramamoorthy A (2005) Membrane permeabilization, orientation, and antimicrobial mechanism of subtilosin A. Chem Phys Lipids 137:38–51

Twomey DP, Wheelock AI, Flynn J, Meaney WJ, Hill C, Ross RP (2000) Protection against *Staphylococcus aureus* mastitis in dairy cows using a bismuth-based teat seal containing the bacteriocin, lacticin 3147. J Dairy Sci 83:1981–1988

Tyopponen S, Petaja E, Mattila-Sandholm T (2003) Bioprotectives and Probiotics for dry sausages. Int J Food Microbiol 83:233–244

Uehara S, Monden K, Nomoto K, Seno Y, Kariyama R, Kumon H (2006) A pilot study evaluating the safety and effectiveness of *Lactobacillus* vaginal suppositories in patients with recurrent urinary tract infection. Int J Antimicrob Agents 28:S30–S34

Urakubo A, Jarskog LF, Lieberman JA, Gilmore JH (2001) Prenatal exposure to maternal infection alters cytokine expression in the placenta, amniotic fluid, and fetal brain. Schizophr Res 47:27–36

Vadyvaloo V, Arous S, Gravesen A, Héchard Y, Chauhan-Haubrock R, Hastings JW, Rautenbach M (2004) Cell-surface alterations in class IIa bacteriocin-resistant *Listeria monocytogenes* strains. Microbiology 150:3025–3033

Vadyvaloo V, Hastings JW, van der Merwe MJ, Rautenbach M (2002) Membranes of class IIa bacteriocin-resistant *Listeria monocytogenes* cells contain increased levels of desaturated and short-acyl-chain phosphatidylglycerols. Appl Environ Microbiol 68:5223–5230

Valore EV, Wiley DJ, Ganz T (2006) Reversible deficiency of antimicrobial polypeptides in bacterial vaginosis. Infect Immun 74:5693–5702

van den Hooven HW, Spronk CAEM, van de Kamp M, Konings RNH, Hilbers CW, van de Ven FJM (1996) Surface location and orientation of the lantibiotic nisin bound to membrane-mimicking micelles of dodecylphosphocholine and of sodium dodecylsulphate. Eur J Biochem 235:394–403 van Reenen CA, Dicks LMT, Chikindas ML (1998) Isolation, purification and partial characterization of plantaricin 423, a bacteriocin produced by *Lactobacillus plantarum*. J Appl Microbiol 84:1131–1137

Velraeds MC, van der Belt B, van der Mei HC, Reid G, Busscher HJ (1998) Interference in initial adhesion of uropathogenic bacteria and yeasts silicone rubber by a *Lactobacillus acidophilus* biosurfactant. J Med Microbiol 49:790–794

Venema K, Kok J, Marugg JD, Toonen MY, Ledeboer AM, Venema G, Chikindas ML (1995) Functional analysis of the pediocin operon of *Pediococcus acidilactici* PAC1.0: PedB is the immunity protein and PedD is the precursor processing enzyme. Mol Microbiol 17:515–522

Verheul A, Russell NJ, van'T Hof R, Rombouts FM, Abee T (1997) Modifications of membrane phospholipid composition in nisin-resistant *Listeria monocytogenes* Scott A. Appl Environ Microbiol 63(9):3451–3457

Wiedemann I, Breukink E, van Kraaij C, Kuipers OP, Bierbaum G, de Kruijff B, Sahl HG (2001) Specific binding of nisin to the peptidoglycan precursor lipid II combines pore formation and inhibition of cell wall biosynthesis for potent antibiotic activity. J Biol Chem 276: 1772–1779

World Health Organization (2009) "Epidemiology". Global tuberculosis control: epidemiology, strategy, financing. pp 1–341

Wright A, Hawkins CH, Änggård EE, Harper DR (2009) A controlled clinical trial of a therapeutic bacteriophage preparation in chronic otitis due to antibiotic-resistant *Pseudomonas aeruginosa*; a preliminary report of efficacy. Clin Otolaryngol 34:349–357

Xie J, Zhang R, Shang C, Guo Y (2009) Isolation and characterization of a bacteriocin produced by an isolated *Bacillus subtilis* LFB112 that exhibits antimicrobial activity against domestic animal pathogens. Afr J Biotechnol 8:5611–5619

Yarin AL, Koombhongse S, Reneker DH (2001) Taylor cone and jetting from liquid droplets in electrospinning of nanofibers. J Appl Phys 90:4836–4846

Yildiz O, Aygen B, Esel D, Kayabas U, Alp E, Sumerkan B, Doganay M (2007) Sepsis and meningitis due to *Listeria monocytogenes*. Yonsei Med J 48:433–439

Zhang R, Xu X, Chen T, Li L, Rao P (2000) An assay for angiotensin-converting enzyme using capillary zone electrophoresis. Anal Biochem 280:286–290

Zipser RD, Laffi G (1985) Prostaglandins, thromboxanes and leukotrienes in clinical medicine. West J Med 143:485–497

Chapter 20
Perspectives and Peptides of the Next Generation

Kim A. Brogden

Abstract Shortly after their discovery, antimicrobial peptides from prokaryotes and eukaryotes were recognized as the next potential generation of pharmaceuticals to treat antibiotic-resistant bacterial infections and septic shock, to preserve food, or to sanitize surfaces. Initial research focused on identifying the spectrum of antimicrobial agents, determining the range of antimicrobial activities against bacterial, fungal, and viral pathogens, and assessing the antimicrobial activity of synthetic peptides versus their natural counterparts. Subsequent research then focused on the mechanisms of antimicrobial peptide activity in model membrane systems not only to identify the mechanisms of antimicrobial peptide activity in microorganisms but also to discern differences in cytotoxicity for prokaryotic and eukaryotic cells. Recent, contemporary work now focuses on current and future efforts to construct hybrid peptides, peptide congeners, stabilized peptides, peptide conjugates, and immobilized peptides for unique and specific applications to control the growth of microorganisms in vitro and in vivo.

Introduction

The study of antimicrobial factors of normal tissues and fluids has been an ongoing endeavor since the 1880s. An excellent review of this early literature was written by Robert C. Skarnes and Dennis W. Watson in 1957 (Skarnes and Watson 1957). Among the antimicrobial substances found in biological secretions, blood, leukocytes, and lymphatic tissues are basic proteins and "tissue basic polypeptides." Descriptions of these peptides in the 1950s, including their amino-acid compositions with high contents of lysine and arginine, their pIs of 10-11, their

K.A. Brogden (✉)
Department of Periodontics and Dows Institute for Dental Research, N447 DSB,
College of Dentistry, The University of Iowa, 801 Newton Road, Iowa City, IA 52242, USA
e-mail: kim-brogden@uiowa.edu

D. Drider and S. Rebuffat (eds.), *Prokaryotic Antimicrobial Peptides: From Genes to Applications*, DOI 10.1007/978-1-4419-7692-5_20,

proposed mechanisms of binding via electrostatic interactions with negatively charged surface constituents of bacteria, and their proposed alterations of bacterial membrane integrity, all suggest that these peptides are among the antimicrobial peptides described in current reviews today. Shortly thereafter, antimicrobial substances were characterized in phagocytic granule extracts by Hirsch (Hirsch 1956) and shown to be correlated with the presence of low molecular weight, highly cationic compounds in granule mixtures (Zeya and Spitznagel 1963; Friedberg et al. 1970; Friedberg and Shilo 1970) including bactericidal/permeability-increasing protein (Weiss et al. 1975). The field expanded further when Hans Boman, Michael Zasloff, and Robert Lehrer independently isolated and purified insect cecropins, amphibian magainins, and mammalian defensins, respectively (Steiner et al. 1981; Zasloff 1987; Ganz et al. 1990). To date, antimicrobial peptides have been reported throughout the Monera (e.g., Eubacteria), Protista (e.g., protozoans and algae), Fungi (yeasts), Plantae (plants), and Animalia (e.g., insects, fish, amphibians, reptiles, birds, and mammals) kingdoms.

Bacteriocins also have a rich history leading to the discovery of extensive families of peptides. The history, current knowledge, and future research on bacteriocins, colicins, and microcins are reviewed in Chapter 1 by Ingolf Nes and Margaret Riley.

Almost from the start, antimicrobial peptides from prokaryotes and eukaryotes were recognized as the next potential generation of pharmaceuticals to treat antibiotic-resistant bacterial infections and septic shock, to preserve food, or to sanitize surfaces. Studies focused on identifying the spectrum of antimicrobial agents in nature; determining the range of antimicrobial activities against bacteria, fungi, and viruses; assessing the antimicrobial activity of synthetic peptides versus their natural counterparts; identifying the mechanisms of antimicrobial peptide activity in model membrane systems; identifying the mechanisms of antimicrobial peptide activity in intact microorganisms; and assessing the cytotoxicity of antimicrobial peptides to eukaryotic cells and erythrocytes. Recent contemporary studies now utilize all this information to design and engineer antimicrobial peptides for a variety of unique and specific applications.

In this chapter, I briefly review the major classes and types of antimicrobial peptides focusing on the parameters that are important for their structure and activity. I then identify the specific antimicrobial events that will serve as targets of future peptide design. Finally, I focus on current and future efforts to construct new antimicrobial peptide analogs or congeners for unique and specific applications. Many of these concepts have also been used to develop other biologically active peptides with enhanced therapeutic potentials (Kluskens et al. 2009; Leupold et al. 2009).

Antimicrobial Peptide Characteristics

In the preceding chapters of this book, the authors provided comprehensive overviews on the history of antimicrobial peptides; classification of prokaryotic antimicrobial peptides; the isolation, purification, and production of antimicrobial peptides; and the biology of prokaryotic antimicrobial peptides. An extensive description and a list of these peptides as well as their mechanisms of action are outside the scope of this chapter. Only a brief description of the families, with a few pertinent examples, is included. Instead, the reader is urged to see the many recent advances described earlier in this book. However, it is worth mentioning that many peptides are tracked in Antimicrobial Sequence Databases (Tossi 2005). These include http://www.bbcm.units.it/~tossi/pag1.htm, http://aps.unmc.edu/AP/main.html, http://defensins.bii.a-star.edu.sg/, and http://faculty.ist.unomaha.edu/chen/rapd/index.php. It is also worth mentioning that other eukaryotic peptides also have antimicrobial activity including some cytokines and chemokines (Cole et al. 2001; Yang et al. 2003), some neuropeptides and peptide hormones (Brogden et al. 2005), and domains in some large proteins, like ovalbumin for example (Pellegrini et al. 2004).

Antimicrobial peptides have a number of unique characteristics that are related to mechanisms of antimicrobial action (Yeaman and Yount 2003; Yount and Yeaman 2004). Generally, antimicrobial peptides are small and range in size from 6 to 100 amino-acid residues. They are often highly charged. Anionic peptides are rich in aspartic and glutamic acids, while cationic peptides are rich in arginine and lysine. Many peptides are amphipathic with hydrophilic and hydrophobic amino-acid residues aligned along the opposite sides of a helical molecule. Some antimicrobial peptides, like the lantibiotics, contain unique amino acids (Willey and van der Donk 2007).

Antimicrobial peptides from prokaryotes and eukaryotes are loosely grouped into families based on their unique amino-acid compositions and secondary structures: first proposed by Hans Boman and later modified by Robert Hancock (Boman 1995; Hancock 1997; Jenssen et al. 2006). One family contains the amphiphilic peptides with two to four beta strands stabilized with 2–3 disulfide bonds. Beta strand peptides include the defensins, thionins, insect defensins, protegrins, and polyphemusin. Another family contains short, amphipathic peptides with alpha helices that lack cysteine residues, and sometimes have a hinge or "kink" in the middle. Alpha helical peptides include magainins, cecropins, melittin, and buforin. Although similar in secondary structure, these peptides have different mechanisms of antimicrobial activity. Melittin, for example, penetrates prokaryotic and eukaryotic membranes, whereas buforin translocates into cells and acts on macromolecular synthesis (Jenssen et al. 2006). A third family of peptides contains peptide loops with a covalently cyclic structure often with a disulfide bond. Peptides structurally arranged in loops include thanatin, bactenecin, gramicidin S, and polymyxin. A fourth family of peptides has extended, coil structures. Family members can contain a high content of select amino acids and include indolicidin and some bactenecins from cattle.

Specific Antimicrobial Events to Target in Design

A specific sequence of events must occur for a peptide to have antimicrobial activity (Matsuzaki et al. 1995). All of these events, outlined below, have proven to be ideal targets to improve the design and antimicrobial activity of peptides.

Antimicrobial peptides must first be attracted to bacterial surfaces likely via an electrostatic interaction among anionic or cationic peptides and the surface structures on microorganisms. Antimicrobial peptides are also attracted to surface lipopolysaccharide (LPS) on gram-negative bacteria (Scott et al. 1999b) or lipoteichoic acid on gram-positive bacteria (Scott et al. 1999a). Hydrophobic interactions among antimicrobial peptides and bacterial membranes may also be a factor.

A popular approach is to increase the number of cationic charges or alter the hydrophobicity of the peptide. Travis and colleagues found a positive correlation between antimicrobial activity and the net positive charge of the peptide but no simple correlation between the calculated average hydrophobicity, amphipathicity of the peptides, or their calculated hydrophobic moment and antibacterial activity (Travis et al. 2000). Another approach is to design a peptide with "carrier" domains or conjugated to larger "carrier" ligands, receptors, or antibodies that are capable of delivering a "killing" domain to the surfaces of specific pathogenic bacteria (Franzman et al. 2009).

Once close to the microbial surface, peptides must traverse capsular polysaccharides and other extracellular matrices and overcome proteolytic degradation before they can interact with the LPS in the outer leaflet of the outer membrane of gram-negative bacteria or the teichoic acid and lipoteichoic acid areas near the outer leaflet of the cytoplasmic membrane in gram-positive bacteria (Hancock and Rozek 2002; Kuo et al. 2007). Once at the cytoplasmic membrane, peptides are free to embed into the lipid bilayer (Huang 2000; Chen et al. 2003). Once peptides are attached, they begin to insert into the membrane in a structured process. "Barrel-stave" (Ehrenstein and Lecar 1977; Yang et al. 2001), "carpet" (Pouny et al. 1992; Oren and Shai 1998; Bechinger 1999; Shai 1999; Ladokhin and White 2001; Yamaguchi et al. 2001), "toroidal pore" (Matsuzaki et al. 1996; Yang et al. 2001; Yamaguchi et al. 2002; Hallock et al. 2003), and "detergent-like" (Bechinger and Lohner 2006) models have all been proposed.

A popular approach is to alter the hydrophobicity or structure of the peptide. Alterations in the flexible secondary structures of peptides increase their ability to insert and permeabilize membranes. Rozek and colleagues and Osapay and colleagues used this approach to change the shape of indolicidin (Osapay et al. 2000; Rozek et al. 2003). Drawing the N terminus and the C terminus closer via a disulfide bond or covalent cross-link between Trp6 and Trp9 increased antimicrobial activity for gram-negative bacteria and decreased protease sensitivity.

Not all peptides kill bacteria by membrane permeabilization, and there is mounting evidence suggesting that some antimicrobial peptides disable the membrane physiology. Under some circumstances, peptides such as cecropin A, magainin 2, and insect Defensin A disrupt the transmembrane potential and dissipate ion gradients, alter osmotic regulation, or uncouple respiration (Cociancich et al. 1993; Matsuzaki et al. 1997; Silvestro et al. 1997).

Other peptides directly translocate to the cytoplasm. Buforin II, arginine-rich peptides such as TAT-related peptides, NLS peptides, RNA-binding peptides, DNA-binding peptides, and polyarginine, and arginine-rich antimicrobial peptides all easily and efficiently translocate across both the cellular and nuclear membranes

(Park et al. 2000; Futaki et al. 2001). Apidaecin, a short, proline-rich antibacterial peptide, is taken up by a permease/transporter-mediated mechanism (Casteels et al. 1993). Once in the cytoplasm, peptides alter the cytoplasmic membrane septum formation, inhibit peptidoglycan biosynthesis, bind DNA, inhibit nucleic acid synthesis, inhibit protein synthesis, or inhibit enzymatic activity.

Modification of peptides can further increase their ability to inhibit metabolic events. Microcin C is actively taken inside sensitive cells and processed. The product, aspartyl-adenylate, inhibits translation by preventing aminoacylation of tRNA(Asp) by aspartyl-tRNA synthetase (Van de Vijver et al. 2009). Chemical synthesis of three microcin C-like compounds containing a terminal aspartate, glutamate, or leucine attached to adenosine all inhibit bacterial growth by targeting aspartyl-tRNA synthetase, glutamyl-tRNA synthetase, and leucine-tRNA synthetase, respectively.

The Rational Design of Antimicrobial Peptides

Early efforts to design more effective peptides simply involved changing a few amino-acid residues to increase or decrease peptide charge, changing a few amino-acid residues to alter the peptide hydrophobicity, or truncating an already effective peptide at the N-terminus and/or C-terminus. The resulting congener peptides were then tested in a battery of antimicrobial and cytotoxic/hemolytic assays with erythrocytes. Results were compared with that of the original peptides to assess efficacy and toxicity. Now, design of more effective peptides can be facilitated with useful Web tools such as http://bioinformatics.biol.rug.nl/websoftware/bagel, http://faculty.ist.unomaha.edu/chen/rapd/about.php, Synthetic Antibiotic Peptide Database (SAPD), and PANAD (Peptides as Novel Anti-Infective Drugs).

Reasons for the design of antimicrobial peptides are varied. Generally, engineered peptides often identify the smallest amino-acid domain that has full antimicrobial activity: an approach that can be cost-effective for commercial development. Engineered peptides often retain antimicrobial activity in adverse environments: an approach that allows their use in complex biological fluids, serum, and high physiologic salt concentrations. Finally, engineered peptides resist degradation by host enzymes and often have reduced broad-spectrum activity: an approach of the latter to enhance specific antimicrobial activity against a select microorganism. Throughout the literature, the development of engineered peptides has resulted in a number of unique groups.

Synthetic Mimics

Synthetic mimics of antimicrobial peptides such as arylamides or oligoacyllysyls have strong potential for future development of antimicrobial agents and antitumor agents (Radzishevsky et al. 2005; Radzishevsky et al. 2007; Rotem et al. 2008;

Held-Kuznetsov et al. 2009; Rotem and Mor 2009). For example, a mimic such as C(12)K-3β(10) is a promising representative and has broad-spectrum activity (MIC(90) = 6.2 μM) and low hemotoxicity (LC(50) > 100 μM). C(12)K-3β(10) kills *E. coli* faster than *S. aureus*. In mice, this mimic is efficacious and reduces bacterial viability upon single-dose systemic treatment (2 mg/kg).

Hybrid Peptides

Hybrid peptides are constructed by combining active regions of existing peptides, often as fusion proteins expressed in recombinant systems. Historically, cecropin A–melittin (CEME) hybrid peptides are good examples of hybrid peptides (Steiner et al. 1988; Wade et al. 1992; Piers et al. 1993; Piers et al. 1994; Piers and Hancock 1994). Cecropin A is a linear 37 amino-acid-residue peptide produced by the cecropia moth, and melittin is a 26 amino-acid-residue peptide from the European honeybee *Apis mellifera* (Boman and Hultmark 1987). CEME is composed of amino acid residues 1–8 of cecropin A and amino acid residues 1–18 of melittin, CEMA is a congener of CEME with a modified C terminus containing two additional cationic charges, CP26 is a congener with additional cationic charges, and CP26 and CP29 are peptides with an increased alpha-helical content. CEME, CEMA, CP26, and CP29 have improved outer membrane-permeabilizing activity of *Pseudomonas aeruginosa*, LPS-binding activity, lipoteichoic acid-binding activity, and CEME, CEMA, and CP29 retain antimicrobial activity in 0.1–0.3 M NaCl. Another hybrid, cecropin A (1–8) and magainin Z (1–12) has high antitumor activity, less hemolytic activity, and induces release of vesicle-entrapped fluorescence probes (Saugar et al. 2002).

Hybrid peptides can specifically target the surface of select microorganisms. Staphylococcal AgrD1 pheromone linked with the channel-forming domain of colicin Ia kills methicillin-sensitive and methicillin-resistant *Staphylococcus aureus*, but not *Staphylococcus epidermidis* or *Streptococcus pneumoniae* (Qiu et al. 2003). Enterococcal cCF109 pheromone fused with the channel-forming domain of colicin Ia kills vancomycin-resistant *Enterococcus faecalis* (Qiu et al. 2005). Vancomycin-resistant-*E. faecalis*-infected mice survive when treated with this hybrid, but untreated control mice do not (Qiu et al. 2005).

A similar approach involves a *s*pecifically *t*argeted *a*nti*m*icrobial *p*eptide (STAMP) (Eckert et al. 2006) and *m*ultiple-*h*eaded *s*pecifically *t*argeted *a*ntimicrobial *p*eptide (MH-STAMP) (He et al. 2009). STAMP utilizes a pheromone produced by *Streptococcus mutans*, the competence stimulating peptide (CSP), as a targeting domain to mediate *S. mutans*-specific delivery of an antimicrobial peptide domain. STAMP is potent against *S. mutans* grown in liquid or biofilm states and does not affect other oral streptococci tested. MH-STAMP displays specific activity against *P. aeruginosa* and *S. mutans* in vitro and can remove both of these species from a mixed planktonic culture with little impact against untargeted bacteria.

A proline-rich peptide, A3-APO, is a synthetic dimer that kills bacteria by a dual mode of action via domains for interaction with both the bacterial membrane and an intracellular target (Rozgonyi et al. 2009). A3-APO disintegrates the bacterial membrane and inhibits the 70-kDa heat-shock protein DnaK alone or in synergy with small molecule antibiotics.

Not all hybrid peptides meet expectations. Recently, we have attached PQGPPQ, a peptide from proline-rich protein 1, to either the N-terminus or the C-terminus of SMAP28 (Bratt et al. 2009). SMAP28 is a sheep myeloid antimicrobial peptide (Brogden et al. 2007). Peptide PQGPPQ has an affinity for fimbriae of *Porphyromonas gingivalis*, and we hypothesized that it would serve as a targeting ligand to deliver SMAP28 to the bacterial surface. Interestingly, attaching PQGPPQ to SMAP28 did not greatly increase the antimicrobial activity of hybrid peptides for *P. gingivalis*, nor did it substantially decrease the antimicrobial activity of hybrid peptides for the four other microorganisms tested. Clearly, additional work is needed to identify other targeting domains that have increased specific antimicrobial activity against *P. gingivalis* and decreased antimicrobial activity against other oral microorganisms.

Peptide Congeners

By definition, a congener is a chemical compound closely related to another in composition and exerting similar or antagonistic effects. Peptide congeners are prepared by (1) "swapping out" specific amino-acid residues within the parent peptide to change either or both the charge and amphipathic characteristics of the molecule, (2) systematically truncating either or both the N-terminus or the C-terminus ends of the parent peptide, or (3) both "swapping out" specific amino-acid residues and simultaneously truncating either or both the N-terminus and the C-terminus ends of the parent peptide. Effective congeners of CAP18, human CAP18/LL-37, SMAP29, HBD3, and enterocin have been found using this approach (Table 20.1).

Rabbit cationic antimicrobial peptide (CAP18) is an 18-kDa protein of 142 amino acids originally isolated from rabbit granulocytes (Larrick et al. 1991). Cloning and sequencing of the cDNA of rabbit CAP18 led to the discovery of a C-terminal, 37-amino-acid fragment designated $CAP18_{106\text{-}142}$. $CAP18_{106\text{-}142}$ and congeners of $CAP18_{106\text{-}142}$ have antimicrobial activity, inhibit the binding of LPS, inhibit LPS activation of mouse macrophages and human monocytes, inhibit LPS-induced release of cytokines and nitric oxide from macrophages, inhibit LPS-induced limulus amebocyte lysate coagulation, and protect mice from LPS lethality (Larrick et al. 1993; Hirata et al. 1994; Larrick et al. 1994; Larrick et al. 1995b; Tasaka et al. 1996; Sawa et al. 1998; Travis et al. 2000).

The homologous domain of rabbit $CAP18_{106\text{-}142}$ in humans is $hCAP18_{104\text{-}140}$, better known as LL-37 (Larrick et al. 1996). Modifications in LL-37 result in congeners with enhanced activities and decreased toxicities (Ciornei et al. 2005). Two

Table 20.1 Examples of the modifications of antimicrobial peptides to improve antimicrobial activity or alter other biological properties

Peptide platform	Congeners	Design parameters
Cecropin and melittin	CEME, CEMA, CP26, CP29	Design parameters. Designed to have an increase in alpha- helical content (CP29 and CP26) and in overall positive charge (CP26) Outcome. Different analogs had improved outer membrane-permeabilizing activity, LPS-binding activity, and retained activity in 0.1–0.3 M NaCl (CP29, CEME, and CEMA) (Steiner et al. 1988; Wade et al. 1992; Piers et al. 1994)
CAP18	CDP	Design parameters. A 32 amino acid C-terminal fragment of CAP18 Outcome. CDP neutralized LPS, attenuated inflammatory cell migration into alveoli, and attenuated lung injury (Hirata et al. 1994; Hirata et al. 1995; Tasaka et al. 1996)
SMAP29	SMAP28, SMAP29, ovispirin-1, novispirin G-10, novispirin T-7	Design parameters. Ovispirin-1 resembles the N-terminal 18 amino acids of SMAP-29. Residue 10 of ovispirin-1 was changed from isoleucine to glycine to make novispirin G-10. Residue 7 of ovispirin-1 was changed from isoleucine to threonine to make novispirin T-7 Outcome. Ovispirin-1 was antimicrobial but cytotoxic/hemolytic to human cells; novispirin T-7 was antimicrobial and less cytotoxic/hemolytic to human cells; novispirin G-10 was antimicrobial and least cytotoxic/hemolytic to human cells (Kalfa et al. 2001; Sawai et al. 2002)
Human β defensin 3 (HBD3)	C2, F2, Y2	Design parameters. Peptides C2, F2, Y2 are C-terminal (R36-K45) analogs of HBD3. Cysteine residues were acetamidomethylated (C2), replaced with phenylalanine (F2), or replaced with tyrosine (Y2) Outcome. Y2 had higher antibacterial activity, lower cytotoxicity against eukaryotic cells compared to the parent HBD3; Y2 oligomerizes and accretes into a unique structure in water and on the surface of lipid membranes (Bai et al. 2009)

congeners, obtained by N-terminal truncation of hydrophobic amino acids and named fragment 106 and fragment 110, inhibit the growth of *Escherichia coli*, *Pseudomonas aeruginosa*, *S. aureus*, and *Candida albicans*, inhibit LPS-induced vascular nitric oxide production, attract neutrophil granulocytes, retain activity in serum, and induce less hemolysis than the parent LL-37. Other congeners of LL-37 also have antimicrobial activity (even in 175-mM NaCl), inhibit the binding of LPS, inhibit LPS-induced release of nitric oxide from macrophages, inhibit LPS-induced generation of tissue factor, and protect mice from LPS lethality (Larrick et al. 1995a; Larrick et al. 1995b; Travis et al. 2000).

Sheep myeloid antimicrobial peptide (SMAP) 29 is also a popular platform for the development of additional congener peptides (Bagella et al. 1995; Mahoney et al. 1995; Skerlavaj et al. 1999; Travis et al. 2000; Brogden et al. 2001; Kalfa et al. 2001; Saiman et al. 2001; Weistroffer et al. 2008). SMAP29 (also known as SC5), SMAP28, thought to be the native form, and SMAP congeners all have varying degrees of antimicrobial activity against gram-negative and gram-positive bacteria and fungi including multiple antibiotic-resistant pathogens from patients with cystic fibrosis and oral bacteria. Congeners are active in both low and high ionic strength conditions, induce significant morphologic alterations in bacterial surfaces, and reduce the concentration of bacteria in both bronchoalveolar lavage fluid and consolidated pulmonary tissues in models of infection. Second- and third-generation congeners are ovispirin and novispirin, respectively.

HBD3 is a human β-defensin and a popular platform for further study (Hoover et al. 2003; Wu et al. 2003; Liu et al. 2008; Taylor et al. 2008; Chandrababu et al. 2009). The disulfide bonding in HBD3 is necessary for binding and activation of cellular receptors for chemotaxis, but not for antimicrobial activity. For example, reordering disulfide bond linkages, substituting the cysteine amino acids with alpha-aminobutyric acid or other amino acid residues, or removing the disulfide bonds completely abolishes the chemotactic activity of the HBD3 congener, diminishes the cytotoxicity of HBD3, but does not alter its antimicrobial activity.

Truncated linear HBD3 fragments have antibacterial and antifungal activities and decreased cytotoxicity for human conjunctival epithelial cells (Taylor et al. 2008; Bai et al. 2009; Krishnakumari et al. 2009). A C-terminal (R36-K45) analog peptide Y2 ,with the two cysteine residues replaced with tyrosines, has high antibacterial activity against *P. aeruginosa* and low cytotoxicity against mammalian cells, compared to the parent HBD3 (Bai et al. 2009).

Melittin has helical regions at the N-terminus and at the C-terminus connected by a hinge region (Saravanan et al. 2009). Deletion of the hinge amino-acid residues along with two C-terminal terminal glutamine residues (Q25 and Q26) yields a peptide analog of 19-amino-acid residues, it does not reduce antibacterial activity, but does reduce hemolytic activity. A diastereomer of Mel-H or Mel-(d)H containing d-amino acids (d)V5, (d)V8, (d)L11, and (d)K16 has similar antibacterial activity but lower hemolytic activity (Saravanan et al. 2009).

The approach of making congeners also works well with bacteriocins. Enterocin CRL35 is a 43-amino-acid-residue peptide with activity against *Listeria* species (Salvucci et al. 2007). A 15-amino-acid-residue congener derived from enterocin CRL35 inhibits the growth of *L. innocua* (10 μM MIC) and *L. monocytogenes* (50 μM MIC). Similarly, 15-amino-acid-residue congeners derived from mesentericin Y105, pediocin PA-1, and piscicolin 126 also have antimicrobial activities.

Stabilized Peptides

Stabilization of biologically active peptides is a major goal in peptide-based drug design, and three recent studies illustrate this point. Cyclization is a popular approach that prevents enzymatic degradation of biologically active peptides and reduces hemolysis to antimicrobial peptides. Levengood and van der Donk used the lantibiotic synthetase LctM to prepare thioether containing analogs of enkephalin, contryphan, inhibitors of human tripeptidyl peptidase II, and inhibitors of spider venom epimerase (Abraham et al. 2007; Levengood and van der Donk 2008; Kluskens et al. 2009). Kluskens and colleagues used plasmid-based nisin modification machinery to produce a thioether-bridged angiotensin (Kluskens et al. 2009). The cyclized angiotensin is fully resistant against purified angiotensin-converting enzyme, has significantly increased stability in homogenates of different organs and in plasma derived from pig, induces relaxation of precontracted SD rat aorta rings in vitro, interacts with the angiotensin receptor, and displays a strongly (34-fold) enhanced survival in Sprague–Dawley rats in vivo. A cyclic diastereomeric lysine ring analog of the antimicrobial peptide gramicidin S exhibits enhanced antimicrobial activity but markedly reduced hemolytic activity compared to gramicidin S itself (Abraham et al. 2007).

Peptide Conjugates

Antimicrobial peptides can be coupled to a ligand, receptor, or antibody to increase its specificity for a bacterial pathogen. Peschen and colleagues used a *Fusarium species*-specific antibody linked to antifungal peptides to protect plants against a *Fusarium oxysporum f.sp. matthiolae* infection (Peschen et al. 2004). We linked SMAP28 to affinity- and Protein-G-purified rabbit IgG antibodies specific to the outer surface of *P. gingivalis* strain 381 and demonstrated that it could selectively kill *P. gingivalis* in an artificially generated microbial community containing *P. gingivalis, Aggregatibacter actinomycetemcomitans,* and *Peptostreptococcus micros* (Franzman et al. 2009). This approach is an initial step for developing selective antimicrobial agents capable of eliminating a specific periodontal pathogen, such as *P. gingivalis,* from patients with periodontal disease, without harming the normal commensal flora.

Peptides can be incorporated in or conjugated to supramolecular structures, particularly micelles and liposomes equipped with uptake-mediating mechanisms. Apolipoprotein E-derived peptide, called A2, efficiently translocates across cell membranes and integrates into lipid bilayers (Leupold et al. 2009). A2, in micelles (colloidal P2A2 micelles) or liposomes (P2A2-tagged liposomes), are taken up differently by endothelial cells of small capillaries (b.End3 cells) and large vascular vessels (BAEC cells). P2A2-tagged liposomes are nonselectively internalized into both b.End3 and BAEC cells via clathrin- and caveolin-independent endocytosis. In contrast, colloidal P2A2 micelles only entered b.End3 cells via clathrin-mediated endocytosis, but not BAEC cells.

Finally, antimicrobial peptides are powerful adjuvants enhancing immune responses alone (Kohlgraf et al. 2010a, b) and in conjugates (containing coupled antimicrobial peptide to antigens) and fusion proteins (containing expressed antimicrobial peptides and antigens). LL-37 enhances an antitumor immune response of M-CSFR$_{J6-1}$, a potential target for tumor immunotherapy, when LL-37 is genetically fused with M-CSFR$_{J6-1}$ and administered to mice (An et al. 2005). Anti-M-CSFR antibody or M-CSF soluble receptor inhibits the growth of leukemia and hepatoma cell lines overexpressing M-CSF and M-CSFR. Mouse β-defensin-based vaccines elicit potent cell-mediated responses and antitumor immunity when genetically fused with another nonimmunogenic tumor antigen (Biragyn et al. 2001). The fusion proteins, consisting of mouse β-defensin linked to a tumor antigen, act directly on immature dendritic cells as an endogenous ligand for TLR4 and upregulate costimulatory molecules, induce dendritic cell maturation, and induce the production of lymphokines.

Immobilized Peptides

Antimicrobial peptides can be immobilized in or to a variety of surfaces where they still retain their antimicrobial activity. Cationic antimicrobial peptides, when attached to resins, are antimicrobial toward *E. coli* and *Bacillus subtilis* (millimolar range) compared to activity of the parent peptides (micromolar range) (Bagheri et al. 2009). Antimicrobial activity depends upon the length of the spacer and the amount of target-accessible peptide. Antimicrobial activity distinctly decreases with reduction of the spacer length. Magainins and structurally related peptides, when conjugated to resins, also arrest the growth and proliferation of *E. coli* (Haynie et al. 1995). Magainin I was incorporated via surface-initiated atom transfer radical polymerization into two coatings of 2-(2-methoxyethoxy) ethyl methacrylate and hydroxyl-terminated oligo(ethylene glycol) methacrylate to prepare antifouling copolymer brushes. The antibacterial activity of the functionalized brushes was successfully tested against two different strains of gram-positive bacteria. (Glinel et al. 2009).

Antimicrobial peptides can be incorporated into films. Nisin in polyethylene or polyethylene oxide polymer films reduce *Brochothrix thermosphacta,* a meat spoilage organism, on beef surfaces up to 21 days to a greater extent than the control plastics (Cutter et al. 2001); defensins in polyelectrolyte multilayer films inhibit the growth of *E. coli* at the surface (Etienne et al. 2004); gramicidin A in polyelectrolyte thin films are active against *E. faecalis* (Guyomard et al. 2008).

Antimicrobial peptides and lysozyme can be incorporated into silica or adsorbed onto solid surfaces (Luckarift et al. 2006). LL-37 in mesoporous silica is slowly released and has high antimicrobial activity against *E. coli* and *S. aureus*, low hemolytic activity for erythrocytes, and low cytotoxicity against keratinocytes (Izquierdo-Barba et al. 2009). Nisin adsorbed to food contact surfaces inhibits the growth of *Enterococcus hirae* (Guerra et al. 2005). Nisaplin, a congener of nisin, adsorbed on surfaces reduces the attachment of *Listeria monocytogenes*. Microbial counts of skim milk in nisin-adsorbed PET bottles are significantly lower after 24 days of refrigerated storage. Magainin I, immobilized on silanized glass slides, binds *Salmonella enterica* serovar Typhimurium and *E. coli* with high affinities (Kulagina et al. 2005).

All of these methods have applications as slow delivery systems, wound dressings, food preservation systems, and coatings for implants, catheters, and toys. Even when immobilized or in complex environments, antimicrobial peptides still retain their antimicrobial activity, which questions earlier work on their mechanisms of antimicrobial activity, particularly for antimicrobial peptides known to form well-structured pores.

Summary

A vast amount of information is known about antimicrobial peptide composition, structure, and activity. Many antimicrobial peptides have unique characteristics related to their mechanisms of antimicrobial action. Contemporary studies now utilize all this information to engineer and design peptide mimics, hybrid peptides, peptide congeners, stabilized peptides, peptide conjugates, and immobilized peptides, all with unique and specific applications.

Acknowledgments This work was supported by NIH/NIDCR R01 DE014390.

References

Abraham T, Marwaha S, Kobewka DM et al (2007) The relationship between the binding to and permeabilization of phospholipid bilayer membranes by GS14dK4, a designed analog of the antimicrobial peptide gramicidin S. Biochim Biophys Acta 1768:2089–2098

An LL, Yang YH, Ma XT et al (2005) LL-37 enhances adaptive antitumor immune response in a murine model when genetically fused with M-CSFR (J6-1) DNA vaccine. Leuk Res 29:535–543

Bagella L, Scocchi M, Zanetti M (1995) cDNA sequences of three sheep myeloid cathelicidins. Fed Eur Biol Sci Lett 376:225–228

Bagheri M, Beyermann M, Dathe M (2009) Immobilization reduces the activity of surface-bound cationic antimicrobial peptides with no influence upon the activity spectrum. Antimicrob Agents Chemother 53:1132–1141

Bai Y, Liu S, Jiang P et al (2009) Structure-dependent charge density as a determinant of antimicrobial activity of peptide analogues of defensin. Biochemistry (Mosc) 48:7229–7239

Bechinger B (1999) The structure, dynamics and orientation of antimicrobial peptides in membranes by multidimensional solid-state NMR spectroscopy. Biochim Biophys Acta 1462:157–183

Bechinger B, Lohner K (2006) Detergent-like actions of linear amphipathic cationic antimicrobial peptides. Biochim Biophys Acta 1758:1529–1539

Biragyn A, Surenhu M, Yang D et al (2001) Mediators of innate immunity that target immature, but not mature, dendritic cells induce antitumor immunity when genetically fused with nonimmunogenic tumor antigens. J Immunol 167:6644–6653
Boman HG (1995) Peptide antibiotics and their role in innate immunity. Annu Rev Immunol 13:61–92
Boman HG, Hultmark D (1987) Cell-free immunity in insects. Annu Rev Microbiol 41:103–126
Bratt CL, Kohlgraf KG, Yohnke K et al (2009) Communication: Antimicrobial activity of SMAP28 with a targeting domain for *Porphyromonas gingivalis* Probiotics and Antimicrobial Proteins 2:21–25
Brogden KA, Guthmiller JM, Salzet M et al (2005) The nervous system and innate immunity: the neuropeptide connection. Nat Immunol 6:558–564
Brogden KA, Kalfa VC, Ackermann MR et al (2001) The ovine cathelicidin SMAP29 kills ovine respiratory pathogens in vitro and in an ovine model of pulmonary infection. Antimicrob Agents Chemother 45:331–334
Brogden KA, Nordholm G, Ackermann M (2007) Antimicrobial activity of cathelicidins BMAP28, SMAP28, SMAP29, and PMAP23 against Pasteurella multocida is more broad-spectrum than host species specific. Vet Microbiol 119:76–81
Casteels P, Ampe C, Jacobs F et al (1993) Functional and chemical characterization of Hymenoptaecin, an antibacterial polypeptide that is infection-inducible in the honeybee (Apis mellifera). J Biol Chem 268:7044–7054
Chandrababu KB, Ho B, Yang D (2009) Structure, dynamics, and activity of an all-cysteine mutated human beta defensin-3 peptide analogue. Biochemistry (Mosc) 48:6052–6061
Chen FY, Lee MT, Huang HW (2003) Evidence for membrane thinning effect as the mechanism for peptide-induced pore formation. Biophys J 84:3751–3758
Ciornei CD, Sigurdardottir T, Schmidtchen A et al (2005) Antimicrobial and chemoattractant activity, lipopolysaccharide neutralization, cytotoxicity, and inhibition by serum of analogs of human cathelicidin LL-37. Antimicrob Agents Chemother 49:2845–2850
Cociancich S, Ghazi A, Hetru C et al (1993) Insect defensin, an inducible antibacterial peptide, forms voltage-dependent channels in Micrococcus luteus. J Biol Chem 268:19239–19245
Cole AM, Ganz T, Liese AM et al (2001) Cutting edge: IFN-inducible ELR- CXC chemokines display defensin-like antimicrobial activity. J Immunol 167:623–627
Cutter CN, Willett JL, Siragusa GR (2001) Improved antimicrobial activity of nisin-incorporated polymer films by formulation change and addition of food grade chelator. Lett Appl Microbiol 33:325–328
Eckert R, He J, Yarbrough DK et al (2006) Targeted killing of Streptococcus mutans by a pheromone-guided "smart" antimicrobial peptide. Antimicrob Agents Chemother 50:3651–3657
Ehrenstein G, Lecar H (1977) Electrically gated ionic channels in lipid bilayers. Q Rev Biophys 10:1–34
Etienne O, Picart C, Taddei C et al (2004) Multilayer polyelectrolyte films functionalized by insertion of defensin: a new approach to protection of implants from bacterial colonization. Antimicrob Agents Chemother 48:3662–3669
Franzman MR, Burnell KK, Dehkordi-Vakil FH et al (2009) Targeted antimicrobial activity of a specific IgG-SMAP28 conjugate against *Porphyromonas gingivalis* in a mixed culture. Int J Antimicrob Agents 33:14–20
Friedberg D, Friedberg I, Shilo M (1970) Interaction of gram-negative bacteria with the lysosomal fraction of polymorphonuclear leukocytes. II. Changes in the cell envelope of Escherichia coli. Infect Immun 1:311–331
Friedberg D, Shilo M (1970) Interaction of gram-negative bacteria with the lysosomal fraction of polymorphonuclear leukocytes. I. Role of cell wall composition of Salmonella typhimurium. Infect Immun 1:305–318
Futaki S, Suzuki T, Ohashi W et al (2001) Arginine-rich peptides. An abundant source of membrane-permeable peptides having potential as carriers for intracellular protein delivery. J Biol Chem 276:5836–5840
Ganz T, Selsted ME, Lehrer RI (1990) Defensins. Eur J Haematol 44:1–8

Glinel K, Jonas AM, Jouenne T et al (2009) Antibacterial and antifouling polymer brushes incorporating antimicrobial peptide. Bioconjug Chem 20:71–77

Guerra NP, Araujo AB, Barrera AM et al (2005) Antimicrobial activity of nisin adsorbed to surfaces commonly used in the food industry. J Food Prot 68:1012–1019

Guyomard A, De E, Jouenne T et al (2008) Incorporation of a hydrophobic antibacterial peptide into amphiphilic polyelectrolyte multilayers: a bioinspired approach to prepare biocidal thin coatings. Adv Funct Mater 18:758–765

Hallock KJ, Lee DK, Ramamoorthy A (2003) MSI-78, an analogue of the magainin antimicrobial peptides, disrupts lipid bilayer structure via positive curvature strain. Biophys J 84:3052–3060

Hancock RE, Rozek A (2002) Role of membranes in the activities of antimicrobial cationic peptides. FEMS Microbiol Lett 206:143–149

Hancock REW (1997) Peptide antibiotics. Lancet 349:418–422

Haynie SL, Crum GA, Doele BA (1995) Antimicrobial activities of amphiphilic peptides covalently bonded to a water-insoluble resin. Antimicrob Agents Chemother 39:301–307

He J, Anderson MH, Shi W et al (2009) Design and activity of a 'dual-targeted' antimicrobial peptide. Int J Antimicrob Agents 33:532–537

Held-Kuznetsov V, Rotem S, Assaraf YG et al (2009) Host-defense peptide mimicry for novel antitumor agents. FASEB J 23:4299–4307

Hirata M, Shimomura Y, Yoshida M et al (1994) Characterization of a rabbit cationic protein (CAP18) with lipopolysaccharide-inhibitory activity. Infect Immun 62:1421–1426

Hirata M, Zhong J, Wright SC et al (1995) Structure and functions of endotoxin-binding peptides derived from CAP18. Prog Clin Biol Res 392:317–326

Hirsch JG (1956) Phagocytin: a bactericidal substance from polymorphonuclear leucocytes. J Exp Med 103:589–611

Hoover DM, Wu Z, Tucker K et al (2003) Antimicrobial characterization of human beta-defensin 3 derivatives. Antimicrob Agents Chemother 47:2804–2809

Huang HW (2000) Action of antimicrobial peptides: two-state model. Biochemistry (Mosc) 39:8347–8352

Izquierdo-Barba I, Vallet-Regi M, Kupferschmidt N et al (2009) Incorporation of antimicrobial compounds in mesoporous silica film monolith. Biomaterials 30:5729–5736

Jenssen H, Hamill P, Hancock RE (2006) Peptide antimicrobial agents. Clin Microbiol Rev 19:491–511

Kalfa VC, Jia HP, Kunkle RA et al (2001) Congeners of SMAP29 kill ovine pathogens and induce ultrastructural damage in bacterial cells. Antimicrob Agents Chemother 45:3256–3261

Kluskens LD, Nelemans SA, Rink R et al (2009) Angiotensin-(1-7) with thioether bridge: an angiotensin-converting enzyme-resistant, potent angiotensin-(1-7) analog. J Pharmacol Exp Ther 328:849–854

Kohlgraf KG, Ackermann A, Lu X et al (2010a) Defensins attenuate cytokine responses yet enhance antibody responses to Porphyromonas gingivalis adhesins in mice. Future Microbiol 5:115–125

Kohlgraf KG, Pingel LC, Dietrich DE et al (2010b) Defensins as anti-inflammatory compounds and mucosal adjuvants. Future Microbiol 5:99–113

Krishnakumari V, Rangaraj N, Nagaraj R (2009) Antifungal activities of human beta-defensins HBD-1 to HBD-3 and their C-terminal analogs Phd1 to Phd3. Antimicrob Agents Chemother 53:256–260

Kulagina NV, Lassman ME, Ligler FS et al (2005) Antimicrobial peptides for detection of bacteria in biosensor assays. Anal Chem 77:6504–6508

Kuo HH, Chan C, Burrows LL et al (2007) Hydrophobic interactions in complexes of antimicrobial peptides with bacterial polysaccharides. Chem Biol Drug Des 69:405–412

Ladokhin AS, White SH (2001) "Detergent-like" permeabilization of anionic lipid vesicles by melittin. Biochim Biophys Acta 1514:253–260

Larrick JW, Hirata M, Shimomoura Y et al (1993) Antimicrobial activity of rabbit CAP18-derived peptides. Antimicrob Agents Chemother 37:2534–2539

Larrick JW, Hirata M, Zheng H et al (1994) A novel granulocyte-derived peptide with lipopolysaccharide-neutralizing activity. J Immunol 152:231–240

Larrick JW, Hirata M, Balint RF et al (1995a) Human CAP18: a novel antimicrobial lipopolysaccharide-binding protein. Infect Immun 63:1291–1297

Larrick JW, Hirata M, Zhong J et al (1995b) Anti-microbial activity of human CAP18 peptides. Immunotechnology 1:65–72

Larrick JW, Lee J, Ma S et al (1996) Structural, functional analysis and localization of the human CAP18 gene. FEBS Lett 398:74–80

Larrick JW, Morgan JG, Palings I et al (1991) Complementary DNA sequence of rabbit CAP18–a unique lipopolysaccharide binding protein. Biochem Biophys Res Commun 179:170–175

Leupold E, Nikolenko H, Dathe M (2009) Apolipoprotein E peptide-modified colloidal carriers: the design determines the mechanism of uptake in vascular endothelial cells. Biochim Biophys Acta 1788:442–449

Levengood MR, van der Donk WA (2008) Use of lantibiotic synthetases for the preparation of bioactive constrained peptides. Bioorg Med Chem Lett 18:3025–3028

Liu S, Zhou L, Li J et al (2008) Linear analogues of human beta-defensin 3: concepts for design of antimicrobial peptides with reduced cytotoxicity to mammalian cells. Chembiochem 9:964–973

Luckarift HR, Dickerson MB, Sandhage KH et al (2006) Rapid, room-temperature synthesis of antibacterial bionanocomposites of lysozyme with amorphous silica or titania. Small 2:640–643

Mahoney MM, Lee AY, Brezinski-Caliguri DJ et al (1995) Molecular analysis of the sheep cathelin family reveals a novel antimicrobial peptide. Fed Eur Biol Sci Lett 377:519–522

Matsuzaki K, Murase O, Fujii N et al (1996) An antimicrobial peptide, magainin 2, induced rapid flip-flop of phospholipids coupled with pore formation and peptide translocation. Biochemistry (Mosc) 35:11361–11368

Matsuzaki K, Murase O, Miyajima K (1995) Kinetics of pore formation by an antimicrobial peptide, magainin 2, in phospholipid bilayers. Biochemistry (Mosc) 34:12553–12559

Matsuzaki K, Sugishita K, Harada M et al (1997) Interactions of an antimicrobial peptide, magainin 2, with outer and inner membranes of Gram-negative bacteria. Biochim Biophys Acta 1327:119–130

Oren Z, Shai Y (1998) Mode of action of linear amphipathic alpha-helical antimicrobial peptides. Biopolymers 47:451–463

Osapay K, Tran D, Ladokhin AS et al (2000) Formation and characterization of a single Trp-Trp cross-link in indolicidin that confers protease stability without altering antimicrobial activity. J Biol Chem 275:12017–12022

Park CB, Yi KS, Matsuzaki K et al (2000) Structure-activity analysis of buforin II, a histone H2A-derived antimicrobial peptide: the proline hinge is responsible for the cell-penetrating ability of buforin II. Proc Natl Acad Sci U S A 97:8245–8250

Pellegrini A, Hulsmeier AJ, Hunziker P et al (2004) Proteolytic fragments of ovalbumin display antimicrobial activity. Biochim Biophys Acta 1672:76–85

Peschen D, Li HP, Fischer R et al (2004) Fusion proteins comprising a *Fusarium*-specific antibody linked to antifungal peptides protect plants against a fungal pathogen. Nat Biotechnol 22:732–738

Piers KL, Brown MH, Hancock RE (1994) Improvement of outer membrane-permeabilizing and lipopolysaccharide- binding activities of an antimicrobial cationic peptide by C-terminal modification. Antimicrob Agents Chemother 38:2311–2316

Piers KL, Brown MH, Hancock REW (1993) Recombinant DNA procedures for producing small antimicrobial cationic peptides in bacteria. Gene 134:7–13

Piers KL, Hancock REW (1994) The interaction of a recombinant cecropin/melittin hybrid peptide with the outer membrane of *Pseudomonas aeruginosa*. Mol Microbiol 12:951–958

Pouny Y, Rapaport D, Mor A et al (1992) Interaction of antimicrobial dermaseptin and its fluorescently labeled analogues with phospholipid membranes. Biochemistry (Mosc) 31:12416–12423

Qiu XQ, Wang H, Lu XF et al (2003) An engineered multidomain bactericidal peptide as a model for targeted antibiotics against specific bacteria. Nat Biotechnol 21:1480–1485

Qiu XQ, Zhang J, Wang H et al (2005) A novel engineered peptide, a narrow-spectrum antibiotic, is effective against vancomycin-resistant *Enterococcus faecalis*. Antimicrob Agents Chemother 49:1184–1189

Radzishevsky IS, Rotem S, Bourdetsky D et al (2007) Improved antimicrobial peptides based on acyl-lysine oligomers. Nat Biotechnol 25:657–659

Radzishevsky IS, Rotem S, Zaknoon F et al (2005) Effects of acyl versus aminoacyl conjugation on the properties of antimicrobial peptides. Antimicrob Agents Chemother 49:2412–2420

Rotem S, Mor A (2009) Antimicrobial peptide mimics for improved therapeutic properties. Biochim Biophys Acta 1788:1582–1592

Rotem S, Radzishevsky IS, Bourdetsky D et al (2008) Analogous oligo-acyl-lysines with distinct antibacterial mechanisms. FASEB J 22:2652–2661

Rozek A, Powers JP, Friedrich CL et al (2003) Structure-based design of an indolicidin peptide analogue with increased protease stability. Biochemistry (Mosc) 42:14130–14138

Rozgonyi F, Szabo D, Kocsis B et al (2009) The antibacterial effect of a proline-rich antibacterial peptide A3-APO. Curr Med Chem 16:3996–4002

Saiman L, Tabibi S, Starner TD et al (2001) Cathelicidin peptides inhibit multiply antibiotic-resistant pathogens from patients with cystic fibrosis. Antimicrob Agents Chemother 45:2838–2844

Salvucci E, Saavedra L, Sesma F (2007) Short peptides derived from the NH2-terminus of subclass IIa bacteriocin enterocin CRL35 show antimicrobial activity. J Antimicrob Chemother 59:1102–1108

Saravanan R, Bhunia A, Bhattacharjya S (2010) Micelle-bound structures and dynamics of the hinge deleted analog of melittin and its diastereomer: Implications in cell selective lysis by d-amino acid containing antimicrobial peptides. Biochim Biophys Acta 1798:128–139

Saugar JM, Alarcon T, Lopez-Hernandez S et al (2002) Activities of polymyxin B and cecropin A-, melittin peptide CA(1–8)M(1–18) against a multiresistant strain of Acinetobacter baumannii. Antimicrob Agents Chemother 46:875–878

Sawa T, Kurahashi K, Ohara M et al (1998) Evaluation of antimicrobial and lipopolysaccharide-neutralizing effects of a synthetic CAP18 fragment against *Pseudomonas aeruginosa* in a mouse model. Antimicrob Agents Chemother 42:3269–3275

Sawai MV, Waring AJ, Kearney WR et al (2002) Impact of single-residue mutations on the structure and function of ovispirin/novispirin antimicrobial peptides. Protein Eng 15:225–232

Scott MG, Gold MR, Hancock RE (1999a) Interaction of cationic peptides with lipoteichoic acid and gram- positive bacteria. Infect Immun 67:6445–6453

Scott MG, Yan H, Hancock RE (1999b) Biological properties of structurally related alpha-helical cationic antimicrobial peptides. Infect Immun 67:2005–2009

Shai Y (1999) Mechanism of the binding, insertion and destabilization of phospholipid bilayer membranes by alpha-helical antimicrobial and cell non-selective membrane-lytic peptides. Biochim Biophys Acta 1462:55–70

Silvestro L, Gupta K, Weiser JN et al (1997) The concentration-dependent membrane activity of cecropin A. Biochemistry (Mosc) 36:11452–11460

Skarnes RC, Watson DW (1957) Antimicrobial factors of normal tissues and fluids. Bacteriol Rev 21:273–294

Skerlavaj B, Benincasa M, Risso A et al (1999) SMAP-29: a potent antibacterial and antifungal peptide from sheep leukocytes. Fed Eur Biol Sci Lett 463:58–62

Steiner H, Andreu D, Merrifield RB (1988) Binding and action of cecropin and cecropin analogues: antibacterial peptides from insects. Biochim Biophys Acta 939:260–266

Steiner H, Hultmark D, Engstrom A et al (1981) Sequence and specificity of two antibacterial proteins involved in insect immunity. Nature 292:246–248

Tasaka S, Ishizaka A, Urano T et al (1996) A derivative of cationic antimicrobial protein attenuates lung injury by suppressing cell adhesion. Am J Respir Cell Mol Biol 15:738–744

Taylor K, Clarke DJ, McCullough B et al (2008) Analysis and separation of residues important for the chemoattractant and antimicrobial activities of beta-defensin 3. J Biol Chem 283:6631–6639

Tossi A (2005) Editorial. Host defense peptides: roles and applications. Curr Protein Peptide Sci 6:1–3

Travis SM, Anderson NN, Forsyth WR et al (2000) Bactericidal activity of mammalian cathelicidin-derived peptides. Infect Immun 68:2748–2755

Van de Vijver P, Vondenhoff GH, Kazakov TS et al (2009) Synthetic microcin C analogs targeting different aminoacyl-tRNA synthetases. J Bacteriol 191:6273–6280

Wade D, Andreu D, Mitchell SA et al (1992) Antibacterial peptides designed as analogs or hybrids of cecropins and melittin. Int J Pept Protein Res 40:429–436

Weiss J, Franson RC, Beckerdite S et al (1975) Partial characterization and purification of a rabbit granulocyte factor that increases permeability of Escherichia coli. J Clin Invest 55:33–42

Weistroffer PL, Joly S, Srikantha R et al (2008) SMAP29 congeners demonstrate activity against oral bacteria and reduced toxicity against oral keratinocytes. Oral Microbiol Immunol 23:89–95

Willey JM, van der Donk WA (2007) Lantibiotics: peptides of diverse structure and function. Annu Rev Microbiol 61:477–501

Wu Z, Hoover DM, Yang D et al (2003) Engineering disulfide bridges to dissect antimicrobial and chemotactic activities of human beta-defensin 3. Proc Natl Acad Sci U S A 100:8880–8885

Yamaguchi S, Hong T, Waring A et al (2002) Solid-state NMR investigations of peptide-lipid interaction and orientation of a beta-sheet antimicrobial peptide, protegrin. Biochemistry (Mosc) 41:9852–9862

Yamaguchi S, Huster D, Waring A et al (2001) Orientation and Dynamics of an Antimicrobial Peptide in the Lipid Bilayer by Solid-State NMR Spectroscopy. Biophys J 81:2203–2214

Yang D, Chen Q, Hoover DM et al (2003) Many chemokines including CCL20/MIP-3alpha display antimicrobial activity. J Leukoc Biol 74:448–455

Yang L, Harroun TA, Weiss TM et al (2001) Barrel-stave model or toroidal model? A case study on melittin pores. Biophys J 81:1475–1485

Yeaman MR, Yount NY (2003) Mechanisms of antimicrobial peptide action and resistance. Pharmacol Rev 55:27–55

Yount NY, Yeaman MR (2004) Multidimensional signatures in antimicrobial peptides. Proc Natl Acad Sci U S A 101:7363–7368

Zasloff M (1987) Magainins, a class of antimicrobial peptides from *Xenopus* skin: Isolation, characterization of two active forms, and partial cDNA sequence of a precursor. Proc Natl Acad Sci U S A 84:5449–5453

Zeya HI, Spitznagel JK (1963) Antibacterial and enzymic basic proteins from leukocyte lysosomes: separation and identification. Science 142:1085–1087

Index

D. Drider and S. Rebuffat (eds.), *Prokaryotic Antimicrobial Peptides: From Genes to Applications*, DOI 10.1007/978-1-4419-7692-5,

D